HEAT EXCHANGER NETWORK SYNTHESIS

Process Optimization by Energy and Resource Analysis

UDAY V. SHENOY

Gulf Publishing Company
Houston, London, Paris, Zurich, Tokyo

To
my parents for their inspiration,
my wife, Maya, for her constant support,
and my three-year-old son, Akshay, for his special company.

Gulf Publishing Company
Book Division
P.O. Box 2608, Houston, Texas 77252-2608

10 9 8 7 6 5 4 3 2 1

Library of Congress Cataloging-in-Publication Data

Shenoy, Uday V.
   Heat exchanger network synthesis : optimization process
by energy and resource analysis / Uday V. Shenoy.
     p. cm.
   Includes bibliographical references and index.
   ISBN 0-88145-319-6 (alk. paper)
   1. Heat exchangers—Design and construction. I. Title.
TJ263.S45   1995
621.402′5—dc20   95-3076
                                           CIP

Printed on Acid-Free Paper (∞)

Contents

Index to Solved Examples

Note: Examples that contain relatively advanced material and that may be omitted on first reading are marked with an asterisk (*) below.

Preface

This book on heat exchanger network synthesis (HENS) and optimization of processes by energy and resource analysis (OPERA) has a three-fold purpose:

- to provide a textbook for a course at the undergraduate or graduate level for students and teachers of chemical engineering, mechanical engineering, and energy systems engineering;
- to serve as a research monograph providing a state-of-the-art review for researchers since more than 350 papers have appeared in the scientific literature on this subject; and
- to act as a guide to the methodology of energy conservation through process integration for practitioners in the industry.

Pinch analysis forms the essence of OPERA; hence, the book emphasizes the concept of establishing targets prior to design. The coverage is not restricted to any particular processing industry for pinch technology, as a thermodynamically-based tool for the synthesis of cost-optimum energy-integrated systems, has proven successful worldwide in a variety of industries (like oil and gas, whisky and brewing, dairy, petrochemical, chemical, cement, steel, pharmaceutical, and pulp and paper) and has resulted in significant cost benefits. However, the technology has not been exploited to its fullest, and applications in many countries are virtually nonexistent. This is probably because the research literature (specially on industrial HENS) does not clearly enunciate the methodology, but only presents the final results. An effort is made in the pages that follow to provide a "how-to-do-it" manual with step-by-step algorithms for targeting and network generation in grassroots as well as retrofit cases. The emphasis is on systematizing the approach to process integration and making the solution procedure transparent. For continuity, a single case study on a simple four-stream problem is used throughout the text (as far as possible) to illustrate various important concepts.

A problem-solving, example-oriented approach is adopted with calculations presented in tabular fashion that lend themselves admirably to computer implementation. Designed to provide a vehicle for learning, through self-study or through a formal course, the book provides a number of case studies from existing literature as problems at the end of each chapter. Consistent with the rapid deployment of computers for design purposes today, the book has accompanying software support (a student version of the PC-based software package, HX) that would reduce tedium in performing repetitive calculations.

The book is at the crossroads of research and industrial practice where researchers can find the important advances achieved in pinch analysis in recent times and process engineers can find an original presentation of the basic notions of heat integration. The book begins with the vital concept in pinch technology of setting targets for energy, units, area, shells, and cost. It introduces a new algorithm for continuous targeting for both minimum approach temperature and minimum flux conditions. The need for globally optimizing energy systems based on the classical energy-capital tradeoff is discussed. Networks for maximum energy recovery are synthesized and further evolved using systematic energy relaxation and considering loop-network interactions. Intricacies like target area distribution, multiple utilities, and forbidden matches are explained. Interfacing network synthesis with detailed exchanger design, using stream pressure drops as the basis, is emphasized in order to achieve consistent values of the heat transfer coefficients. The topic of retrofitting is perhaps the most important from an industrial viewpoint and is treated in detail exploiting tools like remaining problem analysis and driving force plot. Other in-depth topics include dual approach temperature methods, heat and power integration, and flexibility in HENS. Although the emphasis is on application of pinch technology to HENS, the book discusses in detail heat integration in the context of batch processes, distillation systems, evaporation systems, reactor systems, utility systems, reduction of environmental emissions, and total site integration. The extension of the concepts to mass exchanger network synthesis (MENS) and wastewater minimization is also addressed. For completeness, the final chapter presents the alternative approach to HENS, based on mathematical methods (like linear, nonlinear, and integer programming). To gain hands-on experience with mathematical programming formulations, a student version of the GAMS (general algebraic modeling system) software is included with the book through a special arrangement with GAMS Development Corporation (Washington, D.C.).

The book is easier to follow if readers are familiar with basic principles in thermodynamics, heat transfer, and the thermal design of heat exchangers. Chapter 9 on Energy and Resource Analysis of Various Processes requires some prior knowledge of mass transfer and reaction engineering.

The text is divided into ten chapters, each containing numerous illustrative examples. To enhance the educational value of the book, chapters end with a series of questions for discussion. Typically, the answers to these questions will stimulate thinking on the part of the reader. In addition, problems are provided at the end of each chapter to increase understanding through practice. Answers are supplied to the numerical problems.

The coverage in the book is quite comprehensive, and the material is more than can be covered in a typical semester course, so illustrative examples that contain relatively advanced material and that may be omitted on first reading are marked with an asterisk (*) in the index to the solved examples.

Many persons have helped directly or indirectly towards this book. At the start, I must acknowledge my great admiration for Professor Bodo Linnhoff and his ingenious pioneering developments in pinch analysis. I have been inspired by the significant contributions made by Professor Linnhoff and his coworkers at UMIST to process integration technology. Some of the contributors who deserve special mention are R. Smith, G. T. Polley, S. Ahmad, V. R. Dhole, and M. H. Panjeh Shahi (the list is not exhaustive).

My thanks go to those who developed the technology and provided their published material to me on request. These include S. Ahmad, A. Carlsson, V. R. Dhole, Z. Fonyo, D. M. Fraser, P. Glavic, T. Gundersen, S. G. Hall, J. R. Himsworth, H. A. Irazoqui, J. Jezowski, B. Kalitventzeff, I. C. Kemp, R. Lakshmanan, A. J. D. Lambert, B. Linnhoff, M. H. Panjeh Shahi, E. N. Pistikopoulos, G. T. Polley, E. Rev, A. P. Rossiter, N. J. Scenna, R. Smith, B. Sunden, K. K. Trivedi, R. M. Wood, and T. F. Yee.

My gratitude goes to Professor Ranjan K. Malik, for I have learnt much through discussions with him on various aspects of process plant simulation and optimization. Deepak Birewar, Chris Floudas, Ajay Modi, Pratap Nair, Peter Piela, Stratos Pistikopoulos, Sriram Vasantharajan, and Terry Yee may not have realized the influence their friendship has had in generating my interest in computer-aided design.

Thanks are due to GAMS Development Corporation (and Dr. Ramesh Raman, in particular) for granting a special GAMS license to develop the models listed in Chapter 10. Thanks are also due to Bill Lowe, Joyce Alff, Judy King and Gulf Publishing Company for their cooperation.

The quotations that appear at the start of the Chapters 1, 5, and 8 are reproduced by permission from the following sources: Chandra -- A biography of S. Chandrasekhar by Kameshwar C. Wali, The University of Chicago Press, Chicago (1990); How to get control of your time and your life by Alan Lakein, Random House, Inc., New York (1973); and Surely you're joking, Mr. Feynman! by Richard P. Feynman, W. W. Norton & Company, Inc., New York (1985).

I am indebted to IIT-Bombay (and the Chemical Engineering Department and the CAD Centre, in particular) for their support. I am also indebted to Carnegie Mellon University, where I learnt a lot especially from Professors

Herbert L. Toor, Ignacio E. Grossmann, Gary J. Powers, Arthur W. Westerberg, and Larry T. Biegler.

I am deeply grateful to the following individuals for agreeing to review portions of the manuscript and for providing valuable comments:

Prof. Zsolt Fonyo	Prof. Duncan M. Fraser
Prof. Truls Gundersen	Prof. Jacek Jezowski
Mr. Ian C. Kemp	Dr. Kirtan K. Trivedi
Prof. Robert M. Wood	Dr. Terrence F. Yee

I cannot find words to describe the debt I owe to the students at IIT-Bombay for their tireless efforts in developing the computer software and putting together the basic material for the book. Two of them deserve special mention: Yogesh Makwana was responsible in developing Chapters 2 and 10, and Kaushal Patel was the key person in developing the HX software.

In addition, the following students have contributed in a significant way:

Zafar Ali	Rohit Angle	Garima Bhatia
Sandip Bhatwadekar	Sandeep Desai	Amit Duvedi
Sunil Dwivedi	Rahul Ganguli	Smita Golwelker
Pallav Jain	Bikram Kapoor	S. Mohan Krishna
Pranay Narvekar	Shwetal Patel	Y. Purushottam
D. Rajeshkannan	Sameer Ralhan	Kavita Ramanan
Bindu Priya Reddy	Ravi Kishan Reddy	Naveen Sachdeva
Hiren Shethna	Nomula Srinivas	P. Srinivas
Rohit Talwalkar	Kaustubh Tukdev	Ravi Varadachari
Bhushan Vartak	Natarajan Venkatesh	

Last but not the least, sincere thanks to my wife, Maya, for being the first person to read (or should I say, proofread) the entire book.

Uday V. Shenoy
Bombay, India

HEAT EXCHANGER NETWORK SYNTHESIS

CHAPTER 1

From Targeting to Supertargeting

The simple is the seal of the true.
And beauty is the splendor of truth.

Subrahmanyan Chandrasekhar

The words with which S. Chandrasekhar ended his Nobel lecture on 8 December 1983 are appropriate to begin this book. Within the context of Optimization of Processes by Energy and Resource Analysis (OPERA), they apply to the concept of targeting. Targeting, which involves predicting the optimum performance for a process prior to any detailed design, is simple in terms of its usage of fundamental physical principles and has a unique beauty as a tool for preliminary energy and resource analysis. It is the basic philosophy underlying the discussion in the book.

A variety of keywords may be used to describe the subject matter of the book: heat exchanger network synthesis (HENS), energy conservation, process integration, pinch analysis, and mathematical programming. Much of the book focuses on HENS, as it is the topic of process synthesis that is best understood and most developed to date. However, it would be inappropriate not to step beyond the boundaries of HENS to present a glimpse of the future. The term *OPERA* has been coined for this precise purpose, to emphasize the future expanded scope for this technology. First, the focal point of the analysis can be energy, but it must include other resources such as capital equipment (e.g., heat exchanger area), power (e.g., consumption in pumps and compressors through pressure drops), shaftwork (e.g., turbines), refrigerants (e.g., in low temperature processes), water, mass separating agents, waste water, environmental emissions (viewed as negative resources), and capacity (e.g., in debottlenecking). Second, along with the optimal design of heat recovery systems, it is important to consider the flow system, the reactor, and the separation system. Third, integration need not be restricted to a single process, but may need to be viewed in the context of the overall plant. Fourth, the technology must be applicable to continuous processes as well as batch processes (where time, labor, and raw materials tend to be important resources). Finally, being resourceful also implies designing for flexibility, multiple-base cases, safety, good controllability, and operability.

1

Pinch analysis (Linnhoff, 1993a) undoubtedly forms the core of OPERA. It has enjoyed many industrial successes and is therefore the emphasis of the book. However, the structural optimization from pinch analysis may be usefully supplemented by parametric optimization through mathematical programming to create a powerful hybrid design tool.

So, OPERA is the final mission, and the current state of the technology appears to be evolving towards it. As the basic principles from HENS may be extended towards establishing the foundation for OPERA, the thorough understanding of HENS is the logical starting point.

The problem of heat exchanger network synthesis (HENS) may be defined as the determination of a cost-effective network to exchange heat among a set of process streams, where any heating and cooling not satisfied by exchange among these streams must be provided by external utilities (e.g., steam, hot oil, cooling water, and refrigerants). Additional constraints like plant layout, safety, flexibility, operability, and controllability must be accounted for. To synthesize such a network, the following information must be available:

- set of hot streams to be cooled and set of cold streams to be heated;
- inlet (T_{in}) and outlet (T_{out}) temperatures of all these streams;
- flow rates (M) of these streams;
- specific heats (C_p) and heat transfer coefficients (h) of the streams, with their dependence on temperature;
- allowable/available pressure drops (ΔP) for the streams;
- available utilities (with their T_{in}, T_{out}, C_p, h, and costs); and
- cost law for heat exchange area, with plant life and rate of interest.

Pinch technology is one of the tools for solving the above problem. It recognizes the necessity of setting targets, i.e., predicting what is the best performance that can possibly be achieved by the process, before actually attempting to achieve it. Thus, targeting allows the process engineer to determine the minimum utility requirements, area, units, shells, and cost prior to actual design of a heat exchanger network.

The best way to illustrate the basic concepts in pinch technology is to consider a process scheme typically encountered in the chemical industries. Figure 1.1 shows a simplified version of a practical example involving a portion of a petrochemical process. The process involves an exothermic reaction, followed by separation using distillation.

From the viewpoint of HENS, two hot streams and two cold streams may be defined as relevant to heat integration. The feed to the reactor (stream C3) is first heated by the reactor effluent from 20° to 85°C, and then by steam in heater 1 to 155°C. The reactor effluent (stream H1) is cooled by its inlet stream in exchanger 3 from 175° to 45°C. The top product (stream C4) from the distillation column is heated from 40° to 112°C by the bottoms in exchanger 2. The bottoms product (stream H2) is first cooled by the top product from 125° to 98°C, and then by cooling water in cooler 4 to a final temperature of 65°C suitable for onward processing.

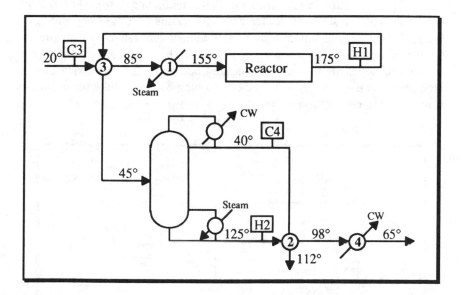

Figure 1.1 The starting point in the application of pinch technology is a simplified flowsheet showing the major unit operations with the heating and cooling duties. (Temperature in °C)

The heat exchanger network for the above flowsheet is conveniently represented using the grid diagram (Linnhoff et al., 1982) as in Figure 1.2. Hot streams are shown by arrows running conventionally from left to right, and cold streams by arrows running right to left. Thus, the temperature axis implicitly runs from right to left, in the direction of the cold streams. Exchangers are drawn as matches between hot and cold streams, i.e., two circles connected by a vertical line with the heat load indicated at the bottom in bold (e.g., units 2 and 3 in Figure 1.2). Usually, heaters appear at the extreme left on the cold streams (e.g., unit 1), and coolers to the extreme right on the hot streams (e.g., unit 4). Temperatures are shown above the "stream arrows" at appropriate points. As will be seen later, the major advantage of this representation in minimum energy designs is that the pinch can be clearly identified as a vertical temperature line dividing the "above pinch" region on the left from the "below pinch" region on the right.

It is important to understand the basic concepts in heat integration before proceeding to the targeting algorithms. Each stream can be represented by its temperature-enthalpy profile (see T-Q plots on the left in Figures 1.3a and 1.3b), with the slope simply given by the reciprocal of the heat capacity flow rate (MC_p as indicated in parenthesis in Figures 1.3a and 1.3b). The profiles will be straight-line segments for streams with constant MC_p, as $dQ = MC_p dT$. The MC_p constancy assumption over finite temperature ranges for all streams is not a serious limitation, as a curved line may be approximated by several straight-line segments to obtain the desired accuracy. The approach to heat integration has traditionally involved overlapping hot and cold stream profiles on temperature-enthalpy diagrams.

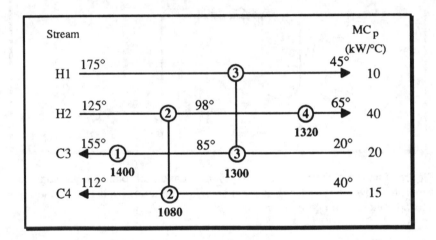

Figure 1.2 The grid representation conveniently displays the hot and cold streams running countercurrently, along with the process exchangers, heater(s), and cooler(s) forming the network. (Temperature in °C and load in kW)

The important step in pinch technology is to combine all hot streams to generate a hot composite curve (HCC as in Figure 1.3a) and all cold streams to generate a cold composite curve (CCC as in Figure 1.3b). The combination is done by summing up the MC_p values of all streams of the same kind (either hot or cold) within each temperature interval. Note that the limits of the temperature intervals correspond to the inlet and outlet temperatures of the streams.

The HCC and CCC (Whistler, 1948; Hohmann, 1971) provide a comprehensive picture of the heat supply and heat demand for the overall process over the entire temperature range. In essence, all process streams may be combined and visualized as a single hot stream and a single cold stream in counterflow (with a piecewise linear variation in their MC_p). When the two composite curves are drawn on the same temperature-enthalpy axis (Figure 1.4), their relative horizontal position is not defined since the enthalpy axis for the hot and cold profiles can be independently (and differently) chosen. Their relative position may, however, be fixed by specifying a minimum temperature difference for heat exchange (ΔT_{min}).

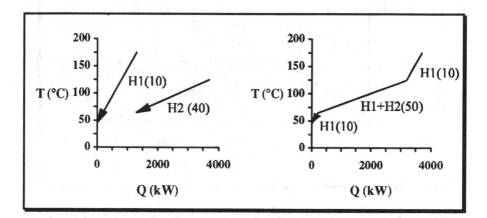

Figure 1.3a The temperature-enthalpy profiles of the individual hot streams (on the left) are combined to obtain the hot composite curve (on the right).

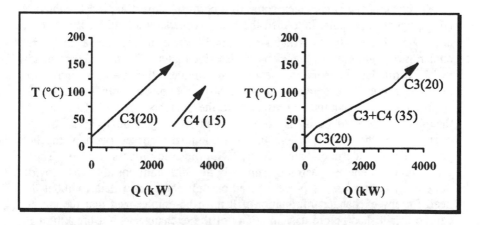

Figure 1.3b The temperature-enthalpy profiles of the individual cold streams (on the left) are combined to obtain the cold composite curve (on the right).

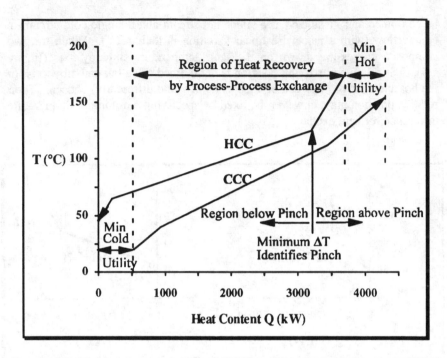

Figure 1.4 The hot composite curve (HCC) and cold composite curve (CCC) respectively show the heat availability and heat requirement for the overall process.

The composite curves then come to a point of closest approach, which is ΔT_{min} apart. This point is termed the pinch (Umeda et al., 1978; Linnhoff et al., 1979; Linnhoff and Hindmarsh, 1983) since it represents the most constrained region for heat recovery as described later. The pinch corresponds to a particular temperature level and divides the process into two thermodynamically separate regions. Above the pinch, only hot utility is required and its magnitude corresponds to the overshoot of the CCC. Below the pinch, only cold utility is necessary and its magnitude is given by the overshoot of the HCC. Where the two composite curves overlap, the hot process streams are in enthalpy balance with the cold process streams. These utility requirements are thermodynamically the minimum possible and provide the energy targets for the process based on only the stream data (without the need for any exchanger information). It must be emphasized that the targets can be established before designing any of the exchangers — a breakthrough that indeed revolutionized the thinking in the area of HENS!

A systematic, tabular approach to targeting is presented in this chapter, where all calculations are done conveniently and compactly through tables.

1.1 ENERGY TARGETING

Maximum energy recovery (MER) implies using the minimum amount of utilities. If Q_{hu} is the heat supplied by hot utility and Q_{cu} is the heat removed by cold utility, then computation of energy targets involves determining the minimum values of Q_{hu} and Q_{cu}. This must be achieved subject to thermodynamic constraints.

1.1.1 Role of Thermodynamic Laws

Consider the four-stream data in Figure 1.2 (see Case Study 4S1 in Appendix B) in terms of the heating and cooling duties. The enthalpy change of a stream may be simply calculated by multiplying its heat capacity flow rate (MC_p) by its change in temperature. Specifically, the enthalpy change for stream H1 is given by $10 (175° - 45°) = 1300$ kW. The maximum amounts of hot and cold utility required may be calculated by summing the enthalpy changes of all the cold streams (ΣQ_c) and those of all the hot streams (ΣQ_h) respectively. Thus,

$$\text{maximum hot utility} = \Sigma Q_c = 2700 + 1080 = 3780 \text{ kW}$$
$$\text{and maximum cold utility} = \Sigma Q_h = 1300 + 2400 = 3700 \text{ kW}.$$

The above solution, which maximizes external utility usage, uses only heaters and coolers in the network. It does not use to advantage the fact that a hot process stream can be used to heat a cold process stream, thus reducing both the hot and cold utility loads by the amount of heat exchanged in the match.

In this context, the first law of thermodynamics may be stated as:

Heat available in hot streams/utilities = Heat required by cold streams/utilities

or $\Sigma Q_h + Q_{hu} = \Sigma Q_c + Q_{cu}.$ (1.1)

Equation 1.1 may be rearranged to give $(Q_{hu} - Q_{cu}) = \Sigma Q_c - \Sigma Q_h = 3780 - 3700 = 80$ kW. The minimum hot and cold utility requirement satisfying this equation is $Q_{hu} = 80$ kW and $Q_{cu} = 0$ kW, which assumes that heat can be transferred from any hot stream to any cold stream.

This is not necessarily feasible as the second law of thermodynamics states that heat transfer can occur only from higher temperatures to lower temperatures. There must be a positive temperature difference to provide an adequate driving force for heat transfer between hot and cold streams. Thus to calculate practical energy targets, the maximum amount of heat transfer possible with a stipulated minimum positive temperature difference (ΔT_{min}) must be determined and the remaining heat must be supplied by external utilities. Hohmann (1971) established a minimum utility target by feasibility table analysis; then, Linnhoff and Flower (1978) developed the problem table algorithm (PTA) to calculate the energy targets algebraically.

1.1.2 Problem Table Algorithm

The algorithm is best illustrated with the help of a typical example.

Example 1.1 Determination of Minimum Utilities
For the stream data in Case Study 4S1, determine the minimum hot utility and minimum cold utility requirements for $\Delta T_{min} = 20°C$.

Solution.   For the calculation of energy targets, only the inlet temperatures, outlet temperatures and heat capacity flow rates are required. The results on implementing the algorithm may be compactly tabulated as in Table 1.1. For each step of the algorithm, the calculation involved is given first and is followed by the rationale.

Step 1. Determination of Temperature Intervals (T_{int} *in Column A*)
$\Delta T_{min}/2$ is subtracted from the hot stream temperatures and $\Delta T_{min}/2$ is added to the cold stream temperatures. These temperatures are then sorted in descending order, omitting temperatures common to both hot and cold streams. These form the limits of the various temperature intervals.

This step ensures that there is an adequate driving force of ΔT_{min} between the hot and cold streams for possible heat transfer within each interval.

Step 2. Calculation of Net MC_p *in Each Interval* ($MC_{p,int}$ *in Column B*)
The sum of the MC_p values of the hot streams is subtracted from the sum of the MC_p values of the cold streams present in each temperature interval and entered in column B against the lower temperature limit of the interval. Thus,

$$MC_{p,int} = \Sigma MC_{p,c} - \Sigma MC_{p,h} \qquad \text{for an interval.} \qquad (1.2)$$

The first interval is between 165° and 122°C and contains streams H1 and C3. As per Equation 1.2, $MC_{p,int} = 20 - 10 = 10$ kW/°C. Similarly, the second interval is between 122° and 115°C and contains streams H1, C3, and C4.

Thus, $MC_{p,int}$ = 20 + 15 − 10 = 25 kW/°C. For convenience in manual determination of the stream populations in each interval, the streams may be represented by vertical bars as in Table 1.1. A zero is placed for the first entry in column B.

Step 3. Calculation of Net Enthalpy in Each Interval (Q_{int} in Column C)
    The $MC_{p,int}$ (calculated in Step 2) is multiplied by the temperature difference for that interval to obtain the heat requirement in the interval (Q_{int}).

For the first interval, Q_{int} = 10 (165° − 122°) = 430 kW.
For the second interval, Q_{int} = 25 (122° − 115°) = 175 kW.

These are the net surplus (Q_{int} < 0) or deficit (Q_{int} > 0) in each interval.

Table 1.1 Implementation of Problem Table Algorithm

Interval i					Col. A T_{int}	Col. B $MC_{p,int}$	Col. C Q_{int}	Col. D Q_{cas}	Col. E R_{cas}
0					165	0	0	0	605
1					122	10	430	−430	175
2					115	25	175	−605	0
3					55	−15	−900	295	900
4					50	25	125	170	775
5					35	10	150	20	625
6					30	20	100	−80	525
Stream	H1	H2	C3	C4					
MC_p	10	40	20	15					

Units: temperature in °C, heat capacity flow rate in kW/°C, and enthalpy in kW.

Step 4. Calculation of Cascaded Heat (Q_{cas} in Column D)
    The net enthalpy in an interval (obtained in Step 3) is subtracted from the cascaded heat in the previous interval to obtain the cascaded heat in that interval. This gives column D by the formula (note that $Q_{cas,i} = 0$ for $i = 0$):

$$Q_{cas,i} = Q_{cas,i-1} - Q_{int,i}$$

(1.3)

Thus, $Q_{cas,1}$ = 0 − 430   = −430 kW,
$Q_{cas,2}$ = −430 − 175 = −605 kW, and so on.

Cascading of heat essentially involves transferring heat from an interval to the next lower temperature interval, starting with the highest interval. Note that in addition to complete heat transfer within an interval, heat can also be transferred from higher temperature intervals to lower ones. This cascaded heat offers the advantage of satisfying some of the heat deficit in the lower intervals.

Step 5. Revision of Cascaded Heat (R_{cas} in Column E)
The most negative Q_{cas} in column D is subtracted from each value in that column to obtain the revised cascaded heat (R_{cas}) in column E. Thus,

$$R_{cas,i} = Q_{cas,i} - min(Q_{cas}) \tag{1.4}$$

provided $min(Q_{cas}) < 0$. In our example, column E is obtained by simply adding 605 to column D.
The cascaded heat needs to be revised since a negative heat transfer (indicated in column D) is thermodynamically infeasible. The negative heat transfer is a consequence of the heat from a higher interval being inadequate to satisfy the requirements of lower intervals. This may be rectified by supplying just enough heat at the highest temperature interval (through a hot utility) to ensure positive heat transfer at every interval. The excess heat at the lowest temperature interval cannot be rejected to any cold stream and thus constitutes the minimum cold utility requirement.
Note that, if all the values in column D are nonnegative, then the cascaded heat requires no revision. Such problems are called threshold problems and are dealt with in detail later in Section 2.1.3.

Step 6. Determination of Energy Targets
The minimum hot utility requirement ($Q_{hu,min}$) and the minimum cold utility requirement ($Q_{cu,min}$) are the first and last values in column E. The temperature in column A that corresponds to zero revised cascaded heat in column E is called the pinch temperature. Its significance is discussed in Section 1.1.4. The energy targets may thus be read off directly from Table 1.1, and these results are summarized in Table 1.2.

1.1.3 Minimum Utility Requirements

The difference between Q_{hu} and Q_{cu} is a constant, given by Equation 1.1 as per the first law of thermodynamics. The minimum Q_{hu} and minimum Q_{cu}

(Table 1.2), in addition to obeying the first law and differing by 80 kW for the case study under consideration, also obey the second law of thermodynamics.

Table 1.2 Results of Energy Targeting

Minimum Hot Utility	= 605
Minimum Cold Utility	= 525
Pinch Temperature	= 115

Units: temperature in °C and utility requirements in kW.

The values of 605 kW and 525 kW are indeed the minimum utility requirements. If the hot utility is further increased, then the extra heat is transferred across the pinch and the cold utility required increases by the same amount in accordance with the first law. Thus, a double energy penalty is incurred, which is referred to as the phenomenon of "more in, more out" in pinch technology parlance.

1.1.4 Concept of Pinch

The pinch temperature (T_p), as determined by the PTA, for Case Study 4S1 is 115°C. On adding $\Delta T_{min}/2$ to T_p, the pinch for the hot streams occurs at 125°C. Similarly, on subtracting $\Delta T_{min}/2$ from T_p, the pinch for the cold streams is at 105°C.

The pinch divides the process into two parts: a high temperature heat sink above the pinch where heat is accepted from only a hot utility and a low temperature heat source below the pinch where heat is rejected to only a cold utility.

From the PTA, it is clear that a cold utility above the pinch will lead to an increase in utility load. If x units of heat from any temperature interval above the pinch are rejected to a cold utility, there will be less heat available for cascading to the lower temperatures, resulting in a negative heat transfer at the pinch. This may be rectified by providing x more units of heat by hot utility. The cooling requirement below the pinch remains unaltered; thus, the overall utility consumption increases by $2x$. An analogous argument holds for a hot utility placed below the pinch.

If x units of heat are transferred across the pinch, then the hot utility required above the pinch as well as the cold utility required below the pinch each increase by x units. As before, a double penalty of $2x$ units is incurred.

To gain a better understanding, consider Figure 1.2 where ΔT_{min} is 13°C (on unit 2). By performing the PTA at ΔT_{min} = 13°C, the reader may verify that the minimum hot and cold utility requirements are 360 kW and 280 kW respectively. The pinch occurs at 125°C for hot streams and at 112°C for cold streams. Now, on Figure 1.2, a portion of the hot utility (namely, 20 x [112° – 85°] = 540 kW) is placed below the pinch. Also, unit 3 transfers 500 kW of heat (namely, 10 x [175° – 125°]) across the pinch. In accordance with the double energy penalty concept, it is observed in Figure 1.2 that the hot utility is 1400 kW (1040 kW above the minimum of 360 kW) and the cold utility is 1320 kW (1040 kW above the minimum of 280 kW).

The pinch temperature is thus crucial for designing MER networks (Linnhoff and Hindmarsh, 1983) that meet energy targets according to the following criteria:

1) no heat transfer must occur across the pinch;
2) no cold utility must be placed above the pinch; and
3) no hot utility must be placed below the pinch.

These form three basic pinch design rules and must be rigorously satisfied since they are based on the second law of thermodynamics (Linnhoff, 1989). For a grassroots design to satisfy the energy targets, there must not be any net heat transfer across the pinch. This provides a powerful design principle -- the complex task of energy optimization of an integrated system is reduced to the relatively simple design objective of zero heat transfer across the pinch. This may be easily implemented for grassroots designs by designing separately above and below the pinch. For retrofitting, the principle indicates that an effort must be made to eliminate or reduce cross-pinch matches.

1.1.5 Grand Composite Curve (GCC)

A plot of column A vs. column E of Table 1.1 yields the GCC (Figure 1.5), which shows the heat available in various temperature intervals. It also is a representation of the net heat flows in the process, which are zero at the pinch. On the GCC, a process sink segment shows an increase in temperature and an increase in enthalpy. A process source segment shows a decrease in temperature and, contrary to convention, an increase in enthalpy. Thus, a process source segment on the GCC uses a sign convention for enthalpy opposite to what is normally used (Linnhoff et al., 1982).

Though the total amounts of the hot and cold utilities are clearly seen on the composite curves in Figure 1.4, their temperature levels are not. These are ideally represented on the GCC, which is particularly useful in deciding the placement of hot and cold utilities (as discussed in detail in Section 1.7.3). The GCC is also useful for profile matching during total energy integration studies (e.g., combined heat and power cycles) as demonstrated in Chapter 8.

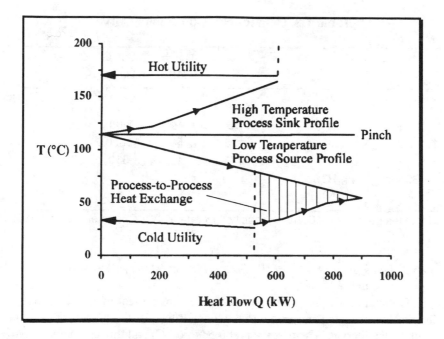

Figure 1.5 The grand composite curve (GCC) shows zero heat flow at the pinch, heating deficit above the pinch, and cooling deficit below the pinch.

1.1.6 The Composite Curves

The plot of the HCC and CCC on a single temperature - enthalpy axis (Figure 1.4) is sometimes called the Hohmann-Lockhart (1976) plot. A method for the algebraic determination of the data for this plot is given in the example below.

Example 1.2 Plotting of the Hot and Cold Composite Curves
For the stream data in Case Study 4S1, plot the hot composite curve (HCC) and the cold composite curve (CCC) for $\Delta T_{min} = 20°C$.

Solution. The plotting of the HCC is conveniently done with the aid of Table 1.3, which is generated by the procedure given below using steps similar to those in the PTA.

Step 1. Sorting of Hot Stream Temperatures (T_h in Column A)
The hot stream temperatures are sorted in ascending order, omitting repeated entries.

Table 1.3 Data for Hot Composite Curve

Interval i			Col. A T_h	Col. B $SumMC_{p,h}$	Col. C $Q_{int,h}$	Col. D $CumQ_h$
0			45	0	0	0
1			65	10	200	200
2			125	50	3000	3200
3			175	10	500	3700
Stream	H1	H2				
MC_p	10	40				

Units: temperature in °C, heat capacity flow rate in kW/°C, and enthalpy in kW.

Step 2. Calculation of MC_p *of Hot Streams in Each Interval (*SumMC$_{p,h}$ *in Column B)*

The sum of the MC_p values of the hot streams present in each temperature interval is calculated and entered in column B against the higher temperature limit of the interval. The first interval (45° to 65°C) and the last interval (125° to 175°C) contain only stream H1 ($SumMC_{p,h} = 10$) while the middle interval (65° to 125°C) contains both streams H1 and H2 ($SumMC_{p,h} = 10 + 40 = 50$). A zero is appropriately placed as the first entry in column B.

Step 3. Calculation of Enthalpy in Each Interval ($Q_{int,h}$ *in Column C)*

The $SumMC_{p,h}$ in each interval (calculated in step 2) is multiplied by the temperature difference for that interval to obtain column C. Thus, for the three intervals, the enthalpies are 10 (65° − 45°) = 200 kW, 50 (125° − 65°) = 3000 kW, and 10 (175° − 125°) = 500 kW.

*Step 4. Calculation of Cumulative Enthalpy (*CumQ$_h$ *in Column D)*

This column is calculated using the formula:

$$CumQ_{h,i} = CumQ_{h,i-1} + Q_{int,hi} \tag{1.5}$$

with $CumQ_{h,i} = 0$ for $i = 0$.

Step 5. Plotting of HCC

Column A is plotted against column D of Table 1.3 to obtain the HCC (Figure 1.4).

Step 6. Generation of CCC Data

The procedure to be followed for the CCC is virtually identical to that adopted for the HCC and is summarized below. The cold stream temperatures are sorted in ascending order, omitting repeated entries, to obtain T_c in column A of Table 1.4. The sum of the MC_p values of the cold streams present in each temperature interval is calculated and entered as $SumMC_{p,c}$ in column B against the higher temperature limit of the interval. A zero is placed for the first entry in column B. The $SumMC_{p,c}$ in each interval is multiplied by the temperature difference for that interval to obtain the enthalpy in each interval ($Q_{int,c}$ in column C). Column D for the cumulative enthalpy ($CumQ_c$) is calculated by using the formula:

$$CumQ_{c,i} = CumQ_{c,i-1} + Q_{int,ci} \tag{1.6}$$

with $CumQ_{c,i} = Q_{cu,min}$ for $i = 0$. This is the only difference between the plotting procedures for the HCC and the CCC. While the HCC starts from zero enthalpy, the CCC is displaced by the cold utility target (for this example, $Q_{cu,min} = 525$ kW from Table 1.2). The above procedure results in Table 1.4, and the CCC (Figure 1.4) is obtained by plotting column A vs. column D.

Table 1.4 Data for Cold Composite Curve

Interval			Col. A	Col. B	Col. C	Col. D
i			T_c	$SumMC_{p,c}$	$Q_{int,c}$	$CumQ_c$
0			20	0	0	525
1			40	20	400	925
2			112	35	2520	3445
3			155	20	860	4305
Stream	C3	C4				
MC_p	20	15				

Units: temperature in °C, heat capacity flow rate in kW/°C, and enthalpy in kW.

To recapitulate, the HCC and CCC merge the effect of all the hot streams and cold streams respectively. When viewed on a single temperature-enthalpy diagram (Figure 1.4), the region of overlap of the two curves represents the extent of heat recovery possible. For a given ΔT_{min}, the overshoot of the HCC on the left side of the figure represents the minimum cooling requirement while the overshoot of the CCC on the right side represents the minimum heating requirement. The temperature difference between the two curves is least at the pinch, where it equals ΔT_{min}.

The greater the value of ΔT_{min}, the larger is the minimum utility consumption. For a larger ΔT_{min}, the composite curves must be further apart. If the CCC is shifted to the right by x units, then both the overshoots increase by that amount. Since the overshoots represent the utility consumption, an increase in either utility can be seen to cause a corresponding increase in the other. The composite curves plot is also useful for area and shell targeting as demonstrated in the next few sections.

1.2 AREA TARGETING

The calculation of $Q_{hu,min}$ and $Q_{cu,min}$ by the PTA is useful in estimating utility costs; similarly, the calculation of areas is useful in estimating capital costs prior to actual network invention.

Area targeting involves calculation of the minimum surface area for heat transfer among hot streams, cold streams, and utilities in a HEN to be yet designed. The area is calculated assuming overall countercurrent heat exchange which manifests itself as vertical heat transfer on the composite curves. Strictly speaking, this is the minimum area only when the heat transfer coefficients (h) of all the streams and utilities are equal.

More complicated targeting procedures (Ahmad, 1985; Colberg and Morari, 1990) consider nonvertical or crisscrossed heat transfer on the composite curves. This sometimes makes better use of the available temperature differences based on the relative heat transfer coefficients of the streams, thus predicting lower area targets. However, according to Linnhoff and Ahmad (1990), the deviation of the target area calculated below (Equation 1.7) from the minimum area is not more than 10% of the entire area even when the heat transfer coefficients differ by one order of magnitude. This accuracy suffices for targeting purposes in most cases.

In essence, the area targeting methods below provide the true minimum area when h-values of all streams are equal and the near-minimum area in other cases. The true minimum area when h-values of various streams differ considerably may be obtained by mathematical programming (see Chapter 10).

1.2.1 Basic Equation for Area Targeting

The composite curves plot is modified to include the utilities and then divided into a number of enthalpy intervals by drawing a vertical line at every vertex (where there is a change in the slope of the T-Q curve) on the HCC as well as the CCC. The area target is basically given by the following equation (Townsend and Linnhoff, 1984; Linnhoff and Ahmad, 1990):

$$A = \sum_i \left(\frac{1}{F\,LMTD}\right)_i \sum_j \left(\frac{Q_j}{h_j}\right)_i \qquad (1.7)$$

where F is the correction factor accounting for noncountercurrent flow, $LMTD$ is the logarithmic mean temperature difference for the interval, Q_j is the enthalpy change of the j-th stream, h_j is the heat transfer coefficient of the j-th stream, subscript i denotes the i-th enthalpy interval, and subscript j denotes the j-th stream. The summation over the streams existing in each enthalpy interval may be split into the two summations, one over the hot streams (jh) and the other over the cold streams (jc). This gives

$$\sum_j \frac{Q_j}{h_j} = \sum_{jh}\left(\frac{Q}{h}\right)_{jh} + \sum_{jc}\left(\frac{Q}{h}\right)_{jc} \qquad (1.8)$$

Within an enthalpy interval, all the hot streams undergo the same temperature change (dT_h) as do all the cold streams (dT_c). Noting that $Q = MC_p dT$, Equation 1.8 yields

$$\sum_j \frac{Q_j}{h_j} = (dT_h)_i \sum_{jh}\left(\frac{MC_p}{h}\right)_{jh} + (dT_c)_i \sum_{jc}\left(\frac{MC_p}{h}\right)_{jc} \qquad (1.9)$$

The use of Equations 1.7 and 1.9 to calculate the near-minimum network area is illustrated next with the help of examples. The calculations are systematically carried out through a series of tables, convenient for computer implementation. The first step in area estimation is the generation of a balanced composites plot that includes utilities.

Example 1.3 Plotting of the Balanced Composite Curves

For the data in Case Study 4S1, plot the balanced hot composite curve (BHCC) and the balanced cold composite curve (BCCC) for $\Delta T_{min} = 20°C$.
Given: hot utility inlet and outlet temperatures are 180°C and 179°C;
  cold utility inlet and outlet temperatures are 15°C and 25°C.

Solution. In addition to the data required for energy targeting, the inlet and outlet temperatures of the utilities are now necessary. The procedure for plotting the BHCC and BCCC is the same as that used in Example 1.2, except that the utilities are treated as additional streams.

Step 1. Calculation of Minimum Hot and Cold Utility
    The energy targets are calculated by the PTA explained in Section 1.1.2. The values of $Q_{hu,min}$ and $Q_{cu,min}$ from Table 1.2 are 605 kW and 525 kW.

Step 2. Calculation of Utility Flow Rates
    The MC_p values of the hot utility (hu) and cold utility (cu) are given by

$$(MC_p)_{hu} = Q_{hu,min}/(T_{in} - T_{out})_{hu} \qquad (1.10a)$$
$$\text{and } (MC_p)_{cu} = Q_{cu,min}/(T_{out} - T_{in})_{cu} \qquad (1.10b)$$

Using the utility inlet and outlet temperatures, the values calculated are $(MC_p)_{hu} = 605/(180° - 179°) = 605$ kW/°C and $(MC_p)_{cu} = 525/(25° - 15°) = 52.5$ kW/°C.

Step 3. Calculation of Balanced Composite Curve Data
    The utilities are treated as additional streams, and the composite curve data are calculated by the same procedure as in Section 1.1.6. In brief, the procedure to generate Table 1.5 is as follows:
  a) The temperatures for the hot streams and hot utility are sorted in ascending order (T_{hb} in column A).
  b) The sum of the MC_p values of the hot streams and utility present in each temperature interval is calculated and entered in column B against the higher temperature limit of the interval ($SumMC_{p,hb}$ in column B). A zero is placed as the first entry in column B.
  c) The $SumMC_{p,hb}$ in each interval (column B) is multiplied by the temperature difference for that interval to obtain $Q_{int,hb}$ in column C.
  d) The cumulative enthalpy in column D is calculated using the formula:

$$CumQ_{hb,i} = CumQ_{hb,i-1} + Q_{int,hbi} \qquad (1.11)$$

with $CumQ_{hb,i} = 0$ for $i = 0$.

    Thus, Table 1.5 is merely an extended version of Table 1.3 with the hot utility information added (intervals 4 and 5).
    The method for generating the BCCC data is identical, and Table 1.6 may be obtained by repeating the procedure given above from a) to d) on the cold stream/utility data. Though Table 1.6 is an extension of Table 1.4 with the cold utility information added, it is different in two ways. First, the cumulative

enthalpy value at the lowest temperature in column D is zero (and not $Q_{cu,min}$ as in Table 1.4). Second, as the cold utility outlet temperature is greater than the lowest inlet temperature of the cold process streams, the utility modifies the shape of the composite curve (rather than merely getting appended to it).

Table 1.5 Data for Balanced Hot Composite Curve

Interval i				Col. A T_{hb}	Col. B $SumMC_{p,hb}$	Col. C $Q_{int,hb}$	Col. D $CumQ_{hb}$
0				45	0	0	0
1				65	10	200	200
2				125	50	3000	3200
3				175	10	500	3700
4				179	0	0	3700
5				180	605	605	4305
Stream	H1	H2	HU				
MC_p	10	40	605				

Units: temperature in °C, heat capacity flow rate in kW/°C, and enthalpy in kW.

Table 1.6 Data for Balanced Cold Composite Curve

Interval i				Col. A T_{cb}	Col. B $SumMC_{p,cb}$	Col. C $Q_{int,cb}$	Col. D $CumQ_{cb}$
0				15	0	0	0
1				20	52.5	262.5	262.5
2				25	72.5	362.5	625
3				40	20	300	925
4				112	35	2520	3445
5				155	20	860	4305
Stream	C3	C4	CU				
MC_p	20	15	52.5				

Units: temperature in °C, heat capacity flow rate in kW/°C, and enthalpy in kW.

Step 4. Plotting of BHCC and BCCC

Column A is plotted against column D of Table 1.5 to obtain the BHCC while column A is plotted against column D of Table 1.6 to obtain the BCCC (Figure 1.6). The balanced composites (as their name suggests) are in perfect enthalpy balance, as they consider both the process as well as utility streams.

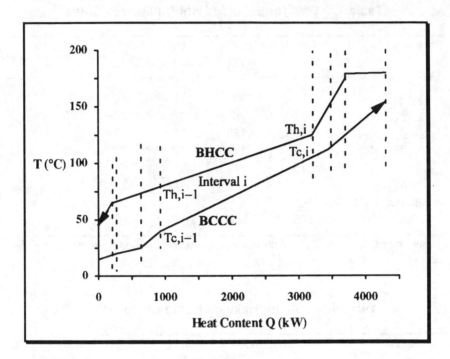

Figure 1.6 The balanced composite curves, on being divided into enthalpy intervals, allow calculation of the area target based on a model of vertical heat transfer.

Example 1.4 Calculation of Target Area for Countercurrent Exchangers

For Case Study 4S1, determine the heat transfer area required for a minimum energy network employing countercurrent exchangers. Assume the heat transfer coefficient (including film, wall, and fouling resistances) to be 0.2 kW/(m^2 °C) for every stream.

Solution. For area targeting, the heat transfer coefficients must be provided. The starting point is the balanced composites plot (Figure 1.6), which must be divided into vertical enthalpy intervals as follows. Table 1.7 tabulates these intermediate results.

Step 1. Determination of Enthalpies for Intervals (CumQ$_i$ *in Column A)*

Columns D of Table 1.5 and Table 1.6 are merged, omitting cumulative enthalpies common to both tables, and the entries are then sorted in ascending order.

This identifies all points where either composite curve has a vertex (change in slope) and thus determines the various enthalpy intervals over which the area is to be summed.

*Step 2. Calculation of Interval Temperatures on BHCC (*T$_{hi}$ *in Column B)*

For each *CumQ$_i$* in column A of Table 1.7, a table lookup is performed on Table 1.5. The least value of *CumQ$_{hb}$* in column D of Table 1.5 that satisfies *CumQ$_{hb}$* ≥ *CumQ$_i$* is identified. Let this value be in row *r*. Then,

$$T_{hi} = T_{hb,row\ r} \quad \text{if } CumQ_{hb,row\ r} = CumQ_i \text{ or } 1/SumMC_{p,hb\ row\ r} = 0$$
or
$$T_{hi} = T_{hb,row\ r} - (CumQ_{hb,row\ r} - CumQ_i)/SumMC_{p,hb\ row\ r}$$
$$\text{in all other cases.} \quad (1.12a)$$

Thus, for *CumQ$_i$* = 0 kW,   *r* = 1, T_{hi} = 45°C
   for *CumQ$_i$* = 200 kW,   *r* = 2, T_{hi} = 65°C
   for *CumQ$_i$* = 262.5 kW, *r* = 3, T_{hi} = 125° – (3200–262.5)/50 = 66.25°C.

Each enthalpy interval is delimited by a vertex on the BHCC or the BCCC. If the vertex is on the BHCC, then the temperature of column A of Table 1.5 corresponding to that cumulative enthalpy is the required interval temperature. If the vertex is on the BCCC, then the required interval temperature can be interpolated from the temperature and enthalpy values of the next vertex on the BHCC, keeping in mind that the slope of the curve below that vertex is given by (1/*SumMC$_{p,hb}$*). In case the BHCC is horizontal in a region indicating a phase change (i.e., 1/*SumMC$_{p,hb}$* = 0), the required temperature will be equal to the temperature at the next vertex on the BHCC.

*Step 3. Calculation of Interval Temperatures on BCCC (*T$_{ci}$ *in Column C)*

These temperatures are calculated in a manner similar to that in step 2. For each *CumQ$_i$* in column A of Table 1.7, the least value of *CumQ$_{cb}$* in column D of Table 1.6 is identified such that *CumQ$_{cb}$* ≥ *CumQ$_i$*. Let this value be in row *r*. Then,

$T_{ci} = T_{cb,row\ r}$ if $CumQ_{cb,row\ r} = CumQ_i$ or $1/SumMC_{p,cb\ row\ r} = 0$
or
$T_{ci} = T_{cb,row\ r} - (CumQ_{cb,row\ r} - CumQ_i)/SumMC_{p,cb\ row\ r}$

in all other cases.   (1.12b)

Thus, for $CumQ_i = 0$ kW, $r = 1, T_{ci} = 15°C$
for $CumQ_i = 200$ kW, $r = 2, T_{hi} = 20° - (262.5 - 200)/52.5 = 18.81°C$
for $CumQ_i = 262.5$ kW, $r = 3, T_{hi} = 20°C$.

Table 1.7 Determination of the Enthalpy Intervals

i					Col. A $CumQ_i$	Col. B T_{hi}	Col. C T_{ci}	Col. D $\Sigma(MC_p/h)_h$	Col. E $\Sigma(MC_p/h)_c$
0					0	45	15	0	0
1					200	65	18.81	50	262.5
2					262.5	66.25	20	250	262.5
3					625	73.5	25	250	362.5
4					925	79.5	40	250	100
5					3200	125	105	250	175
6					3445	149.5	112	50	175
7					3700	175	124.75	50	100
8					3700	179	124.75	0	100
9					4305	180	155	3025	100

H1 H2 C3 C4 HU   CU
MC_p 10 40 20 15 605 52.5

Units: temperature in °C, heat capacity flow rate in kW/°C, enthalpy in kW,
and heat transfer coefficient in kW/(m^2 °C).

*Step 4. Calculation of $\Sigma(MC_p/h)$ in Each Interval ($\Sigma(MC_p/h)_h$ in Column D
and $\Sigma(MC_p/h)_c$ in Column E)*

These columns are calculated in a manner similar to columns B in Tables 1.5 and 1.6. Using the heat transfer coefficients, the sum of the (MC_p/h) values of all the hot process and utility streams present in each temperature interval is calculated and entered in column D against the higher temperature limit of the interval (with a zero placed for the first entry in column D). Similarly, the sum of the (MC_p/h) values of all the cold process and utility streams present in each temperature interval is calculated and entered in column E against the higher temperature limit of the interval (with a zero placed for the first entry in column E). For determining stream populations in each interval, the streams are represented by vertical bars as in Table 1.7.

For example, considering column D, the first interval contains only stream H1 ($\Sigma(MC_p/h)_h = 10/0.2 = 50$), the next four intervals contain both streams H1 and H2 ($\Sigma(MC_p/h)_h = 50/0.2 = 250$), the next two intervals contain only stream H1 ($\Sigma(MC_p/h)_h = 50$), and the last interval contains only the hot utility ($\Sigma(MC_p/h)_h = 605/0.2 = 3025$).

Having characterized the various enthalpy intervals, Equations 1.7 and 1.9 may be readily used after generating Table 1.8. For convenience, columns B and C of Table 1.7 are copied as columns A and B in Table 1.8.

Table 1.8 Calculation of Countercurrent Area

Interval i	Col. A $T_{h,i}$	Col. B $T_{c,i}$	Col. C $Sum(Q/h)$	Col. D $LMTD_i$	Col. E A_i
0	45	15	0	0	0
1	65	18.81	2000	37.51	53.31
2	66.25	20	625	46.22	13.52
3	73.5	25	3625	47.37	76.53
4	79.5	40	3000	43.85	68.42
5	125	105	22750	28.65	794.00
6	149.5	112	2450	27.84	88.01
7	175	124.75	2550	43.56	58.53
8	179	124.75	0	52.22	0
9	180	155	6050	37.76	160.24

Units: temperature in °C, enthalpy in kW,
        heat transfer coefficient in kW/(m^2 °C), and area in m^2.

Step 5. Calculation of Sum(Q/h) *in Each Interval as per Equation 1.9*
        *(*Sum(Q/h) *in Column C)*

The $\Sigma(MC_p/h)_h$ in column D of Table 1.7 for the BHCC is multiplied by the hot composite stream temperature difference from column B of Table 1.7 and then added to the corresponding value for the cold composite curve. As per Equation 1.9 using the values in Table 1.7, interval i of Table 1.8 is given by

$$Sum(Q/h) = \quad (T_{h,i} - T_{h,i-1}) \ \Sigma(MC_p/h)_{h,i}$$
$$+ (T_{c,i} - T_{c,i-1}) \ \Sigma(MC_p/h)_{c,i} \qquad\qquad \text{for } i \geq 1$$

$$Sum(Q/h) = \quad 0 \qquad\qquad\qquad\qquad\qquad \text{for } i = 0. \quad (1.13)$$

Thus, $Sum(Q/h)_{i=1} = (65° - 45°) \ 50 + (18.81° - 15°) \ 262.5 = 2000$
$\quad\quad Sum(Q/h)_{i=2} = (66.25° - 65°) \ 250 + (20° - 18.81°) \ 262.5 = 625$

Step 6. Calculation of Log Mean Temperature Difference in Each Interval
    (LMTD$_i$ in Column D)
This is easily done by the following formula:

$$LMTD_i = \frac{(T_{h,i} - T_{c,i}) - (T_{h,i-1} - T_{c,i-1})}{\ln\left[\dfrac{T_{h,i} - T_{c,i}}{T_{h,i-1} - T_{c,i-1}}\right]} \qquad\qquad \text{for } i \geq 1$$

$$LMTD_i = 0 \qquad\qquad\qquad\qquad\qquad \text{for } i = 0 \quad (1.14)$$

Equation 1.14 yields the following:

$$LMTD_{i=1} = [(65 - 18.81) - (45 - 15)]/\ln(46.19/30) = 37.51°C.$$

Step 7. Calculation of Countercurrent Exchanger Area in Each Interval (A$_i$ in
    Column E)
This is obtained by merely dividing the *Sum(Q/h)* value in column C by
the corresponding *LMTD$_i$* in column D for the interval. Note that $F = 1$ for a
pure countercurrent exchanger. So, the area per interval is given by

$$A_i = [Sum(Q/h)]/LMTD_i. \qquad\qquad\qquad\qquad (1.15)$$

Specifically, $A_i = 0$ (for $i = 0$) and $A_i = 2000/37.51 = 53.31 \text{ m}^2$ (for $i = 1$).

Step 8. Calculation of Countercurrent Exchanger Area Target
On summing the A_i values in column E over all the intervals, the total
countercurrent area is

$$A_c = \Sigma A_i = 1312.57 \text{ m}^2.$$

1.2.2 Understanding the Uniform BATH Formula

Equation 1.7 is referred to as the "Uniform BATH Formula" since it uses a uniform value of ΔT_{min} and was proposed by Townsend and Linnhoff (1984) in a research meeting at Bath (U.K.). The formula does not in general give the absolute minimum total area. The assumption of countercurrency implicit in the composite curves strictly leads to the minimum area only if the overall heat transfer coefficient, U, is constant for all matches (Nishimura, 1980).

In Example 1.4, the U values for the various matches are the same (0.1 kW/m^2 °C) as in the following equation:

$$1/U = 1/h_{jh} + 1/h_{jc} \tag{1.16a}$$

Thus, the target area of 1312.57 m^2 is indeed the minimum for the given stream set. The question to be addressed at this stage is "what is the matching scheme corresponding to this minimum area?"

Overall countercurrency manifests itself as vertical heat transfer on the balanced composite curves and may be ensured by first dividing the plot into vertical enthalpy intervals as in Figure 1.6. Next, a matching scheme may be developed within each interval as follows (Linnhoff and Ahmad, 1990):

- each hot stream is split into as many branches as the number of cold streams within the interval;
- each cold stream is split into as many branches as the number of hot streams within the interval; and
- every hot stream is then matched with every cold stream once.

Every match occurs between the temperature limits of the enthalpy interval (Figures 1.7a and 1.7b) and will therefore necessarily appear vertical on the composite curves. On applying the above procedure to the sixth enthalpy interval in Table 1.7 (where T_{hi} goes from 125° to 149.5°C, and T_{ci} goes from 105° to 112°C), it is seen that two matches are required as in Figure 1.7a. Though the network in Figure 1.7a is simple, this is not the case in general.

For example, the network for the third interval in Table 1.7 requires a number of stream splits and matches (Figure 1.7b). As the design becomes complex and impractical with increasing number of streams within an enthalpy interval, such designs (as in Figure 1.7b) are called "spaghetti" networks (Ahmad and Smith, 1989). The match loads in spaghetti networks can be determined by splitting the heat content of a stream into fractions proportional to the MC_p-values of the streams of the opposite kind. Thus, in Figure 1.7b, the match loads are:

Load on match 1 = 10 (73.5° − 66.25°) x (20/72.5) = 20 kW
Load on match 2 = 10 (73.5° − 66.25°) x (52.5/72.5) = 52.5 kW
Load on match 3 = 40 (73.5° − 66.25°) x (20/72.5) = 80 kW
Load on match 4 = 40 (73.5° − 66.25°) x (52.5/72.5) = 210 kW.

As the $LMTD$ = 47.366°C and U = 0.1 kW/m^2 °C, the areas for the four matches are 4.22, 11.08, 16.89, and 44.34 m^2. This adds to a total area of 76.53 m^2, which is in agreement with the area for the interval in Table 1.7.

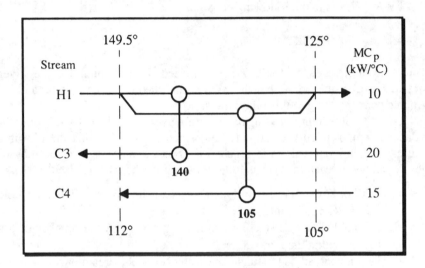

Figure 1.7a The vertical heat transfer model for area targeting implies true countercurrency within each enthalpy interval. Thus, if a single hot stream and two cold streams exist in an enthalpy interval, the hot stream must be split for matching with each of the cold streams. (Temperature in °C and load in kW)

The spaghetti network requires (jh x jc) matches in every enthalpy interval, with every match having precisely the same temperature profiles as those on the composite curves within that enthalpy interval. Thus, the spaghetti design uses the available driving forces to the maximum extent possible for the overall network. As an aside, it may be noted that a minimum of (jh + jc − 1) matches are required for vertical heat transfer in every interval as shown later in Figure 1.8.

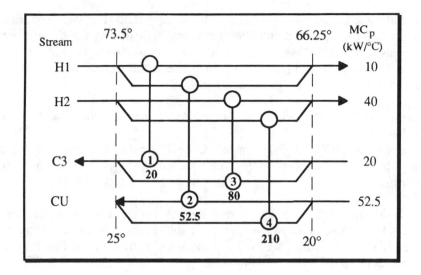

Figure 1.7b If two hot streams and two cold streams exist in an enthalpy interval, then every stream must be split into two to obtain a spaghetti design for vertical heat transfer. (Temperature in °C and load in kW)

For the purposes of the present discussion, consider the rationalization of Equation 1.7 based on the spaghetti network in Figure 1.7b. If the values of U for the various matches are different, then the "countercurrent" area for the interval i is given by

$$A_i = (1/LMTD_i) [(Q_{13}/U_{13}) + (Q_{1U}/U_{1U}) + (Q_{23}/U_{23}) + (Q_{2U}/U_{2U})] \quad (1.16b)$$

The subscripts refer to the match (e.g., 13 refers to the match between streams H1 and C3). On using Equation 1.16a and rearranging,

$$A_i = (1/LMTD_i) [(1/h_1)(Q_{13} + Q_{1U}) + (1/h_2)(Q_{23} + Q_{2U})$$
$$+ (1/h_3)(Q_{13} + Q_{23}) + (1/h_U)(Q_{1U} + Q_{2U})] \quad (1.16c)$$

If Q_1, Q_2, Q_3, and Q_U denote the enthalpy changes of the streams H1, H2, C3, and CU in the temperature interval, then

$$A_i = (1/LMTD_i) [(Q_1/h_1) + (Q_2/h_2) + (Q_3/h_3) + (Q_U/h_U)] \quad (1.16d)$$

Equation 1.16d may be generalized and extended to all the intervals to derive the Uniform BATH formula (Equation 1.7). The discussion so far has focused on counterflow exchangers. The extension to 1-2 shell and tube exchangers (Ahmad and Smith, 1989) requires incorporating the F-correction factor as discussed below. (The reader may wish to browse through Section 6.1 before proceeding to Example 1.5 on 1-2 shell and tube exchangers, if not familiar with the basic design of exchangers.)

Example 1.5   Calculation of Target Area for 1-2 Shell and Tube Exchangers

For Case Study 4S1, determine the heat transfer area required for a maximum energy recovery (MER) network employing 1-2 shell and tube exchangers.

Solution. The major task here is the calculation of the additional area required for an exchanger of the 1-2 type (1 shell pass-2 tube passes) using the F-correction factor which accounts for noncountercurrent flow. This may be done with the aid of Table 1.9 as follows:

Step 1. Calculation of the Temperature Effectiveness for an Interval (P in Column A)

$$P_i = (T_{h,i} - T_{h,i-1}) / (T_{h,i} - T_{c,i-1}) \tag{1.17}$$

A typical entry in column A of Table 1.9 is

$$P_{i=1} = (65° - 45°)/(65° - 15°) = 0.4.$$

The temperature values are listed in columns A and B of Table 1.8.

Equation 1.17 follows from the definition of the heat exchanger temperature effectiveness, which is defined as the ratio of the temperature change in one of the streams (say, the hot one) to the maximum possible temperature difference.

Step 2. Calculation of the Capacity Flow Rate Ratio (R in Column B)

$$R_i = (T_{c,i} - T_{c,i-1}) / (T_{h,i} - T_{h,i-1}) \tag{1.18}$$

A typical entry in column B of Table 1.9 is

$$R_{i=1} = (18.81° - 15°)/(65° - 45°) = 0.1905.$$

Note that R is defined as the ratio of the heat capacity flow rates (say, of the hot streams to the cold streams). By an overall energy balance, R is also the ratio of the temperature change in one of the streams (say, the cold one) to that of the other stream (the hot one).

Table 1.9 Calculation of F-Correction Factor for Noncountercurrent Flow

Interval i	Col. A P	Col. B R	Col. C P_{12}	Col. D S	Col. E F
1	0.4000	0.1905	0.8150	0.2841	0.9908
2	0.0263	0.9524	0.5400	0.0237	0.9999
3	0.1355	0.6897	0.6197	0.1160	0.9973
4	0.1101	2.5000	0.2907	0.2152	0.9921
5	0.5353	1.4286	0.4314	1.7304	0.8244
6	0.5506	0.2857	0.7740	0.5081	0.9612
7	0.4048	0.5000	0.6875	0.3944	0.9706
8	0.0000	0.0000	0.0000	0.0000	1.0000
9	0.0181	30.2500	0.0293	0.3630	0.9963

Step 3. Calculation of the Temperature Effectiveness of an Individual 1-2 Exchanger (P_{12} in Column C)

$$P_{12} = X_p P_{max} \qquad \text{where } P_{max} = 2/(R + 1 + (R^2 + 1)^{1/2}) \qquad (1.19)$$

Column C is obtained by simple substitution in Equation 1.19 (with $X_p = 0.9$),

e.g., $P_{12,i=1} = 0.9 \times 2/(0.1905 + 1 + (0.1905^2 + 1)^{1/2}) = 0.815$.

P_{12} is the temperature effectiveness of each individual 1-2 shell and tube heat exchanger, where a number of such identical shells are connected in series to meet the heat transfer requirements within an interval. While designing such exchangers, a commonly used rule of thumb requires that $F > 0.75$ or 0.8 to ensure that a slight shift in the operating point (due to uncertainties or inaccuracies) does not result in a precipitous drop in the exchanger performance. An alternative criterion proposed by Ahmad et al. (1988) requires that practical designs be limited to some fraction (X_p) of the maximum asymptotic value of P (say, P_{max}) for a specified R. Typically X_p is chosen to be 0.9, as this guarantees that F is greater than 0.75 as well as avoids regions where $F(R, P)$ has a steep slope. Using this criterion to maximize the effectiveness for a feasible 1-2 exchanger, P_{12} may be found from Equation 1.19.

Step 4. Calculation of Number of 1-2 Shells Needed in Series (S in Column D)

$$S = \ln[(1 - RP)/(1 - P)] / \ln[(1 - RP_{12})/(1 - P_{12})] \quad \text{for } R \neq 1$$
$$\text{and } S = [P/(1 - P)] / [P_{12}/(1 - P_{12})] \qquad\qquad \text{for } R = 1 \qquad (1.20)$$

Column D is obtained by merely substituting values in Equation 1.20,

e.g., $S_{i=1} = \ln[(1 - 0.1905 \times 0.4)/(1 - 0.4)]/\ln[(1 - 0.1905 \times 0.815)/(1 - 0.815)]$
$\qquad = 0.2841$.

Multiple shells in series are required when a single shell does not provide a sufficiently high value of F (> 0.75) or the slope of $F(R, P)$ is too steep. S is the number of shells required in series for heat exchange in a particular enthalpy interval. This number as per Equation 1.20 is nonintegral and would need to be rounded off to the next largest integer (say, NS) in actual practice as done below.

Step 5. Calculation of F-Correction Factor on Rounding Off to Integral
       Number of Shells (F in Column E)

a) The following formulae are used to recalculate the effectiveness after the integral number of shells to be used within an interval is decided.

$$P_{12r} = (1 - Y)/(R - Y) \text{ where } Y = [(1 - RP)/(1 - P)]^{1/NS} \text{ for } R \neq 1$$
$$P_{12r} = P/(NS - NS\, P + P) \qquad\qquad\qquad\qquad \text{for } R = 1 \,(1.21)$$

Here, P_{12r} is the effectiveness for a single 1-2 shell (after rounding off S to NS), P is the effectiveness for the overall series of shells (see column A of Table 1.9), R is the capacity flow rate ratio (which is independent of the number of shells for a series circuit), and NS is the integral number of shells required in an enthalpy interval (which is taken to be the least integer greater than S).

b) The number of transfer units, N, is calculated by the expression (Bowman et al., 1940) given below in terms of P_{12r} and R:

$$N = \frac{1}{\sqrt{R^2 + 1}} \ln\left[\frac{2 - P_{12r}(R + 1 - \sqrt{R^2 + 1})}{2 - P_{12r}(R + 1 + \sqrt{R^2 + 1})}\right] \qquad (1.22)$$

c) The F-correction factor (Bowman et al., 1940) may be finally calculated as follows:

$$F = \ln[(1 - P_{12r})/(1 - RP_{12r})] / [N (R - 1)] \qquad \text{for } R \neq 1$$
$$F = P_{12r} / [N (1 - P_{12r})] \qquad\qquad\qquad \text{for } R = 1 \qquad (1.23)$$

It may be noted that $F = 1$ for a pure countercurrent exchanger.

On rounding off the values of S in column D, it is seen that $NS = 1$ for every interval except the fifth interval, where $NS = 2$. Calculations for the first interval are given below as an illustration:

$$Y = [(1 - 0.1905 \times 0.4)/(1 - 0.4)]^{1/1} = 1.5396$$

$$P_{12r} = (1 - 1.5396)/(0.1905 - 1.5396) = 0.4$$
$$\text{(as expected, this equals } P \text{ for } NS = 1)$$

$$N = \frac{1}{\sqrt{0.1905^2 + 1}} \ln\left[\frac{2 - 0.4(1.1905 - \sqrt{0.1905^2 + 1})}{2 - 0.4(1.1905 + \sqrt{0.1905^2 + 1})}\right] = 0.5381$$

$$F = \ln[(1 - 0.4)/(1 - 0.1905 \times 0.4)] / [0.5381 (0.1905 - 1)] = 0.9908.$$

Step 6. Calculation of Noncountercurrent Exchanger Area Target

The noncountercurrent exchanger area in each interval is obtained by dividing A_i in column E of Table 1.8 by the corresponding F in column E of Table 1.9. Therefore, the 1-2 exchanger area per interval is given by

$$A_{12i} = A_i/F \qquad (1.24)$$

On summing the A_{12i} values from Equation 1.24 over all the intervals, the total area for 1-2 shell and tube heat exchangers is obtained as follows:

$$A_{12} \text{ below pinch} = 53.31/0.9908 + 13.52/0.9999 + 76.53/0.9973$$
$$+ 68.42/0.9921 + 794/0.8244 = 1176.16 \text{ m}^2.$$

$$A_{12} \text{ above pinch} = 88.01/0.9612 + 58.53/0.9706 + 160.24/0.9963$$
$$= 312.7 \text{ m}^2.$$

$$A_{12} = \Sigma A_{12i} = 1176.16 + 312.7 = 1488.86 \text{ m}^2.$$

The results from Examples 1.4 and 1.5 for the target areas are summarized in Table 1.10.

Table 1.10 Results of Area Targeting

| | Target Area | |
	Countercurrent	1-2 Shell & Tube
Above Pinch	306.78	312.70
Below Pinch	1005.79	1176.16
Total	1312.57	1488.86

Units: area in m^2.

1.3 UNIT TARGETING

Unit targeting involves calculation of the minimum overall units, the minimum units for an MER network, the minimum units with vertical heat transfer, and the number of loops in the network.

The calculations are based on Euler's Network Theorem from graph theory, which in the context of HENS may be stated as (Linnhoff et al., 1979)

$$N_u = N_s + N_l - N_p \tag{1.25}$$

where N_u is the number of units, N_s is the number of streams (both process and utility), N_l is the number of loops, and N_p is the number of independent problems or subsystems. A loop in a network is any path that starts at a point and returns to the same point. An independent problem or subsystem (also referred to as a subset equality) is a set of streams which is perfectly matched (i.e., the streams are in enthalpy balance with each other).

1.3.1 Overall Unit Target

The minimum overall units target (also referred to as the Euler minimum unit target) is given by

$$N_{u,min} = N_s - 1. \tag{1.26}$$

As loops increase the number of units, N_l is set to zero for a design with minimum units. As the number of subset equalities in a particular problem cannot be determined a priori, the number of independent problems is taken to be unity (i.e., $N_p = 1$) for targeting purposes.

1.3.2 Unit Target for an MER Network

The target for the minimum number of units (or heat exchange matches) in an MER network is given by

$$N_{u,mer} = (N_a - 1) + (N_b - 1) \tag{1.27}$$

where N_a and N_b are the number of streams (both process and utility) above and below the pinch respectively. $N_{u,mer}$ is commonly referred to as simply the "units target."

An MER network recognizes a division at the pinch. Since there is no heat flow across the pinch, there is an independent problem above the pinch and another below it. Applying Equation 1.26 to the two subsystems above and below the pinch separately yields Equation 1.27 (Linnhoff et al., 1979). Equation 1.27 applies to the case of a single pinch (multiple pinches are discussed in Chapter 2).

1.3.3 Unit Target for Vertical Heat Transfer

The minimum number of units for a vertical heat transfer model is

$$N_{u,vht} = \Sigma\,(N_{si} - 1) \tag{1.28}$$

where N_{si} is the number of streams present in the i-th interval.

Here, the system is divided into several independent problems equal in number to the enthalpy intervals. Equation 1.26 is applied to each interval and the summation is carried out over all the intervals. This target may not be of practical significance since it leads to a network with a prohibitively large number of units (though less than in a spaghetti network).

A possible network that meets the vertical heat transfer unit target is shown in Figure 1.8. This network provides an alternative to the spaghetti design in Figure 1.7b. The driving forces for the enthalpy interval are the same as those in the spaghetti design. Matches are chosen so that their MC_p-ratio (of hot stream to cold stream) is identical to that on the composite curves for that interval. Also the match must satisfy the enthalpy change of the stream in the interval. On placing a match, one stream is eliminated; simultaneously, the MC_p-ratio of the remaining streams continues to be the same as that for the interval on the composite curves. This ensures that the $(N_{si} - 1)$ target can always be satisfied.

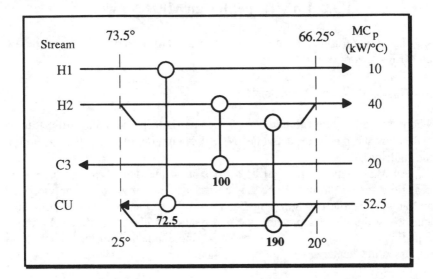

Figure 1.8 An alternative network to the spaghetti design in Figure 1.7b can be generated that meets the unit target for vertical heat transfer. (Temperature in °C and load in kW)

For Figure 1.8, the MC_p-ratio based on the composite curves is 50/72.5. Consider splitting of streams with higher MC_p. Stream H1 (with $MC_p = 10$) requires that it be matched with a cold stream of $MC_p = 14.5$ while stream C3 (with $MC_p = 20$) requires that it be matched with a hot stream of $MC_p = 13.793$. The final result is the network shown in Figure 1.8, where stream H2 is split into two branches (of $MC_p = 13.793$ and 26.207) and stream CU also needs to be split into two branches (of $MC_p = 14.5$ and 38). The reader may attempt to generate another network that satisfies the "vertical heat transfer" unit target (i.e., a further alternative to Figure 1.8), which involves splitting stream C3 rather than stream CU.

1.3.4 Number of Loops in a Network

The number of loops in any network is given by

$$N_l = N_u - N_{u,min} \tag{1.29}$$

where N_u is the number of units in the network and $N_{u,min}$ is the overall unit target from Equation 1.26. Equation 1.29 is obtained by combining Equations 1.25 and 1.26 with $N_p = 1$.

Example 1.6 Calculation of Unit Targets

For Case Study 4S1, determine the minimum overall units. Also, find the minimum units for an MER network, the minimum units with vertical heat transfer, and the number of loops in an MER network for $\Delta T_{min} = 20°C$.

Solution. Equations 1.26 through 1.29 yield the following results: As there are four process streams and two utilities in this case study,

$$N_{u,min} = 6 - 1 = 5.$$

As per Table 1.1, there are three process streams and one utility (hot) above the pinch. Below the pinch, there are four process streams and one utility (cold).

Thus, $N_{u,mer} = (4 - 1) + (5 - 1) = 7.$

The following observations are readily made on the basis of Table 1.7:

Interval i	1	2	3	4	5	6	7	9
Number of streams N_{si}	2	3	4	3	4	3	2	2

$$N_{u,vht} = 1 + 2 + 3 + 2 + 3 + 2 + 1 + 1 = 15.$$

The number of loops that cross the pinch in an MER network is given by

$$N_l = N_{u,mer} - N_{u,min} = 7 - 5 = 2.$$

The results for unit targets are summarized in Table 1.11.

Table 1.11 Results of Unit Targeting

Overall Unit Target = 5
MER Unit Target = 7
Vertical Heat Transfer Unit Target = 15
Loops in MER Network = 2

1.4 SHELL TARGETING

A single match in a heat exchanger network may require more than one shell. This may be due to the use of noncountercurrent exchangers or a restriction on the maximum area for a shell.

Shell targeting involves calculating the minimum number of shells below and above the pinch separately and then simply adding them to obtain the total number of shells for an optimal network. An optimal network makes the best use of the overall temperature differences between the streams, with the widest overall distribution being found on the composite curves (Nishimura, 1980). Thus, in accordance with the procedure used for area targeting, the number of shells required for a design that exactly obeys the vertical temperature differences on the composite curves is calculated. Although the vertical heat transfer model does not guarantee the true minimum for area targeting, it may yet provide the best model for shell targeting purposes (Ahmad and Smith, 1989). It may be noted that Trivedi et al. (1987a) proposed a "stepping-off" method on the composite curves (similar to the McCabe-Thiele construction) for shell targeting. It is not discussed here because it appears to be less reliable than the procedure of Ahmad and Smith (1989) and may significantly underpredict the shell target at times.

1.4.1 Basic Equations for Shell Targeting

An estimate of the minimum number of shells may be obtained from (Ahmad and Smith, 1989)

$$S_{min} = \Sigma \, S_{mi} \, (N_{si} - 1) \qquad\qquad (1.30)$$

where $(N_{si} - 1)$ is the minimum number of matches and S_{mi} is the number of shells required per match in the i-th enthalpy interval. Since shell targeting assumes vertical heat transfer on the composite curves, each interval forms an independent subsystem (i.e., the streams in each interval are perfectly matched) and Equation 1.26 may be applied to each interval separately to obtain the minimum number of matches $(N_{si} - 1)$. This is multiplied by the number of shells required per match, S_{mi}, and summed over all the enthalpy intervals to obtain Equation 1.30. In summing the real number of shells over all the intervals, an additivity property of shells from one interval to another is assumed, which may be easily verified (Ahmad and Smith, 1989). The prohibitively large number of matches implied in Equation 1.30 (see Section 1.3.3) does not lead to an impractical shell target because of the additivity property in terms of the real number of shells. Note that Equation 1.30

appropriately reduces to Equation 1.28 when every match requires only a single shell.

For an MER design, Equation 1.30 is applied separately below and above the pinch. The number of shells in the two regions separated by the pinch division must be integral; so, rounding off to the next highest integer is necessary to get the final shell targets.

Equation 1.30 may be rewritten as (Ahmad and Smith, 1989)

$$S_{min} = \Sigma (S_{mi} \cdot N_{si}) - \Sigma S_{mi} = \Sigma S_j - \Sigma S_{mi} \qquad (1.31)$$

where S_j represents the shell contribution of the j-th stream and is given by summing S_{mi} over all the enthalpy intervals in which stream j exists. Equation 1.31 holds because the shell contribution can be added either on an enthalpy-interval basis or on a streamwise basis, when the problem as a whole is considered.

Example 1.7 Rough Estimation of Shells

For Case Study 4S1, obtain a rough estimate of the number of shells required for a minimum energy network.

Solution. The enthalpy intervals in Table 1.7 provide a useful starting point and are used in Table 1.12 for the rough estimation of the shells.

Step 1. Characterization of Enthalpy Intervals (T_{hi}* in Column A and* T_{ci} *in Column B)*

Columns A and B are obtained by merely copying the T_{hi} and T_{ci} values listed in Table 1.7.

Step 2. Calculation of Number of Shells Required per Match (S_{mi}* in Column C)*

The number of shells required in each interval is obtained by merely copying the S values listed in Table 1.9. It may be noted that the number of shells per match for a countercurrent exchanger is one.

Importantly, each match in an interval requires the same number of shells because it has the same end temperatures (see Figures 1.7 and 1.8).

Step 3. Calculation of Number of Streams in an Interval (N_{si}* in Column D)*

The number of streams in each interval is counted after representing the streams by vertical bars as done earlier. For example, the first interval contains two streams (H1 and CU). In fact, these calculations were performed in Example 1.6 when calculating the unit target for vertical heat transfer.

Table 1.12 Approximate Determination of Shells

i							Col. A T_{hi}	Col. B T_{ci}	Col. C S_{mi}	Col. D N_{si}	Col. E $S_{mi}(N_{si}-1)$
0							45	15	0	0	0
1							65	18.81	0.2841	2	0.2841
2							66.25	20	0.0237	3	0.0474
3							73.5	25	0.1160	4	0.3481
4							79.5	40	0.2152	3	0.4305
5							125	105	1.7304	4	5.1912
6							149.5	112	0.5081	3	1.0163
7							175	124.75	0.3944	2	0.3944
8							179	124.75	0	0	0
9							180	155	0.3630	2	0.3630
	H1	H2	C3	C4	HU	CU					

Units: temperature in °C.

Step 4. Calculation of Number of Shells in an Interval (S_{mi} [N_{si} − 1] *in Column E*)

The minimum number of shells in an enthalpy interval is S_{mi} (N_{si} − 1). Using columns C and D, the third entry in column E is $0.0237 \times 2 = 0.0474$.

Step 5. Calculation of Rough Estimate of Shell Targets

The number of shells in enthalpy intervals below and above the pinch are summed separately. They are rounded off to the next highest integer before being added to get the total shell targets.

The pinch occurs at 125°C/105°C. So,

$$\text{shells below pinch} = 0.2841 + 0.0474 + 0.3481 + 0.4305 + 5.1912$$
$$= 6.3013, \text{ which may be rounded off to 7; and}$$
$$\text{shells above pinch} = 1.0163 + 0.3944 + 0.3630$$
$$= 1.7737, \text{ which may be rounded off to 2.}$$

Thus, the total number of shells required is 9.

Example 1.8 Accurate Determination of Shells

For Case Study 4S1, determine accurately the number of shells required for a minimum energy network. Given: maximum area per shell = 250 m^2.

Solution. This is primarily done by use of Equation 1.31. The calculations are summarized in Table 1.13a.

Table 1.13a Detailed Shell Targets

Row	Col. A *Stream Label*	Col. B *SumSj,bp*	Col. C *SumSj,ap*
1	H1	2.3695	0.9026
2	H2	2.0854	0
3	HU	0	0.3630
4	C3	2.0617	1.2655
5	C4	1.7304	0.5081
6	CU	0.4239	0

Step 1. Calculation of Shell Contribution on a Streamwise Basis (Stream Label in Column A, $SumS_{j,bp}$ in Column B, and $SumS_{j,ap}$ in Column C)

The number of shells required per match in an enthalpy interval is given in column C in Table 1.12. To calculate the shells required by any stream, it is sufficient to sum the shells required per match over the enthalpy intervals in which the stream exists. For an MER network, the shell contribution below the pinch ($SumS_{j,bp}$) and that above the pinch ($SumS_{j,ap}$) are calculated separately and entered against the appropriate stream label (column A). The calculations may be done easily with the help of Table 1.12. The sample calculation for stream H1 is as follows:

$SumS_{j,bp}$ = 0.2841 + 0.0237 + 0.1160 + 0.2152 + 1.7304 = 2.3695
(note that temperature range for stream H1 below pinch is 45° to 125°C)
$SumS_{j,ap}$ = 0.5081 + 0.3944 = 0.9026
(note that temperature range for stream H1 above pinch is 125° to 175°C)

Step 2. Sum Streamwise Shell Contributions after Appropriate Revision

If the shell contribution from a stream is less than 1, it is revised to unity. Then, the streamwise shell contributions below and above the pinch are summed separately to give $\Sigma S_{j,bp}$ and $\Sigma S_{j,ap}$ respectively.

Thus, $\Sigma S_{j,bp}$ = 2.3695 + 2.0854 + 2.0617 + 1.7304 + 1 = 9.2470
and $\Sigma S_{j,ap}$ = 1 + 1 + 1.2655 + 1   = 4.2655.

The rationale for the revision to unity is that the streams contributing a small fractional number of shells (i.e., less than one shell) will require at least one whole shell in the design. So, if the revision to unity is not done, then a gross underestimate may result for the shell target in practice, especially when the number of streams with such fractional contributions is large. This aspect was ignored in Example 1.7 and is the primary conceptual difference between the approximate and accurate methods for the determination of shell targets.

Step 3. Sum Enthalpywise Shell Contributions per Match
    The shells required per match (S_{mi} in column C of Table 1.12) in enthalpy intervals below and above the pinch are summed to get $\Sigma S_{m,bp}$ and $\Sigma S_{m,ap}$.

Thus, $\Sigma S_{m,bp}$ = 0.2841 + 0.0237 + 0.1160 + 0.2152 + 1.7304 = 2.3695
and $\quad \Sigma S_{m,ap}$ = 0.5081 + 0.3944 + 0.3630 $\quad\quad\quad\quad\quad$ = 1.2655.

Step 4. Calculation of Number of Shells
    As per Equation 1.31, the number of shells below and above the pinch may be calculated. These are rounded off to the next largest integer to give practical values.

$$S_{min,bp} = \Sigma S_{j,bp} - \Sigma S_{m,bp} = 9.2470 - 2.3695 = 6.8775 \text{ (rounded off to 7)}$$
and $\quad S_{min,ap} = \Sigma S_{j,ap} - \Sigma S_{m,ap} = 4.2655 - 1.2655 = 3.0000 \text{ (rounded off to 3)}$.

Step 5. Checking for Shell Size Violation
    If the maximum area per shell is specified (e.g., in this case, MaxArea = 250 m^2), then a check must be performed at this stage. From the area targets, the areas below and above the pinch can be calculated. Dividing the areas below and above the pinch (from step 6 in Example 1.5) by the corresponding number of shells (as calculated in step 4 above), the area per shell is

Area per shell below pinch = 1176.16/7 = 168.02 m^2
Area per shell above pinch =   312.70/3 = 104.23 m^2.

These are within the specified limit of 250 m^2. If these areas had exceeded the maximum stipulated area, the number of shells would have to be increased. This is done by determining the ratio of the target area to the maximum area per shell and then rounding this value off to the next largest integer.

Step 6. Calculation of Accurate Shell Targets
    The total number of shells is merely the sum of the minimum shells required below and above the pinch. Thus, the total shell target = 7 + 3 = 10.

The results from Examples 1.7 and 1.8 for shell targeting are summarized in Table 1.13b.

Table 1.13b Results of Shell Targeting

| | Number of Shells | | Area per shell |
	Approximate	Accurate	
Above Pinch	2 (1.7737)	3 (3.0000)	104.23
Below Pinch	7 (6.3013)	7 (6.8775)	168.02
Total	9	10	148.86

Units: area in m^2. Note: Values in parenthesis indicate number of shells before rounding off.

1.5 COST TARGETING

The cost of a network basically comprises the operating cost and the capital cost. These cost contributions are discussed below.

1.5.1 Operating Cost

The operating cost (OC) is a function of the energy requirements and is given by:

$$OC = (C_{hu} \cdot Q_{hu,min}) + (C_{cu} \cdot Q_{cu,min}) \qquad (1.32)$$

where C_{hu} and C_{cu} are the cost of unit load of hot and cold utility respectively and $Q_{hu,min}$ and $Q_{cu,min}$ are the minimum requirements of hot and cold utility respectively.

1.5.2 Capital Cost

The starting point is an expression for the capital cost (CC) of a single heat exchanger. If A is the surface area, then a simple cost law typically used has the form:

$$CC = a + b A^c \qquad \text{for a single exchanger} \qquad (1.33)$$

where a, b, and c are the cost law coefficients which depend on the material of construction, the pressure rating, and the type of heat exchanger. Note that the coefficient a represents the fixed costs, and CC in Equation 1.33 is the installed capital cost.

When establishing capital cost targets for a network, the area distribution among the individual exchangers comprising the network (yet to be designed) is not known. Consequently, it is simplest to assume each individual exchanger to have the same area. With this assumption that appears to give good capital cost predictions (Ahmad et al., 1990), the expressions for the network capital cost based on Equation 1.33 are

$$CC = N_{u,mer} [a + b (A_c/N_{u,mer})^c]$$

for a network of countercurrent exchangers   (1.34a)

$$CC = a N_{u,mer} + b S_{min} (A_{12}/S_{min})^c$$

for a network of 1-2 shell & tube exchangers   (1.34b)

where $N_{u,mer}$ is the minimum number of units in an MER network, A_c and A_{12} are the appropriate area targets, and S_{min} is the minimum shell target. Note that all these are targets established in the previous sections. For a network of 1-2 exchangers, Equation 1.34b appropriately accounts for two effects: the increased area ($A_{12} > A_c$) and the difference between the numbers of shells and units ($S_{min} \neq N_{u,mer}$, in general). Note that $S_{min} = N_{u,mer}$ for truly countercurrent exchangers, provided the maximum area per shell specification is not violated.

In addition to assuming uniform area distribution among the various exchangers, Equation 1.34 assumes that all exchangers use the same cost coefficients. These coefficients will vary for exchangers with different specifications, in which case the assumption of identical a, b, and c values for all exchangers will lead to unrealistic cost targets. The exchanger specifications (i.e., material of construction, pressure rating, and equipment type) are dictated by the chemical nature and operating conditions of the process streams. Methods which incorporate the effects of nonuniform exchanger specifications and nonuniform area distribution have been developed (Hall et al., 1990; Jegede and Polley, 1992b) and are discussed in the examples below.

1.5.3 Total Annual Cost

Since the energy cost is a recurring expense whereas the capital cost is a one-time investment, the expected life of the plant has to be considered while calculating annual costs. Thus, the total annual cost (*TAC*) is given by

$$TAC = OC + CC \cdot A_f \qquad \text{where } A_f = (1 + r)^t / t \qquad (1.35)$$

where A_f is the annualization factor, r is the rate of return of capital interest, and t is the expected plant life.

The calculation of cost targets for uniform exchanger specifications is straightforward and involves mere substitution in the relevant formulae, provided the cost data (C_{hu}, C_{cu}, a, b, c, r, and t) and other targets (energy, area, unit, and shell) are known. So, an example for cost targeting is included only for the case of nonuniform exchanger specifications.

Example 1.9 Cost Target for Nonuniform Exchanger Specifications

For the stream and utility data in Case Study 4S1 with $\Delta T_{min} = 20°C$, determine the total annual cost target for countercurrent flow. The following specifications need to be adhered to (note that SS stands for stainless steel and CS for carbon steel):

Stream	Exchanger Specification
H1	shell & tube (SS)
H2	spiral (SS)
C3	plate & frame (SS)
C4	shell & tube (SS)
HU	shell & tube (CS)
CU	shell & tube (CS)

Given: Annual cost of unit duty in $/(kW.yr): 120 (for HU) and 10 (for CU)
    Plant lifetime: 5 yr   Rate of interest: 10%
Assume cost data for different exchanger specifications as given in Table 1 14.

Table 1.14 Cost Data for Various Exchanger Specifications

Exchanger Specification	Capital Cost ($)	Identifier
spiral (SS-SS)	$30000 + 19700 \, A^{0.59}$	HX1
plate & frame (SS-SS)	$30000 + 1900 \, A^{0.78}$	HX2
shell & tube (CS-CS)	$30000 + 750 \, A^{0.81}$	HX3
shell & tube (SS-SS)	$30000 + 1650 \, A^{0.81}$	HX4
shell & tube (CS-SS)	$30000 + 1350 \, A^{0.81}$	HX5

Note: A is exchanger area in m^2; SS is stainless steel and CS is carbon steel.

Solution. The cost targeting is done here using the approach developed by Hall et al. (1990), which accounts for nonuniform exchanger specifications but assumes equal area distribution. Further, heat transfer coefficients are assumed to be match-independent. The basis of the method is discussed first.

Equation 1.7 provides an expression for the area target by summing the area contributions for each enthalpy interval. An equivalent expression based on summing the area contribution for each stream is (Ahmad et al., 1990)

$$A = \sum_j \left[\sum_i \left(\frac{1}{F \, LMTD}\right)_i \left(\frac{Q_j}{h_j}\right)_i \right] = \sum_j A_j \tag{1.36}$$

This form has the advantage that it clearly shows that each stream contributes to the total area, which in turn contributes to the total cost. The contribution of each stream depends on the specification it requires in terms of exchanger type, pressure rating, and material of construction. For example, a corrosive stream requiring a special material would contribute more to the capital cost. This was accounted for by Hall et al. (1990) in terms of an increased area contribution for the "special" stream by weighting its heat transfer coefficient. Note that this area (say, A^*) is a fictitious one and should not be confused with the actual area target. The purpose of this fictitious area is to yield a practical cost target when used with a single set of cost coefficients (a, b, and c). The method for determining the factor ϕ_j used for weighting heat transfer coefficients is explained next.

In the discussion below, specification refers to the material of construction, or the pressure rating, or the exchanger type. The specification that is used for the majority of streams is chosen to be the reference specification, and quantities associated with it are denoted by the subscript r. Therefore, the capital cost of a single exchanger using the reference specification is $CC_r = a_r + b_r A^{cr}$. Denoting the special specification by the subscript s, the capital cost for a single "special" exchanger is $CC_s = a_s + b_s A^{cs}$. To use the reference cost law for the special exchanger, a fictitious area A^* is defined that satisfies the following:

$$A^* = [(b_s/b_r)^{1/cr} A^{cs/cr - 1}] A \tag{1.37a}$$

Equation 1.37a is simply derived from $CC_s = a_s + b_s A^{cs} = a_r + b_r (A^*)^{cr}$, assuming $a_s = a_r$. Hall et al. (1990) have argued that the exchanger cost data can typically be adjusted in a manner which prevents coefficient a from changing with exchanger specification.

From $Q = U A (F\ LMTD)$, it is seen that A is proportional to $1/U$ because $Q/(F\ LMTD)$ is constant for a particular exchanger. So, Equation 1.37a, in terms of the overall heat transfer coefficient, is

$$1/U^* = [\,(b_s/b_r)^{1/cr}\ A^{cs/cr\,-1}\,]\,(1/U) \tag{1.37b}$$

The overall coefficient is defined in terms of the thermal resistance contributions of the two streams (each including wall and fouling) and is approximately given by $1/U = 1/h_{hot} + 1/h_{cold}$ (see Equation 1.16a). Thus, Equation 1.37b may be decomposed into individual stream heat transfer coefficients to yield

$$h_j^* = \phi_j\,h_j \qquad \text{where } \phi_j = (b_r/b_s)^{1/cr}\ A^{1-cs/cr} \tag{1.37c}$$

Equation 1.37c provides the definition of the stream-dependent weighting factor, ϕ_j. The above arguments can be extended to define the weighting factors, ϕ_j, in an entire heat exchanger network as

$$\phi_j = (b_r/b_s)^{1/cr}\ (A_c/N_{u,mer})^{1-cs/cr}$$
for a network of countercurrent exchangers $\hspace{1em}$ (1.38a)

or $\quad \phi_j = (b_r/b_s)^{1/cr}\ (A_{12}/S_{min})^{1-cs/cr}$
for a network of 1-2 shell & tube exchangers $\hspace{1em}$ (1.38b)

In Equation 1.38, A_c and A_{12} are the actual target areas, $N_{u,mer}$ is the minimum number of units in an MER network, and S_{min} is the minimum shell target.

With the necessary expressions derived, two comments are in order at this stage. If h_j^* values for the various streams differ by more than one order of magnitude, a more accurate value of A^* may be obtained by mathematical programming. A simpler version of Equation 1.38 (namely, $\phi_j = (b_r/b_s)^{1/cr}$) may be used if $c_s = c_r$; thus, the cost laws may be manipulated a priori as done by Hall (1986) for shell and tube exchangers to ensure that c does not change for different specifications.

The costing procedure is outlined below. First, calculate ϕ_j using Equation 1.38 for streams with a special specification. Next, calculate $h_j^* = \phi_j\,h_j$. Then compute A^* from Equation 1.7 using h_j^* in place of h_j. Finally, calculate the target capital cost for the mixed-specification network from Equation 1.34 using A^* in place of A. The stepwise calculations are now given for Case Study 4S1.

Step 1. Selection of Reference Specification
Here, the shell and tube (CS) exchanger may be chosen as the reference material. Thus, $a_r = 30000$, $b_r = 750$, and $c_r = 0.81$.

Step 2. Calculation of Weighting Factors for Special Streams
All process streams need to be treated as special streams. Use of Equation 1.38a gives

$$\phi_1 = \phi_4 = (750/1650)^{1/0.81} = 0.3778$$
$$\phi_2 = (750/19700)^{1/0.81} (1312.57/7)^{1-0.59/0.81} = 0.0733$$
$$\phi_3 = (750/1900)^{1/0.81} (1312.57/7)^{1-0.78/0.81} = 0.3853$$

Obviously, the value of ϕ_j for the reference streams (in this case, the utility streams HU and CU) is unity.

Step 3. Calculation of Weighted Heat Transfer Coefficients for Special Streams
Using $h_j^* = \phi_j h_j$, it is seen that

$$h_1^* = h_4^* = 0.3778 \times 0.2 = 0.07556 \text{ kW/(m}^2 \text{ °C)};$$
$$h_2^* = 0.0733 \times 0.2 = 0.01466 \text{ kW/(m}^2 \text{ °C)}; \text{ and}$$
$$h_3^* = 0.3853 \times 0.2 = 0.07706 \text{ kW/(m}^2 \text{ °C)}.$$

For the utility streams (i.e., HU and CU), $h_j^* = h_j = 0.2 \text{ kW/(m}^2 \text{ °C)}$.

Step 4. Calculation of Fictitious Area A*
The calculation of the area target is repeated as in Example 1.4 (left as an exercise for the reader), with the difference that the heat transfer coefficients for process streams are taken as calculated in step 3 rather than 0.2 kW/(m^2 °C). This yields the fictitious area A^* as 7411.01 m^2 for $\Delta T_{min} = 20$°C.

Step 5. Calculation of Capital Cost Target
The installed capital cost of the network is determined using Equation 1.34 and the coefficients for the reference cost law. For a network of countercurrent exchangers,

$$CC = N_{u,mer} [a + b (A^*/N_{u,mer})^c]$$
$$= 7 [30000 + 750 (7411.01/7)^{0.81}] = 1689.89 \times 10^3 \text{ \$}.$$

Step 6. Calculation of Total Annual Cost Target
The operating cost is calculated using Equation 1.32 as

$$OC = (C_{hu} \cdot Q_{hu,min}) + (C_{cu} \cdot Q_{cu,min})$$
$$= (120)(605) + (10)(525) = 77.85 \times 10^3 \text{ \$/yr.}$$

Finally, the total annual cost is determined using Equation 1.35 as

$$TAC = OC + CC(1 + r)^t/t$$
$$= 77850 + (1689.89 \times 10^3)(1 + 0.1)^5/5 = 622.17 \times 10^3 \text{ \$/yr.}$$

For a network using all CS exchangers in counterflow, the reader may verify that the TAC is 262.79×10^3 \$/yr.

1.6 SUPERTARGETING OR ΔT_{min} OPTIMIZATION

The targeting thus far has assumed a particular value of ΔT_{min} in order to provide a minimum driving force for heat transfer. The choice of a reasonable value of ΔT_{min} has traditionally been based on intuition and experience to account for the capital cost (e.g., 20°C in above-ambient processes, 5°C in refrigeration systems, and so on, as indicated by Townsend and Linnhoff [1984]). However, a poor choice could result in a nonoptimal network.

The goal of HENS is to design a cost-efficient network. In other words, the network must feature minimum utility requirements, minimum area, and minimum number of units/shells. However, tradeoffs between these desired features are often necessary and, consequently, the need for optimization arises. This may be simply explained as follows. The total cost of a network is the sum of the operating cost and the capital cost (Equation 1.35), both of which are functions of ΔT_{min}. The higher the value of ΔT_{min}, the higher will be the energy requirements and, consequently, the operating costs. On the other hand, with increasing ΔT_{min}, the lower will be the area requirements and, consequently, the capital costs. Thus, there will be an optimum value of ΔT_{min} for which total annual cost is a minimum. The process of pre-design optimization for ΔT_{min} based on total cost (with no piece of equipment specified) is called supertargeting (Linnhoff and Ahmad, 1989).

The Hohmann-Lockhart plot shows the basis of supertargeting. On this plot, the value of ΔT_{min} determines the relative positions of the composite curves. An increase in ΔT_{min} corresponds to a horizontal shift of the CCC to the right and results in an increase in the overshoots of the HCC and CCC. Thus, the utility requirements increase (in a piecewise linear manner) with increasing ΔT_{min}. On the other hand, the area requirement decreases with increasing ΔT_{min}. This is due to the increase in the temperature differences that provide the driving forces for heat transfer on the composite curves. The dependence of the number of units/shells on ΔT_{min} is more complicated. For example, the units target for an MER network changes as a function of ΔT_{min} as the stream population at the pinch varies. The variations of these targets with ΔT_{min} will be examined in detail in Chapter 2 under continuous targeting. For the present, it may be noted that the procedures in the previous sections of this chapter may be used to obtain the energy, area, unit, and shell targets at various values of ΔT_{min}. These may then be converted into operating cost and capital cost profiles using Equations 1.32 and 1.34. Summing operating and annualized capital costs as per Equation 1.35 gives the total annual cost profile. Typical *TAC* vs. ΔT_{min} profiles are shown in Figure 1.9. They go through a minimum, as expected, and show discontinuities wherever there is a change in number of units.

A HEN designed by the conventional pinch design method at an arbitrary (experience-based) value of ΔT_{min} does not guarantee cost-optimality. This is because the traditional approach involved two separate activities: first, synthesis for energy efficiency (see Chapter 3), and then optimization of the capital cost on the actually designed HEN by tedious evolution methods (as discussed in Chapter 4). This is in contrast to supertargeting, where the optimization is performed based on operating and capital cost targets together. This practically guarantees near global optimality ahead of design (Linnhoff and Ahmad, 1989). Thus, supertargeting ensures an initial design at an optimal ΔT_{min} that requires minimal evolution and saves considerable effort in post-design optimization.

Example 1.10 Pre-design Optimization of ΔT_{min}

Determine the optimum value of ΔT_{min} (to an accuracy of 1°C) for the Case Study 4S1, which results in the minimum total annual cost for the following cases:

countercurrent exchangers of carbon-steel; and
nonuniform exchanger specifications as given in Example 1.9.

Solution. The procedure to be adopted is summarized below.

Step 1. Choose a value of ΔT_{min}.

Step 2. Determine the energy, area, units, and cost targets as in Examples 1.1, 1.4, 1.6, and 1.9 respectively.

Step 3. Repeat steps 1 and 2 for various values of ΔT_{min}.

Step 4. Plot the curves of *TAC* vs. ΔT_{min} for the two cases and determine the optimum ΔT_{min} corresponding to the minimum cost.

The results (ideally obtained through a computer program like HX) are summarized in Table 1.15, and the total cost target profiles are plotted in Figure 1.9 for the two cases.

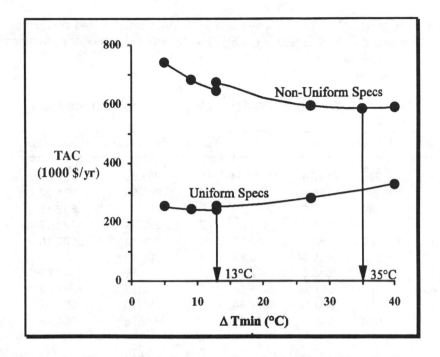

Figure 1.9 The total annual cost profile provides the optimum ΔT_{min} on applying the cost data to the energy, area, and units targets for different ΔT_{min}.

For the case of countercurrent CS exchangers, the minimum cost is observed to be 240.42×10^3 \$/yr. Thus, the optimum ΔT_{min} is seen to be 13°C rather than the experience-based value of 20°C. For the case of nonuniform exchanger specifications, the minimum cost is 586.06×10^3 \$/yr and the optimum ΔT_{min} is 35°C. This is because the cost of area for the special exchangers is relatively more than that for energy; so, it is preferable to use more energy and less area (which is the case at 35°C compared to 13°C).

The procedure for supertargeting is tedious and not suited to manual calculations. Table 1.15 shows that the targets are first established in the 5° - 50°C range in steps of 5°C. Based on these initial results, targets are generated in steps of 1°C in the 5° - 15°C range (for the uniform specifications case) and in the 30° - 40°C range (for the nonuniform specifications case). A more efficient search technique (e.g., golden section search) could be used to obtain the optimum ΔT_{min}. However, the discontinuities (due to the sudden change in the units target) could be many in number and cause difficulties. These difficulties may be circumvented by bracketing the minimum within an interval, in which the *TAC* function is smooth, differentiable, and continuous. Such a procedure is discussed in Chapter 2 within the framework of continuous targeting methodologies.

Table 1.15 Targets for Various ΔT_{min} Used for Optimization

ΔT_{min}	$Q_{hu,min}$	A_c	A^*	$N_{u,mer}$	$TAC_{uniform}$	$TAC_{nonuniform}$
5	200	2157.44	11412.94	6	253.55	739.92
10	300	1788.87	9701.01	6	241.90	671.92
15	430	1517.99	8633.17	7	254.70	662.15
20	605	1312.57	7411.01	7	262.79	622.17
25	780	1184.51	6560.52	7	276.18	600.10
30	955	1099.00	5936.22	7	292.58	589.25
35	1130	1040.02	5462.50	7	310.89	586.06
40	1305	998.99	5095.64	7	330.52	588.42
45	1480	970.60	4807.66	7	351.10	594.97
50	1655	951.42	4579.58	7	372.38	604.76
6	220	2057.08		6	249.70	
7	240	1972.48		6	246.82	
8	260	1899.55		6	244.66	
9	280	1835.62		6	243.05	
10	300	1778.87		6	241.90	
11	320	1727.97		6	241.11	
12	340	1681.95		6	240.63	
13	360	1640.06		6	240.42	
14	395	1574.89		7	254.15	
31	990		5831.14	7		588.08
32	1025		5731.58	7		587.19
33	1060		5637.18	7		586.57
34	1095		5547.59	7		586.20
35	1130		5462.50	7		586.06
36	1165		5381.65	7		586.14
37	1200		5304.76	7		586.42
38	1235		5231.61	7		586.91
39	1270		5161.97	7		587.58

Units: temperature in °C, utility in kW, area in m^2, and cost in 10^3 \$/yr.

1.7 SOME INTRICACIES IN TARGETING

In this section, some subtleties and advances in targeting procedures are presented. First, the distribution of area (both streamwise and matchwise) is discussed. This distribution is used to obtain better estimates for the capital cost in the case of nonuniform exchanger specifications. Finally, the important problem of multiple utilities is addressed.

1.7.1 Target Area Distribution

In addition to knowing the total area target, it is useful to know how this total area is distributed amongst the various streams. This may be determined from Equation 1.36 as demonstrated below. The streamwise area distribution may then be used to compute the matchwise area distribution in the spaghetti design for the vertical heat transfer model.

Example 1.11 Determination of the Area Distribution Matrices
Compute the distribution for the target area among the four process streams and the two utility streams for Case Study 4S1 with $\Delta T_{min} = 20°C$. Further, establish the matchwise area distributions in the case of the spaghetti design for vertical heat transfer. Assume countercurrent exchangers.

Solution. The first objective is to determine A_j, starting with Equation 1.36. Noting that $Q_j = (MC_p)_j\, dT_j$ gives

$$A_j = \frac{(MC_p)_j}{h_j} \sum_i \left(\frac{dT_j}{F\,LMTD}\right)_i \tag{1.39a}$$

The terms that appear in the above equation are readily available in Table 1.8.

Step 1. Calculation of an Element in the Streamwise Area Distribution Matrix
Each element in the streamwise area distribution matrix (Table 1.16) corresponds to the area contribution of a stream j within an enthalpy interval i. Sample calculations for matrix elements using Equation 1.39a are given below:

Stream j	Interval i	$(MC_p)_j$	h_j	dT_j	$LMTD_i$	Area
H1	1	10	0.2	$(65° - 45°) = 20°$	$37.51°$	26.66
H1	4	10	0.2	$(79.5° - 73.5°) = 6°$	$43.85°$	6.84
H2	4	40	0.2	$(79.5° - 73.5°) = 6°$	$43.85°$	27.37
C3	4	20	0.2	$(40° - 25°) = 15°$	$43.85°$	34.21

Note that dT_j and $LMTD_i$ are simply read from Table 1.8, and F is 1 (for counterflow). The entire matrix in Table 1.16 is filled in a similar manner.

Table 1.16 Streamwise Area Distribution Matrix

Interval	Stream						A_i
	H1	H2	C3	C4	HU	CU	
1	26.66					26.66	53.31
2	1.35	5.41				6.76	13.52
3	7.65	30.61	10.56			27.71	76.53
4	6.84	27.37	34.21				68.42
5	79.40	317.6	226.86	170.14			794.00
6	44.00		25.14	18.86			88.01
7	29.27		29.27				58.53
8							
9			80.12		80.12		160.24
A_j	195.17	380.99	406.16	189.00	80.12	61.13	1312.57

Units: area in m².

Step 2. Calculation of Streamwise Area Contributions

Summing the elements in each column of Table 1.16 gives the streamwise area contributions, i.e., A_j. Summing the elements in each row of the matrix yields the intervalwise area contributions, i.e., A_i. This is identical to column E in Table 1.8. As a check, both ΣA_j as well as ΣA_i must tally with the area target of 1312.57 m² obtained earlier.

Step 3. Calculation of an Element in the Matchwise Area Distribution Matrix

To determine the area contributions of the matches that constitute the spaghetti design (see Section 1.2.2), each element in Table 1.16 must be multiplied by two factors. The first factor converts the heat transfer coefficient for the stream into an overall heat transfer coefficient for the match as per Equation 1.16a. The second factor recognizes that the heat content of a stream must be split into fractions proportional to the MC_p-values of the streams of the opposite kind within an interval to obtain the match loads in spaghetti networks. Thus, each element in the matchwise area distribution matrix (Table 1.17) may be computed using the following formula based on hot streams:

$$A_{mh} = \frac{(MC_p)_{jh}}{h_{jh}} \left(\frac{dT_{jh}}{F\,LMTD} \right)_i h_{jh} \left(\frac{1}{h_{jh}} + \frac{1}{h_{jc}} \right) \frac{(MC_p)_{jc}}{\displaystyle\sum_{jc} (MC_p)_{jc}} \qquad (1.39b)$$

Though an analogous equation based on cold streams may be written down, it is unnecessary since the matrix elements (being match-dependent) can be generated using either hot streams or cold streams as the basis.

For Case Study 4S1, the heat transfer coefficients are the same for all streams; hence, h_{jh} $(1/h_{jh} + 1/h_{jc}) = 2$. So, each matrix element from Table 1.16 must be multiplied by $2(MC_p)_{jc}/\Sigma(MC_p)_{jc}$. Sample calculations for some matrix elements (Table 1.17) on using Equation 1.39b are given below:

Match	Interval i	$(MC_p)_{jc}$	$\Sigma(MC_p)_{jc}$	Element$_{Table\ 1.16}$	Element$_{Table\ 1.17}$
H1-CU	1	52.5	52.5	26.66	53.32
H1-C3	3	20	72.5	7.65	4.22
H1-CU	3	52.5	72.5	7.65	11.08
H2-C3	3	20	72.5	30.61	16.89
H2-CU	3	52.5	72.5	30.61	44.34

Table 1.17 Matchwise Area Distribution Matrix

| Interval | Match | | | | | | | |
	H1-C3	H1-C4	H1-CU	H2-C3	H2-C4	H2-CU	HU-C3	HU-C4
1			53.31					
2			2.70			10.82		
3	4.22		11.08	16.89		44.34		
4	13.68			54.74				
5	90.74	68.06		362.97	272.23			
6	50.29	37.72						
7	58.53							
8								
9							160.24	
A_m	217.46	105.78	67.09	434.60	272.23	55.16	160.24	
HX type	HX2	HX4	HX5	HX1	HX1	HX1	HX2	HX5

Units: area in m^2.

The area contributions of the four matches in the third interval agree with those obtained in Section 1.2.2 for Figure 1.7b.

The area distributions in each column of Table 1.17 are added to get total match areas (A_m), which are fruitfully exploited in the next subsection to obtain an improved capital cost estimate for nonuniform exchanger specifications.

1.7.2 Cost Targets for Nonuniform Exchanger Specifications

The approach recently proposed by Jegede and Polley (1992b) for nonuniform exchanger specifications provides an improvement over the method of Hall et al. (1990) discussed in Section 1.5. Its implementation is facilitated by the use of the matchwise area distribution matrix and is demonstrated below.

Example 1.12   Improved Cost Targets for Nonuniform Exchanger Specifications

Rework Example 1.9 using the approach proposed by Jegede and Polley (1992b). The goal as before is to determine the total annual cost target for the stream and utility data in Case Study 4S1 with $\Delta T_{min} = 20°C$.

Solution.   The method proposed by Jegede and Polley (1992b) determines the area and unit targets for each specification separately. It proceeds then to determine the cost contributions of the various specifications and adds them up to obtain the cost of the network.

Step 1. Classification of Matches According to the Exchanger Specifications

As suggested by Jegede and Polley (1992b), the priority assignment for the various exchanger types is as follows: spiral > plate and frame > shell and tube. The identifiers in Table 1.14 may be used for convenience to abbreviate the exchanger types. With the above priorities and convention, the matches may be assigned exchanger types as shown at the bottom of Table 1.17.

For a match between streams H1 and C3, the plate and frame exchanger assumes priority over the shell and tube; so, HX2 is entered in the table. An SS shell and tube exchanger (HX4) is the choice for a match between streams H1 and C4 whereas a CS-SS shell and tube exchanger (HX5) is the choice for a match between stream H1 and CU. For a match between stream H2 and any cold stream, the spiral exchanger assumes priority; so, HX1 is entered for matches involving stream H2. With similar reasoning, exchanger classification may be systematically determined for all matches.

Step 2. Computation of Target for Area Distribution

The area distribution among the different units affects the capital cost. For example, if a special unit has a high area, then it should have a high contribution to the capital cost. However, this factor was not accounted for in the method of Hall et al. (1990). The area contribution of only those units having the same specifications should be computed together. Thus, according to Jegede and Polley (1992b), for a specification *L*,

$$A_L = \sum_i \left(\frac{1}{F\ LMTD} \right)_i \left[\sum_{jh} \sum_{jc} Q_{jh-jc} \left(\frac{1}{h_{jh-jc}} + \frac{1}{h_{jc-jh}} \right) \right]_i \quad \text{and} \quad \sum_{L=1}^{N_{spec}} A_L = A$$

(1.39c)

Equation 1.39c allows the use of match-dependent heat transfer coefficients (Jegede and Polley, 1992b) and is used to calculate the area for all the different specifications. If N_{spec} is the total number of specifications, then the sum of the areas (A_L) obtained from Equation 1.39c would result in the same area target as given by Equation 1.7.

The matchwise area distribution matrix (Table 1.17) provides an elegant representation for readily calculating the values of A_L. On summing up the values of A_m belonging to the same specification, the A_L row in Table 1.18 may be filled. For example, $A_{HX1} = 434.60 + 272.23 + 55.16 = 761.99$ m².

Step 3. Computation of Target for Unit Distribution

Just like the area distribution, the unit distribution affects the capital cost. A higher number of units of a special match would result in a higher capital cost contribution of that particular specification. The total area for each specification is constant. Therefore, the cost contribution of each specification depends entirely on the number of units belonging to that specification. Hence, the area and unit distribution together give a proper estimate of the capital cost. An assumption of uniform unit distribution among all specifications would be inaccurate.

The unit distribution between different specifications is obtained by the following procedure. The vertical targeting model matches pairs of streams in each enthalpy interval. Thus, for an enthalpy interval, the total number of vertical matches (*V*) is known. While area targeting for a specification *L*, the number of vertical matches for that specification (V_L) is also known. Now, the minimum number of units in an MER network is given by Equation 1.27. This equation may simply be extended (Jegede and Polley, 1992b) to yield an expression for the unit target for a specification *L*:

$$N_{u,mer,L} = (N_a - 1)(V_{La}/V_a) + (N_b - 1)(V_{Lb}/V_b) \tag{1.40}$$

where the subscripts a and b denote the above and below pinch regions respectively. The number of units having specification L ($N_{u,mer,L}$ from Equation 1.40) will typically be a fraction (not a whole number). The factor (V_L/V) is the ratio between the spaghetti matches of specification L to the total spaghetti matches. It is assumed that this ratio also exists in the case of the actual MER network, i.e., if a specification L has a higher number of units in a spaghetti network, then the same would be true in the MER network. In essence, $N_{u,mer,L}$ is not the actual number of units of specification L, but provides a factor that allows accurate calculation of capital cost targets accounting for the unit distribution.

For Case Study 4S1, intervals 1 through 5 are below the pinch and intervals 6 through 9 are above the pinch. From Example 1.6, $N_a = 4$ and $N_b = 5$. On counting the non-zero entries in the matrix in Table 1.17, $V_a = 4$ and $V_b = 13$. The thirteen matches below the pinch are distributed among the various specifications as follows: six of HX1; three of HX2; one of HX4; and three of HX5. Similarly, there are three matches of type HX2 and one match of type HX4 above the pinch. Using Equation 1.40 gives

$$N_{u,mer,HX1} = (4-1)(0/4) + (5-1)(6/13) = 1.8462$$
$$N_{u,mer,HX2} = (4-1)(3/4) + (5-1)(3/13) = 3.1731, \text{and so on.}$$

Step 4. Computation of Capital Cost Target

The capital cost target for specification L (denoted by CC_L) is given by Equation 1.34 using the area target (A_L calculated in step 2), the unit target ($N_{u,mer,L}$ calculated in step 3), and the cost coefficients (a, b, and c) for the particular specification (given in Table 1.14). The annualization factor, A_{fL}, is given by $(1 + r)^{t_L}/t_L$ as in Equation 1.35 with the difference that t_L now refers to the useful lifespan of the equipment of specification L. The total installed capital cost of the network is given by the summation of CC_L after amortization. Thus,

$$TAC = OC + \sum_{L=1}^{N_{spec}} (CC_L)(A_{fL})$$

$$= OC + \sum_{L=1}^{N_{spec}} N_{u,mer,L}[a_L + b_L(A_L / N_{u,mer,L})^{c_L}](1+r)^{t_L} / t_L . \tag{1.41}$$

For the present example, $A_{fL} = 0.3221$ and the values of CC_L are tabulated in Table 1.18. So, the annualized capital cost is given by $(1845.2 \times 10^3)(0.3221) = 594.34 \times 10^3$ \$/yr. Adding this to the operating cost of 77.85×10^3 \$/yr gives the TAC to be 672.19×10^3 \$/yr, which is 7.4% higher than the value of 622.17×10^3 \$/yr obtained in Example 1.9 by the method of Hall et al. (1990).

Table 1.18 Targets According to Exchanger Specification

	Exchanger Classification					
	HX1	HX2	HX3	HX4	HX5	
A_L	761.99	377.70	0	105.78	67.09	
V_{La}	0	3	0	1	0	$V_a = 4$
V_{Lb}	6	3	0	1	3	$V_b = 13$
$N_{u,mer,L}$	1.85	3.17	0	1.06	0.92	
CC_L	1327.12	345.80	0	104.59	67.69	

Units: area in m^2 and cost in 10^3 \$.

Conceptually speaking, the method of Jegede and Polley (1992b) is more correct because it accounts for specifications in terms of matches rather than in terms of streams. Note that the specification of a CS-SS shell and tube exchanger (identified by HX5 in Table 1.14) plays an appropriate role in the H1-CU match (see Table 1.17) in the Jegede and Polley (1992b) method whereas it does not appear at all in the Hall et al. (1990) approach (see step 2 in Example 1.9).

1.7.3 Targeting for Multiple Utilities

The problem arises when more than one hot utility and/or cold utility is available, as is typically the case in the industrial scenario. These utilities may be point utilities (i.e., those undergoing negligibly small or no temperature changes such as steam at various pressure levels from a boiler or a steam turbine, and refrigerants at various levels) or nonpoint utilities (i.e., those undergoing significant temperature changes such as cooling water, hot water, hot oil, and flue gas from a furnace or a gas turbine exhaust). In addition, opportunities may exist for generating utilities (e.g., low pressure steam and air preheating). The selection of utilities depends on their relative unit costs and the associated capital equipment. The problem involves determining the optimum usage levels of the various available utilities so that their sum equals the overall minimum utility requirements established in Section 1.1.2.

An analytical approach (Sachdeva, 1993) is presented below based on the graphical method discussed by Linnhoff et al. (1982). It uses the GCC and a simple cost-based heuristic. The results are identical to those obtained by a method recently introduced by Jezowski and Friedler (1992), based on a five-component flow model. Qassim and Silveira (1988) studied the multiple utilities problem using a goal-programming approach and a prioritization sequence using special weighting factors applicable to only point utilities. The approach below has the following advantages: it is analytical; it is based on the universally-used GCC; it uses a commonly-used prioritization scheme and a simple cost-based heuristic (Linnhoff et al., 1982; Jezowski and Friedler, 1992); and it is easily extended to both point and nonpoint utilities as well as variable utility outlet temperatures.

Example 1.13 Allocation of Multiple Utilities Based on a Simple Heuristic
For the stream data in Case Study 4S1 with ΔT_{min} = 20°C, determine the optimum usage levels for each of the following utilities.

Utility Label	CU1	CU2	HU1	HU2
Inlet Temperature T_{in} (°C)	15	100	180	135
Outlet Temperature T_{out} (°C)	25	101	179	134

If another portion of the plant requires very low pressure (VLP) steam at 100°C, what is the maximum amount of steam-raising possible against the process source profile? It may be noted that the purpose of utility CU2 is specifically to raise VLP steam at 100°C from feed water at the same temperature. Assume the global ΔT_{min} of 20°C applies to utilities also.

Solution. The allocation is done starting with the cheapest utility and proceeding to the most expensive. The usage of the cheapest utility available is maximized at each stage. In the absence of cost data, the simple heuristic is to assume that the hottest cold utility and the coldest hot utility are the cheapest. The inlet temperatures of the utilities are considered in deciding the hottest/coldest utilities. The calculations are compactly tabulated in Table 1.19 and may be better understood by referring to Figure 1.10.

Step 1. Generation of GCC Information
Using the PTA, the GCC information may be obtained as in Example 1.1. The GCC (Figure 1.5) is a plot of shifted stream temperatures vs. revised cascaded heat (i.e., column A vs. column E of Table 1.1). The overall minimum hot and cold utilities required are 605 kW and 525 kW respectively.

These utility loads assume that the hot utility is sufficiently hot to satisfy the heating needs and the cold utility is sufficiently cold to meet the cooling needs at any required temperature levels. In practice, the utility temperature levels are not arbitrary, and these set constraints on the utility usage (load).

Step 2. Shifting Utility Temperatures and Prioritizing Utilities
    Because determination of the individual optimum contributions of the different utilities is based on the GCC, which is in terms of shifted temperatures, the end temperatures of the hot utility must be decreased by $\Delta T_{min}/2$ and those of the cold utility increased by $\Delta T_{min}/2$. Using the "hottest cold utility and the coldest hot utility" heuristic, the cold utilities must be prioritized in decreasing order of their inlet temperatures and the hot utilities in increasing order. In this case, CU2 assumes precedence over CU1 (bearing in mind the demand for steam generation at the plant site) and HU2 assumes priority over HU1. The shifted temperatures and priorities appear in the first four rows of Table 1.19.

Table 1.19 Usage Levels of Multiple Utilities

Utility Label	CU1	CU2	HU1	HU2
Shifted Inlet Temperature T_{int}	25	110	170	125
Shifted Outlet Temperature	35	111	169	124
Priority Index	2	1	2	1
$R_{cas,x}$ (kW)	525	75	605	205
$R_{cas,o}$ (kW)	525	525	605	605
Utility Usage Level	450	75	400	205

Note: temperature in °C and utility usage in kW.

Step 3. Intersection on GCC
    The point on the GCC in terms of R_{cas} is determined corresponding to the shifted inlet temperatures of the utilities. This is done by linear interpolation using column E of Table 1.1. For CU2, $T_{int} = 110°C$ which lies in the 55°-115°C interval in Table 1.1. Thus,

$$R_{cas,x} = 900 \times (115° - 110°)/(115° - 55°) = 75 \text{ kW}.$$

If T_{int} is beyond the temperature range of the GCC, then the R_{cas} value at the appropriate end of the GCC is used. For HU1, $T_{int} = 170°C$ which lies beyond 165°C in Table 1.1. So,

$$R_{cas,x} = R_{cas} \text{ (at 165°C on GCC)} = 605 \text{ kW.}$$

Step 4. Determination of Minimum R_{cas} *outside GCC Interval*

The minimum R_{cas} value is determined outside the GCC interval in which the shifted inlet temperature of the utility lies. Only the region of the GCC below the lower limit is considered for a cold utility whereas only the region above the upper limit is considered for a hot utility.

For CU2, the minimum R_{cas} value in the region $T_{int} \leq 55°C$ is 525 kW from Table 1.1. For HU2, the minimum R_{cas} in the region $T_{int} \geq 165°C$ is 605 kW. These are listed as $R_{cas,o}$ in Table 1.19.

Step 5. Determination of Individual Usage Levels of Various Utilities

The optimum usage level of each utility is determined starting with the utility having top priority (namely, 1) using the following formula:

$$Q_{u,pi} = \min(R_{cas,x}, R_{cas,o}) - \sum_{pj=1}^{pi-1} Q_{u,pj} \tag{1.42}$$

where *pi* denotes the priority index. The formula is applied separately for hot and cold utilities as shown below:

For CU2, $Q_{cu,p1} = \min(75, 525) - 0$ $= 75$ kW.
For CU1, $Q_{cu,p2} = \min(525, 525) - 75$ $= 450$ kW.
For HU2, $Q_{hu,p1} = \min(205, 605) - 0$ $= 205$ kW.
For HU1, $Q_{hu,p2} = \min(605, 605) - 205$ $= 400$ kW.

Step 6. Plotting Utility Profiles on GCC

The utility profiles are plotted on the GCC (Figure 1.10) using the results of the shifted temperatures and usage levels in Table 1.19.

Step 7. Post-analysis

The following observations may be made based on Figure 1.10. The 450 kW rejected to CU1 (with an inlet temperature of 15°C) may be given to a cold utility at a higher temperature level. On shifting the profile segment corresponding to CU1 vertically upwards till it touches the GCC, it is seen that the maximum cold utility inlet temperature at which 450 kW can be rejected is

70°C (i.e., a shifted temperature of 80°C given by 115° − [115°−55°] 525/900). If the cold utility temperature is increased beyond 70°C, then only a portion of the 450 kW can be rejected to this utility and another cold utility at a lower temperature is required.

What is the maximum outlet temperature for this cold utility entering at 70°C? At best, the end point (corresponding to the outlet temperature) of this cold utility profile can touch the process GCC. In this case, the maximum outlet temperature is 100°C, after which steam-raising by CU2 occurs. The cold utility profile segment (running from 80° to 110°C) will be parallel to the process source profile, and the flow rate of the cold utility will be minimized. However, this gain in the operating cost obtained by minimizing utility flow rate is offset by the increased capital cost since this process-utility heat exchange occurs at the minimum driving force.

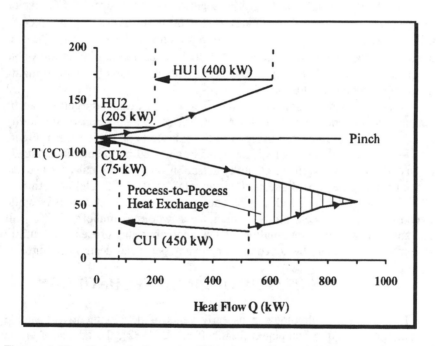

Figure 1.10 The selection of utilities is conveniently done on the grand composite curve rather than on the composite curves.

From the previous example, it may be observed that the GCC provides a more convenient representation than the composite curves plot for the selection of utility loads and temperature levels. This is especially true when the utility temperature levels lie within the range of the process streams and would cause disruptions on the composite curves. In other words, on incorporating multiple utilities, the resulting balanced composites will have different shapes depending on the load and temperature level of the utilities. However, the process GCC in Figure 1.10 is sacrosanct and need not be redrawn since it is independent of the utility loads/levels.

This aspect further manifests itself in Figure 1.10 in terms of two distinct types of heat exchange: process-to-utility exchange and process-to-process exchange. The shaded region in Figure 1.10 (often referred to as the "pocket" or nonmonotonic part of the GCC) is a portion in perfect enthalpy balance, where the process streams can exchange heat among themselves to satisfy their heat requirements effectively. It is essentially outside such regions that hot utility is necessary for exchanging heat with the process sinks and cold utility is required for the process sources. Further, consider the region below the pinch. Though the 30°-55°C profile segment of the process GCC (which represents a local heat sink) can accept heat from any portion of the process GCC between 115° and 55°C (which represents a heat source), heat is transferred from the segment between 80° to 55°C (at the minimum allowable temperature driving force) to be consistent with the heuristic of maximizing usage of the hottest cold utility.

Finally, it must be pointed out that the concept of supertargeting can be extended to include multiple utilities as follows. On selection of the utility levels, the GCC is converted to a plot of the balanced composites (as in Figure 1.6) from which capital cost targets may be established by the methods already discussed. Thus, the effects of utility selection on total annual cost and the optimum may be explored ahead of design (Hall et al., 1992; Hui and Ahmad, 1994b). Supertargeting, with this broader scope, considers the energy-capital tradeoff for the process-utility system as a whole (Linnhoff, 1986). It guarantees near-global optimality by not analyzing the process and utility systems separately and by not committing to any particular design structure.

1.8 OVERALL LOGIC OF PINCH TECHNOLOGY

Though pinch technology in its early years of development was simple enough not to require computers (Linnhoff et al., 1982), the advances in the last decade make software programs not merely convenient but essential. This is certainly true with regard to targeting (especially supertargeting), where the calculations tend to be repetitive and tedious. Thus, with the exception of energy and unit targeting for small-sized problems, the other targeting calculations are best done with the help of a computer package such as HX (see Appendix A).

The logic used in designing such a software based on pinch technology is presented in Figure 1.11.

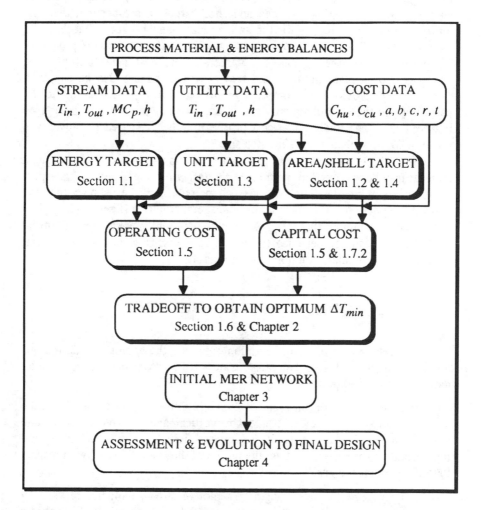

Figure 1.11 The logic of pinch technology for grassroots design is summarized in the above information flow diagram, which shows that cost optimization may be effectively performed before design.

The targeting module is conceptually based on the composite curves and requires the specification of ΔT_{min} to fix their relative positions. It assumes that the base-case material and energy balances for the process plant are available (say, through an external simulation). The data extraction may then be performed and the streams relevant to heat integration specified in terms of their T_{in}, T_{out}, MC_p, and h values. The stream data are sufficient to perform energy and unit targeting. If area and shell targeting are desired, then the utility data must also be specified in an identical manner, except that the MC_p values of the utilities are not required (they will be calculated by the program). To perform cost targeting and supertargeting, the cost data (utility costs, exchanger costs, plant lifetime, and rate of interest) must also be supplied. These would provide the complete inputs and allow calculations to be performed by the algorithms provided in earlier sections of this chapter (see Figure 1.11). The total annual cost profile may be generated for any given range of ΔT_{min} with any specified step size.

Figure 1.11 highlights the fact that the classical energy-capital cost tradeoff may be exploited to optimize at the targeting stage itself. This aspect is treated in more detail in the next chapter, where it is shown that the optimization may be in terms of a minimum temperature difference (ΔT_{min}) or in terms of a minimum flux (Q''_{min}). At this optimum specification, an initial MER network may be designed (using methods discussed in Chapter 3) that would require minimal evolution (see Chapter 4).

QUESTIONS FOR DISCUSSION

1. In the context of heat integration, what are the appropriate definitions for a hot stream and a cold stream? Can a cold stream have a higher inlet temperature than a hot stream?

2. Draw the network in Figure 1.2 using a "cross hatched grid" in which hot and cold streams are represented by vertical and horizontal arrows and matches by circles at the points of intersection. In terms of a general representation for HENs, what are the merits of the "parallel grid" with respect to the "cross hatched grid"?

3. Draw a heat-content diagram (a plot of stream temperature vs. heat capacity flow rate) corresponding to the network in Figure 1.2. Note that a stream is then represented by a rectangle, whose area corresponds to the enthalpy content of the stream. Show that horizontal divisions of these rectangles may be used to represent matches (and vertical divisions to represent stream splits).

4. Show how the composite curves may be obtained from the temperature-enthalpy profiles of the individual streams (see Figure 1.3) by a

vector type addition of appropriate stream segments and a horizontal sliding process.

5. Will the following conventions for temperature shifting yield identical results from the PTA?

- Hot stream temperatures are decreased by $\Delta T_{min}/2$ and cold stream temperatures are increased by $\Delta T_{min}/2$.
- Hot stream temperatures are decreased by ΔT_{min} and cold stream temperatures are left unchanged.
- Hot stream temperatures are left unchanged and cold stream temperatures are increased by ΔT_{min}.

6. How can streams that change phase be handled in pinch technology? How do such streams affect the definition of the temperature intervals in the PTA? Is it convenient to approximate isothermal phase changes for streams by a 1° temperature change? Can this cause significant errors?

7. Discuss the rationale for the arrow convention used on the GCC (see Figure 1.5).

8. For the case where the overall heat transfer coefficient, U, is constant, show that Equation 1.7 simplifies to $A = 1/U \ \Sigma Q_i/(F \ LMTD)_i$ where the summation runs over all the enthalpy intervals and Q_i is the enthalpy width of interval i.

9. Do the area and shell targeting formulae give correct results when applied to a single exchanger (which can be viewed as a network with a single hot stream and a single cold stream in perfect enthalpy balance)?

10. The targets for Case Study 4S1 indicate the need for three units above the pinch and four units below the pinch (Example 1.6), but Table 1.13b shows the approximate shell target to be two (above the pinch) and seven (below the pinch). The number of shells cannot be lower than the number of units (above the pinch). Discuss this limitation of the approximate shell targeting method.

11. Equation 1.32 is valid for a single hot utility and a single cold utility. Generalize the equation to determine the operating cost for the case of multiple utilities.

12. Equation 1.34a is obtained by application of Equation 1.33 for the overall network (and assuming equi-distribution of area). Would it be better to predict capital costs for the above- and below-pinch regions separately, and then sum them to obtain the total network capital cost?

13.   Capital cost estimates from Equations 1.34a and 1.34b appear reasonably good in practice, inspite of the following assumptions/approximations:

- The uniform BATH formula overpredicts the area for the case of unequal heat transfer coefficients.
- The number of units considered in Equations 1.34a and 1.34b corresponds to an MER network and not a vertical heat transfer model.
- The equal distribution of area among the units does not consider economy of scale.

Discuss the effects of the above approximations on the capital cost. Could the effects cancel one another?

14.   Can A^* (see Equation 1.37a) be obtained by using ϕ_j-values along with the streamwise area distribution matrix?   Show that the value of A^* (7411.01 m^2) in Example 1.9 is obtainable from Table 1.16.

15.   Because ΔT_{min} is continuously increased during the supertargeting in Table 1.15, ΔT_{min} for the process-to-utility heat transfer must necessarily be chosen differently from that for the process-to-process heat transfer.   What should these ΔT_{min} values be for the process-to-hot-utility and process-to-cold-utility heat exchanges in Table 1.15?

16.   Consider that a plant is to be built in a country L where the cost of energy is low.   The optimum ΔT_{min} through supertargeting is found to be $\Delta T_{min,L}$.   If the same plant (with no change in the stream data) is to be set up in country H where the cost of energy is high, then would the optimum ΔT_{min} for this case be expected to be higher or lower than $\Delta T_{min,L}$?   State any assumptions made.

17.   What modifications are necessary in the analytical approach outlined in Example 1.13 if nonpoint utilities are to be used with variable outlet temperatures?   In other words, only the minimum (for hot utility) and maximum (for cold utility) values are specified for the utility outlet temperatures.

PROBLEMS

1.A Energy and Area Targeting

Determine the minimum utility requirements, pinch temperature, and counterflow area target for the following case studies discussed by Colberg and Morari (1990):

a) Case Study 4C1 with $\Delta T_{min} = 10$ K;

b) Case Study 4G1 with $\Delta T_{min} = 20$ K;

c) Case Study 5N1 with $\Delta T_{min} = 10$ K; and

d) Case Study 7C1 with $\Delta T_{min} = 20$ K.

Answers:

a) $Q_{hu,min} = 620$ kW; $Q_{cu,min} = 230$ kW; $T_p = 358$ K; $A_c = 295.78$ m^2

b) $Q_{hu,min} = 1075$ kW; $Q_{cu,min} = 400$ kW; $T_p = 353$ K; $A_c = 2898.01$ m^2

c) $Q_{hu,min} = 0$ kW; $Q_{cu,min} = 0$ kW; $A_c = 47.71$ m^2

d) $Q_{hu,min} = 244.13$ kW; $Q_{cu,min} = 172.6$ kW;
$T_p = 507$ K; $A_c = 227.1$ m^2

1.B Area and Shell Targeting

Determine the energy, area, unit, and shell targets for the following case studies discussed by Ahmad and Smith (1989), assuming $\Delta T_{min} = 20°C$ and 1-2 shell and tube exchangers:

a) Case Study 4A1   b) Case Study 4A2

c) Case Study 9A1   d) Case Study 28A1

Answers:

a) $Q_{hu,min} = 1380$ kW; $T_p = 90°C$; $N_{u,mer} = 7$;
$A_{12} = 591.62$ m^2; $S_{ap} = 4$ (4.00); $S_{bp} = 4$ (3.32)

b) $Q_{hu,min} = 4000$ kW; $T_p = 150°C$; $N_{u,mer} = 7$;
$A_{12} = 2027.01$ m^2; $S_{ap} = 7$ (6.43); $S_{bp} = 5$ (4.70)

c) $Q_{hu,min} = 21580$ kW; $T_p = 150°C$; $N_{u,mer} = 14$;
$A_{12} = 5769.89$ m^2; $S_{ap} = 14$ (13.93); $S_{bp} = 17$ (16.83)
This requires assuming $\Delta T_{min} = 15°C$ for the cold utility.

d) $Q_{hu,min} = 6196$ kW; $T_p = 150°C$; $N_{u,mer} = 31$;
$A_{12} = 2868.76$ m^2; $S_{ap} = 29$ (28.94); $S_{bp} = 20$ (19.87)

1.C Cost Targeting for Countercurrent and 1-2 Exchangers

Determine the total annual cost target for the case study discussed by Ahmad et al. (1990) under the following conditions (Note that the data are identical to those in Case Study 9H1 except that the MC_ps of all process streams are halved and $C_{hu} = 110$ $/[kW.yr]):

a) countercurrent exchangers with $\Delta T_{min} = 10°C$; and

b) 1-2 shell and tube exchangers with $\Delta T_{min} = 17°C$.

Answers:

a) $Q_{hu,min} = 9225$ kW; $A_c = 6570.63$ m^2; $TAC = 1601.31 \times 10^3$ $/yr.

b) $Q_{hu,min} = 10125$ kW; $N_{u,mer} = 12$; $A_{12} = 5369.80$ m^2;
$S_{ap} = 13$ (12.92); $S_{bp} = 5$ (4.06); $TAC = 1657.13 \times 10^3$ $/yr.

1.D Capital Cost Targeting for Nonuniform Specifications

Determine the capital cost target for the Case Study 9H1 discussed by Hall et al. (1990) under the following conditions:

a) all streams use CS;

b) streams H3, H4, C6, and C9 use SS, whereas the rest use CS;

c) streams H3, H4, C6, and C9 use Ti, whereas the rest use CS;

d) streams H3 and H4 use SS, streams C6 and C9 use Ti, whereas the rest use CS;

e) stream H3 uses SS; streams H4, C6, and C9 use Ti; and the rest use CS;

f) pressure ratings of streams H4 and C9 are 35 bar, of streams H3 and C7 are 60 bar, and the rest are 10 bar;

g) streams H3 and C7 require spiral exchangers, whereas the rest require plate and frame exchangers.

Assume $\Delta T_{min} = 20°C$ and counterflow. Use the ϕ-factor approach.

Note: CS stands for carbon-steel, SS for stainless-steel, and Ti for titanium.

Assume cost data for different exchanger specifications as given below.

Cost Data for Various Exchanger Specifications (Case Study 9H1)

Exchanger Specification	Capital Cost ($)
spiral (SS-SS)	$19687\ A^{0.59}$
plate & frame (SS-SS)	$1905\ A^{0.78}$
shell & tube (CS-CS)	$30800 + 750\ A^{0.81}$
shell & tube (SS-SS)	$30800 + 1644\ A^{0.81}$
shell & tube (CS-SS)	$30800 + 1339\ A^{0.81}$
shell & tube (Ti-Ti)	$30800 + 4407\ A^{0.81}$
shell & tube (CS-Ti)	$30800 + 3349\ A^{0.81}$
shell & tube (SS-Ti)	$30800 + 3749\ A^{0.81}$
shell & tube (10/10 bar)	$30800 + 750\ A^{0.81}$
shell & tube (10/35 bar)	$30800 + 890\ A^{0.81}$
shell & tube (35/35 bar)	$30800 + 1089\ A^{0.81}$
shell & tube (10/60 bar)	$30800 + 983\ A^{0.81}$
shell & tube (60/60 bar)	$30800 + 1438\ A^{0.81}$
shell & tube (35/60 bar)	$30800 + 1201\ A^{0.81}$

Note: A is exchanger area in m^2.
Data taken from Hall et al. (1990)

Answers:

$Q_{hu,min}$ = 20950 kW; $Q_{cu,min}$ = 7000 kW; T_p = 125°C; $N_{u,mer}$ = 13
a) A_c =   9523.77 m²; CC = 2440.00 x 10³ $
b) A^* = 12438.58 m²; CC = 2932.47 x 10³ $
c) A^* = 23610.49 m²; CC = 4655.65 x 10³ $
d) A^* = 16371.66 m²; CC = 3563.59 x 10³ $
e) A^* = 18453.35 m²; CC = 3885.63 x 10³ $
f) A^* = 12587.78 m²; CC = 2957.04 x 10³ $
g) A^* = 16175.69 m²; CC = 6425.03 x 10³ $

1.E Pre-design Optimization through Supertargeting

Determine the optimum value of ΔT_{min} (to an accuracy of 1°C) for the Case Study 9H1 discussed by Hall et al. (1990) under the following conditions:
a) all streams use CS;
b) all streams use Ti;
c) streams H4, H5, C6, and C7 use Ti, whereas the rest use CS;
d) all streams require plate and frame exchangers;
e) all streams require spiral exchangers;
Assume counterflow and use the ϕ-factor approach.
Given: Annual cost of unit duty in $/(kW.yr):
120 (for HU) and 10 (for CU)
Plant lifetime: 6 yr
Rate of interest: 16% (for spiral, and plate and frame cases) and
10% (for other cases)
Use cost data from Problem 1.D.

Answers:
a) $\Delta T_{min,opt}$ = 9°C; A_c =   13923.36 m²; TAC = 3157.34 x 10³ $/yr (CS)
b) $\Delta T_{min,opt}$ = 27°C; A_c =   8281.04 m²; TAC = 6125.82 x 10³ $/yr (Ti)
c) $\Delta T_{min,opt}$ = 18.2°C; A^* = 36158.39 m²;
   TAC = 4373.20 x 10³ $/yr (CS/Ti)
d) $\Delta T_{min,opt}$ = 18.2°C; A_c = 10010.84 m²;
   TAC = 4279.51 x 10³ $/yr (P&F)
e) $\Delta T_{min,opt}$ = 28°C; A_c = 8160.88 m²; TAC = 7535.22 x 10³ $/yr (Sp)

1.F Allocation of Multiple Utilities

Determine the usage levels of the three cold utilities and the single hot utility (given below) for the Case Study 5J1 discussed by Jezowski and Friedler (1992). Assume ΔT_{min} = 10°C.

Utility Label	CU1	CU2	CU3	HU
Inlet Temperature T_{in} (°C)	150	130	38	290
Cost ($/kW.yr)	0.4	0.5	1.0	--

Answers:

   Q_{hu1} = 127.68 kW;
   Q_{cu1} = 321.63 kW; Q_{cu2} = 65.16 kW; Q_{cu3} = 622.35 kW

CHAPTER 2

Continuous Targeting

If you're headed in the right direction,
each step, no matter how small,
is getting you closer to your goal.

Anonymous

The problem table algorithm (PTA), introduced in Chapter 1, allows calculation of energy targets, but only at a particular value of ΔT_{min} specified a priori. A new algorithm (Makwana and Shenoy, 1993) is presented in the first section of this chapter that permits continuous determination of the targets over the entire allowable range of ΔT_{min}. As discussed later, another major advantage of the continuous targeting algorithm is the accurate analytical computation of the ΔT_{min} values at which multiple pinches (including topology traps) and the sensitivity threshold occur. The proposed algorithm uses a novel pinch tableau representation to track the horizontal shifts of the composite curve systematically on the temperature-enthalpy diagram.

The second section of the chapter introduces the diverse pinch concept useful when the stream heat transfer coefficients differ by orders of magnitude. The minimum flux (Q''_{min}) rather than the minimum approach temperature (ΔT_{min}) is the appropriate design specification in this case. The last section attempts to look at these two design specifications in a unified framework by performing continuous targeting over the entire range of ΔT_{min} as well as Q''_{min}.

2.1 THE CONTINUOUS TARGETING ALGORITHM

The minimum utility requirement as well as the pinch temperature can be calculated as targets before inventing the actual heat exchanger network using the PTA, provided ΔT_{min} is prespecified. Multiple pinches and topology traps can occur at certain values of ΔT_{min}. A topology trap as indicated by Linnhoff and Ahmad (1990) is a multipinched situation where a pinch swap occurs and across which evolution of networks is not possible due to major changes in the network topology. The ΔT_{min} value at which a topology trap exists cannot be accurately determined by the PTA since it can locate pinches only at discrete

71

values of ΔT_{min} and would consequently need to adopt a trial-and-error procedure. An analogous argument holds for the determination of the threshold ΔT_{min}, below which only a single utility (either hot or cold, but not both) is required.

This section provides an accurate analytical method for (a) calculation of minimum utility requirements and pinch temperatures as piecewise linear functions of ΔT_{min}, (b) computation of the threshold ΔT_{min}, (c) determination of multiple pinches/topology traps, and (d) calculation of the ΔT_{min} for a specified utility consumption. The method assumes significance in the context of supertargeting, where the continuous variation of ΔT_{min} is important in deciding the optimum (see Section 1.6). The method extensively depends on the composite curves on the conventional temperature-enthalpy diagram. As discussed in Chapter 1, a feature of the composite curves is a change in the slope, caused by a stream population change. These points where the slope changes will hereafter be referred to as vertices. Each vertex on a composite curve may be described by its enthalpy-temperature coordinates, (Q, T). The T coordinate of a vertex is unique. A change in ΔT_{min} is typically made by shifting the cold composite curve (CCC) horizontally, keeping the hot composite curve (HCC) fixed. This implies that each vertex on the HCC (or hot vertex) has a unique Q coordinate whereas each vertex on the CCC (or cold vertex) has a unique Q coordinate only for a given value of ΔT_{min}. The proposed algorithm tracks the locations of the composite curves on a tableau in terms of the coordinates of the vertices and simultaneously determines the position of the pinch(es) where the two composite curves are ΔT_{min} apart.

The following observations (Makwana and Shenoy, 1993) are central to the development of the continuous targeting algorithm (CTA):

a) For a given ΔT_{min}, there exists a unique set of composite curves and unique minimum utility requirements, and vice versa.

b) A pinch occurs necessarily at a vertex, i.e., at the pinch, the slope of either the hot or the cold composite curve (or both) changes. Hence, if the ΔT at all the vertices is known, the minimum will correspond to the pinch. In a multipinched situation, the minimum ΔT will occur at more than one vertex.

c) The ΔT_{min} may be increased by moving the CCC to the right (a positive Q-shift, hereafter referred to as forward tracking) or decreased by moving the CCC to the left (a negative Q-shift, referred to as backward tracking).

d) As the CCC is moved horizontally, the ΔT at the vertices changes. The rate of change of ΔT at a vertex depends on the slope of the line segment facing it (vertically above on the HCC for a cold vertex, and vertically below on the CCC for a hot vertex). From Chapter 1, it may be noted that the slope is given by the reciprocal of the sum of the heat capacity flow rates of the streams within the temperature interval. The rates remain unchanged over certain ranges when a vertex faces the same line segment during the horizontal shifting process. These ranges are demarcated by what are termed as "significant shifts."

e) A significant shift may be defined as a Q-shift which brings a cold vertex directly below a hot vertex. Beyond a significant shift, the two vertices involved will have different rates of change of ΔT.

f) A vertex can exist with a lower rate of change of ΔT than at the pinch. Then, it is possible that, for a Q-shift less than the significant shift, the ΔT at that vertex and at the pinch become equal. This would result in a multipinched situation. For further shifts, the pinch swaps to the new vertex. This sudden change in pinch location is called a topology trap (see Figure 2.6).

g) The energy targets (namely, $Q_{hu,min}$, $Q_{cu,min}$, and T_p) as well as the units target ($N_{u,mer}$) vary linearly with ΔT_{min}. Linear interpolation is possible between two consecutive significant shifts after taking topology traps into account. The CTA determines the end points within which such an interpolation is permissible.

h) The pinch tableau is a useful representation of the coordinates of the vertices on the composite curves that enables direct calculation (without resorting to graphical procedures) of the minimum utility and the pinch.

i) Every pinch tableau corresponds to a unique position of the composite curves on the temperature-enthalpy diagram (i.e., to a particular ΔT_{min}).

j) The basic structure of a pinch tableau (see Tables 2.3 and 2.5) is as follows. Every column in the pinch tableau corresponds to a vertex on either the HCC or CCC. The first row contains the Q coordinates (in ascending order) of the vertices of the composite curves. The second and third rows contain the temperatures on the HCC and CCC respectively. In each column, the vertex temperature itself is not underlined although its projection on the opposite curve is. The fourth row conveniently stores information on the ΔT at all the vertices by simply subtracting the third row from the second.

k) Importantly, given a tableau corresponding to a particular location of the composite curves, a new tableau may easily be generated for *any* specified shift of the composite curve in terms of the Q coordinate. So, if the amount of utility available is known, then the ΔT_{min} for operation may be determined (which is the inverse problem to that solved by the PTA).

l) The algorithm is designed to handle complexities such as several topology traps in succession between two significant shifts, and multipinched situations (with two or more pinches).

2.1.1 The Pinch Tableau Approach

The continuous targeting algorithm (CTA) is illustrated with an example below.

Example 2.1 Continuous Determination of Energy and Unit Targets

For the stream data in Case Study 4S1, determine how the utility requirements, the pinch temperature, and the minimum number of units in an MER network vary with ΔT_{min}.

Solution. Only the inlet temperatures, outlet temperatures, and heat capacity flow rates are necessary. The steps below may be better understood by referring to Figure 2.1.

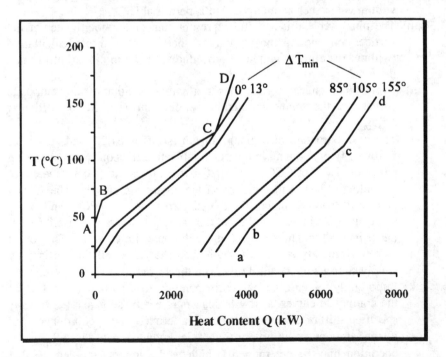

Figure 2.1 The cold composite curve may be shifted horizontally to vary ΔT_{min} and study its effect on the energy targets.

Step 1. Generation of Preliminary Composite Curve Data

It is clear from the first law of thermodynamics (see Section 1.1.1) that there is a heat deficiency of 80 kW; so, let $Q_f = -80$.

Next, the data for the preliminary HCC and the preliminary CCC are generated as in Example 1.2. The end temperatures of the hot and cold streams are separately arranged in ascending order (column A). The sum of the MC_p values of the streams present in each temperature interval is calculated and entered in column B against the higher temperature limit of the interval. A zero is appropriately placed as the first entry in column B. The value in column B is multiplied by the temperature difference for that interval to obtain the enthalpy associated with each interval (column C). The cumulative enthalpies are finally calculated by using the formula:

$$(CumQ)_i = (CumQ)_{i-1} + (Q_{int})_i \qquad (2.1)$$

with $(CumQ)_{i=0} = 0$ for preliminary HCC,

$(CumQ)_{i=0} = 0$ for preliminary CCC if $Q_f \leq 0$;

and $(CumQ)_{i=0} = Q_f$ for preliminary CCC if $Q_f > 0$.

It may be noted that the above procedure generates Table 1.3 (for the preliminary HCC) as well as columns A through C of Table 1.4 (for the preliminary CCC) with no changes. The only change occurs in the calculation of the cumulative enthalpy column for the preliminary CCC, where it must be noted that $(CumQ_c)_{i=0} = 0$ for Case Study 4S1. The preliminary composite curve data are summarized in Table 2.1, along with the labels for convenient identification of the vertices. The data on plotting yield the composite curves in Figure 1.3.

Table 2.1 Preliminary Composite Curve Data

T_h	$CumQ_h$	Hot Vertex	T_c	$CumQ_c$	Cold Vertex
45	0	A	20	0	a
65	200	B	40	400	b
125	3200	C	112	2920	c
175	3700	D	155	3780	d

Units: temperature in °C and enthalpy in kW.

Step 2. Generation of an Initial Pinch Tableau

The initial pinch tableau (Table 2.2), which is important in determining the sensitivity threshold, is generated as follows. The Q coordinates of the vertices of the two composite curves (i.e., the *CumQ* entries in Table 2.1) are merged and then arranged in ascending order. In the pinch tableau, the first column will correspond to a hot vertex and the last column to a cold vertex. Hot vertices are labeled A, B, C, etc. while cold vertices are labeled a, b, c, etc. For hot vertices, temperatures are simply copied from the T_h column of Table 2.1 into the next row of the tableau. In a similar fashion, the cold stream temperatures are copied from the T_c column of Table 2.1 into the third row of the tableau. The "missing" temperatures in the second and third rows of the tableau can now be entered by simple linear interpolation. The interpolated temperatures, which correspond to the projections of vertices on the "facing" line segments, are shown underlined for convenience. The fourth row of the tableau contains ΔT, namely $T_h - T_c$.

The least value of ΔT in the fourth row of the pinch tableau directly gives the ΔT_{min}. If this ΔT_{min} in the initial pinch tableau is nonnegative, then it is the sensitivity threshold ΔT_{min} (as discussed later). If the ΔT_{min} is negative, then it must be set to zero as described in the next step since it indicates this is not a threshold problem.

Table 2.2 Initial Pinch Tableau

Vertices	A	a	B	b	c	C	D	d
Q	0	0	200	400	2920	3200	3700	3780
T_h	45	<u>45</u>	65	<u>69</u>	<u>119.4</u>	125	175	--
T_c	--	20	<u>30</u>	40	112	<u>126</u>	<u>151</u>	155
ΔT	--	25	35	29	7.4	−1	24	--

$\Delta T_{min} = -1°C$

Units: temperature in °C, and enthalpy in kW.

Step 3. Setting ΔT_{min} to Zero

Physically, it is impossible to have a negative ΔT_{min}. So, Table 2.2 must be revised and ΔT_{min} set to zero by shifting the CCC using the procedure described below.

Note that every column in the pinch tableau corresponds to a vertex either on the HCC or CCC. In Table 2.2, there is a vertex (C) with coordinates (3200, 125) on the HCC where the ΔT is −1. For a T-value of 125, the Q-value on the CCC is 3180 by linear interpolation. Thus, the CCC needs to be given a Q-shift of 20 (i.e., 3200 − 3180) to make ΔT zero. All vertices with negative

ΔT must be similarly examined and the maximum value of the Q-shifts used to give a horizontal shift to the CCC so that no vertex has a negative ΔT.

Step 4. Revision of a Pinch Tableau for a Specified Composite Curve Shift

In the previous step, it was found that a Q-shift of 20 was required to make ΔT_{min} zero. Given a pinch tableau (as in Table 2.2), generation of a new tableau (Table 2.3) to reflect a specified horizontal shift in the CCC is a straightforward task. First, the value of the Q-shift is added to all the cold vertex entries in the first row of the existing tableau in order to appropriately revise the Q coordinates of the vertices of the CCC. The Q-values in the first row are then arranged in ascending order. Note that if the Q coordinate for a hot vertex and a cold vertex is the same, then the entry is repeated with the hot vertex entry preceding the cold one. This is done because the hot vertex will precede the cold vertex on the composite curves plot for further positive Q-shifts. The remainder of the procedure is analogous to that described in Step 2. The hot and cold stream temperatures are copied from Table 2.1 into the next two rows of the tableau. "Missing" temperatures can be now entered by linear interpolation and ΔT values can then be calculated and entered in the fourth row of the tableau.

Table 2.3 Reference Pinch Tableau (for ΔT_{min} = 0°C)

Vertices	A	a	B	b	c	C	D	d
Q	0	20	200	420	2940	3200	3700	3800
T_h	45	<u>47</u>	65	<u>69.4</u>	<u>119.8</u>	125	175	--
T_c	--	20	<u>29</u>	40	112	<u>125</u>	<u>150</u>	155
ΔT	--	27	36	29.4	7.8	0	25	--

ΔT_{min} = 0°C, $Q_{cu,min}$ = 20 kW, $Q_{hu,min}$ = 100 kW, and T_p = 125°C.

Units: temperature in °C and enthalpy in kW.

Importantly, the energy targets may be obtained directly from a pinch tableau as follows. The least ΔT value in the fourth row specifies the ΔT_{min}. The pinch is obviously given by the column where ΔT_{min} occurs. The minimum cold utility requirement ($Q_{cu,min}$) is given by the Q coordinate of the first cold vertex (namely, the first cold vertex entry in the first row of the pinch tableau). The minimum hot utility requirement ($Q_{hu,min}$) is given by the difference in the Q coordinates of the last vertices on the CCC and HCC (namely, on subtracting the last hot vertex entry from the last entry in the first row of the pinch tableau). For Case Study 4S1, Table 2.3 yields $\Delta T_{min} = 0°C$, $Q_{cu,min} = 20$ kW, $Q_{hu,min} = 3800 - 3700 = 100$ kW, and $T_p = 125°C$.

Step 5. Determination of Significant Composite Curve Shifts
    The lower limit on ΔT_{min} is zero. The upper limit is given by the difference between the highest hot stream temperature and the lowest cold stream temperature in the stream data. For Case Study 4S1, the upper limit on ΔT_{min} is $175° - 20° = 155°$. This value corresponds to the case where there is no overlap between the HCC and the CCC. As the composite curve can be horizontally shifted through any arbitrary amount within this allowable ΔT_{min} range of $0° - 155°$, the total number of possible shifts is infinite. However, the number of significant shifts is limited because it is possible meaningfully to interpolate between these shifts to obtain the energy targets. It may be noted that $Q_{hu,min}$, $Q_{cu,min}$, and T_p are all piecewise linear functions of ΔT_{min}. Interpolation is valid when a vertex on one composite curve "sees" a line segment (and not a vertex) on the other composite curve during the horizontal sliding process. (An exception occurs when a topology trap exists, as discussed later). Hence, a significant shift may be defined as the Q-shifts of the CCC (with respect to a reference position) that result in a hot vertex and a cold vertex having the same Q coordinate. Thus, linear interpolation for $Q_{hu,min}$, $Q_{cu,min}$, and T_p is possible between two consecutive significant shifts (after sorting in ascending order) when no topology trap exists in between.
    Significant shifts (Table 2.4) may be determined by the following procedure using the pinch tableau for $\Delta T_{min} = 0°$ (Table 2.3) as the reference.

Table 2.4 Determination of Significant Q-Shifts

Cold Vertex	a	a	a	b	b	c	c
Shift to Hot Vertex	B	C	D	C	D	C	D
Significant Q-Shifts	180	3180	3680	2780	3280	260	760

Units: enthalpy in kW.

    Consider the first cold vertex (a), whose coordinates are (20, 20) in Table 2.3. Its horizontal distance from all hot vertices to its right is simply determined by subtracting 20 from the Q coordinates of these hot vertices (namely, 200, 3200, and 3700). These distances (180, 3180, and 3680) along with the distances for all other cold vertices (calculated in the same manner) are entered in Table 2.4 as the significant shifts.

Step 6. Generation of Energy Targets for the Complete ΔT_{min} Range

Given the reference pinch tableau, new tableaux can be generated for the various significant shifts in Table 2.4 using the procedure discussed in step 4. As an example, the updated tableau for a significant shift of 180 is given in Table 2.5. It is observed that $Q_{cu,min} = 200$ kW and $Q_{hu,min} = 280$ kW (both being merely incremented by the value of Q-shift, namely, 180).

Table 2.5 Pinch Tableau for $\Delta T_{min} = 9°C$ and Q-Shift = 180

Vertices	A	B	a	b	c	C	D	d
Q	0	200	200	600	3120	3200	3700	3980
T_h	45	65	<u>65</u>	<u>73</u>	<u>123.4</u>	125	175	--
T_c	--	--	20	40	112	<u>116</u>	<u>141</u>	155
ΔT	--	--	45	33	11.4	9	34	--

$\Delta T_{min} = 9°C$, $Q_{cu,min} = 200$ kW, $Q_{hu,min} = 280$ kW, and $T_p = 120.5°C$.

Units: temperature in °C and enthalpy in kW.

On creating tableaux analogous to Table 2.5 for all the other significant shifts in Table 2.4, the energy targets for the entire range of ΔT_{min} from 0° to 155°C may be obtained. The final results are summarized in Table 2.6, and the composite curves for some of these significant ΔT_{min} values are shown in Figure 2.1.

Table 2.6 Energy Targets for Complete ΔT_{min} Range

ΔT_{min}	T_{ph}	T_{pc}	$Q_{cu,min}$	$Q_{hu,min}$	Q-shift	$N_{u,mer}$
0	125	125	20	100	Ref	6
9	125	116	200	280	(R) 180	6
13	125	112	280	360	260	6
27.286	125	97.714	780	860	(R) 760	7
85	125	40	2800	2880	2780	6
105	125	20	3200	3280	3180	5
115	135	20	3300	3380	(R) 3280	5
155	175	20	3700	3780	3680	4

Units: temperature in °C and enthalpy in kW.

Table 2.6 contains a comprehensive picture of the energy targets and may be used to obtain this information at any value of ΔT_{min}. For example, at ΔT_{min} = 20°C, straightforward linear interpolation between 13° and 27.286°C in Table 2.6 yields $Q_{cu,min}$ = 525 kW, $Q_{hu,min}$ = 605 kW, and T_{ph} = 125°C and T_{pc} = 105°C. This is in agreement with the results in Table 1.2.

The variation in the hot utility requirement and the pinch temperature may be plotted (as in Figure 2.2) using Table 2.6. The variation in the cold utility will be identical to that of the hot utility. It will appear as a parallel curve (not shown) in Figure 2.2, except that it will be vertically displaced by 80 kW downwards. The variation of pinch temperature (based on the average of hot and cold stream temperatures) in Figure 2.2 is observed to have a slope of –0.5 when the pinch is at a hot vertex and a slope of +0.5 when at a cold vertex. For Case Study 4S1, the pinch is at hot vertex (C) for $0° \leq \Delta T_{min} \leq 105°$ and at cold vertex (a) for $105° \leq \Delta T_{min} \leq 155°$.

Some entries in Table 2.6 are marked (R) to indicate redundant. These correspond to shifts at which the trends in Figure 2.2 for the minimum utility vs. ΔT_{min} and T_p vs. ΔT_{min} do not change. Though the number of redundant shifts in this example is three, the number can be very large for some problems and the effort in generating redundant tableaux (specially when calculating manually) is in vain. Makwana and Shenoy (1993) present a method for elimination of these redundant shifts. However, this method is not discussed here since a priori elimination leads to difficulties in supertargeting by the extended pinch tableau (see Section 2.1.2) and in determination of topology traps (see Section 2.1.3).

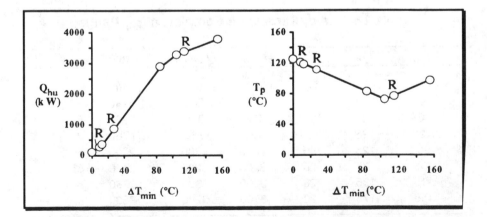

Figure 2.2 The minimum utility requirements and the pinch temperature show a piecewise linear variation with ΔT_{min}. Redundant shifts are indicated by R.

Step 7. Generation of Unit Targets for the Complete ΔT_{min} *Range*

The final step is to establish unit targets for maximum energy recovery (MER) networks over the entire range of ΔT_{min} starting with Equation 1.27 and the results obtained from the CTA thus far (see Table 2.6). Assuming that there is one utility (hot) above the pinch and one utility (cold) below the pinch, $N_{u,mer}$ as per Equation 1.27 is obtained by summing the number of process streams above and below the pinch separately.

Consider the problem of determining $N_{u,mer}$ at ΔT_{min} = 13°C, for example. From Table 2.6, T_{ph} is found to be 125°C. From the stream data for Case Study 4S1, only one hot stream (H1) is found to be above the pinch and two hot streams (H1 and H2) below the pinch. Similarly, using T_{pc} = 112°C, one cold stream (C3) is found to exist above the pinch and two cold streams (C3 and C4) below the pinch. Thus, $N_{u,mer}$ = 2 + 4 = 6. By following the above procedure for each of the ΔT_{min} values listed in Table 2.6, the $N_{u,mer}$ column may easily be calculated.

The change in the number of units always occurs at the points given by Table 2.6. However, an additional calculation for the unit target is necessary at any value of ΔT_{min} within each interval to obtain the variation of $N_{u,mer}$ with ΔT_{min} over the entire range (see Fig. 2.3). This is due to two reasons: the number of units within an interval cannot be determined easily from the values at the end-points; on the other hand, the number of units does not change within an interval (excluding end-points) and is unique. On performing this additional calculation within each ΔT_{min} interval, the final results for the unit targets are obtained as in Figure 2.3.

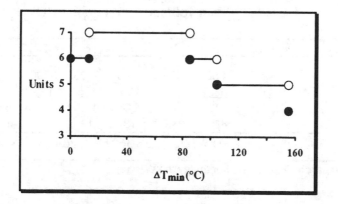

Figure 2.3 The variation of the unit targets with ΔT_{min} appears as horizontal segments (solid dots denote that the end-point of the segment is included, and open circles denote that the end-point is excluded).

2.1.2 The Extended Pinch Tableau

The pinch tableau from the previous subsection may be extended to obtain information on the GCC, the area targets, the shell targets, and the cost targets. The conversion of the pinch tableau (essentially containing composite curve information) to the GCC is straightforward and is left as an exercise for the reader. In this subsection, the extension of the tableau for area and cost targeting is discussed as it provides a sound methodology for supertargeting.

Example 2.2 Supertargeting Using the Pinch Tableau Approach
For the stream data in Case Study 4S1, determine the optimum value of ΔT_{min} (to an accuracy of 0.1°C) that yields the minimum total annual cost for the two cases discussed in Example 1.10.

Solution. Area and shell targeting may conveniently be performed on the pinch tableau, as the information stored is enthalpy-interval-based. The procedure for developing the extended pinch tableau will be illustrated for $\Delta T_{min} = 13$°C (for concreteness).

Step 1. Generation of the Basic Pinch Tableau
The first step is to develop the basic pinch tableau as in Example 2.1. The result is shown in Table 2.7 for $\Delta T_{min} = 13$°C, which is generated by revising the reference pinch tableau (Table 2.3) for a Q-shift of 260 kW using the procedure outlined in step 4 of the previous example.

Table 2.7 Pinch Tableau for $\Delta T_{min} = 13$°C

Vertices	A	B	a	b	C	c	D	d
Q	0	200	280	680	3200	3200	3700	4060
T_h	45	65	66.6	74.6	125	125	175	--
T_c	--	--	20	40	112	112	137	155
ΔT	--	--	46.6	34.6	13	13	38	--

$\Delta T_{min} = 13$°C, $Q_{cu,min} = 280$ kW, $Q_{hu,min} = 360$ kW, and $T_p = 118.5$°C.

Units: temperature in °C, and enthalpy in kW.

Step 2. Incorporation of the Utilities
Area targeting requires the composite curves to be balanced as in Example 1.3 by incorporation of the utilities as additional streams.

The problem now is to incorporate an additional stream into a given composite curve. The MC_p values of these additional streams as per Equation 1.10 are $(MC_p)_{hu} = 360/(180° - 179°) = 360$ kW/°C and $(MC_p)_{cu} = 280/(25° - 15°) = 28$ kW/°C. As there is no hot stream in the existing tableau between 179° and 180°C, the incorporation of hot utility is straightforward. The Q coordinate of one end-point of the hot utility segment will be 4060 kW (see Table 2.7). The other end-point will be at $Q = 4060 - 360 (180° - 179°) = 3700$ kW.

The incorporation of the cold utility is more involved. The cold utility segment extends from 15° to 25°C; however, stream C3 (with $MC_p = 20$ kW/°C) also exists over the 20° to 25°C range and the overlap must be carefully accounted for. One end-point of the cold utility segment will be at $Q = 0$ kW. The other end-point will be at $Q = 380$ kW as per the calculations below:

In 15° to 20°C range (stream CU): $Q = 28 (20° - 15°) = 140$ kW
In 20° to 25°C range (streams CU and C3): $Q = 140 + 48 (25° - 20°)$
$= 380$ kW

Table 2.8 shows the pinch tableau with the utilities. It includes the three vertices calculated above, namely, (140, 20), (380, 25), and (3700, 179).

Table 2.8 Extended Pinch Tableau with Utilities for $\Delta T_{min} = 13°C$

Vertices	A	a	B	V_{CU}	b	C	c	D	V_{HU}	d
Q	0	140	200	380	680	3200	3200	3700	3700	4060
T_h	45	59	65	68.6	74.6	125	125	175	179	180
T_c	15	20	21.25	25	40	112	112	137	137	155
ΔT	30	39	43.75	43.6	34.6	13	13	38	42	25
$\Sigma(MC_p/h)_h$	0	50	50	250	250	250	250	50	0	1800
$\Sigma(MC_p/h)_c$	0	140	240	240	100	175	175	100	100	100
$Sum(Q/h)$	0	1400	600	1800	3000	25200	0	5000	0	3600
$LMTD_i$	0	34.30	41.33	43.67	38.93	22.07	13	23.31	39.97	32.77
A_i	0	40.81	14.52	41.21	77.07	1142	0	214.53	0	109.86

$\Delta T_{min} = 13°C$, $Q_{cu,min} = 280$ kW, $Q_{hu,min} = 360$ kW, and $T_p = 118.5°C$.
$N_{u,mer} = 6$, $A_c = 1640.06$ m^2,
$OC = 46 \times 10^3$ \$/yr, $CC = 603.59 \times 10^3$ \$, and $TAC = 240.42 \times 10^3$ \$/yr.

Units: temperature in °C, enthalpy in kW, heat capacity flow rate in kW/°C,
    heat transfer coefficient in kW/(m^2 °C), and area in m^2.

Step 3. Calculation of Area Target Profile

The first three rows of Table 2.8 correspond to the first three columns of Table 1.7. These are merely the enthalpy intervals in a horizontal format rather than in the vertical format of Table 1.7. At this stage, the area target can be determined in the same manner as in Example 1.4 by calculating $\Sigma(MC_p/h)_h$, $\Sigma(MC_p/h)_c$, $Sum(Q/h)$, $LMTD_i$, and A_i. On summing the values of A_i, the countercurrent area target is 1640.06 m^2. In essence, the methodology from Chapter 1 holds for area, shell, and cost targeting; however, the extended pinch tableau provides a convenient representation for these calculations. As an exercise, the reader may attempt the calculation of the *F*-correction factor as in Table 1.9 in the pinch tableau format and verify the area target for 1-2 shell and tube exchangers to be 1883 m^2 at $\Delta T_{min} = 13°C$.

It is possible to generate continuous area targets in principle by developing an extended pinch tableau that holds between two significant *Q*-shifts. The procedure just used may be followed, except that entries in the tableau will be functions of ΔT_{min} rather than constant values. As a matter of fact, a formula for the area target may be written (Bhatwadekar, 1993) valid between two significant *Q*-shifts; however, such an expression is unwieldy. So, it may be advisable to pursue with area targets at particular ΔT_{min} values.

To obtain the area target profile as a function of ΔT_{min}, the area targets may first be determined at the significant shifts themselves (Table 2.9).

Table 2.9 Targets at the Significant Shifts

ΔT_{min}	OC	$N_{u,mer}$	A_c	$TAC_{uniform}$	A^*	$TAC_{nonuniform}$
1	14.8	6	3037.78	297.57	14905.50	888.07
9	35.6	6	1835.62	243.05	9970.36	682.23
13	46.0	6	1640.06	240.42	9018.50	646.68
13+	46.0	7	1640.06	254.14	9018.50	672.47
27.286	111.0	7	1141.36	283.39	6252.61	594.01
85–	373.6	7	941.55	530.87	3767.70	716.82
85	373.6	6	941.55	518.62	3767.70	699.20
105–	425.6	6	959.75	571.98	3600.84	741.56
105	425.6	5	959.75	559.31	3600.84	723.11
115	438.6	5	965.09	572.69	3595.64	735.82
155–	490.6	5	987.75	626.32	3450.54	779.65
155	490.6	4	987.75	613.03	3450.54	760.00

Units: temperature in °C, area in m^2, and cost in 10^3 \$/yr.

The area target is continuous, smooth, and differentiable between two significant shifts; so, it may suffice to calculate the area at a few points within such an interval and then draw a smooth curve between every two significant shifts. Figure 2.4 shows the area target profile obtained following such a procedure.

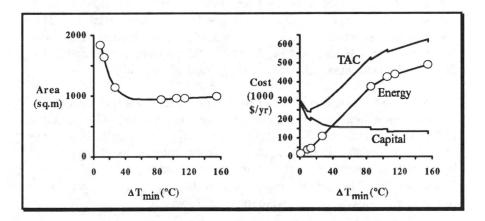

Figure 2.4   The area and cost targets are continuous, smooth, and differentiable only between two significant shifts (shown by open circles above).

Step 4. Calculation of Cost Target Profiles

The operating cost as per Equation 1.32 for Case Study 4S1 for the entire ΔT_{min} range is (assuming annual cost of unit duty to be 120 \$/(kW.yr) for HU and 10 \$/(kW.yr) for CU):

$$OC = 120\,(Q_{hu,min}) + 10\,(Q_{hu,min} - 80) = 130\,(Q_{hu,min}) - 800$$

Thus, the hot utility requirement profile in Figure 2.2 can be transformed into the energy cost profile (Figure 2.4) using the above equation.

It is more complex to obtain the capital cost profile since it requires combining the unit target profile (Figure 2.3) and the area target profile (Figure 2.4) as per Equation 1.34a. The important aspect to remember is that discontinuities and nonsmooth behavior can occur only at the significant shifts; so the procedure used to obtain the area target profile may be followed. Capital cost calculations need to be performed at the significant shifts (Table 2.9) and at a few points between every two of them. At the discontinuities in the unit target in Figure 2.3, additional calculations for the capital cost need to be performed using a different value of $N_{u,mer}$ in Equation 1.34a. These additional calculations at ΔT_{min} values of 13+, 85−, 105−, and 155− °C are shown in Table 2.9. The resulting capital cost profile is shown in Figure 2.4.

Using the annualization factor (in this case, 0.3221 as in Example 1.9) and Equation 1.35, the total annual cost profile in Figure 1.9 may be obtained. The above procedure requires less computational effort and ensures that the *TAC* profile is fundamentally accurate (especially at the discontinuities and the topology traps discussed in the next section). The procedure used in Section 1.6 with calculations at 1°C intervals of ΔT_{min} requires relatively more computational effort and yet does not necessarily capture the intricacies in the behavior of the *TAC* profile. In Figure 1.9, the procedure in Example 1.10 does not yield the actual behavior for ΔT_{min} between 13° and 14°C.

Step 5. Continuous Determination of Targets for Nonuniform Specifications

For nonuniform specifications, the targeting procedure of Hall et al. (1990) may be used as in Example 1.9. The energy, units, and operating cost targets do not change from those obtained in the uniform specifications case. The area targets need to be recalculated using the appropriate φ-factors from Equation 1.38a. The fictitious area A^* and the $TAC_{nonuniform}$ obtained by this method for various ΔT_{min} values are shown in Table 2.9.

Step 6. Determination of Optimum ΔT_{min} *by Golden Section Search*

The *TAC* profile is continuous, smooth, and differentiable between every two significant shifts; so any standard search technique may be used between any two significant shifts to determine the optimum within that range.

On examining Table 2.9, it is observed that the minimum *TAC* for the uniform specifications case lies in the interval 9° - 13°C. On using the golden section search technique, the optimum ΔT_{min} is found to be 13°C. Golden section search is a region elimination method. If L_0 is the original length of the region and L is its length after k iterations, then $L = [(\sqrt{5} - 1)/2]^k L_0$ (Edgar and Himmelblau, 1989). So, if the optimum ΔT_{min} is desired to an accuracy of 0.1°C within the range 9° - 13°C, the number of iterations required will be eight on rounding off $\ln(0.1/4)/\ln[(\sqrt{5} - 1)/2]$. If the accuracy desired is 1°C, then only three iterations are necessary.

For the nonuniform specifications, a search in the range 13° - 27.286°C shows the minimum *TAC* to be 594.01 x 10^3 $/yr at a ΔT_{min} of 27.286°C. Another search in the range 27.286° - 85°C shows the minimum *TAC* is 586.05 x 10^3 $/yr at a ΔT_{min} of 35.12°C (which is the global optimum for the nonuniform specifications case).

2.1.3 Topology Traps and Sensitivity Threshold

Continuous targeting allows determination of the minimum utility requirements and the pinch temperature over the entire range of ΔT_{min}. In Example 2.1, both hot and cold utilities were required over the entire ΔT_{min} range. This may not necessarily be the case. There is a class of examples (Linnhoff et al., 1982) called threshold problems, where only a single utility (either hot or cold) is required below a certain value of ΔT_{min}. This value of ΔT_{min} is referred to as the threshold ΔT_{min} or the sensitivity threshold.

Another aspect not observable in Example 2.1 is multiple pinches. At certain values of ΔT_{min}, there may be more than one pinch point. In such a case, the plot of pinch temperature as a function of ΔT_{min} will show discontinuities. Both these features are demonstrated in the next example, in which a small modification is made in Case Study 4S1. The example also highlights the drastic effect a single change in the stream data can cause.

Example 2.3 Determination of Sensitivity Threshold and Multiple Pinches
Consider the stream data in Case Study 4S1 with the following modification: the MC_p-value of stream H2 is 15 kW/°C (and not 40 kW/°C). Let this modified stream data be referred to as Case Study 4S1t. Determine how the utility requirements and pinch temperature vary with ΔT_{min}. Indicate clearly the sensitivity threshold and multiple pinches, if any exist.

Solution. The procedure adopted in Example 2.1 will be followed.

Step 1. Generation of Preliminary Composite Curve Data
From the first law of thermodynamics (see Section 1.1.1), it can be easily shown that there is a heat deficiency of 1580 kW. In other words,

$$Q_f = \Sigma Q_h - \Sigma Q_c = 1300 + 900 - 2700 - 1080 = -1580.$$

The data for the preliminary CCC does not change from that obtained in Table 2.1. However, the change in the heat capacity flow rate of stream H2 will cause some changes in the preliminary HCC. In Table 2.1, the $CumQ_h$ entries of the hot vertices C and D will be 1700 and 2200 respectively (and not 3200 and 3700). Except for this change, the rest of Table 2.1 holds.

Step 2. Generation of Initial Pinch Tableau
The initial pinch tableau is generated by the procedure used in step 2 of Example 2.1. The sensitivity threshold ΔT_{min} is given by the value of ΔT_{min} in the initial pinch tableau, provided it is nonnegative. Table 2.10 shows that

the threshold ΔT_{min} is 25°C for Case Study 4S1t. The physical explanation of the sensitivity threshold will be offered at the end of this problem.

Step 3. Generation of Energy Targets for the Complete ΔT_{min} *Range*

The significant Q-shifts are next determined, and the revised pinch tableaux are obtained for shifts of the CCC. As in Example 2.1, the energy targets for the entire range of ΔT_{min} from 0° to 155°C are obtained (Table 2.11) For $\Delta T_{min} \geq 85°C$, the two case studies (4S1 and 4S1t) show no significant change in behavior.

Table 2.10 Initial Pinch Tableau Showing Sensitivity Threshold

Vertices	A	a	B	b	C	D	c	d
Q	0	0	200	400	1700	2200	2920	3780
T_h	45	45	65	73	125	175	--	--
T_c	--	20	30	40	77.14	91.43	112	155
ΔT	--	25	35	33	47.86	83.57	--	--

$\Delta T_{min} = 25°C$, $Q_{cu.min} = 0$ kW, and $Q_{hu.min} = 1580$ kW.
Units: temperature in °C and enthalpy in kW.

Table 2.11 Energy Targets for Entire Range for Case Study 4S1t

ΔT_{min}	T_{ph}	T_{pc}	$Q_{cu,min}$	$Q_{hu,min}$	Q-shift	$N_{u,mer}$
25	45	20	0	1580	0	4
41	81	40	200	1780	200	7
85	125	40	1300	2880	1300	6
105	125	20	1700	3280	1700	5
115	135	20	1800	3380	1800	5
155	175	20	2200	3780	2200	4
Topology Trap:						
38.33	58.33	20	133.33	1713.33	133.33	7
	78.33	40				

Units: temperature in °C and enthalpy in kW.

Step 4. Detection of Topology Traps

It must be realized that the pinch can occur only at a vertex. With a small shift (between two significant shifts), the pinch will usually continue to be at the same vertex. However, as a rule, when the CCC is slid horizontally to the right, ΔT at each vertex increases at a rate dependent upon the slope of the line segment facing it. So, it is possible that another vertex exists with a lower rate of increase of ΔT compared to that at the pinch. In such an event, ΔT at this vertex will become equal to that at the pinch after a certain shift, leading to a multipinched situation. If the composite curve is further slid, the pinch will shift to the other vertex provided the rate of increase of ΔT at both vertices remains unchanged. This situation where the pinch moves to an entirely new vertex for an infinitesimally small shift is called a topology trap, and the value of ΔT_{min} at which this pinch swap occurs is of interest.

Further, it may be argued that a multipinched situation without a pinch swap (see Problem 2.A) can occur only at a significant shift (because it must be accompanied by a change in the rate of increase of ΔT on at least one of the two pinches involved). It may be noted that the rate of increase of ΔT remains the same, provided the vertices are facing line segments (and not vertices) on the opposite curve. However, a topology trap (a multipinched situation with a pinch swap) can occur at a significant shift or between two consecutive significant shifts. When the occurrence is at a significant shift, it is automatically detected by the algorithm outlined in Example 2.1.

Specifically, a topology trap exists between two consecutive significant shifts if both T_{ph} and T_{pc} change. From Table 2.11, a topology trap is observed for ΔT_{min} between 25° and 41°C. The pinch is at cold vertex (a) for $\Delta T_{min} = 25$°C, but is at cold vertex (b) for $\Delta T_{min} = 41$°C. The question addressed below is "what is the pinch location for $25 \leq \Delta T_{min} \leq 41$?"

The following procedure (which is explained later) may be adopted for the accurate determination of the topology trap. The topology trap may be determined by backtracking from the pinch tableau for ΔT_{min} of 41°C (Table 2.12) or by forward-tracking from the pinch tableau for ΔT_{min} of 25°C. Backtracking is used below, as forward-tracking in general causes difficulties when redundant shifts are eliminated (Makwana and Shenoy, 1993). In backtracking, a minor change is required in the ordering of entries in the pinch tableau for the case of a hot vertex and a cold vertex having the same Q coordinate. The cold vertex now precedes the hot vertex (as opposed to the convention adopted earlier). This can be achieved simply by interchanging the vertex labels (see vertices [a] and [B] with a Q coordinate of 200 in Table 2.12).

Table 2.12 Pinch Tableau for ΔT_{min} = 41°C and Q-Shift = 200

Vertices	A	a	B	b	C	D	c	d
Q	0	200	200	600	1700	2200	3120	3980
T_h	45	<u>65</u>	65	<u>81</u>	125	175	--	--
T_c	--	20	<u>20</u>	40	<u>71.43</u>	<u>85.71</u>	112	155
ΔT	--	45	45	41	53.57	89.29	--	--
$\Sigma(MC_p)_h$	--	10	10	25	25	10	--	--
$\Sigma(MC_p)_c$	--	--	20	20	35	35	35	20
Q_{tt}	--	-66.67	-400	--	--	--	--	--

ΔT_{min} = 41°C, $Q_{cu,min}$ = 200 kW, $Q_{hu,min}$ = 1780 kW, and T_p = 60.5°C.

Units: temperature in °C, enthalpy in kW, and heat capacity flow rate in kW/°C.

The basic tableau is extended by adding three more rows. The $\Sigma(MC_p)_h$ and $\Sigma(MC_p)_c$ rows can be filled, bearing in mind the streams that exist in each temperature interval. As before, the $\Sigma(MC_p)$ values for the various temperature intervals are appropriately written in the tableau against the higher temperature limit of the interval. The Q_{tt} row, useful in determining the topology trap, is obtained from the following equation:

$$Q_{tt} = (\Delta T - \Delta T_{min}) / (1/\Sigma(MC_p)_p - 1/\Sigma(MC_p)) \qquad (2.2)$$

where
ΔT = ΔT at the vertex under consideration
$\Sigma(MC_p)$ = $\Sigma(MC_p)_c$ for hot vertex and $\Sigma(MC_p)_h$ for cold vertex
$\Sigma(MC_p)_p$ = $\Sigma(MC_p)_c$ at pinch if it is a hot vertex,
        $\Sigma(MC_p)_h$ at pinch if it is a cold vertex,
        $\min[\Sigma(MC_p)_h$ one interval below pinch,
          $\Sigma(MC_p)_c$ one interval above pinch]
          if pinch features two vertices and backtracking is used
        $\max[\Sigma(MC_p)_h$ one interval above pinch,
          $\Sigma(MC_p)_c$ one interval below pinch]
          if pinch features two vertices and forward-tracking is used.

In backtracking, the Q_{tt} calculation can be skipped for points with $\Sigma(MC_p) \geq \Sigma(MC_p)_p$, as it results in a positive shift (or an infinite shift in the case of an equality). For Table 2.12, the pinch at ΔT_{min} = 41°C features cold vertex (b); consequently, $\Sigma(MC_p)_p$ = 25. Now, for vertex (a), Q_{tt} = (45 − 41)/(1/25 −

1/10) = −66.67 and for vertex (B), $Q_{tt} = (45 − 41)/(1/25 − 1/20) = −400$. The Q_{tt} with the minimum absolute value signifies the backward (negative) Q-shift required (−66.67 in this case to be given to Table 2.12) for the occurrence of the topology trap. Without generating a new tableau, the ΔT_{min} at which the topology trap occurs can be directly determined from

$$\Delta T_{min,tt} = \Delta T_{min} + Q_{tt}/\Sigma(MC_p)_p. \qquad (2.3)$$

Using the above formula, $\Delta T_{min,tt} = 41 − 66.67/25 = 38.33°C$. The pinch will be at cold vertex (a) for ΔT_{min} slightly less than 38.33°C and abruptly changes to cold vertex (b) for ΔT_{min} slightly greater than 38.33°C. The energy targets at this topology trap ($\Delta T_{min} = 38.33$ and Q-shift = −66.67) are $Q_{cu,min} = 200 −$ 66.67 = 133.33 kW and $Q_{hu,min} = 1780 − 66.67 = 1713.33$ kW. At $\Delta T_{min} = 38.33°C$ (the topology trap itself), there are two pinches: one at cold vertex (a) with $T_{pc} = T_{c,vertex\ a} = 20°C$, $T_{ph} = T_{pc} + \Delta T_{min} = 58.33°C$; another at cold vertex (b) with $T_{pc} = T_{c,vertex\ b} = 40°C$, $T_{ph} = T_{pc} + \Delta T_{min} = 78.33°C$. As $T_{pc} = 20°C$ agrees with the T_{pc} at $\Delta T_{min} = 25°C$ and $T_{pc} = 40°C$ agrees with that at $\Delta T_{min} = 41°C$, there is only one topology trap between these two significant shifts. If two or more topology traps occurred in succession, the procedure would need to be repeated. By including the results for the topology trap at the bottom of Table 2.11, the variation in the utility requirements and the pinch temperature with ΔT_{min} may be plotted (Figure 2.5). In addition to the features seen in Figure 2.2, the effect of the sensitivity threshold and the topology trap are clearly observed in Figure 2.5, as emphasized by Kemp (1991).

Step 5. Generation of Unit Targets for the Complete ΔT_{min} Range
     The procedure in Example 2.1 (step 7) may be adopted to determine the variation of the unit targets except for the following modification. For the case of multiple pinches (including topology traps), the unit target needs to be carefully determined as the problem gets partitioned into more than two subsystems. The number of units in each subsystem must be separately determined using Equation 1.26, and then summed to obtain $N_{u,mer}$. For example, at $\Delta T_{min} = 38.33°C$, there are four process streams with one hot utility above the first pinch (at 78.33°C/40°C), one process stream with one cold utility below the second pinch (at 58.33°C/20°C), and three streams with no utility in the region between these two pinches; so, $N_{u,mer} = 4 + 1 + 2 = 7$. Using Table 2.11, the reader should attempt to draw the analog of Figure 2.3 for Case Study 4S1t. It will be seen that $N_{u,mer}$ is 4 for $0 \leq \Delta T_{min} \leq 25$, is 5

for $25 < \Delta T_{min} < 38.33$, and is 7 for $38.33 \le \Delta T_{min} < 85$. For $\Delta T_{min} \ge 85$, the plot is identical to Figure 2.3.

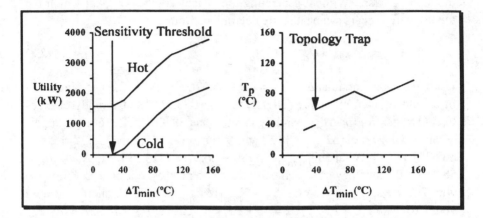

Figure 2.5 A constant amount of one kind of utility (either hot or cold, but not both) is required below the threshold ΔT_{min} (plot on left). A discontinuity occurs in the pinch temperature vs. ΔT_{min} curve at a topology trap (plot on right).

Before proceeding further, it would be useful to understand the physical implications of the sensitivity threshold and the topology trap.

In Case Study 4S1, both hot and cold utility were required even at $\Delta T_{min} = 0°C$ (see Table 2.6). The area and consequently the capital cost would be infinite for $\Delta T_{min} = 0°C$; hence, Table 2.9 started with $\Delta T_{min} = 1°C$. This may be contrasted with the situation in Case Study 4S1t, where the need for both utilities exists only above a certain value of ΔT_{min} (called the sensitivity threshold). Below the threshold ΔT_{min}, only a single utility (either hot or cold) is required and its amount is not affected by a change in ΔT_{min} (Figure 2.5). For Case Study 4S1t, 1580 kW of hot utility needs to be supplied for $\Delta T_{min} \le 25°C$. It may be visualized through the dotted line on Figure 2.6 (by shifting the HCC to the right, keeping the CCC fixed at its threshold position) that a portion of this 1580 kW of hot utility can be supplied at the lower temperature end if $\Delta T_{min} < 25°C$ for the system. Note that, at the sensitivity threshold, *either* the HCC has an overshoot at the left end *or* the CCC has an overshoot at the right end (Figure 2.6) with the remainder of the composite curves in perfect enthalpy balance.

Equation 2.1 ensures that if the overall system is heat deficient, then the preliminary CCC starts at $Q = 0$ which automatically places the hot utility to account for the right overshoot in the CCC. On the other hand, in the case of an overall heat excess, the CCC starts at $Q = Q_f$, which appropriately accounts for the left overshoot in the HCC in terms of a cold utility. Thus, in the case of threshold problems, the initial pinch tableau corresponds to the position of the composite curves at the sensitivity threshold. Essentially, the sensitivity threshold divides the problem into two regions in the ΔT_{min} space. If $\Delta T_{min} \leq$ threshold ΔT_{min}, no tradeoff exists between energy and capital costs due to the invariance of the utility requirements. If $\Delta T_{min} >$ threshold ΔT_{min}, utility requirements vary with ΔT_{min} and optimization is meaningful because the classical tradeoff between operating and capital costs exists. In the latter case, supertargeting may be performed and then the network may be generated by the pinch design method (Chapter 3). In the former case where $\Delta T_{min} \leq$ threshold ΔT_{min}, the thermodynamic constraint imposed by the pinch does not exist; so, the pinch design method is not used and special methods (discussed in Chapter 3) are employed. Further, as there is no pinch, $N_{u,mer} = N_{u,min}$ and typically no evolution (Chapter 4) is necessary.

Next, attention is focused on multiple pinches and topology traps. Consider a typical pair of hot and cold composite curves (see Figure 2.6). Each composite curve is piecewise linear, and the slope of a line segment on it is given by the reciprocal of the $\Sigma(MC_p)$ in that temperature interval. The vertices, where the slopes change, indicate a change in $\Sigma(MC_p)$ as caused by a change in the stream population.

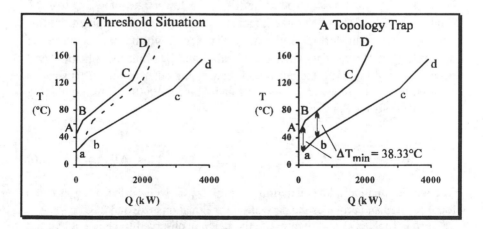

Figure 2.6 Only one kind of utility (hot utility for plot on left) is required for the case of a threshold problem. A multipinched situation occurs at a topology trap (plot on right): the pinch remains with vertex (a) if the CCC is moved slightly to the left and with vertex (b) if the CCC is moved slightly to the right.

The rate of change of ΔT at any vertex, for a shift in the CCC of Q (with a positive sign if to the right), depends on the slope of the line segment it faces and is given by

$$(d\Delta T/dQ_{shift})_{vertex} = 1/\Sigma(MC_p)_{facing\ line\ segment}. \tag{2.4a}$$

The equation in a more useful form may be rewritten as

$$(\Delta T)_{vertex,after\ shift} = (\Delta T)_{vertex,before\ shift} + Q_{shift}/\Sigma(MC_p)_{facing\ line\ segment}. \tag{2.4b}$$

For example with $\Delta T_{min} = 85°C$ in Figure 2.1, vertex (D) faces line segment (bc), whose $\Sigma(MC_p)$ must be used in the above relation. While forward-tracking, the $\Sigma(MC_p)$ of segment (ab) needs to be used beyond the shift, which brings vertex (b) below vertex (D). By similar reasoning, the $\Sigma(MC_p)$ of segment (ab) is to be used for vertex (C) during forward-tracking and the $\Sigma(MC_p)$ of segment (bc) is to be used during backtracking from $\Delta T_{min} = 85°C$.

The above arguments hold for all vertices (hot or cold). Applying the equation to the pinch point gives

$$(\Delta T_{min})_{after\ shift} = (\Delta T_{min})_{before\ shift} + Q_{shift}/\Sigma(MC_p)_p \tag{2.5}$$

where $\Sigma(MC_p)_p$ is the $\Sigma(MC_p)$ of the line segment facing the vertex at which the pinch occurs (as defined in Equation 2.2). There can exist a situation (topology trap) where some other vertex has a lower rate of change of ΔT (higher $\Sigma[MC_p]$ of the line segment facing it) than at the current pinch. In this case, after a certain shift (Q_{tt}), the ΔT at that vertex will become equal to the new ΔT_{min} at the current pinch point and a multiple-pinched situation will occur. Beyond this shift, the pinch will jump to the other vertex. This swap is referred to as a topology trap (Linnhoff and Ahmad, 1990). Thus, combining Equations 2.4b and 2.5 gives

$$(\Delta T_{min})_{current\ position} + Q_{tt}/\Sigma(MC_p)_p =$$
$$(\Delta T)_{other\ vertex} + Q_{tt}/\Sigma(MC_p)_{line\ segment\ facing\ other\ vertex} = \Delta T_{min,tt}. \tag{2.6}$$

The above equation allows determination of Q_{tt} as well as the ΔT_{min} for the topology trap and provides the derivation for Equations 2.2 and 2.3.

From Table 2.11 and Figure 2.5, the pinch is observed to be at cold vertex (a) for $25 \le \Delta T_{min} \le 38.33$, at cold vertex (b) for $38.33 \le \Delta T_{min} \le 85$, at hot vertex (C) for $85 \le \Delta T_{min} \le 105$, and at cold vertex (a) for $105 \le \Delta T_{min} \le 155$. These ranges of ΔT_{min} define four topology regions, within each of which the network structure does not change significantly and evolution (as discussed in Chapter 4) is possible. In other words, if the optimum ΔT_{min} is in the same topology region as the initial design (at a nonoptimal ΔT_{min}), then network optimization and evolution will prove successful in locating the global optimum (Linnhoff and Ahmad, 1990). This is otherwise not possible, simply because the pinch location and consequently the stream populations on either side of the pinch are substantially different in the different topology regions leading to different matching options during network design.

It is possible to have multipinched situations without pinch swaps (Makwana and Shenoy, 1993). This occurs when a shift causes the ΔT at another vertex to become equal to the ΔT at the pinch vertex, and the rate of change of ΔT at this other vertex is more than that at the pinch for further shifts. Such a situation, however, can occur only at the significant shifts.

Finally, a pinch occurs necessarily at a vertex except when two line segments (one on HCC and another on CCC) are parallel to each other and ΔT_{min} apart. In such a case, the whole region is pinched (see Ponton and Donaldson [1974] problem discussed by Makwana and Shenoy [1993]).

2.2 DIVERSE PINCH FOR DIFFERENT HEAT TRANSFER CONDITIONS

The diverse pinch concept (Fraser, 1989; Rev and Fonyo, 1991) is useful when the stream heat transfer coefficients differ considerably. In this case, it is more meaningful to use minimum flux (Q''_{min}) as a design specification rather than minimum approach temperature (ΔT_{min}). In what follows, energy and area targeting for the diverse pinch concept are explained.

2.2.1 Energy Targeting for Minimum Flux Specification

The ideas presented in Section 1.1 for the conventional pinch apply for the case of the diverse pinch, except that the temperature shifting of the streams is not by a uniform amount (namely, $\Delta T_{min}/2$). The PTA from Section 1.1.2 is applicable as illustrated in the example below, with the modification suggested by Townsend and Linnhoff (1983) for stream-dependent ΔT_{min} contributions. As argued by Fraser (1989), the temperature shift may be appropriately chosen to be (Q''_{min}/h_j), where h_j is the stream-dependent heat transfer coefficient.

Example 2.4 Determination of Diverse-Type Heat Cascade
Consider the stream data in Case Study 4S1t with the following heat transfer coefficients in kW/(m^2 °C) for the various streams: 0.2 (for H1 and C3); 2.0 (for H2, C4, and CU); and 4.0 (for HU). Determine the minimum

utility requirements for a minimum flux specification $Q''_{min}= 90/11 = 8.1818$ kW/m^2.

Solution. In addition to the inlet temperatures, outlet temperatures, and heat capacity flow rates, the heat transfer coefficients are required while performing energy targeting for minimum flux specification.

The diverse-type heat cascade (Table 2.13) is similar to the conventional-type heat cascade (Table 1.1) and is developed as follows. The temperature intervals (T_{int} in column A of Table 2.13) are carefully determined first. All the hot stream temperatures are decreased by (Q''_{min}/h_j) while all the cold stream temperatures are increased by (Q''_{min}/h_j). For hot stream H1, the inlet and outlet temperatures of 175°C and 45°C are decreased by 40.91° (namely, 8.1818/0.2) to yield 134.09°C and 4.09°C respectively. For cold stream C4, the inlet and outlet temperatures of 40°C and 112°C are increased by 4.09° (namely, 8.1818/2) to yield 44.09°C and 116.09°C respectively. These shifted temperatures are then sorted in descending order, omitting temperatures common to both hot and cold streams. These form the limits of the temperature intervals.

Table 2.13 Diverse-Type Heat Cascade

Interval i					Col. A T_{int}	Col. B $MC_{p,int}$	Col. C Q_{int}	Col. D Q_{cas}	Col. E R_{cas}
0				|	195.91	0	0	0	1980
1	|			|	134.09	20	1236.36	−1236.36	743.64
2	|	|		|	120.91	10	131.82	−1368.18	611.82
3	|	|		| |	116.09	−5	−24.09	−1344.09	635.91
4	|	|	|	|	60.91	10	551.82	−1895.91	84.09
5	|			|	44.09	5	84.09	−1980	0
6	|				4.09	−10	−400	−1580	400
Stream	H1	H2	C3	C4					
MC_p	10	15	20	15					
h_j	0.2	2	0.2	2					

Units: temperature in °C, heat capacity flow rate in kW/°C, enthalpy in kW, and heat transfer coefficient in kW/(m^2 °C).

The rest of the procedure does not change from the PTA described in steps 2 through 6 in Example 1.1. As before, the minimum hot utility ($Q_{hu,min}$) and the minimum cold utility ($Q_{cu,min}$) are the first and last values in column E. The temperature in column A that corresponds to zero revised cascaded heat in column E is brought back to its unshifted value and this is called the pinch temperature. In this example, the pinch temperature is 40°C (given by T_{int} − Q''_{min}/h_j = 44.09 − 8.1818/2. Note that the pinch is caused by the inlet temperature of stream C4). This is in contrast to the conventional-type heat cascade where the pinch temperature is often reported in terms of the shifted value, which is meaningful since the temperature shift is uniform (namely, $\Delta T_{min}/2$) for all streams. The energy targets are summarized in Table 2.14.

Table 2.14 Results of Energy Targeting for Diverse Pinch Case

Minimum Hot Utility	= 1980
Minimum Cold Utility	= 400
Pinch Temperature	= 40

Units: temperature in °C and utility requirements in kW.

Hot and cold composites for the diverse pinch problem are plotted in a manner similar to that in Example 1.2, except that shifted temperatures are used on the ordinate. The procedure will be clear when the balanced composites are plotted in the next subsection.

2.2.2 Area Targeting for Minimum Flux Specification

The area targeting procedure in Section 1.2.1 based on Equation 1.7 (referred to as the uniform BATH area formula) yields the true minimum area only when the heat transfer coefficients of all the streams are equal. For the case where heat transfer coefficients differ considerably, a diverse BATH area formula based on a minimum flux specification has been suggested by Rev and Fonyo (1991). They claim that areas calculated from the diverse BATH formula are closer to those obtained by a more rigorous model based on a linear programming formulation and a crisscross spaghetti network.

As before, the plot of the composite curves is modified to include the utilities and then divided into a number of enthalpy intervals. Equation 1.7 applies, but needs a minor modification to account for the fact that the composite curves for the diverse pinch case are based on shifted temperatures. The modification is essentially in the manner by which $LMTD$ in Equation 1.7 is calculated. The temperature difference at each end of the enthalpy interval must be calculated using (Rev and Fonyo, 1991):

$$\Delta T = \Delta T_{sf} + \frac{1}{Q_i} \sum_j \left| Q_j \Delta T_{j,sf} \right| \tag{2.7}$$

where ΔT_{sf} is the temperature difference between the shifted hot and cold curves at the end of the interval, Q_i is the length of the enthalpy interval, Q_j is the enthalpy change of the j-th stream in the interval, and $\Delta T_{j,sf}$ is the individual ΔT shift contribution of the j-th stream. Noting that $\Delta T_{j,sf} = Q''_{min}/h_j$, Equation 2.7 may be rewritten in a more convenient form:

$$\Delta T = \Delta T_{sf} + SCF \quad \text{where} \quad SCF = \frac{Q''_{min}}{Q_i} \sum_j \left(\frac{Q_j}{h_j} \right)_i. \tag{2.8}$$

The second term in the above equation may be appropriately referred to as the shift compensating factor. The summation term involved in the SCF appears in Equation 1.9 and hence may simply be evaluated.

If all the stream heat transfer coefficients are equal, then the conventional definition of $LMTD$ based on unshifted temperatures and used in Example 1.4 must be recovered. In other words, ΔT in Equation 2.8 must be the temperature difference between the *unshifted* hot and cold curves at the end of the interval for the case of equal h_j. It is left to the reader to prove this (see Problem 2.C).

Given that utility and area targets can be established for the minimum flux specification, supertargeting can be performed to determine an optimum Q''_{min} rather than an optimum ΔT_{min} (Fraser, 1989). If the stream-dependent ΔT_{min} contributions are specified from experience and are not related to a single minimum flux value, then the utility-area tradeoff is not easily evaluated since the optimization problem becomes multivariable.

Example 2.5 Determination of Area Using Diverse BATH formula

For the stream data in Case Study 4S1t, determine the area target for the case of countercurrent exchangers and a minimum flux specification of Q''_{min} = 90/11 = 8.1818 kW/m^2.

Solution. The procedure is essentially along the lines of Examples 1.3 and 1.4. First, a balanced composite curves plot that includes utilities is generated; subsequently, it is divided into vertical enthalpy intervals and the area calculated using Equations 1.7, 1.9, and 2.8.

Step 1. Plotting of Balanced Hot and Cold Composite Curves

The steps are identical to those in Example 1.3, except that shifted temperatures are used throughout. Note that the hot and cold utilities undergo temperature changes of 1°C and 10°C respectively; therefore, the utility flow rates for the utilities (in terms of MC_p) may be calculated from Equations 1.10a and 1.10b as

$$(MC_p)_{hu} = Q_{hu,min}/(T_{in} - T_{out})_{hu} = 1980/1 \quad = 1980 \text{ kW/°C}$$
$$\text{and } (MC_p)_{cu} = Q_{cu,min}/(T_{out} - T_{in})_{cu} = \quad 400/10 = 40 \text{ kW/°C}.$$

The data for the BHCC and the BCCC are shown in Tables 2.15 and 2.16 (which are the analogs of Tables 1.5 and 1.6). Note that the HCC and CCC (not including utilities) may be directly obtained by deleting the last two rows of Table 2.15 and the first two rows of Table 2.16 in this case since the shifted utility temperatures do not have any overlap with the shifted process temperatures.

Step 2. Determination of Enthalpy Intervals

On following steps 1 through 4 in Example 1.4, Table 2.17 may be easily constructed. The only difference between Table 2.17 and Table 1.7 (which is its analog) is that columns B and C now contain shifted temperatures.

Table 2.15 Balanced Hot Composite for Q''_{min} = 90/11 kW/m²

Interval i				Col. A T_{hb}	Col. B $SumMC_{p,hb}$	Col. C $Q_{int,hb}$	Col. D $CumQ_{hb}$
0				4.09	0	0	0
1				60.91	10	568.18	568.18
2				120.91	25	1500	2068.18
3				134.09	10	131.82	2200
4				176.95	0	0	2200
5				177.95	1980	1980	4180
Stream	H1	H2	HU				
MC_p	10	15	1980				
h_j	0.2	2	4				

Units: temperature in °C, heat capacity flow rate in kW/°C, enthalpy in kW, and heat transfer coefficient in kW/(m² °C).

Table 2.16 Balanced Cold Composite for $Q''_{min} = 90/11$ kW/m^2

Interval				Col. A T_{cb}	Col. B $SumMC_{p,cb}$	Col. C $Q_{int,cb}$	Col. D $CumQ_{cb}$
0					19.09	0	0	0
1					29.09	40	400	400
2					44.09	0	0	400
3					60.91	15	252.27	652.27
4					116.09	35	1931.37	2583.64
5					195.91	20	1596.36	4180
Stream	C3	C4	CU				
MC_p	20	15	40				
h_j	0.2	2	2				

Units: temperature in °C, heat capacity flow rate in kW/°C, enthalpy in kW, and heat transfer coefficient in kW/(m^2 °C).

Having characterized the various enthalpy intervals, Equations 1.9, 2.8, and 1.7 must be used in that order. For convenience, columns A, B, and C of Table 2.17 are copied into Table 2.18.

Step 3. Calculation of Sum(Q/h) *in Each Interval as per Equation 1.9*
   Column D in Table 2.18 is obtained using Equation 1.13 (step 5 of Example 1.4).

Step 4. Calculation of Log Mean Temperature Difference in Each Interval (LMTD in Column F of Table 2.18)
   This is the key difference in area targeting between the conventional pinch and diverse pinch cases. Equation 2.8 must be used at this stage. The shift compensating factors in column E of Table 2.18 are obtained from

$$SCF_i = \frac{Q''_{min}}{CumQ_i - CumQ_{i-1}} \, Sum(Q/h)_i \qquad \text{for } i \geq 1$$
$$SCF_i = 0 \qquad \text{for } i = 0. \qquad (2.9)$$

Equation 2.9 yields the following:

$$SCF_{i=1} = (90/11)(2200)/(400 - 0) = 45°C;$$
$$SCF_{i=4} = (90/11)(235.45)/(652.27 - 568.18) = 22.91°C; \text{ etc.}$$

Table 2.17 Determination of the Enthalpy Intervals for Diverse BATH Area Formula

i							Col. A $CumQ_i$	Col. B T_{hi}	Col. C T_{ci}	Col. D $\Sigma(MC_p/h)_h$	Col. E $\Sigma(MC_p/h)_c$
0	|					|	0	4.09	19.09	0	0
1	|					|	400	44.09	29.09	50	20
2	|			|			400	44.09	44.09	50	0
3	|	|		|			568.18	60.91	55.30	50	7.5
4	|	|	|	|			652.27	64.27	60.91	57.5	7.5
5	|	|	|	|			2068.18	120.91	101.36	57.5	107.5
6	|		|	|			2200	134.09	105.13	50	107.5
7			|	|	|		2200	176.95	105.13	0	107.5
8			|	|	|		2583.64	177.15	116.09	495	107.5
9				|	|		4180	177.95	195.91	495	100

	H1	H2	C3	C4	HU	CU
MC_p	10	15	20	15	1980	40
h_j	0.2	2	0.2	2	4	2

Units: temperature in °C, heat capacity flow rate in kW/°C, enthalpy in kW, and heat transfer coefficient in kW/(m^2 °C).

Table 2.18 Calculation of Area Using Diverse BATH Formula

Int i	Col. A $CumQ_i$	Col. B $T_{h,i}$	Col. C $T_{c,i}$	Col. D $Sum(Q/h)$	Col. E SCF_i	Col. F $LMTD_i$	Col. G A_i
0	0	4.09	19.09	0	0	0	0
1	400	44.09	29.09	2200	45	43.28	50.83
2	400	44.09	44.09	0	--	--	0
3	568.18	60.91	55.30	925	45	47.75	19.37
4	652.27	64.27	60.91	235.45	22.91	27.38	8.60
5	2068.18	120.91	101.36	7605.45	43.95	55.01	138.27
6	2200	134.09	105.13	1063.96	66.04	90.21	11.79
7	2200	176.95	105.13	0	--	--	0
8	2583.64	177.15	116.09	1274.22	27.18	93.51	13.63
9	4180	177.95	195.91	8380.91	42.95	55.42	151.22

Units: temperature in °C, enthalpy in kW, heat transfer coefficient in kW/(m^2°C), and area in m^2.

Next, the *LMTD is* obtained from the following formula:

$$LMTD_i = \frac{(T_{h,i} - T_{c,i}) - (T_{h,i-1} - T_{c,i-1})}{\ln\left[\dfrac{(T_{h,i} - T_{c,i}) + SCF_i}{(T_{h,i-1} - T_{c,i-1}) + SCF_i}\right]} \qquad \text{for } i \geq 1$$

$$LMTD_i = 0 \qquad\qquad\qquad\qquad \text{for } i = 0. \quad (2.10)$$

From Equation 2.10,

$LMTD_{i=1} = [(44.09-29.09) - (4.09-19.09)]/\ln[(15+45)/(-15+45)] = 43.28°C$
$LMTD_{i=4} = [(64.27-60.91) - (60.91-55.3)]/\ln[(3.36+22.91)/(5.61+22.91)]$
$$= 27.38°C.$$

Step 5. Calculation of Countercurrent Exchanger Area Target
   The countercurrent exchanger area in each interval (A_i in column G) is obtained by merely dividing the *Sum(Q/h)* value in column D by the corresponding $LMTD_i$ in column F for each interval, so the area per interval is given by Equation 1.15 as before. On summing the A_i values in column G over all the intervals, the total countercurrent area for the diverse case is

$$A_c = \Sigma A_i = 393.71 \text{ m}^2.$$

2.3 CONTINUOUS HEAT CASCADES FOR DIVERSE AND CONVENTIONAL PINCH CONCEPTS

   So far, it has been seen that energy targeting of HENS may be done with the PTA using a minimum temperature difference (ΔT_{min}) or a minimum flux (Q''_{min}) specified a priori. The PTA provides information on the minimum utility requirements, the pinch temperature, and the GCC, but only at a particular value of ΔT_{min} or Q''_{min}. The pinch tableau approach allows continuous energy targeting for the entire range of ΔT_{min}. However, it cannot be readily extended to perform continuous energy targeting for the entire range of Q''_{min} since the composite curves for the diverse pinch concept must be in terms of shifted temperatures (Rev and Fonyo, 1991) and the horizontal shifting of the CCC (as in Figure 2.1) is not meaningful in this case. In what follows, an algorithm is presented that permits continuous determination of the energy targets and the GCC over the entire allowable range of ΔT_{min} as well as

Q''_{min}. This continuous targeting automatically establishes the equivalence between ΔT_{min} and Q''_{min} for which the minimum utility requirements are identical. The equivalence is important because it obviates the iterative procedure (Rev and Fonyo, 1991) during comparison of the area estimates from the diverse BATH and uniform BATH formulae. Furthermore, the concepts of multiple pinches (including topology traps) and the sensitivity threshold as discussed in Section 2.1.3 hold for the diverse pinch case. The accurate analytical computation of the ΔT_{min} and Q''_{min} values at which these occur is an additional major advantage provided by the algorithm.

In essence, this section provides an accurate analytical method for (a) calculation of minimum utility requirements and pinch temperatures as piecewise linear functions of ΔT_{min} and Q''_{min}, (b) continuous determination of the GCC in every interval over the entire possible range of ΔT_{min} and Q''_{min}, (c) computation of sensitivity thresholds, (d) detection of multiple pinches /topology traps, and (e) establishment of the mapping from ΔT_{min} to Q''_{min} such that both specifications result in the same minimum utility requirements. Continuous targeting is important in the context of supertargeting, where ΔT_{min} or Q''_{min} must be varied continuously to establish their optimum values.

Example 2.6 Continuous Energy Targeting for Diverse and Conventional Pinch Concepts

For the stream data in Case Study 4S1t, determine how the utility requirements and the pinch temperature vary with Q''_{min} and ΔT_{min}.

Solution. The PTA may be adapted to perform continuous energy targeting within an interval of Q''_{min} and ΔT_{min}. The computational effort may be reduced by judiciously choosing the intervals so that the sensitivity (slope) of the linear variation of the energy targets within the interval is constant. In other words, the algorithm below exploits the property of piecewise linear variation of the energy targets with Q''_{min} and ΔT_{min}. It first determines the intervals in which the targets are linear, and then performs the PTA only at the delimiting values of the intervals.

Step 1. Performing Stream-dependent Temperature Shifting

The first step is to shift the hot streams down and the cold streams up by their ΔT_j values. This is achieved by simply subtracting ΔT_j from the end temperatures of the hot streams and adding ΔT_j to the end temperatures of the cold streams. The stream-dependent shifting values are calculated using

$$\Delta T_j = \kappa \, h_j^{-1} = Q''_{min} h_j^{-1} \tag{2.11}$$

where κ is the notation used by Rev and Fonyo (1991) for the factor in the diversity relation and h_j is the stream heat transfer coefficient. The shifted temperatures may then be arranged in descending order (leftmost column in Table 2.19), assuming κ to be small.

Table 2.19 Determination of κ-Shifts

Shifted Temps.	A	B	C	D	E	F	G	H
A 175 – 5κ	--	2	100/9	126/11	220/9	--	270/11	155/10
B 155 + 5κ	--	--	--	--	--	--	--	--
C 125 – κ/2			--	13	--	--	85	210/11
D 112 + κ/2				--	--	--	--	184/9
E 65 – κ/2					--	--	25	90/11
F 45 – 5κ						--	10/11	5/2
G 40 + κ/2							--	40/9
H 20 + 5κ								--

Units: temperature in °C and κ in kW/m^2.

Step 2. Determining κ-Shifts

Arranging the shifted temperatures in decreasing order is useful in determining the heat cascade. The order of the shifted temperatures (labeled A through H) in Table 2.19 holds for a small enough value of κ. As κ is changed (increased), the shifted temperatures swap positions, thus affecting the subsets of the stream population within each temperature interval and in turn the final heat cascade. The values of κ at which such temperature swaps occur are of importance since the minimum utility required is a linear function of κ within such a κ-shift. The κ-shifts may be determined readily by equating two shifted temperatures as tabulated in Table 2.19. For example, shifted temperatures (A) and (B) will swap when $175 - 5\kappa = 155 + 5\kappa$ or $\kappa = 2$. Similar calculations are performed for all possible swaps to obtain the entire matrix of κ-shifts in Table 2.19. Only finite positive values of κ-shifts are shown in Table 2.19, as these correspond to physically realistic κ-shifts.

Step 3. Generating Energy Targets for the Complete κ Range

The κ-shifts in Table 2.19 are arranged in ascending order in Table 2.20 for convenience. Then, the minimum heating and cooling for any given value of κ may be determined easily by constructing a diverse-type heat cascade using the method adopted in Example 2.4. By constructing cascades similar to the one in Table 2.13, values of the minimum utility may be obtained for all the possible κ-shifts. The final results are summarized in Table 2.20.

Table 2.20 Energy Targets for Entire Range for Case Study 4S1t for Diverse and Conventional Pinch Cases

Utility	κ	Diverse Pinch Vertices	ΔT_{min}	Conventional Pinch Vertices
1580	0		0	
1580	10/11, 2		5,13,20	
1580	5/2	H	25	f, h
1774.44	40/9	G, H		
1880			45	g
1980	90/11	G		
2232.78	100/9	H		
2272.27	126/11	H		
2330			63	g
2450	13	H		
2737.5	155/10	A, H		
2791.36	210/11	A, C, H		
2811.67	184/9	A, C		
2874.44	220/9	G		
2880	270/11	A, G		
2880	25	A, C, E, G	85	c, g
3280			105	c, h
3580			135	h
3780	85	C, G	155	h
Topology Trap:				
1713.33			115/3	g, h
2057.08	115/12	G, H		
2870.63	195/8	A, C, G		

Units: utility in kW, κ in kW/m², and temperature in °C.

Step 4. Developing Continuous Heat Cascades

The objective is to develop a heat cascade as in Table 2.13, which is valid between two consecutive κ-shifts given in Table 2.20. Such a continuous heat cascade is useful in determining topology traps and the sensitivity threshold.

Consider the development of a continuous heat cascade for $\kappa \geq 90/11$. The procedure is similar to that used in Example 2.4, except that κ is treated as an unknown variable leading to the shifted temperatures being linear functions of κ rather than constant values. The order of the shifted temperatures is identical to that in Table 2.13. The resulting continuous cascade is shown in Table 2.21.

Table 2.21 Continuous Diverse-Type Heat Cascade for $\kappa \geq 90/11$

Int. i	Col. A T_{shift}	Col. B $MC_{p,int}$	Col. C Q_{int}	Col. D Q_{cas}	Col. E κ_{tt}
0	B $155 + 5\kappa$	0	0	0	--
1	A $175 - 5\kappa$	20	$20(-20 + 10\kappa)$	$400 - 200\kappa$	13.31
2	C $125 - \kappa/2$	10	$10(50 - 4.5\kappa)$	$-100 - 155\kappa$	14.3
3	D $112 + \kappa/2$	-5	$-5(13 - \kappa)$	$-35 - 160\kappa$	14.24
4	H $20 + 5\kappa$	10	$10(92 - 4.5\kappa)$	$-955 - 115\kappa$	9.58
5	E $65 - \kappa/2$	-10	$-10(-45+5.5\kappa)$	$-1405 - 60\kappa$	25
6	G $40 + \kappa/2$	5	$5(25 - \kappa)$	$-1530 - 55\kappa$	--
7	F $45 - 5\kappa$	-10	$-10(-5 + 5.5\kappa)$	-1580	0.91

Str.	H1	H2	C3	C4
MC_p	10	15	20	15
h_j	0.2	2	0.2	2

Units: temperature in °C, heat capacity flow rate in kW/°C, enthalpy in kW, heat transfer coefficient in kW/(m^2 °C), and κ in kW/m^2.

Step 5. Detecting Topology Traps

By definition, the pinch corresponds to that point where the cascaded heat is most negative (e.g., vertex G in Table 2.21 corresponding to the Q_{cas} entry of -1980 in Table 2.13). Further, it is observed that all entries in the Q_{cas} column of a continuous-type heat cascade (see Table 2.21) are linear functions of κ and of the form $a + b\kappa$ (where a and b are constants in a temperature interval). Importantly, the rate of change of utility requirement with κ depends on the value of b at the most negative entry. There can exist an entry in the Q_{cas} column with a value of b lower than that at the current pinch (vertex G in Table 2.21). In this case, as κ is increased to a certain value (say, κ_{tt}) less than the next κ-shift, it is possible for two Q_{cas} entries in Table 2.21 to become equal. Then, beyond this value of κ_{tt}, the pinch will jump to a new vertex and the utilities will change at a rate dependent on the b-value of the new vertex. This phenomenon may be termed a topology trap (see Section 2.1.3).

All topology traps can be identified from the table of continuous energy targets (i.e., Table 2.20). A topology trap exists between two consecutive κ-shifts if the pinch vertices at these two κ-shifts do not have a single entry in common. Examination of Table 2.20 reveals that two topology traps exist: one between the κ-shifts of 90/11 and 100/9 and another between the κ-shifts of 184/9 and 220/9. The exact location of the first topology trap may be determined as follows, using the continuous heat cascade for $\kappa \geq 90/11$ shown in Table 2.21.

The rate of decrease of Q_{cas} at the current pinch (vertex G for $\kappa = 90/11$) depends on its b-value of -55. There are several other Q_{cas} entries (column D in Table 2.21) with a b-value lower than -55 which have the potential of becoming the pinch vertex after a certain increase in κ (say, κ_{tt}). A column of candidate κ_{tt} values can be generated using the following formula:

$$\kappa_{tt} = [a_{current\ pinch} - a]/[b - b_{current\ pinch}]. \qquad (2.12)$$

Noting that the current pinch is at vertex G, the entry in the candidate κ_{tt} column for the case of vertex H (for example) may be calculated as $\kappa_{tt} = [-1530 - (-955)]/[-115 - (-55)] = 9.583$. The entries that lie between 8.182 and 11.111 (see κ-shifts in Table 2.20) in the candidate column give the κ-shifts at which the topology traps occur. From Table 2.21, it is seen that a topology trap occurs for a κ-shift of 9.583, where the utility requirement is given by $1530 + 55 (9.583) = 2057.08$ kW. It must be emphasized that, for $\kappa \leq 9.583$ the pinch is at vertex G whereas, for $\kappa \geq 9.583$, it is at vertex H. When the procedure is repeated starting with a continuous diverse-type heat cascade for $\kappa \geq 184/9$, the second topology trap may be shown to lie at a κ-shift of 24.375. The results on the two topology traps for the diverse pinch case are summarized at the bottom of Table 2.20.

Step 6. Determining Threshold κ and Maximum κ

Analogous to the concept of the threshold ΔT_{min} (see Section 2.1.3), the threshold κ is that value of κ below which the utility requirement remains unaffected by a change in κ. From Table 2.20, it is observed that the utility requirement remains unchanged at 1580 kW for $\kappa \leq 2.5$. However, in principle, the threshold κ can lie in the interval [5/2, 40/9] and its exact value may be determined by starting with the continuous diverse-type heat cascade for this interval (Table 2.22).

Table 2.22 Continuous Diverse-Type Heat Cascade for $\kappa \geq 2.5$

Int. i		Col. A T_{shift}	Col. B $MC_{p,int}$	Col. C Q_{int}	Col. D Q_{cas}	Col. E κ_{thres}
0		B $155 + 5\kappa$	0	0	0	--
1		A $175 - 5\kappa$	20	$20(-20 + 10\kappa)$	$400 - 200\kappa$	9.90
2		C $125 - \kappa/2$	10	$10(50 - 4.5\kappa)$	$-100 - 155\kappa$	9.55
3		D $112 + \kappa/2$	-5	$-5(13 - \kappa)$	$-35 - 160\kappa$	9.66
4		E $65 - \kappa/2$	10	$10(47 + \kappa)$	$-505 - 170\kappa$	6.32
5		G $40 + \kappa/2$	25	$25(25 - \kappa)$	$-1130-145\kappa$	3.10
6		H $20 + 5\kappa$	10	$10(20 - 4.5\kappa)$	$-1330-100\kappa$	2.50
7		F $45 - 5\kappa$	-10	$-10(-25 + 10\kappa)$	-1580	--

Str. H1 H2 C3 C4
MC_p 10 15 20 15
h_j 0.2 2 0.2 2

Units: temperature in °C, heat capacity flow rate in kW/°C, enthalpy in kW, heat transfer coefficient in kW/(m^2 °C), and κ in kW/m^2.

The aim is to determine that value of κ for which at least one of the Q_{cas} entries becomes less than -1580. This may be achieved by equating the entries in column D of Table 2.22 to -1580. The resulting values of κ are shown in column E. The lowest of these κ-values gives the threshold κ. In the example considered, the threshold κ is 2.5, which happens to be a κ-shift itself.

What is the maximum value of κ? The highest value of the κ-shift in Table 2.19 at which a hot stream shifted temperature (with a negative κ-coefficient) swaps with a cold stream shifted temperature (with a positive κ-coefficient) is the maximum κ. In Table 2.19, this maximum κ is observed to be 85 (for a swap between shifted temperatures C and G). The κ-shifts above this value (none exist in this example) correspond to swaps between two hot or cold stream shifted temperatures and may be ignored, as the utility requirements do not change above the maximum κ.

Step 7. Determining Energy Targets for the Complete ΔT_{min} Range
The methodology for minimum approach temperature specification is analogous to that adopted in steps 1 through 6 for the minimum flux specification. The first step is to shift the hot streams down and the cold streams up using the specified stream-independent ΔT_{min} value. This is

achieved by simply subtracting $\Delta T_{min}/2$ from the end temperatures of the hot streams and adding $\Delta T_{min}/2$ to the end temperatures of the cold streams. The shifted temperatures (labeled a through h for consistency with Table 2.19) are then arranged in descending order (leftmost column in Table 2.23), assuming ΔT_{min} to be small.

Table 2.23 Determination of ΔT_{min} - Shifts

Shifted Temps.	a	b	c	d	e	f	g	h
a $175 - \Delta T_{min}/2$	--	20	--	63	--	--	135	155
b $155 + \Delta T_{min}/2$		--	--	--	--	--	--	--
c $125 - \Delta T_{min}/2$			--	13	--	--	85	105
d $112 + \Delta T_{min}/2$				--	--	--	--	--
e $65 - \Delta T_{min}/2$					--	--	25	45
f $45 - \Delta T_{min}/2$						--	5	25
g $40 + \Delta T_{min}/2$							--	--
h $20 + \Delta T_{min}/2$								--

Units: temperature in °C.

As ΔT_{min} is increased, the shifted temperatures swap positions, affecting the subsets of the stream population within each temperature interval and consequently the final heat cascade. The values of ΔT_{min} at which such temperature swaps occur are important, as the minimum utility is a linear function of ΔT_{min} within a ΔT_{min}-shift. As before, the ΔT_{min}-shifts may be determined by equating two shifted temperatures. For example, shifted temperatures a and b will swap when $175 - \Delta T_{min}/2 = 155 + \Delta T_{min}/2$ or ΔT_{min} = 20. Similar calculations, when performed for all possible swaps, yield the matrix of ΔT_{min}-shifts in Table 2.23. Blank entries in Table 2.23 indicate negative or zero ΔT_{min}-shifts. The minimum utility requirements for a given value of ΔT_{min} are easily determined by constructing a conventional-type heat cascade using the PTA. By constructing cascades similar to the one in Table 1.1, the minimum utilities are obtained for all ΔT_{min}-shifts. Results for ΔT_{min}-shifts are merged with those obtained earlier for κ-shifts (Table 2.20).

Step 8. Detecting Topology Traps and Determining Threshold ΔT_{min}
    Continuous heat cascades valid between two consecutive ΔT_{min}-shifts are useful for this purpose. A continuous heat cascade for $\Delta T_{min} \geq 25$ (Table 2.24)

may be generated by treating ΔT_{min} as an unknown variable and shifted temperatures as linear functions of ΔT_{min}.

Table 2.24 Continuous Conventional-Type Heat Cascade for $\Delta T_{min} \geq 25$

Int. i				Col. A T_{shift}	Col. B Q_{int}	Col. C Q_{cas}	Col. D ΔT_{tt}	Col. E ΔT_{thr}
0				b 155 + ΔT_m/2	0	0	--	--
1				a 175 − ΔT_m/2	20(−20+ΔT_m)	400−20ΔT_m	173	99
2				d 112 + ΔT_m/2	10(63−ΔT_m)	−230−10ΔT_m	--	135
3				c 125 − ΔT_m/2	25(−13+ΔT_m)	95−35ΔT_m	57	47.9
4				g 40 + ΔT_m/2	10(85−ΔT_m)	−755−25ΔT_m	38.33	33
5				e 65 − ΔT_m/2	−5(−25+ΔT_m)	−880−20ΔT_m	45	35
6				h 20 + ΔT_m/2	10(45−ΔT_m)	−1330−10ΔT_m	--	25
7				f 45 − ΔT_m/2	−10(−25+ΔT_m)	−1580	25	--

H1 H2 C3 C4
10 15 20 15

Units: temperature in °C and enthalpy in kW. (Note: ΔT_{min} is abbreviated as ΔT_m.)

The topology trap is determined by the procedure used earlier in step 5 with the help of Table 2.24. A column of candidate ΔT_{tt} is generated using the formula in Equation 2.12. Since the current pinch is at vertex h, the entry in the candidate ΔT_{tt} column for the case of vertex g (for example) may be calculated as ΔT_{tt} = [−1330 − (−755)]/[−25 − (−10)] = 38.33. The entries that lie between 25 and 45 (see ΔT_{min}-shifts in Table 2.20) in the candidate column give the ΔT_{min}-shifts for the topology traps. From Table 2.24, it is seen that a topology trap occurs for a ΔT_{min}-shift of 38.33, where the utility requirement is given by 1330 + 10 (38.33) = 1713.33 kW. For $\Delta T_{min} \leq 38.33$°C, the pinch is at vertex h whereas, for $\Delta T_{min} \geq 38.33$, it is at vertex g. This topology trap for the conventional pinch case is added at the bottom of Table 2.20.

From Table 2.20, it is observed that the utility requirement remains unchanged at 1580 kW for $\Delta T_{min} \leq 25$. In principle, the threshold ΔT_{min} can lie in the interval [25, 45] and its exact value may be determined by starting with the continuous heat cascade for this interval (Table 2.24). The entries in column C of Table 2.24 are equated to −1580 in order to determine that value of ΔT_{min} for which at least one of the Q_{cas} entries becomes less than −1580. The resulting values of ΔT_{thr} are shown in column E. The lowest of these ΔT-values gives the threshold ΔT_{min}. In this example, the threshold ΔT_{min} is 25°C, which happens to be a ΔT_{min}-shift itself.

Both the topology trap and the sensitivity threshold obtained are in agreement with those determined earlier (Table 2.11) by the pinch tableau approach. The advantage of Table 2.20 is that it provides an equivalence between ΔT_{min} and κ, which may be visualized through a plot as in Figure 2.7.

Step 9. Comparison of Area Estimates from Various Targeting Formulae

The equivalence allows comparison of the estimates provided by the various formulae for area targeting. Table 2.25 compares the area estimates from the diverse BATH formula and the uniform BATH formula to those from the NLP minimum area. The uniform BATH and diverse BATH formulae give similar results for this case study; however, it is seen that, for the case of significantly different heat transfer coefficients, both formulae overestimate the true minimum area obtained from a nonlinear programming (NLP) formulation (see Chapter 10). Initially, Rev and Fonyo (1991) argued that the diverse BATH estimation is much nearer to the true minimum area compared to the uniform BATH estimation. Later, they clarified (Rev and Fonyo, 1993) that the purpose of the diverse BATH formula is to provide a consistent methodology with the diverse pinch concept and not really a better area approximation. This conclusion is borne out by the present case study also.

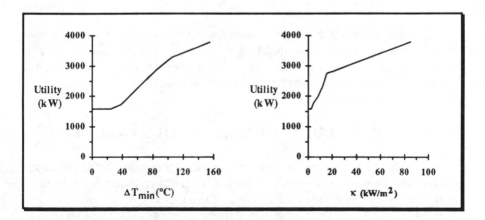

Figure 2.7 The variation of the utility requirements with ΔT_{min} and κ provides an equivalence between these parameters.

Table 2.25 Comparison of Estimated Areas from Uniform BATH and Diverse BATH formulae to the NLP Minimum

Utility	Uniform BATH Area	Diverse BATH Area	NLP Minimum Area
1580	484.53	482.34	434.42
1713.33	442.76	440.08	393.90
1774.44	428.69	426.44	381.48
1880	408.64	408.04	365.51
1980	393.26	393.71	354.39
2057.08	383.21	384.92	347.41
2232.78	364.69	368.22	335.46
2272.27	361.20	365.09	332.94
2330	356.49	360.84	330.03
2450	347.96	353.14	324.94
2737.5	332.92	340.11	316.82
2791.36	330.77	339.05	315.77
2811.67	330.00	336.07	315.41
2870.63	327.90	330.31	314.45
2874.44	327.78	329.93	314.34
2880	327.59	329.40	314.31
3280	316.82	312.78	310.68
3580	313.56	311.05	310.68
3780	313.43	313.90	311.00

Units: utility in kW and area in m^2.

The point to note is that the minimum flux specification may be preferable to the minimum approach temperature specification for the case of widely differing heat transfer coefficients. With the targeting methodology established for the diverse pinch concept, its utility at the synthesis stage will be examined in Section 4.4.4.

2.4 OVERALL LOGIC OF CONTINUOUS TARGETING

The logic flow diagram for the continuous targeting algorithm is shown in Figure 2.8. The concept of significant shifts is central to continuous targeting as it leads to reduced computational effort. Three types of significant shifts are important: Q-shifts (in the case of conventional pinch problems) for the pinch tableau approach and κ-shifts (in the case of diverse pinch problems) and ΔT_{min}-shifts (in the case of conventional pinch problems) for the continuous cascade approach. Exact determination of multipinched situations and the sensitivity threshold is an important benefit during continuous targeting, which requires only the basic stream data (T_{in}, T_{out}, MC_p, and h) as the input.

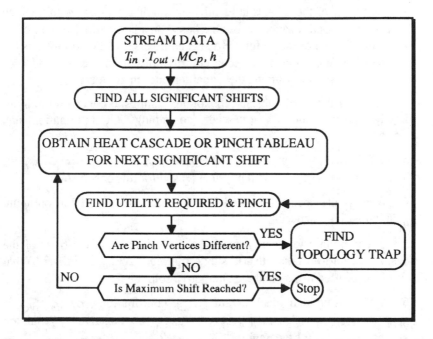

Figure 2.8 The above algorithm allows continuous targeting with reduced computational effort by determining significant shifts.

The ability to do energy and area targeting using the diverse pinch concept is useful for problems with significantly differing heat transfer coefficients. Here, the input must include the minimum flux specification (Q''_{min} or κ).

Armed with the targets established in the first two chapters, networks for maximum energy recovery will be designed in Chapter 3.

QUESTIONS FOR DISCUSSION

1. Comment on the validity of the following statements:
- The pinch temperature is often determined by the inlet temperature of the stream having the largest MC_p.

- The pinch in industrial situations is often determined by the condensation/vaporization requirements of distillation columns.
- It is not desirable to design the actual network at a topology trap or a significant shift.
- The power of the CTA is in identifying topology regions within which parametric optimization is possible based on a good initial structure.
- A nice feature of the uniform BATH and diverse BATH formulae is the effective separation of the area contributions from the hot and cold streams, without assuming the actual matches in an interval.

2. How can the pinch tableau (containing composite curve information) be directly converted into a heat cascade (containing GCC information) and vice versa?

3. In Table 2.9, it is observed that the area (A_c) increases with ΔT_{min} for very high values of ΔT_{min}. Give a physical reason for this observation.

4. What is the accuracy desired for the determination of the optimum ΔT_{min} in industrial practice?

5. Is it important to identify quasi-optimal solutions on the basis of the TAC vs. ΔT_{min} profile rather than a single optimum ΔT_{min} solution? What criteria could be used to screen these alternative solutions?

6. While using individual stream ΔT_{min} contributions, $\Delta T_{min,match} = \Delta T_{min,jh} + \Delta T_{min,jc}$. Can $\Delta T_{min,jh}$ or $\Delta T_{min,jc}$ be negative? What are $\Delta T_{min,jh}$ and $\Delta T_{min,jc}$ for the uniform ΔT_{min} case?

7. Is it appropriate to use a global (match-independent) ΔT_{min} in practice? What would be appropriate assignments for the ΔT_{min} contributions of gas streams, liquid streams, and condensing/vaporizing streams?

8. What are the advantages in choosing the individual stream ΔT_{min} contributions in accordance with Equation 2.11?

9. Discuss the rationale for the pseudo BATH formula (Ahmad, 1985) given below:

$$A = \sum_i \sum_{jh} \sum_{jc} \left(\frac{1}{LMTD_{jh-jc}} \right)_i \left(\frac{Q_{jh-jc}}{h_{jh}} + \frac{Q_{jc-jh}}{h_{jc}} \right)_i \quad \text{where} \quad Q_{jh-jc} = Q_{jh} \frac{Q_{jc}}{Q_i}$$

How does the above formula account for individual stream ΔT_{min} contributions? Does it assume a spaghetti network?

10. Data extraction is a critical step prior to targeting. In this context, comment briefly on the following aspects:

- Why should the data in the pinch region be accurate? Is it a good strategy initially to locate the pinch region approximately from coarse data and then fine-tune the critical data?
- Would it be reasonable to define a stream for heat integration studies as one which changes in heat content but not in composition? When should a stream be included/excluded in a process integration study?
- Does the designer need to be cautious in defining streams for heat integration? Specifically, what would happen if stream H2 in Figure 1.1 was defined as two streams (i.e., stream H2a from 125° to 98°C and stream H2b from 98° to 65°C) rather than a single stream?
- Why is it important to double-check that the data have been properly extracted at the end of the targeting phase?
- When MC_ps depend on temperature, the composite curves may be approximated by several straight line segments. Why should these straight line segments be placed below the HCC and above the CCC for predicting realizable targets?
- When isothermal phase changes occur, why should the dew point and bubble point be chosen as vertex temperatures on the composites?
- What can be done if the data from a plant flowsheet is incomplete and/or conflicting? How can heat losses and direct utility heating/cooling be accounted for? How can mass exchange be accounted for?

PROBLEMS

2.A Continuous Energy and Unit Targeting by Pinch Tableau Approach

Determine how the utility requirements, the pinch temperature, and the unit targets vary with ΔT_{min} for Case Study 6M1 from Makwana and Shenoy (1993). Indicate clearly any multiple pinches (including topology traps).

Answers:

Q-shift	Ref	50	65	104	150	169	170	235	270	335
ΔT_{min}	0	10	13	23.5	35	41.33	41.67	63	71.75	88
$Q_{hu,min}$	5	55	70	109	155	174	175	240	275	340
T_{ph}	100	110	113	60	60	66.33	66.67	113	113	113
T_{pc}	100	100	100	36.5	25	25	25	50	41.25	25
$N_{u,mer}$	8	10	7	8	7	7	7	7	7	6

Multiple Pinch (no swap) at $\Delta T_{min} = 10°C$ ($Q_{hu,min} = 55$ kW;
  $N_{u,mer} = 10$; $T_{ph} = 110°C$ & $T_{pc} = 100°C$; $T_{ph} = 60°C$ & $T_{pc} = 50°C$)

Topology Trap at $\Delta T_{min} = 15.14°C$ ($Q_{hu,min} = 75.57$ kW; $N_{u,mer} = 8$;
$T_{ph} = 113°C$ & $T_{pc} = 97.86°C$; $T_{ph} = 60°C$ & $T_{pc} = 44.86°C$)
Topology Trap at $\Delta T_{min} = 62.5°C$ ($Q_{hu,min} = 237.5$ kW; $N_{u,mer} = 9$;
$T_{ph} = 87.5°C$ & $T_{pc} = 25°C$; $T_{ph} = 112.5°C$ & $T_{pc} = 50°C$)

2.B Topology Traps and Sensitivity Threshold
Verify the topology traps and the threshold for the following case studies:

Case Study (Reference)	ΔT_{min} for topology trap(s)	ΔT_{min} for threshold
4L1 (Lee et al., 1970)	58.987°C	--
4L2 (Linnhoff et al., 1982)	--	5.556°C
4P1 (Ponton & Donaldson, 1974)	46.429-375°F*	46.429°F
5M1 (Masso & Rudd, 1969)	83.175°F	42.857°F
6C1 (Ciric & Floudas, 1989)	27.778°C	20°C
6L1 (Lee et al., 1970)	--	65.25°F
7M1 (Masso & Rudd, 1969)	--	49.401°F
7M2 (Masso & Rudd, 1969)	58.878°F, 227.925°F	50°F
7T1 (Trivedi et al., 1990a)	5.101°C	1.004°C
9A1 (Ahmad & Smith, 1989)	40.625°C	--
9A2 (Ahmad & Linnhoff, 1989)	2.462°C, 19.231°C	--
9H1 (Hall et al., 1990)	18.214°C	--
10P1 (Pho & Lapidus, 1973)	--	71.616°F

* denotes multiple-pinched region (not a topology trap)

2.C Reduction of Diverse BATH Formula to Uniform BATH Formula
Show that the diverse BATH area formula reduces to the uniform BATH area formula when the heat transfer coefficients are equal. It is sufficient to prove that the SCF in Equation 2.8 equals ΔT_{min} for these conditions.

2.D Calculation of Diverse BATH Areas
For the stream data in Case Study 6G1,
  a) determine the values of Q''_{min} (equivalent) that yield the same minimum utility requirements as at $\Delta T_{min} = 12°, 16°, 20°, 24°, 28°C$.
  b) determine the area targets using the diverse BATH formula at values of the equivalent Q''_{min} established in part a) above. Verify your answers with the results of Rev and Fonyo (1993).

Answers:

ΔT_{min} (°C)	12	16	20	24	28
Q''_{min} (kW/m^2)	1.2727	1.9697	3.2609	3.5507	4.643
$A_{diverse}$ (m^2)	1261.54	813.88	555.5	458.26	390.27

CHAPTER 3

Networks for
Maximum Energy Recovery

Energy is for the mechanical world
what consciousness is for the human world.
If energy fails, everything fails.

E. F. Schumacher

The pinch design method (PDM), proposed by Linnhoff and Hindmarsh (1983), recognizes that no heat must cross the pinch and develops the design for two separate problems (namely, one above the pinch and another below it). Recognizing the pinch division allows the PDM to generate maximum energy recovery (MER) networks that meet the energy targets. These were previously thought to be close to the optimum solution but were proved otherwise (Gundersen and Naess, 1988). However, combining the PDM with recently developed supertargeting procedures (see Chapter 1) and remaining problem analysis (RPA) gives a network that is near optimal. This is because supertargeting ensures that the design starts at the optimum ΔT_{min} based on the energy-capital tradeoff whereas RPA (Ahmad and Smith, 1989) attempts to achieve the area and shell targets in addition to the energy targets. The basic PDM is discussed in the first part of this chapter whereas RPA is dealt with in the next chapter. The latter part of this chapter presents various methods for designing MER networks in some special cases like multiple utilities, multiple pinches, threshold problems, and forbidden matches.

3.1 BASIC PINCH DESIGN METHOD

A major difference between other published methods and the PDM is that it identifies certain matches between hot and cold streams which are essential for an MER network. This helps in dramatically reducing the search space of possible design solutions. At the same time, it identifies options whenever they exist and thereby provides the designer with some degree of control over the design decisions as is essential in practical situations. Furthermore, the PDM identifies situations where stream splitting is necessary to obtain an MER design. The identification of essential matches, matching options, and the requirement of stream splitting are decided by certain feasibility criteria that are explained below. It must be emphasized that the criteria are applicable only at the pinch and not away from it. This is because the pinch is the most constrained region, a fact that has serious implications and leads to the PDM starting the design at the pinch.

3.1.1 Important Pinch Design Criteria

The basic theory for the PDM in terms of design rules (Linnhoff and Hindmarsh, 1983) is briefly covered first, and then illustrated by an example problem.

The Number Criterion

The first feasibility criterion concerns the number of process streams at the pinch. To obtain an MER design, no utility cooling should be provided above the pinch. This implies that every hot stream must be brought to its pinch temperature with the help of a cold stream. In other words, the number of hot streams at the pinch cannot be greater than the number of cold streams. The reverse is obviously true below the pinch.

Thus, the first feasibility criterion (or the number criterion) at the pinch is

$$N_{hp} \leq N_{cp} \quad \text{above the pinch}$$
$$\text{and } N_{hp} \geq N_{cp} \quad \text{below the pinch} \tag{3.1}$$

where N_{hp} = number of hot streams at the pinch
and N_{cp} = number of cold streams at the pinch.

If the condition in Equation 3.1 is not satisfied, then stream splitting is necessary (i.e., to ensure that the inequalities are satisfied, cold stream[s] will have to be split above the pinch and hot stream[s] will have to be split below the pinch).

The MC_p Criterion

The second feasibility criterion concerns the approach temperature for pinch matches. Because the approach temperature is a minimum at the pinch, the temperature driving force (ΔT) in a pinch match cannot decrease as one moves further away from the pinch. On a T-Q plot, the straight line segments representing the hot and cold streams will be closest at the pinch (being ΔT_{min} apart) and must not converge with movement away from the pinch. Since the slope on a T-Q plot is given by $1/MC_p$, the second feasibility criterion (or the MC_p criterion) at the pinch is

$$MC_{p,hp} \leq MC_{p,cp} \quad \text{above the pinch}$$
$$\text{and } MC_{p,hp} \geq MC_{p,cp} \quad \text{below the pinch} \tag{3.2}$$

where $MC_{p,hp}$ = heat capacity flow rate of hot stream at the pinch
   and $MC_{p,cp}$ = heat capacity flow rate of cold stream at the pinch.

If a design is not possible where all pinch matches satisfy the inequalities in Equation 3.2, stream splitting is necessary. Either hot streams or cold streams may be split, and this depends on the stream data. However, keeping the first feasibility criterion in mind, it is preferable to split cold stream(s) above the pinch and hot stream(s) below the pinch.

The Tick Off Heuristic
   The tick off heuristic attempts to choose the heat loads of the exchangers in such a manner that the capital cost of the network is minimum. In a sense, it tries to keep the number of matches to a minimum in accordance with the unit target in Section 1.3.2. It recognizes that, if one of the two streams in a match is brought to its target temperature, then it may be ticked off (i.e., it need not be considered any longer in the design to follow). This is simply achieved by choosing each exchanger load to be the smaller of the heat loads of the two streams being matched. However, it must be emphasized that this is merely a heuristic (though a useful one) and can occasionally penalize the design by necessitating the targets set in the last chapter to be exceeded. To ensure that such penalties are not excessive, RPA will be introduced in the next chapter.

Example 3.1 Pinch Design Method for MER Networks
   For the stream data in Case Study 4S1, design an MER network for a ΔT_{min} of 13°C using the pinch design method.

   Solution. The design is performed at the optimum ΔT_{min} of 13°C, obtained by supertargeting (see Example 1.10). The design procedure may be followed easily through Figure 3.1, where the matches on the network are numbered in the sequence in which they are generated.

Step 1. Determination of Pinch Temperature
   Using the PTA (Section 1.1.2) or the pinch tableau approach (Section 2.1.1), it is found that the pinch temperature is 125°C (for hot streams) and 112°C (for cold streams). Furthermore, the minimum utility requirements are 360 kW (for hot utility) and 280 kW (for cold utility). These targets are available in Table 2.6.

Step 2. Calculation of Stream Heat Loads

The problem is divided as in Figure 3.1 into two regions (one above the pinch and another below it). Above the pinch, only two streams exist and their heat loads are 500 kW (i.e., 10[175 – 125] for stream H1) and 860 kW (i.e., 20[155 – 112] for stream C3). Below the pinch, all streams exist and their heat loads are 800 kW (i.e., 10[125 – 45] for stream H1), 2400 kW (i.e., 40[125 – 65] for stream H2), 1840 kW (i.e., 20[112 – 20] for stream C3) and 1080 kW (i.e., 15[112 – 40] for stream C4). These loads appear in Figure 3.1.

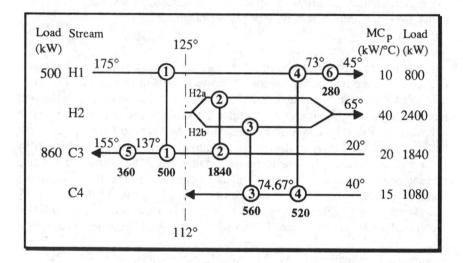

Figure 3.1 A network for maximum energy recovery that meets the energy targets can always be synthesized by designing separately above and below the pinch using the pinch design method. (Temperature in °C and load in kW)

Step 3. Matches above the Pinch

Above the pinch, there is one hot stream (H1) and one cold stream (C3) at the pinch; thus, the number criterion in Equation 3.1 is satisfied. Also, the MC_p criterion in Equation 3.2 is satisfied as $MC_{p,H1} \le MC_{p,C3}$ (since 10 < 20). The tick off heuristic suggests that the heat load for this match be 500 kW (minimum of 500 kW and 860 kW). No further matches are possible. The remaining heat of 360 kW on stream C3 must be supplied by an external hot utility. This completes the design above the pinch, which in this case is very simple. However, the design below the pinch is more challenging, as seen below.

Step 4. Matches below the Pinch

Below the pinch, there are two hot streams and two cold streams at the pinch; thus, the number criterion in Equation 3.1 is satisfied. The MC_p criterion in Equation 3.2 requires that every pinch match satisfy $MC_{p,hp} \geq MC_{p,cp}$. It is not possible to match stream H1 with either of the cold streams (since the MC_p of stream H1, being 10, is lower than the MC_p of stream C3 and stream C4). On the other hand, it is possible to match stream H2 with either of the cold streams (since the MC_p of stream H3, being 40, is higher than the MC_p of 20 for stream C3 and 15 for stream C4). If we proceed to match H2 with either stream C3 or stream C4, then there will be no possible match for the other (remaining) cold stream that meets a ΔT_{min} of 13°C. The situation arises because no pinch matches are possible for stream H1 based on the MC_p criterion. This means, in terms of matching options below the pinch, there is effectively only one hot stream and two cold streams at the pinch, so this violates the number criterion. The only option is to resort to stream splitting. Stream H2 is split into two branches: one branch (say, H2a) to match with stream C3 and another branch (say, H2b) to match with stream C4.

Depending on which cold stream is ticked off, two possible cold-end designs are possible. If stream C3 is ticked off, then unit 2 (match between split-stream H2a and C3) will have a load of 1840 kW (minimum of 1840 kW and 2400 kW). The other split-stream H2b with a load of 560 kW (i.e., 2400 – 1840) can be matched with stream C4 to obtain unit 3 (see Figure 3.1). Another possible cold-end design (see Figure 3.2) is obtained by ticking off stream C4 through unit 2 (match between split-stream H2b and C4) with a load of 1080 kW (minimum of 1080 kW and 2400 kW). The other split-stream H2a with a load of 1320 kW (i.e., 2400 – 1080) can be matched with stream C3 to obtain unit 3 (see Figure 3.2). All pinch matches are completed at this stage.

Step 5. Design Away from the Pinch

The remaining heat of 520 kW on stream C4 (in Figure 3.1) and on stream C3 (in Figure 3.2) is supplied by matching with stream H1 through unit 4. Note that the feasibility criteria (Equations 3.1 and 3.2) need not be satisfied away from the pinch. In fact, the MC_p criterion in Equation 3.2 is violated by unit 4 since $MC_{p,H1}$ is lower than the MC_p of either cold stream. However, this does not make the match infeasible. This is because an away-pinch match does not have the driving force equal to ΔT_{min} at its terminal. The temperatures at the ends of unit 4 need to be calculated and a check needs to be performed to ensure that ΔT_{min} has not been violated (see step 6 below).

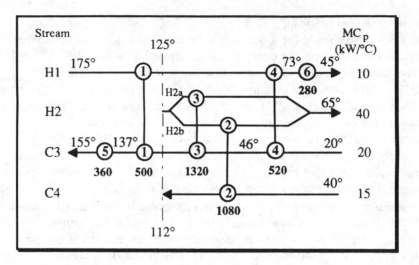

Figure 3.2 An alternative to the network in Figure 3.1 is possible with a different cold-end design. (Temperature in °C and load in kW)

Step 6. Adding Heaters/Coolers and Checking Final Design

A heater of 360 kW is placed on stream C3 and a cooler of 280 kW is placed on stream H1 to complete the design.

Next, all intermediate temperatures are systematically calculated. For example, the calculations for stream H1 starting from the hot end of the network will be as follows:

$$175 - 500/10 = 125; \quad 125 - 520/10 = 73; \quad 73 - 280/10 = 45.$$

Finally, a check is performed on the driving forces at the two ends of each unit to verify that they are not less than the ΔT_{min} specified (in this case, 13°C). For example, the ΔT at the two ends of unit 1 are: $\Delta T_{hot\ end} = 175 - 137 = 38°C$; and $\Delta T_{cold\ end} = 125 - 112 = 13°C$.

Step 7. Defining the Stream Split

For the split branches in Figures 3.1 and 3.2, only the heat loads have been specified. The MC_ps and end temperatures of the split branches have intentionally not been labeled since a number of possibilities exist and a representation to include them all is given below.

If $MC_{p,a}$ is the heat capacity flow rate of the split branch H2a in Figure 3.1, then the split may be defined in general terms as

	Heat Capacity Flow Rate	Temperature at Cold End
Branch H2a	$MC_{p,a}$	$125 - 1840/MC_{p,a}$
Branch H2b	$40 - MC_{p,a}$	$125 - 560/(40-MC_{p,a})$

Appropriately, the heat capacity flow rates add to 40 kW/°C, and the temperature after mixing of the two split branches is 65°C. The legal range for $MC_{p,a}$ may be determined using the feasibility criterion in Equation 3.2. As H2a is the branch that matches with stream C3, $MC_{p,a} \geq 20$. Similarly, as H2b is the branch that matches with stream C4, $40 - MC_{p,a} \geq 15$ or $MC_{p,a} \leq 25$.

For an isothermal mixing junction, $125 - 1840/MC_{p,a} = 125 - 560/(40-MC_{p,a})$ or $MC_{p,a} = 30.667$. As this is outside the legal range ($20 \leq MC_{p,a} \leq 25$), it is not possible for the network in Figure 3.1 to have an isothermal mixing junction.

The stream split in Figure 3.2 may be represented as

	Heat Capacity Flow Rate	Temperature at Cold End
Branch H2a	$MC_{p,a}$	$125 - 1320/MC_{p,a}$
Branch H2b	$40 - MC_{p,a}$	$125 - 1080/(40-MC_{p,a})$

where $20 \leq MC_{p,a} \leq 25$, as before. The network in Figure 3.2 can have an isothermal mixing junction if $MC_{p,a}$ is chosen to be 22.

It is clear that the MER network is not unique. Two possible topologies are shown in Figures 3.1 and 3.2, and each includes a number of designs depending on the stream split ratio chosen. Other designs are also possible. For instance, either cold stream may be split below the pinch to allow a match with stream H1. However, such a design will require two stream splits (one for the cold stream and another for stream H2). Figure 3.3 shows an MER network where stream C3 is split.

The stream splits in Figure 3.3 may be represented as

	Heat Capacity Flow Rate	Temperature at Cold End
Branch H2a	$MC_{p,h}$	$125 - 1040/MC_{p,h}$
Branch H2b	$40 - MC_{p,h}$	$125 - 1080/(40-MC_{p,h})$

	Heat Capacity Flow Rate	Temperature at Hot End
Branch C3a	$MC_{p,c}$	$20 + 800/MC_{p,c}$
Branch C3b	$20 - MC_{p,c}$	$20 + 1040/(20-MC_{p,c})$

where $MC_{p,c} = 8.696$ kW/°C and $11.304 \leq MC_{p,h} \leq 25$. In order not to have any ΔT_{min} violations at the hot end of units 2 and 3, the mixing junction on stream C3 must be isothermal with both the branches at the cold pinch temperature of 112°C. An MER network (similar to Figure 3.3) involving two stream splits may also be obtained by splitting stream C4 and is left as an exercise to the reader.

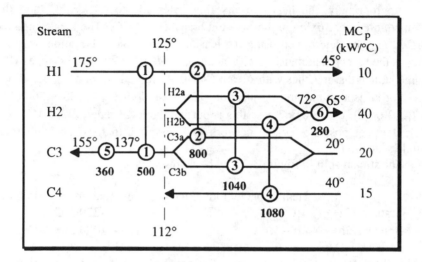

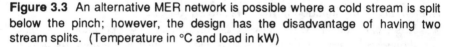

Figure 3.3 An alternative MER network is possible where a cold stream is split below the pinch; however, the design has the disadvantage of having two stream splits. (Temperature in °C and load in kW)

For all the three MER networks (Figures 3.1 through 3.3), only hot utility (360 kW) is needed above the pinch and cold utility (280 kW) below the pinch. Also, there are two units above the pinch and four units below the pinch. This is in accordance with the energy and unit targets established in Table 2.6.

In the above example, there was only one hot end design and a number of cold end designs. In general, there will be many hot end designs and many cold end designs for a problem; thus, a large number of final MER networks are possible based on different combinations of hot and cold end designs. All these networks maximize energy recovery and, hence, the "best" network among these must be determined based on evaluation criteria like area, complexity (in terms of stream splits and number of units), cost, flexibility, controllability, and safety. This discussion is reserved for the next chapter, as the issue in this chapter is to design networks that maximize energy recovery.

3.1.2 Understanding the Pinch Design Feasibility Criteria

Before proceeding any further, a second look at the feasibility criteria (Equations 3.1 and 3.2) may be worthwhile.

Consider the below-pinch design (in Figure 3.1). For the sake of argument, let stream H1 be matched with one of the cold streams. Then, for a pinch match of load Q, $\Delta T_{hot\ end} = 125 - 112 = 13°C$ and $\Delta T_{cold\ end} = (125 - Q/10) - (112 - Q/MC_{p,cp}) = 13 - Q(1/10 - 1/MC_{p,cp})$. For $\Delta T_{cold\ end} \geq 13$, $MC_{p,cp} \leq 10$. The MC_p criterion (Equation 3.2) is a generalization of this concept which simply ensures that the driving force at the end of an exchanger (match) is not less than ΔT_{min}.

Consider next a pinch match for stream H2, but without stream splitting. Let stream H2 be matched with one of the cold streams (say, stream C3 for concreteness). Although stream H2 has heat remaining in it after this match, it cannot supply the heat to the other cold stream (C4) as the minimum ΔT is not available. From the earlier discussion, a pinch match for stream H1 is not possible because of the MC_p criterion. As no hot utility can be supplied below the pinch, there is no matching option whatsoever for cold stream C4. The problem may be solved if there is another hot stream with an appropriate MC_p. Such a stream is created by stream splitting (see Figures 3.1 through 3.3) in a manner that both feasibility criteria (the number criterion in Equation 3.1 and the MC_p criterion in Equation 3.2) are simultaneously satisfied.

3.2 MER NETWORKS FOR MULTIPLE UTILITIES AND MULTIPLE PINCHES

The use of multiple utilities causes multiple pinches, and this section looks at designs for these cases. Three methods proposed for designing MER networks in the case of multiple pinches are outlined below.

Trivedi et al. (1989a) defined a new concept of an inverse pinch point (between two pinches) such that

$$\Sigma MC_{p,h} < \Sigma MC_{p,c} \quad \text{just below the inverse pinch}$$
$$\text{and } \Sigma MC_{p,h} > \Sigma MC_{p,c} \quad \text{just above the inverse pinch} \tag{3.3}$$

Equation 3.3 provides a necessary and sufficient condition for an inverse pinch. It may be simply stated as that point where the temperature approach between the composite curves reaches a maximum. The inverse pinch divides the problem into two regions: a region between the higher pinch and the inverse pinch (where PDM heuristics for below pinch are applicable) and another region between the lower pinch and the inverse pinch (where PDM heuristics for above pinch are applicable).

Jezowski (1992a) argued that it is unnecessary to evaluate the inverse pinch and design the network for the two separate regions. He suggested that synthesis should start from both the pinches simultaneously. PDM heuristics for above the pinch should be applied for designing from the lower temperature pinch, and PDM heuristics for below the pinch should be applied for designing from the higher temperature pinch. According to Jezowski (1992a), the PDM rules are valid for a region between two pinches and their simultaneous application should not lead to any clashes. The approach appears elegant and seems to lead to simple networks. The method in principle is identical to that described by Linnhoff et al. (1982). The only difference is that Linnhoff et al. (1982) suggest initializing the design from the more constrained pinch when designing between two pinches.

The approach based on simultaneous synthesis from both pinches (Jezowski, 1992a), combined with the heuristic of designing from the more constrained pinch (Linnhoff et al., 1982) in the case of a clash, provides a simple attractive option and is discussed in detail through an example below. Before proceeding to the network design itself, it must be emphasized that a multiple utilities problem is in effect a problem of multiple pinches. It is obvious that, wherever the utility profile touches the GCC (Figure 1.10), a new pinch (which may be termed the utility pinch) is created. However, utility pinches may occur in a more subtle way (see Problem 3.H), and a definite procedure (see Example 3.2) is helpful in determining all utility pinches.

Example 3.2 Determination of Utility Pinches

For the stream data in Case Study 4S1, determine all the pinches (process as well as utility) for a ΔT_{min} of 20°C, when multiple utilities are used as in Example 1.13.

Solution. The utility GCC (obtained by connecting the utility profile segments) and the process GCC in Figure 1.10 are in perfect enthalpy balance and together produce the balanced GCCs (Hall et al., 1992). The points where the two GCCs touch each other yield the utility pinches in this graphical construction. However, if the utilities are treated as dummy process streams, then the utility pinches may be analytically found as described below.

Step 1. Merging Utility Streams and Process Streams

The usage level of the various utilities is determined as in Example 1.13, and then converted to MC_p values using Equation 1.10. The four utility streams are then merged with the four process streams to obtain the following stream data:

Stream	H1	H2	C3	C4	CU1	CU2	HU1	HU2
T_{in} (°C)	175	125	20	40	15	100	180	135
T_{out} (°C)	45	65	155	112	25	101	179	134
MC_p (°C)	10	40	20	15	45	75	400	205

Step 2. Determination of Pinches Using PTA

Use of the PTA (as in Example 1.1) on the above eight-stream data results in the following heat cascade at $\Delta T_{min} = 20°C$:

T_{int} (°C)   170 169 165 125 124 122 115 111 110 55 50 35 30 25
R_{cas} (kW)   0 400 400 0 195 175 0 60 0 825 700 550 225 0

Three pinches are observed (corresponding to the three zeroes in the R_{cas} row above, when the zeroes at the ends of the cascade are ignored). There is a process pinch at 115°C (as in Table 1.2), and there are two utility pinches at 125° and 110°C.

It is noticed that the intermediate utilities, which fall within the range of the process temperatures, lead to additional pinches. The PDM in Section 3.1 needs to be adapted appropriately to design such multipinched problems.

Example 3.3 MER Network Design for Multiple Pinches

Design an MER network for a ΔT_{min} of 20°C for the multiple-pinched problem in Example 3.2.

Solution. It must be emphasized that, when utility temperatures fall within the temperature range of the process streams, the MER design is ideally performed on a balanced grid (Linnhoff, 1986). This merely implies that the utility streams are drawn as if they are (dummy) process streams in the classical grid representation for networks (Figure 3.4). This inclusion of utilities as process streams has an important implication, for it allows the process and utility systems to be designed simultaneously (on a balanced grid). This is in contrast to conventional practice where the process is designed first and the utility system later. When the design is sequential (process first and utility afterwards), commitments and design decisions made in the early stages may adversely affect the final design (see Example 3.4).

Step 1. Conventional PDM for Normal Above- and Below-Pinch Regions

Given the three pinches at 125°, 115°, and 110°C, it is clear that the region above 125°C is a "normal above-pinch" region and the region below 110°C is a "normal below-pinch" region. The conventional PDM (discussed in Section 3.1) is used to design these two regions.

Consider the region above the 125°C pinch (Figure 3.4a). The only near-pinch match is between streams H1 and C3, which ticks off stream H1 (with a load of 400 kW). The remaining load (of 400 kW) on stream C3 is taken care of by the hot utility HU1 (match 2).

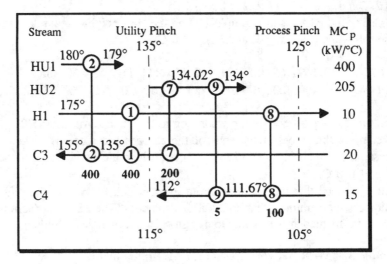

Figure 3.4a For a multiple utility - multiple pinch design, an MER network is synthesized above the process pinch with the hot utilities represented as dummy process streams on a balanced grid. (Temperature in °C and load in kW)

Next, consider the region below the 110°C pinch (Figures 3.4b and 3.4c). No pinch matches are possible for stream H1 because of the MC_p criterion. Given the number criterion, stream H2 (with a load of 2200 kW) must be split and then used to tick off either stream C3 (of load 1600 kW) or stream C4 (of load 900 kW). The remaining load (of 300 kW) on the cold process stream is satisfied using stream H1 (an away-pinch match). The design is completed through match 6 of 450 kW between streams H1 and CU1. Possible cold-end designs are shown in Figure 3.4b (where stream C4 is ticked off using a split stream H2b) and Figure 3.4c (where stream C3 is ticked off with split stream H2a). The stream split on stream H2 in Figure 3.4b is represented as

	Heat Capacity Flow Rate	Temperature at Cold End
Branch H2a	$MC_{p,a}$	$120 - 1300/MC_{p,a}$
Branch H2b	$40 - MC_{p,a}$	$120 - 900/(40 - MC_{p,a})$

The stream split on stream H2 in Figure 3.4c may be represented in the same manner, except that the loads 1300 and 900 above should be replaced by 1600 and 600 respectively. In both cases, $20 \leq MC_{p,a} \leq 25$.

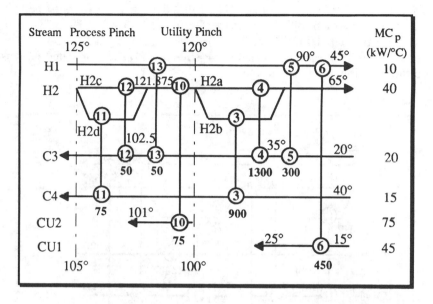

Figure 3.4b For a multiple utility - multiple pinch design, an MER network is synthesized below the process pinch with the cold utilities represented as dummy process streams on a balanced grid. (Temperature in °C and load in kW)

Step 2. Designing between Two Pinches

Consider the region between the two pinches (125° and 115°C) in Figure 3.4a. Above the 115°C pinch, stream H1 can be matched with either stream C3 or C4. Below the 125°C pinch, stream H1 has no match possible because of its low MC_p. The only option is to match stream HU2 with stream C3. Given this constraint at the 125°C pinch, match 7 of load 200 kW may be made to tick off stream C3. Then, at the 115°C pinch, stream H1 is ticked off using stream C4. Match 9 of heat load 5 kW allows completion of the design. As an aside, the reader may attempt to obtain another possible design wherein stream H1 (of load 100 kW) is first ticked off with stream C3. The remaining load of 100 kW on stream C3 is then satisfied by matching with stream HU2. In this case, match 9 ends up with a load of 105 kW (instead of 5 kW as in Figure 3.4a).

Next, consider the region between the two pinches (115° and 110°C). Above the 110°C pinch, stream H1 can be matched with any cold stream whereas stream H2 can be matched only with stream CU2 on the basis of the MC_p criterion. Below the 115°C pinch, stream H1 has no possible match. Given the number criterion, stream H2 (with a load of 200 kW) must be split and then used to tick off either stream C4 of load 75 kW (Figure 3.4b) or stream C3 of load 100 kW (Figure 3.4c). The remaining load (of 50 kW) on the cold process stream is satisfied using stream H1. The completed designs are shown in Figures 3.4b and 3.4c.

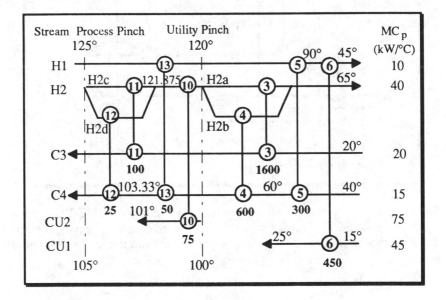

Figure 3.4c Alternative designs are possible for both the regions in Figure 3.4b, so these designs can be combined in various ways for the final network. (Temperature in °C and load in kW)

Combining Figure 3.4a (design above the process pinch) with Figure 3.4b or 3.4c (designs below the process pinch) gives various possibilities for the final network, which comprises thirteen matches in any case. These thirteen units in the MER design are consistent with the units target (Equation 1.26) and occur because of the three pinches. Given the complexity of the final network (thirteen units and two stream splits), some simplification would be desirable to arrive at a practical design. This is indeed possible through evolution (as shown in the next chapter in Example 4.2).

3.3 BALANCED GRID NETWORKS FOR APPROPRIATE UTILITY PLACEMENT

The use of a balanced grid (Linnhoff, 1986), with utilities drawn as dummy process streams in the network, is not limited to multiple utility designs. It may be profitably used to decide on the appropriate placement of a utility, even in the case where a single utility is used. Instead of treating the utility on a stand-alone basis, the process and utility may be considered as a unified system with their design done simultaneously as illustrated below.

Example 3.4 Appropriate Location of a Utility

Imagine that you have been invited as a consultant to analyze the MER network design in Figure 3.1. The basic idea is to use flue gas (instead of steam at 180°C) with a theoretical flame temperature (T_{tf}) of 1500°C to meet the hot utility demand of 360 kW and extract the maximum possible heat from the flue gas. The lowest temperature for the flue gas is the acid dew point temperature $(T_{adp}$, given as 160°C). To compensate for the poor heat transfer coefficient, it is desirable to have a ΔT_{min} of 50°C between the utility (flue gas) and the process. How should the furnace be located in order to minimize fuel consumption? Given: ambient temperature $T_a = 20°C$.

Solution. While designing the MER network in Example 3.1, the hot utility was assumed to be hot enough to satisfy the heating need at any temperature and the cold utility was assumed cold enough to meet the cooling need at any temperature. The validity of this assumption must indeed be checked. For steam (180° to 179°C), the 360 kW heating from 137° to 155°C for stream C3 has the required ΔT_{min} of 13°C. However, in the case of flue gas, the specification on ΔT_{min} of 50°C requires its outlet temperature not to be below 187°C. Now,

$$MC_{p,flue} = Q_{hu,min}/(T_{tf} - T_{flue,out}). \tag{3.4a}$$

With $MC_{p,flue} = 360/(1500 - 187) = 0.2742$ kW/°C, the fuel consumption is calculated from

$$Q_{fuel} = Q_{hu,min} + MC_{p,flue} (T_{flue,out} - T_a) \tag{3.4b}$$

Thus, $Q_{fuel} = 360 + 0.2742 (187 - 20) = 405.79$ kW. This results in a furnace efficiency $(Q_{hu,min}/Q_{fuel})$ of 88.7%. The stack losses are given by

$$Stack\ losses = MC_{p,flue} (T_{flue,out} - T_{stack}) \tag{3.4c}$$

where T_{stack} is the minimum allowable stack temperature (typically chosen slightly above the acid dew point temperature to minimize stack corrosion). Therefore, if the allowance for corrosion is neglected ($T_{stack} \approx T_{adp}$), the stack losses are 7.4 kW, i.e., 0.2742 (187 − 160). Can these losses be minimized in a systematic manner by appropriately locating the furnace?

Step 1. Matching Utility Profile with Process GCC
    The GCC for the process at ΔT_{min} = 13°C obtained from the PTA is plotted (Figure 3.5, not to scale).

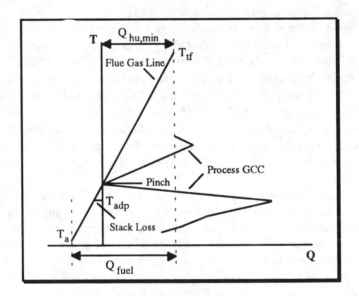

Figure 3.5 A fuel target may be determined by representing the flue gas as a straight line (of constant heat capacity flow rate), starting from the theoretical flame temperature and ending at the ambient temperature.

As the ΔT_{min} for the utility-process transfer is 50°C, the ΔT_{min} contribution for the flue gas is taken to be 43.5°C (i.e., 50 − 13/2). Given the maximum temperature change for the flue gas from T_{tf} to T_{adp} (i.e., from 1500° to 160°C in this case), the utility profile must go from 1456.5° to 116.5°C. This is not allowed since it cuts the process GCC (not shown in figure) because the shifted T_{adp} is less than the pinch temperature. This implies that, at best, the utility profile can just touch the GCC (at the pinch of 118.5°C) in order to minimize the flue gas flow rate. This determines that the outlet temperature of the flue gas is 162°C (namely, 118.5 + 43.5) with a MC_p value of 0.269 kW/°C (i.e., 360/[1500 − 162]). Thus, Q_{fuel} = 360 + 0.269 (162 − 20) = 398.21 kW. The furnace efficiency is increased to 90.4%, and the stack losses are reduced to 0.54 kW, i.e., 0.269 (162 − 160). The requirement of the cold utility (15° − 25°C) is 280 kW, as before.

Step 2. Appropriate Utility Placement on a Balanced Grid
    The next step is to design a network that meets the fuel target calculated in step 1. The balanced GCCs show only a single pinch at 118.5°C (i.e., the earlier process pinch remains unchanged and no utility pinches are added). Since the temperatures and the MC_p for the flue gas and the cold utility are known, the utilities may be drawn as dummy process streams on a balanced grid. Figure 3.6 shows a possible MER network obtained by the PDM, where the design below the pinch is identical to that in Figure 3.1.

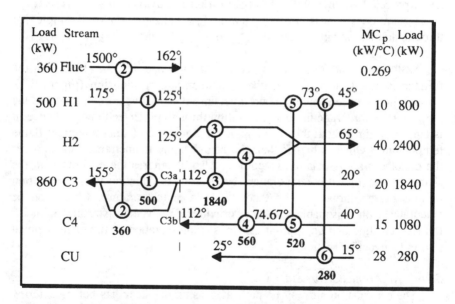

Figure 3.6 The design in Figure 3.1 may be appropriately modified if heating demand above the pinch is to be met by flue gas. (Temperature in °C and load in kW)

    The interesting aspect obviously is the incorporation of the furnace above the pinch. At the pinch, there are two hot streams (H1 and flue gas) and one cold stream (C3) above the pinch. The number criterion requires that the cold stream be split. The C3 split may then be represented as

	Heat Capacity Flow Rate	Temperature at Hot End
Branch C3a	$MC_{p,a}$	$112 + 500/MC_{p,a}$
Branch C3b	$20 - MC_{p,a}$	$112 + 360/(20 - MC_{p,a})$

where $10 \leq MC_{p,a} \leq 19.731$. Though the above-pinch design has only two matches (the same as in the case where steam is used), it requires a stream split. The design would not have been easy without the use of the PDM and the balanced grid.

Significant advantages may result by preheating the combustion air entering the furnace (Linnhoff and de Leur, 1988) because it increases furnace efficiency and reduces fuel consumption. This is explored next.

Example 3.5 Minimizing Fuel Consumption in Furnace by Air Preheat

Determine the possible reduction in fuel consumption in Example 3.4 through air preheating. Given: $MC_{p,air}/MC_{p,flue}$ (in furnace) = 0.87.

Solution. The model, proposed by Linnhoff and de Leur (1988), for the furnace is based on a linear enthalpy-temperature relationship (Figure 3.5). Though the furnace phenomena are oversimplified in this model, it serves the purpose for appropriate furnace integration through pinch technology; hence, it is employed throughout this discussion. The model uses the theoretical flame temperature (which is higher than the actual flame temperature) and ignores the endothermic dissociation reactions. Also, it ignores the difference in the shape of the enthalpy-temperature curve in the radiation and convection regions of the furnace. The theoretical flame temperature, which can be calculated with reasonable accuracy, provides a convenient starting point for the flue gas line and an appropriate linear representation in the critical pinch region (convection section).

Step 1. Air Preheating by Stack Gas

The combustion air may be preheated using the stack gas before entering the furnace. Consider the design in Figure 3.1, where there is no stream split and the flue gas leaves the furnace at a temperature of 187°C. The enthalpy lost by the stack gas equals the enthalpy gained by the air; thus, it is seen that

$$MC_{p,flue}(T_{flue,out} - T_{stack}) = MC_{p,air}(T_{preheat} - T_a), \tag{3.4d}$$

i.e., $(187-160) = 0.87(T_{preheat}-20)$ or $T_{preheat} = 51°C$ (letting $T_{stack} \approx T_{adp}$).

The new theoretical flame temperature is increased to

$$T_{tf,new} = T_{tf,old} + (MC_{p,air}/MC_{p,flue})\,(T_{preheat} - T_a),\qquad(3.4e)$$

so $T_{tf,new} = 1500 + 0.87\,(51 - 20) = 1527°C$.

With $MC_{p,flue}$ reduced to 0.2687 kW/°C (i.e., 360/[1527 − 187] as per Equation 3.4a), the fuel requirement in the presence of air preheating reduces to

$$Q_{fuel} = Q_{hu,min} + MC_{p,flue}\,(T_{flue,out} - T_a) - MC_{p,air}\,(T_{preheat} - T_a),\qquad(3.4f)$$

i.e., $Q_{fuel} = 360 + 0.2687\,(187 - 20) - 0.87 \times 0.2687\,(51 - 20) = 397.61$ kW.

With this arrangement, there are virtually no stack losses and the furnace efficiency is 90.54% (a marginal increase over the design in Figure 3.6). This is the conventional approach to air preheating. Can pinch technology offer something more attractive?

Step 2. Air Preheating Using Process Streams below the Pinch

The combustion air may be preheated using heat available in the process streams. In that sense, it may be looked upon as a cold utility. Pinch technology dictates that the air be introduced below the pinch. The air enters at 20°C (corresponding to a shifted temperature of 63.5°C, i.e., 20 + 43.5). The maximum temperature to which it may be preheated is the pinch temperature (118.5°C), which corresponds to a final temperature for the preheated air of 75°C (i.e., 118.5 − 43.5). The calculations for this new situation may be performed as follows:

$T_{tf,new} = 1500 + 0.87\,(75 - 20) = 1547.85°C$ (using Equation 3.4e);
$MC_{p,flue} = 360/(1547.85 - 162) = 0.2598$ kW/°C   (using Equation 3.4a);
$Q_{fuel} = 360 + 0.2598\,(162 - 20) - 0.87 \times 0.2598\,(75 - 20) = 384.46$ kW
(using Equation 3.4f).

In the above calculations, the design in Figure 3.6 (with the furnace appropriately placed) is used as the starting point. This is then further improved through air preheating by process streams to yield a final design with a furnace efficiency of 93.64% (i.e., 360/384.46) and a negligible stack loss (0.52 kW, i.e., 0.2598 [162 − 160]). Note that the air can be preheated to 75°C when process streams are used and to only 51°C when the stack gas is used. Also, the cold utility requirement is decreased by 12.43 kW, as the combustion air acts like a cold utility stream ($MC_{p,air}$ = 0.226 kW/°C) below the pinch with inlet and outlet temperatures of 20° and 75°C respectively.

Step 3. Choosing Process-Process and Process-Utility Matches Simultaneously
    The balanced GCCs may be obtained next by merging the four process streams with the three utility streams, using the following stream data:

Stream	H1	H2	C3	C4	CU	Air	Flue
T_{in} (°C)	175	125	20	40	15	20	1547.85
T_{out} (°C)	45	65	155	112	25	75	162
MC_p (kW/°C)	10	40	20	15	26.76	0.226	0.2598
ΔT_c (°C)	6.5	6.5	6.5	6.5	6.5	43.5	43.5

    The last row above shows the ΔT_{min} contribution associated with each individual stream. While shifting the end temperatures of the streams, these ΔT_c values must be used (instead of $\Delta T_{min}/2$ in the PTA detailed in Section 1.1.2). The resulting heat cascade shows only a single pinch at 118.5°C. Using the PDM, an MER network may be designed using the balanced grid as in Figure 3.7. The design is straightforward; it involves only minor changes from the network in Figure 3.6 to incorporate the combustion air stream. The fuel demand as well as the requirement for the cold utility (CU) is reduced, but at the expense of an additional unit. The network in Figure 3.7 is converted into a process scheme and represented in Figure 3.8.

3.4 THE FAST MATCHING ALGORITHM

    In addition to the PDM, another useful method for designing HENs is the fast matching algorithm of Ponton and Donaldson (1974). It is also referred to as the hottest/highest matching heuristic and allows networks to be generated with relative ease and rapidity. Typically, it may be used to synthesize near-optimal solutions to threshold problems. Unlike the PDM (which generates many alternative networks), this heuristic method will generate only one network. It does not specify stream splits; however, it produces cyclic networks that are often much cheaper than acyclic ones.
    Though the discussion here focuses on the proposal of Ponton and Donaldson (1974), similar thermodynamic matching rules have been presented by other workers (Nishida et al., 1971; Nishida et al., 1977; Pehler and Liu, 1981). One such recommendation, to approach the most efficient or nearly reversible exchange of heat, suggests matching the hot process and utility streams with the cold process and utility streams consecutively in a decreasing order of their average stream temperatures (Liu, 1987).

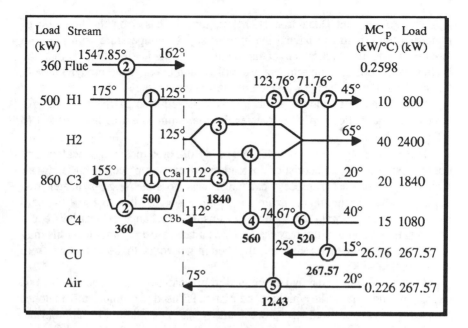

Figure 3.7 In the above network, preheating of the combustion air to the cold pinch temperature helps reduce both fuel consumption and the cold utility required. (Temperature in °C and load in kW)

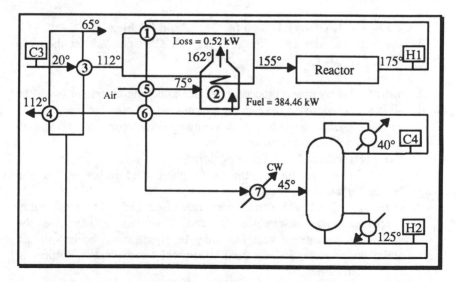

Figure 3.8 The flowsheet for the network in Figure 3.7 shows the appropriate location of the furnace with air preheating. (Temperature in °C)

The Ponton and Donaldson (1974) method is based on the heuristic of matching the hot stream having the highest supply temperature with the cold stream having the highest target temperature. The justification is that the highest target cold stream temperature will be achieved (if thermodynamically feasible) through exchange with the hottest stream available.

Once a match is chosen, some heuristic is required to fix its load. An appropriate heuristic may be to transfer the maximum possible heat subject to ΔT_{min} constraints.

However, Ponton and Donaldson (1974) did not locate the heaters and coolers using the same heuristic. Instead, a utility was placed in the network whenever no process-to-process exchange was possible (e.g., a heater was inserted if the maximum hot stream temperature was too low to allow the cold stream to reach its target temperature). Thus, the exchange of heat between a process stream and a utility was used only as a last resort. Sometimes this may cause cheaper and thermodynamically better networks to be excluded from consideration.

In what follows, the balanced grid representation (see Section 3.3) is exploited and the hottest/highest heuristic is used to guide all matches (process-process as well as process-utility). This helps in appropriate heater/cooler placement and multiple utility selections. With this modification to the Ponton and Donaldson (1974) matching method, the philosophy of simultaneous design of process and utility systems (from the previous sections) is continued in this section.

Example 3.6 MER Network Design for Threshold Problem
Use the fast matching algorithm to design an MER network for the Case Study 4S1t at $\Delta T_{min} = 13°C$.

Solution. The synthesis procedure by fast matching algorithm (with the modification for appropriate heater/cooler placement) is presented first.
1. Perform energy targeting to determine utility requirements and to ascertain a threshold problem.
2. Use a balanced grid as the starting point.
3. Select the hottest hot stream and highest cold target as per the hottest/highest (H/H) heuristic.
4. Consider a match with known hot stream inlet and cold stream outlet temperatures (as determined in step 3 above). Determine the maximum quantity of heat that may be transferred, the hot stream outlet temperature, and the cold stream inlet temperature, subject to ΔT_{min} constraints.
5. Repeat steps 3 and 4 on the remaining problem (excluding match[es] made).

Consider the application of the above algorithm on Case Study 4S1t.

Step 1. Determining the Energy Targets
At $\Delta T_{min} = 13°C$, the hot utility requirement is 1580 kW. There is no cold utility requirement; thus, this is a threshold problem. Note that ΔT_{min} of 13°C is below the sensitivity threshold of 25°C (see Table 2.11).

Step 2. Drawing the Balanced Grid
The hot utility (180° to 179°C) is treated as a stream with an MC_p of 1580 kW/°C. Figure 3.9 shows the balanced grid.

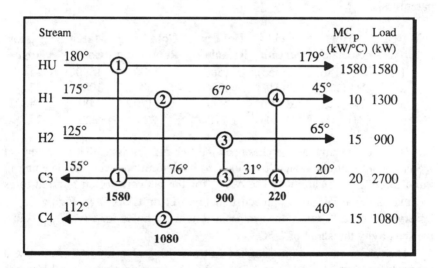

Figure 3.9 The application of the hottest/highest heuristic on a balanced grid allows rapid generation of a near-optimal network for a threshold problem. (Temperature in °C and load in kW)

Step 3. Applying the Hottest/Highest Heuristic
The hot stream with the highest inlet temperature is the hot utility (180°C). The cold stream with the highest cold target temperature is stream C3. As per the H/H heuristic, a match between streams HU and C3 is recommended.

Step 4. Determining the Match Load
For a match between stream HU (with a load of 1580 kW) and C3 (with a load of 2700 kW), the maximum possible load is 1580 kW. This would tick

off stream HU and result in an inlet temperature of 76°C (i.e., 155 − 1580/20) for stream C3. This gives a ΔT_{min} of 103°C at the cold end of match 1, which is above the specified ΔT_{min} of 13°C. This completes the first match.

Step 5. Applying the H/H Heuristic Repeatedly for Remaining Problem

On excluding match 1, the H/H heuristic indicates a match between stream H1 (inlet temperature of 175°C) and stream C4 (outlet temperature 112°C). Match 2 between streams H1 and C4 can have a maximum load of 1080 kW and helps tick off stream C4. The procedure is repeated to obtain the design in Figure 3.9. The calculations for every match are summarized below, with the hottest hot stream temperature and the highest cold target temperature in parenthesis.

Match	Hot Stream	Cold Stream	Hot Load Remaining	Cold Load Remaining	Match Load	ΔT_{min} for Match
1	HU (180)	C3 (155)	1580	2700	1580	25
2	H1 (175)	C4 (112)	1300	1080	1080	27
3	H2 (125)	C3 (76)	900	1120	900	34
4	H1 (67)	C3 (31)	220	220	220	25

In this case study, all matches obey the tick off heuristic. This is often not the case, and then the match loads need to be determined carefully keeping in mind the ΔT_{min} constraints. The ΔT_{min} for the network (often referred to as the exchanger minimum approach temperature, or EMAT) in Figure 3.9 is observed to be 25°C; thus, the synthesized network holds for any ΔT_{min} below the sensitivity threshold of 25°C.

The reader can attempt using the fast matching algorithm to synthesize a network for Case Study 4S1 at ΔT_{min} = 13°C. For this pinched problem, the fast matching algorithm runs into difficulties and the PDM provides a superior methodology.

3.5 CONSTRAINED HEAT EXCHANGER NETWORKS

Just thermodynamic feasibility may not be sufficient to match two streams. In industrial problems, matches between particular streams are often not allowed due to practical constraints like plant layout, safety, operability, product contamination, corrosion, fouling, materials of construction, and area integrity. Simultaneous satisfaction of these constraints and the energy targets in Section 1.1 is then possible only if sufficient design options exist. More often than not, this is not the case. The constraints interfere with the matching options preferred in the unconstrained case, and this leads to an increase in the minimum utility consumption.

This section contains a detailed treatment of constrained heat exchanger networks (CHENs). Before one proceeds to design CHENs, it is important to establish new targets for this situation of forbidden matches.

3.5.1 The Dual Cascade Approach

Energy targeting for the case of forbidden matches may be done using dual heat cascades: a constrained cascade and an unconstrained one. The dual cascade approach (Ali, 1993; Sachdeva, 1993) is based on the heat flow model originally proposed by Jezowski and Friedler (1992). However, it modifies the Jezowski-Friedler approach by converting their heat flow model into a heat cascade representation, which is consistent with that used in the previous chapters. The dual cascade approach is introduced below through an example.

Example 3.7 Energy Targeting for Forbidden Matches by the Dual Cascade Approach

Consider the Case Study 4S1 with the constraint that streams H2 and C3 cannot be matched. Let this modified problem be referred to as 4S1c. Obtain the minimum utility requirement for this constrained problem at $\Delta T_{min} = 20°C$.

Solution. The targeting procedure divides the problem into the conventional temperature intervals and then works out the possible heat flows and the need for utility within each interval.

The heat flow model for any given temperature interval (say, the i-th interval) may be represented as in Figure 3.10.

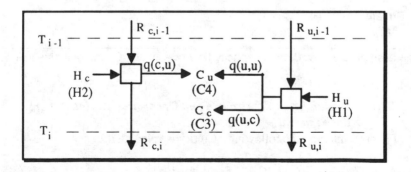

Figure 3.10 The model for each temperature interval maintains the heat flows for the constrained and unconstrained portions separately.

For the present case study, streams H2 and C3 are the constrained streams (denoted by c) whereas streams H1 and C4 are the unconstrained ones (denoted by u). In Figure 3.10, R is the residual heat cascaded between intervals; q is the heat that may be transferred between the two streams in parenthesis. For example, $q(c, u)$ is the heat that may be transferred between the constrained hot and unconstrained cold streams.

Step 1. Enthalpy Content of a Stream in Each Interval

The temperature intervals are generated in the same manner as in the PTA. The intervals generated in Table 1.1 for $\Delta T_{min} = 20°C$ are reproduced in Table 3.1. Then, the enthalpy content of each stream in each interval is computed. In the first interval, streams H1 and C3 exist. So, $QH1 = 10 (165 - 122) = 430$ kW and $QH2 = 20 (165 - 122) = 860$ kW. Table 3.1 may be easily filled in this manner.

Table 3.1 Enthalpy of Streams in Various Temperature Intervals

Interval i				T_{int}	QH_u $QH1$	QH_c $QH2$	QC_c $QC3$	QC_u $QC4$
0				165	0	0	0	0
1				122	430	0	860	0
2				115	70	0	140	105
3				55	600	2400	1200	900
4				50	50	0	100	75
5				35	150	0	300	0
6				30	0	0	100	0
Stream	H1	H2	C3	C4				
MC_p	10	40	20	15				

Units: temperature in °C, heat capacity flow rate in kW/°C, and enthalpy in kW.

Step 2. Heat Availability in Constrained and Unconstrained Hot Streams for Each Interval

For each interval, the following quantity is calculated.

$$\Delta_c = R_{c,i-1} + QH_c - QC_u \qquad \text{for } i \geq 1 \qquad (3.5a)$$

where $R_{c,0} = 0$. Δ_c is the difference between the heat available in the constrained hot stream and the heat required by the unconstrained cold stream. Note that the constrained hot stream can give heat to only the unconstrained cold stream whereas the unconstrained hot stream can give heat to any cold stream. If there is an excess of heat in the constrained hot stream ($\Delta_c > 0$), then it supplies all the heat required by the unconstrained cold stream and cascades the remainder. If there is a deficit ($\Delta_c < 0$), then all the heat available in the constrained hot stream is given to the unconstrained cold stream with nothing to be cascaded. In mathematical terms,

$$\text{If } \Delta_c \geq 0, \quad R_{c,i} = \Delta_c \quad \text{and} \quad q(c, u) = QC_u. \tag{3.5b}$$

$$\text{If } \Delta_c < 0, \quad R_{c,i} = 0 \quad \text{and} \quad q(c, u) = R_{c,i-1} + QH_c. \tag{3.5c}$$

The remaining heat for the unconstrained cold stream comes from the unconstrained hot stream, which also provides heat to the constrained cold stream. Thus,

$$q(u, u) = QC_u - q(c, u) \tag{3.5d}$$

$$q(u, c) = QC_c \tag{3.5e}$$

$$R_{u,i} = QH_u + R_{u,i-1} - q(u, u) - q(u, c) \tag{3.5f}$$

where $R_{u,0} = 0$. Equations 3.5a - 3.5f may be used (in that order) to obtain the values for each row (interval) of Table 3.2.

Table 3.2 Constrained and Unconstrained Cascade Calculations

Int i	Δ_c Eq 3.5a	R_c Eqs 3.5b & 3.5c	$q(c, u)$ Eq 3.5d	$q(u, u)$ Eq 3.5e	$q(u, c)$ Eq 3.5e	R_u Eq 3.5f	Revised R_c	Revised R_u
0	0	0	0	0	0	0	0	1505
1	0	0	0	0	860	−430	0	1075
2	−105	0	0	105	140	−605	0	900
3	1500	1500	900	0	1200	−1205	1500	300
4	1425	1425	75	0	100	−1255	1425	250
5	1425	1425	0	0	300	−1405	1425	100
6	1425	1425	0	0	100	−1505	1425	0

Units: enthalpy in kW.

A sample calculation for the third interval is given below:

$$\Delta_c = 0 + 2400 - 900 = 1500 \qquad \text{(using Equation 3.5a)}$$
$$R_{c,i} = 1500 \quad \text{and} \quad q(c, u) = 900 \qquad \text{(using Equation 3.5b)}$$
$$q(u, u) = 900 - 900 = 0 \qquad \text{(using Equation 3.5d)}$$
$$q(u, c) = 1200 \qquad \text{(using Equation 3.5e)}$$
$$R_{u,i} = 600 + (-605) - 0 - 1200 = -1205 \qquad \text{(using Equation 3.5f)}$$

Step 3. Removal of Infeasibilities to Obtain Revised Cascades

The R_c and R_u columns must be revised if they contain any infeasibilities (negative entries). The revised R_c column is identical to the unrevised one as it contains no infeasibilities. On the other hand, the infeasibilities in the R_u column may be removed by adding 1505 to every entry. The revised R_u column appears as the last column in Table 3.2. Thus, with the constraint that streams H2 and C3 cannot be matched, the minimum hot utility requirement is the sum of the first entries in the revised R_c and R_u columns. The minimum cold utility requirement is the sum of the last entries in these columns. For Case Study 4S1c, the hot utility required is 1505 kW and the cold utility required is 1425 kW. The equivalent quantities for the unconstrained problem were 605 kW and 525 kW (see Table 1.2); thus, an energy penalty of 900 kW is incurred on account of the forbidden match.

The dual cascade approach provides an elegant way for energy targeting under constraints. However, its extension for more than one forbidden match in a temperature interval appears difficult. The approach of O'Young et al. (1988) based on the enthalpy matrix provides an alternative method. It is capable of handling more than one constraint and is easy (though at times tedious) to perform manually. However, it appears difficult to program. Importantly, note that targeting for CHENs may be conveniently done by mathematical programming (see Chapter 10).

The rest of this section illustrates the approach of O'Young et al. (1988).

3.5.2 The Enthalpy Matrix Approach

The constrained problem table algorithm (CPTA) proposed by O'Young et al. (1988) is essentially an extension of the original PTA. The two important thermodynamic principles on which the unconstrained PTA is based, and which continue to form basic tenets in the CPTA, are

- maximization of process-process heat transfer within each temperature interval (before using external hot utility or cascading surplus heat to lower temperature intervals); and

- usage of excess hot stream enthalpy at the highest temperature level possible.

On the other hand, the CPTA differs from the PTA in two fundamental ways (O'Young and Linnhoff, 1989):

- requirement of both hot and cold utility within the same temperature interval; and
- maximization of process-process heat transfer within a particular temperature interval is achievable in various ways, some of which may have adverse effects in recovering heat at the lower temperature levels.

The appropriate maximization scheme within each interval may be chosen, without adverse effects on the later heat recovery, by the maximum possible remaining (MPR) concept discussed in the next subsection. An example to explain the CPTA and the enthalpy matrix representation is presented first.

Example 3.8 Energy Targeting for Forbidden Matches by the Enthalpy Matrix Approach

Rework Example 3.7 using the CPTA. The goal as before is to determine the minimum utility requirements for Case Study 4S1c at ΔT_{min} = 20°C when the match between streams H2 and C3 is forbidden.

Solution.   The temperature intervals are first constructed as in the conventional PTA (see Table 3.1), and calculations are then performed starting with the highest temperature interval.

Step 1. Construction of the Enthalpy Matrix

The enthalpy matrix representation is a two-dimensional array, where hot streams form columns and cold streams form rows. The enthalpy contribution of each stream within each temperature interval is shown in parenthesis. Each element in the matrix represents a possible match between a hot and a cold stream in that interval. The element may be filled with a cross (when a match is forbidden), a tick (when a match is made), a blank (when a stream does not exist in the temperature interval), or match indices (as explained below). The element corresponding to streams H2 and C3 will always have a cross to indicate the forbidden match.

The enthalpy matrix for the first temperature interval (165° - 122°C) is shown in Figure 3.11a. The interval enthalpy contribution is 430 kW for stream H1, 860 kW for stream C3, and 0 kW for streams H2 and C4 (see Table 3.1). As stream C4 does not exist in this interval, the elements in that row are left blank.

	H1 (430)	H2 (0)
C3 (860)	1.1	X
C4 (0)	--	--

Match H1 & C3 ↓

	H1 (0)	H2 (0)
C3 (430)	√	X
C4 (0)	--	--

(a) Interval 1 (165° - 122°C)

	H1 (70)	H2 (0)
C3 (140)	1.2	X
C4 (105)	1.2	--

Match H1 & (C3, C4) ↓

	H1 (0)	H2 (0)
C3 (140)	√	X
C4 (105)	√	--

MPR(C3, C4) = 175 kW

(b) Interval 2 (122° - 115°C)

	H1 (600)	H2 (2400)
C3 (1200)	1.2	X
C4 (900)	2.2	2.1

Match H1 & C3 ↓

	H1 (0)	H2 (2400)
C3 (600)	√	X
C4 (900)	--	1.1

Match H2 & C4 ↓

	H1 (0)	H2 (1500)
C3 (600)	√	X
C4 (0)	--	√

(c) Interval 3 (115° - 55°C)

	H1 (50)	H2 (1500)
C3 (100)	1.2	X
C4 (75)	2.2	2.1

Match H1 & C3 ↓

	H1 (0)	H2 (1500)
C3 (50)	√	X
C4 (75)	--	1.1

Match H2 & C4 ↓

	H1 (0)	H2 (1425)
C3 (50)	√	X
C4 (0)	--	√

(d) Interval 4 (55° - 50°C)

	H1 (150)	H2 (1425)
C3 (300)	1.1	X
C4 (0)	--	--

Match H1 & C3 ↓

	H1 (0)	H2 (1425)
C3 (150)	√	X
C4 (0)	--	--

(e) Interval 5 (50° - 35°C)

	H1 (0)	H2 (1425)
C3 (100)	--	X
C4 (0)	--	--

No Match Possible

(f) Interval 6 (35° - 30°C)

Figure 3.11 The enthalpy matrix representation allows the constrained problem table algorithm to be implemented conveniently. The enthalpy matrices and their revisions are shown above for the various temperature intervals. The match with the highest priority is underlined.

The match indices for a potential match (between streams H1 and C3) are obtained by counting the number of match options available to the cold stream (C3) and the hot stream (H1). As there is only one match choice for stream H1 and one match choice for stream C3, the match indices are written as 1.1 for the top left element in Figure 3.11a.

Step 2. Revision of the Enthalpy Matrix

On filling the entire enthalpy matrix, the highest priority is given to the element with the lowest match index for a cold stream. A match is completed by ticking off the stream with the lower enthalpy content; then, the enthalpy matrix is appropriately revised. In Figure 3.11a, there is no option but to tick off the match between streams H1 and C3 (with a heat load of 430 kW). The enthalpy contents in the revised matrix will be 0 kW (i.e., 430 − 430, for stream H1) and 430 kW (i.e., 860 − 430, for stream C3).

Step 3. Tabulation of the Match Information

For each interval, the match information may be tabulated (as in Table 3.3) for ready reference. For the first interval, the match of 430 kW between streams H1 and C3 leaves 430 kW still unsatisfied on the cold streams. There is no surplus heat in the interval.

Table 3.3 Match Information from Constrained PTA

Interval	Match	Load	Unsatisfied Cold	Surplus Hot
1	H1 & C3	430	430	0
2	H1 & (C3, C4)	70	175	0
3	H1 & C3	600		
3	H2 & C4	900	600	1500
4	H1 & C3	50		
4	H2 & C4	75	50	1425
5	H1 & C3	150	150	1425
6	No Match	--	100	1425

Units: load/enthalpy in kW.

Step 4. Establishment of Enthalpy Matrices for All Intervals

In the second interval, both the cold streams have the same priority (Figure 3.11b). Without deciding the actual match, a notional match between streams H1 and (C3, C4) may be made. On ticking off stream H1, the unsatisfied cold enthalpy is found to be 175 kW (i.e., 140 + 105 − 70). If streams H1 and C3 are matched, then the remaining loads will be 70 kW (on stream C3) and 105 kW (on stream C4). On the other hand, if streams H1 and C4 are matched, then the remaining loads will be 140 kW (on stream C3) and 35 kW (on stream C4). The enthalpies indicated in parenthesis in the revised matrix (after streams H1 and [C3, C4] are notionally matched) in Figure 3.11b are the maximum possible remaining.

In the third interval (Figure 3.11c), a match is first made between streams H1 and C3 (of 600 kW) and then between streams H2 and C4 (of 900 kW). There remains 600 kW of unsatisfied enthalpy on cold stream C3 (which must be satisfied by an external hot utility) and 1500 kW of surplus heat on hot stream H2 (which is cascaded to interval 4) because the match between streams H2 and C3 is forbidden.

The fourth interval (Figure 3.11d) is similar (qualitatively) to the third interval. Note that the 1500 kW cascaded from the third interval are included in the enthalpy of stream H2. In the fifth interval (Figure 3.11e), there is a straightforward match possible between streams H1 and C3 (of 150 kW). However, in the sixth interval (Figure 3.11f), there is no match possible (inspite of the 100 kW in stream C3 and the 1425 kW in stream H2 because of the constraint).

Step 5. Determination of the Minimum Utility Requirements

The last entry (1425 kW) in the "surplus hot" column in Table 3.3 gives the minimum cold utility required. Note that the "surplus hot" column is truly a cascade; on the other hand, the "unsatisfied cold" column is not a cascade and represents the enthalpies in each temperature interval. Therefore, the entries in the "unsatisfied cold" column (Table 3.3) must be summed to obtain the minimum hot utility requirement as 1505 kW.

3.5.3 The Maximum Possible Remaining Concept

The maximum possible remaining (MPR) concept was introduced, in passing, in Example 3.8 (see step 4, second temperature interval). This subsection dwells on the concept and its importance in detail. In addition to the two principles for the CPTA mentioned in Section 3.5.2, there are two more conditions (O'Young et al., 1988) which govern the MPR concept. They are

- heat must not be committed in any interval unless not doing so leads to the use of utility in that interval; and
- heat must be cascaded in a manner not to prejudice maximization of process-process heat transfer in the remaining temperature intervals.

When no match options exist, the MPR concept commits to matches in order to avoid usage of external hot utility. When match options exist, the MPR concept commits to match loads (and not particular matches). It does careful bookkeeping of the maximum possible heat remaining in each stream that may be cascaded to the other temperature levels, after maximization of heat recovery within the interval.

Instead of starting at the highest temperature interval and proceeding all the way to the lowest (as in Example 3.8), it is advantageous (O'Young and Linnhoff, 1989) to compute the MPR surplus for the region above and below the unconstrained pinch separately. The procedure to calculate the MPR hot surplus (for the above-pinch region) and the MPR cold surplus (for the below-pinch region) is demonstrated next.

Example 3.9 Calculation of MPR Surplus above and below Pinch

Rework Example 3.8 by calculating the MPR hot surplus and the MPR cold surplus separately. As before, the aim is to determine the minimum utility requirements for Case Study 4S1c at ΔT_{min} = 20°C when the match between streams H2 and C3 is forbidden.

Solution. Starting from the highest temperature interval, the above-pinch imbalance is computed to obtain the MPR hot surplus (called MPRH). Then, starting from the lowest temperature interval, the below-pinch imbalance is calculated to obtain the MPR cold surplus (called MPRC). Note that when doing MPRH calculations with the temperature-decreasing approach, prioritization is based on cold streams. However, when doing MPRC calculations with the temperature-increasing approach, prioritization is based on hot streams.

Step 1. Calculation of the Unconstrained Pinch

The pinch for the unconstrained problem at ΔT_{min} = 20°C is at 115°C (see Table 1.2). Thus, intervals 1 and 2 lie above the pinch and the remaining intervals below it.

Step 2. Calculation of MPR Hot Surplus (MPRH)

The calculations for the above-pinch region (i.e., the first two intervals) remain unchanged from the ones performed in Example 3.8 (Figures 3.11a and 3.11b), so MPRH = 0 kW as per the "surplus hot" column in Table 3.3 when only the first two temperature intervals are considered.

Step 3. Calculation of MPR Cold Surplus (MPRC)

The calculations for the below-pinch region (i.e., intervals 3 - 6) must be started from the lowest temperature interval.

In the sixth temperature interval, no match is possible (see Figure 3.12a). The cold surplus of 100 kW is cascaded to stream C3 in the fifth interval.

The fifth interval (Figure 3.12b) comprises only one possible match between streams H1 and C3 (of 150 kW). On ticking off stream H1, the cold surplus is 250 kW, which is cascaded to stream C3 in the fourth interval.

Both the cold streams have the same priority in the fourth interval (Figure 3.12c). As there is more one match choice available (for stream H1), the MPR concept is used. Considering its importance, a clear explanation for the MPR is provided separately in the next step.

	H1 (0)	H2 (0)
C3 (100)	--	X
C4 (0)	--	--

No Match Possible

(a) Interval 6 (35° - 30°C)

	H1 (150)	H2 (0)
C3 (400)	1.1	X
C4 (0)	--	--

Match H1 & C3 ↓

	H1 (0)	H2 (0)
C3 (250)	√	X
C4 (0)	--	--

(b) Interval 5 (50° - 35°C)

	H1 (50)	H2 (0)
C3 (350)	1.2	X
C4 (75)	1.2	--

Match H1 & (C3, C4) ↓

	H1 (0)	H2 (0)
C3 (350)	√	X
C4 (75)	√	--

MPR (C3, C4) = 375

(c) Interval 4 (55° - 50°C)

	H1 (600)	H2 (2400)
C3 (1550)	1.2	X
C4 (975)	2.2	2.1

MPR (C3, C4) = 2475
Match H2 & C4 ↓

	H1 (600)	H2 (1425)
C3 (1500)	1.1	X
C4 (0)	--	√

Match H1 & C3 ↓

	H1 (0)	H2 (1425)
C3 (900)	√	X
C4 (0)	--	√

(d) Interval 3 (115° - 55°C)

Figure 3.12 The enthalpy matrices (and their revisions) for determining the MPR cold surplus use a temperature-increasing approach. They start at the lowest temperature interval and proceed towards the unconstrained pinch, with the prioritization based on hot streams. The highest priority match is underlined.

Step 4. The Maximum Possible Remaining Concept

The concept ensures that no individual match commitments are made when options exist. This least commitment strategy, where the actual distribution of loads between the equal-priority streams is not predetermined, ensures that prejudices do not get built in and heat recovery does not get hindered in the other temperature intervals. Instead, the maximum possible heat that can remain for cascading is calculated for each stream (hot during MPRH calculations and cold during MPRC calculations) by a simple energy balance, after grouping the equal-priority streams. In this case (for the fourth interval), a notional match of 50 kW is made between streams H1 and (C3, C4), so MPR(C3, C4) = 350 + 75 − 50 = 375. If streams H1 and C3 are matched, then the remaining loads will be 300 kW (on stream C3) and 75 kW (on stream C4). On the other hand, if streams H1 and C4 are matched, then the remaining loads will be 350 kW (on stream C3) and 25 kW (on stream C4). The enthalpies indicated in parenthesis in the revised matrix in Figure 3.12c are the maximum possible remaining. Thus, only the upper limit on the heat that can be cascaded by each stream (hot in the case of MPRH calculations and cold in the case of MPRC calculations) is obtained. It turns out to be the minimum of the MPR and the enthalpy of that stream (i.e., 350 kW for stream C3 and 75 kW for stream C4). The MPR(C3, C4) value of 375 kW must be cascaded to the third interval. Thus, the cold streams in the MPR set are available in the next step of the matching process with maximum possible heat loads.

Now, consider the third interval. The enthalpies associated with the cold streams are 1550 kW (i.e., 20 [115 − 55] + 350 for C3) and 975 kW (i.e., 15 [115 − 55] + 75 for C4). Note that MPR (C3, C4) = 20 (115 − 55) + 15 (115 − 55) + 375 = 2475 kW. Using prioritization based on hot streams, stream H2 has the highest priority (Figure 3.12d). On matching it with stream C4, the value of MPR (C3, C4) becomes 1500 kW (i.e., 2475 − 975). The enthalpy associated with stream C3 also becomes 1500 kW (as no individual stream enthalpy can be higher than the MPR). Now, the only match possible is between streams H1 and C3 (of 600 kW). At this stage, the unsatisfied hot enthalpy is 1425 kW (on stream H2) and the surplus cold enthalpy is 900 kW (on stream C3). The analog of Table 3.3 is shown in Table 3.4, but for the region below the unconstrained pinch and by using the temperature-increasing approach (Figure 3.12). Therefore, MPRC = 900 kW as per the "surplus cold" column in Table 3.4.

The minimum hot utility requirement obtained by summing MPRC and the unsatisfied cold enthalpies above the pinch is 1505 kW (i.e., 900 + 430 + 175). The minimum cold utility requirement obtained by summing MPRH and the unsatisfied hot enthalpies below the pinch is 1425 kW (i.e., 0 + 1425).

It is observed that the utility targets may be obtained by starting either at the highest temperature interval (and using a temperature-decreasing approach) or at the lowest temperature interval (and using a temperature-increasing approach). This duality (O'Young et al., 1988) is important in multiple utility design and MER design for CHENs. This is discussed next, along with the significance of the MPRH and the MPRC.

Table 3.4 Match Information for Region Below Pinch

Interval	Match	Load	Unsatisfied Hot	Surplus Cold
6	No Match	--	0	100
5	H1 & C3	150	0	250
4	H1 & (C3, C4)	50	0	375
3	H2 & C4	975		
3	H1 & C3	600	1425	900

Units: load/enthalpy in kW.

3.5.4 Three Constrained Energy Penalty Components

It must be clear by now that the minimum utility requirements increase when constraints are imposed. In our example, the hot as well as the cold utility have increased by 900 kW. This difference in utility requirements, with reference to the unconstrained case, may be called (O'Young et al., 1988) the constrained energy penalty, QP.

The penalty occurs when the hot streams above the pinch are not allowed fully to heat the cold streams above the pinch. The resulting hot imbalance above the pinch is called the MPRH. A penalty also occurs when the cold streams below the pinch are not allowed to be heated completely by the hot streams below the pinch. The resulting cold imbalance below the pinch is called the MPRC. The MPRH may be satisfied either by a cold utility above the pinch or by cold process streams below the pinch. Similarly, the MPRC may be satisfied either by a hot utility below the pinch or by hot process streams above the pinch. To minimize the total utility requirements in CHENs, the feasible heat recovery between the MPRH and the MPRC must be maximized. The sacrosanct condition in unconstrained HENs of no heat transfer across the pinch does not hold for a maximum energy recovery CHEN.

In any case, both the MPRH and the MPRC ultimately lead to cross-pinch heat transfer, which is inevitable in CHENs. The total amount of cross-pinch heat flow exactly equals the increase in utility consumption (i.e., the penalty). The constrained energy penalty is thus made up of three components (O'Young and Linnhoff, 1989):

- QP_c -- penalty due to cold utility used above the pinch to satisfy the MPRH
- QP_h -- penalty due to hot utility used below the pinch to satisfy the MPRC
- QP_p -- penalty due to surplus hot streams above the pinch used to satisfy unsatisfied cold streams below the pinch (i.e., process-process heat transfer across the pinch).

The three energy penalty components are depicted on a general network in Figure 3.13. Note that $Q_{hu,min}$ and $Q_{cu,min}$ are the minimum hot and cold utility requirements for the unconstrained case.

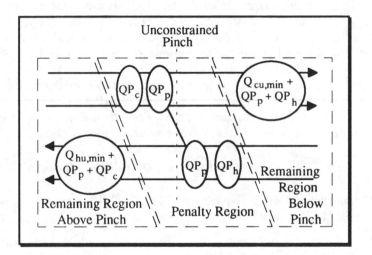

Figure 3.13 A schematic diagram to show the three constrained energy penalty components and the decomposition of the problem into three regions during synthesis of CHENs for maximum energy recovery.

A concise mathematical statement of the constrained energy penalty is

$$QP \equiv Q_{constrained} - Q_{unconstrained} = \text{Total cross-pinch heat flow}$$
$$= QP_p + QP_c + QP_h \tag{3.6a}$$

Linnhoff and O'Young (1987) have established upper and lower bounds for each of the three components. It may be argued that there are only two degrees of freedom available, if the total energy penalty (QP) is fixed as per Equation 3.6a. Consequently, QP_p may be traded off against QP_h. Alternatively, QP_p may be traded off against QP_c. In other words, the penalty components can contribute to varying extents to the total cross-pinch heat flow even at the same utility consumption level. This provides a vital tool in the hands of the designer: flexibility in design is available in terms of appropriate choice of utility and network capital cost. There are limiting values to these tradeoffs, for which the following relations hold:

$$QP = QP_{p,min} + QP_{c,max} + QP_{h,max} \qquad (3.6b)$$
$$QP = QP_{p,max} + QP_{c,min} + QP_{h,min} \qquad (3.6c)$$

Equation 3.6b may be referred to as the $QP_{p,min}$ condition, and Equation 3.6c as the $QP_{p,max}$ condition. Note that Trivedi (1988) has used the transshipment model to establish the upper and lower bounds on the energy penalty components by introducing a dummy cold utility stream and a dummy hot utility stream above and below the unconstrained pinch respectively.

A number of MER designs are possible between these two limiting conditions. More details on these tradeoffs and limits for the three component flows are covered in Section 4.5 on the evolution of CHENs.

Having established the constrained energy penalty and its causes, a CHEN for maximum energy recovery may now be systematically synthesized.

3.5.5 MER Networks with Forbidden Matches

The inevitability of cross-pinch heat flow in CHENs has two important ramifications in terms of MER network synthesis:

- A design for minimum utility should, in principle, simultaneously maximize the process-process heat exchange above the pinch, below the pinch, and across the pinch. Fortunately, this is not necessary and a sequential maximization strategy may be adopted because of the independence of the above- and below-pinch maximizations (see Linnhoff and O'Young [1987] for proof).

- The problem must be decomposed into three thermally-independent regions (not two, as in the unconstrained case). All the cross-pinch heat flow is concentrated in the penalty region (O'Young and Linnhoff, 1989), which comprises the MPRH and the MPRC identified in Example 3.9. The remaining regions (Figure 3.13) contain no cross-pinch heat transfer and may consequently be designed by the conventional PDM.

Example 3.10 Designing a CHEN for Maximum Energy Recovery

For Case Study 4S1c, design an MER network for a ΔT_{min} of 20°C. Note that the match between streams H2 and C3 is forbidden.

Solution. The general procedure for CHEN synthesis (O'Young and Linnhoff, 1989) is as follows. The process-process heat exchange above and below the pinch is separately maximized to obtain the MPRH and the MPRC (see Example 3.9). This is followed by the maximization of the process-process heat exchange between the MPRH and the MPRC. The enthalpy matrix at the pinch may be used for this purpose to identify the cross-pinch heat flows and to obtain a design for the penalty region. The remaining regions above and below the pinch may finally be designed by conventional techniques. The final result is an MER network (out of the many possible designs) at the $QP_{p,min}$ condition. In Chapter 4, the $QP_{p,min}$ design is used as a starting point to generate other possible designs and determine the optimum one. In fact, the $QP_{p,min}$ design provides an appropriate initialization as the cross-pinch heater is at its maximum load and lowest temperature level whereas the cross-pinch cooler is at its maximum load and highest temperature level. These are favorable conditions for evolution without causing ΔT_{min} violations (Linnhoff and Flower, 1978).

Step 1. Determination of the MPRH and the MPRC

The MPRH and the MPRC were determined in Example 3.9 using the CPTA and the MPR concept. The results obtained were 0 kW (for MPRH, on maximizing the heat recovery above the pinch) and 900 kW (for MPRC on stream C3, on maximizing the heat recovery below the pinch).

Step 2. Determination of the $QP_{p,min}$ Condition Using the Enthalpy Matrix for the Pinch Interval

The enthalpy matrix for the pinch interval (MPRH = 0, MPRC = 900) is shown below.

	H1 (0)	H2 (0)
C3 (900)	--	X
C4 (0)	--	--

$QP_{p,min}$ is given by the maximum heat recovery possible between MPRH and MPRC. In the above enthalpy matrix, there are no match options and consequently no heat recovery possible. Therefore, the $QP_{p,min}$ condition is defined by $QP_{p,min} = 0$, $QP_{c,max} = 0$, and $QP_{h,max} = 900$ (on stream C3).

Step 3. Design for the Penalty Region

All the cross-pinch heat flow must occur in the penalty region. One possible design for the penalty region is clearly and simply identified in terms of $QP_{p,min}$, $QP_{c,max}$, and $QP_{h,max}$ in step 2. For this example, the design is trivial. All it involves is a heater of 900 kW on stream C3 whose temperature changes from 60° to 105°C (see Figure 3.14).

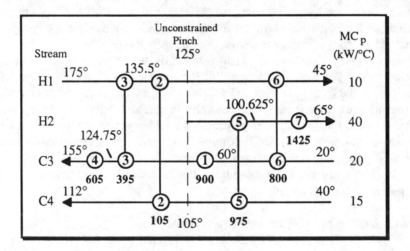

Figure 3.14 The design for the penalty portion may be combined with those for the remaining portions above and below the pinch to get a possible CHEN for maximum heat recovery. The above network is at the $QP_{p,min}$ condition. (Temperature in °C and load in kW)

Step 4. Design for the Remaining Regions

After the stream segment(s) involved in the penalty region are eliminated, the remaining regions are obtained. These have no cross-pinch heat transfer. If the PTA (in Section 1.1.2) is applied to the stream data for the remaining regions, the pinch is found to be unchanged in position (at 115°C). The utility values are consistent with those indicated in Figure 3.13 (i.e., hot utility = $Q_{hu,min} + QP_p + QP_c = 605 + 0 + 0 = 605$ kW and cold utility = $Q_{cu,min} + QP_p + QP_h = 525 + 0 + 900 = 1425$ kW).

The basic PDM (see Section 3.1) may be used to synthesize a network that meets these altered energy targets. The matches must be selected, keeping the forbidden ones in mind. The enthalpy matrix (Section 3.5.2), augmented with stream MC_p information, may be used for this purpose especially in the case of complex problems. The MC_p information allows quick checking of the feasibility criteria. The augmented enthalpy matrix for the remaining region above the pinch is given below:

MC_p		10	40
		H1 (500)	H2 (0)
20	C3 (1000)	1.2	X
15	C4 (105)	1.2	--

There are two equal-priority options: stream H1 may be ticked off against stream C3 or stream C4 may be ticked off against stream H1. The resulting enthalpy matrices for the two options are shown below. The MC_p information is not necessary any further, as what remains are away-pinch matches.

Match H1 & C3 ↓

	H1 (0)	H2 (0)
C3 (500)	√	X
C4 (105)	--	--

Match H1 & C4 ↓

	H1 (395)	H2 (0)
C3 (1000)	1.1	X
C4 (0)	√	--

Match H1 & C3 ↓

	H1 (0)	H2 (0)
C3 (605)	√	X
C4 (0)	√	--

For the case of a match of 500 kW between streams H1 and C3, the design is completed by placing two heaters of 500 kW (on stream C3) and 105 kW (on stream C4). For the case of a match of 105 kW between streams H1 and C4, the design is completed by using a match of 395 kW (between streams H1 and C3) and a heater of 605 kW (on stream C3). Thus, the remaining region above the pinch has two possible designs, both requiring 605 kW of hot utility as per the target.

Next, consider the augmented enthalpy matrix for the region below the pinch. It suggests a match between streams H2 and C4 (of load 975 kW) and a match between streams H1 and C3 (of load 800 kW). The design is completed with a cooler of 1425 kW (as targeted) on stream H2. The enthalpy matrices are shown below.

MC_p		10	40
		H1 (800)	H2 (2400)
20	C3 (800)	1.1	X
15	C4 (975)	--	1.1

Match H2 & C4 ↓

	H1 (800)	H2 (1425)
C3 (800)	1.1	X
C4 (0)	--	√

Match H1 & C3 ↓

	H1 (0)	H2 (1425)
C3 (0)	√	X
C4 (0)	--	√

Finally, the penalty portion is combined with the remaining portion above and below the pinch to yield the MER network (Figure 3.14). This is not the only possible network. Other networks are generated in Section 4.5 by systematic evolution exploiting the QP_p-QP_h and QP_p-QP_c tradeoffs.

For this example, the CHEN design is quite straightforward and could have been generated by inspection. However, the procedure above is systematic and powerful, especially for complicated problems.

3.6 OVERALL LOGIC OF MER NETWORK SYNTHESIS

The logic flow diagram for the synthesis of MER networks is shown in Figure 3.15. The synthesis procedure essentially depends on the type of problem. If there are forbidden matches, the constrained energy penalty must be determined and the enthalpy matrix approach used. If only one kind of utility (either hot or cold) is required, a network for the threshold problem may be synthesized by the fast matching algorithm using the hottest/highest heuristic. If the problem involves multiple utilities, then the utility pinches must be determined using balanced GCCs. A balanced grid must be utilized during network design if the temperatures of the utilities overlap those of the process streams. Finally, the problem is decomposed into separate subproblems recognizing the pinch division(s) across which there is no heat transfer. The design always begins at the pinch (the most constrained region), where the feasibility criteria are applicable. Stream splitting may be necessary to satisfy the number criterion and the MC_p criterion. In the region above the pinch, only hot utility is used. In the region below the pinch, only cold utility is used. For a multipinched problem, the region between the two pinches is in perfect enthalpy balance and requires no utility.

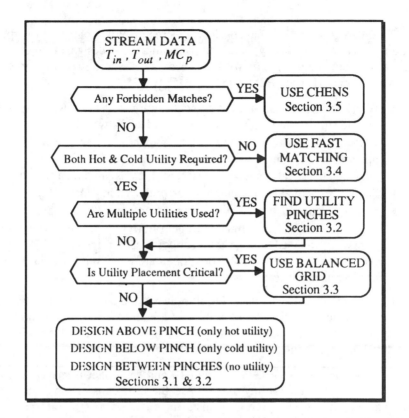

Figure 3.15 Depending on the type of problem (constrained, threshold, multiple utilities, or pinched), a suitable synthesis strategy must be adopted.

The designs in this chapter have focused on maximum energy recovery and have been based only on the energy targets. In the next chapter, the remaining targets (unit, area, shell, and cost) are utilized to evolve the MER design and obtain improved networks. As a large number of designs emerge for a given set of stream data, these need to be evaluated (see Chapter 4) using some criteria to decide on the final network.

QUESTIONS FOR DISCUSSION

1. Discuss how stream splits can be represented on a temperature-enthalpy plot by a parallelogram (whose diagonal corresponds to the stream before splitting and whose sides correspond to the split branches).

2. Do area targets (without considering stream splits) hold good for split-stream designs? Specifically, consider the split-stream design in Figure 3.2 and perform area targeting on a five-stream data set where stream H2 is split into two streams (H2a and H2b) with $20 \leq MC_{p,a} \leq 25$.

3. With regard to stream splits, comment briefly on the following aspects:
- Should stream splitting in general be avoided? Why?
- In industrial practice, how many split branches are reasonable from an operational and control viewpoint?
- How could a stream split be represented if it was more than two-way?
- Should isothermal mixing junctions be preferred in general?
- How can optimal branch flow rates be computed? For example in Figure 3.1, can the optimum value of $MC_{p,a}$ be determined that minimizes the total area (or total capital cost) for units 2 and 3? Does it matter whether split-stream flow rate optimization is done on an area basis or a capital cost basis?

4. In Example 3.1, heat loads for all matches are calculated first and the temperature differences later (in step 6). Is it a better strategy to calculate the temperature differences on both sides of the exchanger simultaneously with the heat duty? If ΔT_{min} violations occur, how can they be handled?

5. Draw the utility GCC in Figure 1.10 and show that it snugly fits the process GCC at optimum utility usage. Are the directions of the source and sink segments in the utility GCC opposite in direction to the corresponding segments in the process GCC? On this balanced GCCs plot, show that the utility pinches correspond to the points where the two GCCs touch.

6. With regard to multiple pinch designs, what are the advantages and disadvantages of the inverse pinch approach? Does it place matches that exhibit vertical heat transfer? Does it increase the subnetworks to be designed and require more evolution?

On the other hand, does examining match options simultaneously at both pinches (see Example 3.3) become difficult as the number of streams increases? Does this procedure place matches that exhibit greater crisscross heat transfer?

7. In Figure 3.4a, show that the inverse pinch point is at 134°C/108°C. Show that the design by the inverse pinch approach (Trivedi et al., 1989a) is as given below:

Below the utility pinch: HU2-C3 (140 kW) & HU2-C4 (65 kW).

Above the process pinch: H1-C3 (60 kW) & H1-C4 (40 kW).

The above design requires slipping stream H1 totally into the region between the inverse pinch and the process pinch. Why is this done? How can the additional unit be eliminated?

8. Define clearly the legal range of split ratios for the stream splits in Figures 3.4b and 3.4c. Can the mixing junctions be isothermal?

9. Discuss the tradeoffs in Example 3.4. For example, the reduction in stack losses is achieved through an additional stream split.

10. In Figure 3.5, how does the flue gas line change when
a) the preheated air temperature is changed, keeping the fuel flow constant; and
b) the fuel flow is changed, keeping the preheated air temperature constant?

11. Draw the balanced GCCs corresponding to the use of flue gas with air preheating (say, for Figure 3.7).

12. In Figure 3.8, the stack losses are negligible. If they were large, then the combustion air could be further preheated using the stack gas. Draw the utility GCC for this case. Does it cross the temperature axis and lie partly in the second quadrant of the temperature-enthalpy diagram? What are the tradeoffs (in terms of area, units, and fuel consumption)?

13. A widespread design practice is to use the same flowsheet after structural optimization in various countries. Of course, parametric optimization is performed to adapt it to local conditions. Is this a healthy practice? Can the structure of the heat recovery network change significantly from one location to another?

14. As a rule of thumb, the number of pinches in a general problem may be estimated to be one less than the number of utilities. Is this an expected relation between multiple pinches and multiple utilities?

15. Instead of using the hottest/highest heuristic from the hot end of the network, can the following (coldest/lowest) heuristic be used from the cold end of the network: match the cold stream having the lowest supply temperature with the hot stream having the lowest target temperature?

16. Give a proof/derivation for Equations 3.6b and 3.6c.

PROBLEMS

3.A MER Networks Using the Pinch Design Method
Design an MER network for the following case studies by recognizing the pinch division:

a) Case Study 4L2 at ΔT_{min} of 10°C
b) Case Study 4D1 at ΔT_{min} of 10°C
c) Case Study 4A2 (with temperatures reduced by 20°C) at ΔT_{min} of 20°C
d) Case Study 9A2 at ΔT_{min} of 19°C

Answers:

a) A possible MER design appears in Figure 2.22 (Linnhoff et al., 1982, p. 37).

$T_p = 85°C$	*Above Pinch*	*Below Pinch*
	H2 - C3 (90 kW)	H1 - C3 (90 kW)
	H1 - C4 (240 kW)	H2 - C3 (30 kW)
	HU - C3 (20 kW)	H2 - CU (60 kW)

b) A possible MER design appears in Figure 8.4-11 (Douglas, 1988, p. 244).

$T_p = 135°\ C$	*Above Pinch*	*Below Pinch*
	H1 - C3 (60 kW)	H2 - C3 (120 kW)
	H2 - C4 (240 kW)	H1 - CU (20 kW)
	H1 - C4 (50 kW)	H2 - CU (40 kW)
	HU - C4 (70 kW)	

c) A possible MER design appears in Figure 4 (Trivedi et al., 1990a, p. 605).

$T_p = 130°C$	*Above Pinch*	*Below Pinch*
	H1 - C3 (1600 kW)	H2 - C3 (4000 kW)
	H2 - C4 (1800 kW)	H1 - CU (2400 kW)
	H1 - C4 (2600 kW)	H2 - CU (1400 kW)
	HU - C4 (4000 kW)	

d) A possible MER design appears in Figure 5 (Ahmad & Linnhoff, 1989, p. 134).

$T_p = 150.5°C$	*Above Pinch*	*Below Pinch*
	H1 - C5 (6 MW)	H4 - C9 (0.2 MW)
	H2 - C9 (9.6 MW)	H1 - C5 (4.1 MW)
	H3 - C8 (1.74 MW)	H3 - C8 (4.86 MW)
	H3 - C6 (1.61 MW)	H4 - C6 (3.86 MW)
	H3 - C9 (0.25 MW)	H1 - C6 (3.56 MW)
	H1 - C9 (10.7 MW)	H4 - C7 (18.55 MW)
	HU - C5 (9.9 MW)	H1 - CU (4.34 MW)
	HU - C9 (11.25 MW)	H3 - CU (1.14 MW)
		H4 - CU (17.39 MW)

Note that the above-pinch match H1-C5 (max load = 15.9 MW) and the below-pinch match H4 -C6 (max load = 7.42 MW) have not been ticked off.

3.B MER Networks with Stream Splits

Design an MER network for the following case studies by recognizing the pinch division. Employ stream splits wherever necessary.

a) Case Study 4A1 at ΔT_{min} of 20°C
b) Case Study 4T1 at ΔT_{min} of 10°C

Answers:

a) A possible MER design appears in Figure 11a (Ahmad & Smith, 1989, p. 492).

$T_p = 90°C$

Above Pinch	Below Pinch
H1 - C4a (160 kW)	H2 - C3 (140 kW)
H2 - C4b (600 kW)	H1 - CU (320 kW)
HU - C3 (40 kW)	H2 - CU (460 kW)
HU - C4 (1340 kW)	

b) Two possible below-pinch designs appear in Figures 6a & 6c (Trivedi et al., 1990a, p. 606). $T_p = 165°C$

Above Pinch	Below Pinch	Below Pinch
H1 - C4 (280 kW)	H2a - C4 (200 kW)	H2a - C3 (500 kW)
HU- C4 (120 kW)	H2b - C3a (150 kW)	H2b - C4a (25 kW)
	H1 - C3b (350 kW)	H1 - C4b (175 kW)
	H2 - CU (250 kW)	H1 - CU (175 kW)
		H2 - CU (75 kW)

3.C MER Design for Multiple Utilities

Design an MER network for Case Study 4L3 at ΔT_{min} of 10°C. It is desired to raise VLP steam at 110°C from feed water at the same temperature (which acts as a cold utility). Assume that there is a hot utility available at a high enough temperature (say, 200°C) and there is a cold utility available at a low enough temperature (say, 15°C).

Answers:

The process pinch is at 145°C, and the utility pinch is at 115°C.

A possible MER design appears in Figure 2.39 (Linnhoff et al., 1982, p. 63).

Above Pinch (145°C)	Below Pinch (115°C)	Between Pinches
H1 - C3 (600 kW)	H1 - C4a (1600 kW)	H2 - C3 (900 kW)
HU - C3 (600 kW)	H2a - C4b (160 kW)	H2 - VLP (300 kW)
	H2b - C3 (1500 kW)	H1 - VLP (160 kW)
	H2 - CU (1540 kW)	H1 - C4 (440 kW)

3.D Balanced Grid for Appropriate Placement of Utility

Design MER networks for Case Study 5L1 at ΔT_{min} of 50°C. The hot utility is flue gas at 1500°C, which may be cooled to a minimum temperature (acid dew point) of 160°C. The fuel consumption must be minimized by extracting the maximum heat from the flue gas, i.e., by minimizing stack losses. The cold utility is cooling water (15° - 20°C), for which the ΔT_{min} may be 35°C. To understand the advantage of using a balanced grid,

   a) design an unbalanced network with utility streams not included; and
   b) design a balanced network with utility streams explicitly included.

Answers:

   a) A possible MER design (unbalanced grid) appears in Figure 7 (Linnhoff, 1986).

   $T_p = 125°C$ *Above Pinch*   *Below Pinch*

                      H1 - C4 (260 kW)   H1 - CU (70 kW)

                      H2 - C5 (275 kW)   H2 - CU (100 kW)

                      H2 - C3 (115 kW)

                      H1 - C3 (335 kW)

                      HU - C4 (300 kW)

   The flue gas can cool to only 336°C in the above design; thus, stack loss = 300 (336 − 160)/(1500 − 336) = 45.36 kW

   b) A possible MER design (balanced grid) appears in Figure 15 (Linnhoff, 1986). There are no new utility pinches.

   $T_p = 125°C$ *Above Pinch*   *Below Pinch*

                      H2 - C4 (390 kW)   H1 - CUa (70 kW)

                      H1 - C5a (145 kW)   H2 - CUb (100 kW)

                      HU - C5b (130 kW)

                      H1 - C3 (450 kW)

                      HU - C4 (170 kW)

   The flue gas can cool all the way to 160°C, resulting in no stack losses.

3.E Calculation of Efficiencies for Integrated Furnaces

Consider a furnace integrated with a background process (Linnhoff and de Leur, 1988). The air (not fuel) is preheated to the cold pinch temperature using the process heat below the pinch. The flue gas itself is cooled to the hot pinch temperature using the process heat above the pinch. Both the air and fuel are initially at ambient temperature. A global value of ΔT_{min} is specified (i.e., the same ΔT_{min} also applies to the flue gas and air). Assume $MC_{p,air}/MC_{p,flue}$ (in furnace) = R (given constant).

   a) Sketch qualitatively the GCC for the background process along with the profiles for the flue gas and air.
   b) Derive an equation for the fuel consumption.
   c) Calculate the integrated furnace efficiencies ($Q_{hu,min}/Q_{fuel}$) for pinch temperatures of 100°C, 300°C, and 500°C.
      Given: $T_{tf} = 1800°C$; $\Delta T_{min} = 20°C$; $T_a = 20°C$; $R = 0.9$.

Answers:

b) $Q_{fuel} = Q_{hu,min} \{1 + [\Delta T_{min} + (1-R)(T_p - \Delta T_{min}/2 - T_a)]/$
$$[T_{tf} - T_p - \Delta T_{min}/2]\}$$

c) 98.43% (for T_p= 100°C), 96.94% (for T_p= 300°C),
95.06% (for T_p= 500°C)

3.F Fast Matching Algorithm for Threshold Problem

Design an MER network for Case Study 4P1 at ΔT_{min} of 20°C (threshold problem) using the fast matching algorithm.

Answers:

A possible design appears in Figure 2B (Ponton & Donaldson, 1974, p. 2377).

Match 1: H1 - C4 (1680000 Btu/hr)   Match 6: H1 - C4 (3405500 Btu/hr)
Match 2: H2 - C4 (2450000 Btu/hr)   Match 7: H3 - C4 (6967500 Btu/hr)
Match 3: H1 - C4 (980000 Btu/hr)   Match 8: H2 - C4 (768750 Btu/hr)
Match 4: H3 - C4 (1732500 Btu/hr)   Match 9: H1 - C4 (1734500 Btu/hr)
Match 5: H2 - C4 (6781250 Btu/hr)   Match 10: HU - C4 (1150000 Btu/hr)

3.G Constrained Heat Exchanger Network

Design an MER network for Case Study 10O1 at ΔT_{min} of 20°C by decomposing the problem into a penalty region, a remaining problem above the pinch and a remaining problem below the pinch. The forbidden matches are H1- C5, H1 - C9, H1 - C10, H2 - C6, H2 - C9, H2 - C10, H4 - C5, and H4 - C10. The network may correspond to the $QP_{p,min}$ condition.

Answers:

The unconstrained energy targets are: $Q_{hu,min}$ = 34.58; T_p = 150°C.
Possible MER designs appear in Figure 24 (O'Young & Linnhoff, 1989).

Above Pinch	*Below Pinch*	*Penalty Region*
H3 - C9 (1.2)	H1 - C6 (7.35)	H1 - CU (13.22)
H2 - C5 (12)	H4 - C8 (4.8)	H2 - C5 (3)
H1 - C6 (1.68)	H4 - C7 (7.875)	HU - C5 (5 or 4.2)
H1 - C8 (1.8)	H3 - C10 (2.0)	HU- C10 (13 or 13.8)
HU - C5 (20)	H1 - CU (5.65)	2 possibilities here
HU - C9 (30.8)	H4 - CU (26.425)	

3.H MER Design with Stream Split and Multiple Utilities

Consider the stream data for Case Study 4S1 with the following utilities:

Utility	T_{in} (°C)	T_{out} (°C)	Cost ($/(kW.yr))
HU	200	199	130
CU1	25	100 (max)	8
CU2	15	25 (max)	10

Other Cost Data: Capital cost of exchanger (\$) = 30000 + 750 $A^{0.81}$
        Plant life: 5 yr   Rate of interest: 10%
Let ΔT_{min} = 30°C and h = 0.2 kW/(m^2 °C) for every stream.

a) Determine the usage levels of the three utilities for minimum operating cost. Assume the global ΔT_{min} of 30°C applies to utilities also.

b) Determine all the pinches (process as well as utility).

c) Obtain the total annual cost target for the case of countercurrent exchangers.

d) The following are the boundary temperatures for one of the enthalpy intervals used during area targeting: T_{hi} goes from 125° to 73.536°C, and T_{ci} goes from 40° to 95°C. Develop a spaghetti design for vertical heat transfer and establish the matchwise area distribution in this interval.

e) The spaghetti design in d) has too many units. Develop a design for the interval that has the minimum number of units as well as the minimum area.

f) Design a network for maximum energy recovery for the multiple utility - multiple pinch problem under consideration.

Answers:

a) 955 kW of HU; 825 kW (MC_p = 11.786 kW/°C) of CU1 (25° to 95°C); 50 kW (MC_p = 10 kW/°C) of CU2 (15° to 20°C)

b) Process pinch is at 110°C. Utility pinch is at 30°-35°C (pinched region).

c) A_c = 1246.2 m^2; TAC = 323.92 x 10^3 \$/yr

d) H1 - C3 (220 kW and 69.32 m^2), H2 - C3 (880 kW and 277.30 m^2),
   H1 - C4 (165 kW and 51.99 m^2), H2 - C4 (660 kW and 207.97 m^2),
   H1 - CU1 (129.64 kW and 40.85 m^2),
   H2 - CU1 (518.57 kW and 163.41 m^2)

e) Design 1: H1 - CU1 (514.64 kW)   H2 - C3 (1100 kW)
              H2 - C4 (825 kW)   H2 - CU1 (133.57 kW).
   Design 2: H1 - C3 (514.64 kW)   H2 - C4 (825 kW)
              H2 - C3 (585.36 kW)   H2 - CU1 (648.21 kW).
   Design 3: H1 - C4 (514.64 kW)   H2 - C3 (1100 kW)
              H2 - C4 (310.36 kW)   H2 - CU1 (648.21 kW).

f) A possible MER design featuring a three-way split on stream H2 and a two-way split on stream C3 is given below.

Above Pinch (110° C)	*Below Pinch (35° C)*	*Between Pinches*
H1 - C3 (500 kW)	H1 - CU2 (50 kW)	H2a - C3a (750 kW)
HU - C3 (700 kW)		H2b - C4 (825 kW)
HU - C4 (255 kW)		H2c - CU1 (825 kW)
		H1 - C3b (750 kW)

CHAPTER 4

Network Evolution and Evaluation

*What the caterpillar
calls the end of the world,
the master calls a
butterfly.*

Richard Bach

In the first two chapters, targets for energy, units, area, shells, and cost were developed. Based on the energy targets set by the closest approach temperature, maximum energy recovery (MER) networks were synthesized in Chapter 3. This chapter further evolves these networks to make them compatible with the remaining targets (units, area, shells, and cost).

The MER network is the most energy optimum network; however, it does not feature the minimum number of units (exchangers). The difference between the number of units in an MER design and the minimum possible number of exchangers gives the number of independent loops that cross the pinch in an MER network (see Section 1.3.4). However, the maximum number of loops present in an MER network is given by (Gibbs, 1969)

$$N_{l,max} = 2^{N_l} - 1 \tag{4.1}$$

where N_l is the number of independent loops as per Equation 1.29. These $N_{l,max}$ loops are combinations of the N_l independent loops.

Identification of loops in a complex network can be a challenging task. Su and Motard (1984) defined levels for loops, where nth level loops involve n hot and n cold streams. Obviously, the highest possible level is given by the minimum of the number of hot or cold streams. As loop identification is an important step in network evolution, it will be discussed later in a separate subsection.

As per Equation 1.25, each independent loop implies the existence of an additional exchanger in the network. So, if loops are identified, then the number of exchangers in the network may be reduced by loop breaking. However, an exchanger can be eliminated from a network only at the expense of an increased energy penalty (when the loop crosses the pinch). This leads to the well-known "capital-energy tradeoff," which must be evaluated.

The tradeoff, in fact, is more complicated as the capital cost involves units/shells as well as area (Equation 1.34). As pointed out by Colberg and Morari (1990), a three-way tradeoff between energy, area and units must be considered. A sophisticated tool called remaining problem analysis (RPA) is introduced in this chapter. RPA attempts to determine the match loads during network design consistent with the various targets.

When the heat transfer coefficients differ considerably, Rev and Fonyo (1991) suggest the use of the diverse pinch concept (see Section 2.2). The design of a network that meets the energy and area targets for such a minimum flux condition is also discussed in this chapter.

As a number of designs are possible for a given problem, evaluation of the various networks according to some criterion is necessary. The criterion considered here is the total annual cost. In reality, flexibility, controllability, operability, resiliency, and safety must also be considered.

4.1 LOOP BREAKING AND PATH RELAXATION

Classically, evolution of HENs is done by procedures like loop breaking and path relaxation that attempt to achieve the unit target by reducing the number of exchangers in the network. The result is a network where the heat transfer area is concentrated in fewer matches. Such networks often have lower capital costs since larger exchangers are cheaper on a per-unit-area basis.

4.1.1 Loops and Paths

A loop (see also Section 1.3) is a set of connections that can be traced through a network (via streams and units) that starts at one exchanger and returns to the same exchanger. A loop may pass through utilities of the same kind (which is obvious on a balanced grid). The definition of a loop (and a path) will be further clarified through Example 4.1.

A path is a connection between a heater and a cooler in a network. Heat loads can be "shifted" along a path as follows. Let a certain amount of heat be added to the hot utility. Then, the same amount must be subtracted from an exchanger on the same stream. As a result, the heat load on the other stream passing through that exchanger is reduced. This difference in heat load may be added to another exchanger. The process of shifting heat loads by addition and subtraction is continued till the cooler is reached and the same amount of heat is added to it. A similar "shifting" of heat loads is possible for loops and will be illustrated in Example 4.1.

4.1.2 Reducing Exchangers by Eliminating Loops

A number of heuristics have been proposed for breaking loops (Shah and Westerberg, 1975; Nishida et al., 1977; Linnhoff et al., 1982; Su and Motard, 1984; Pehler and Liu, 1984; Trivedi et al., 1990a). Attention is focused here on the three simple design heuristics suggested by Linnhoff et al. (1982) and Linnhoff and Hindmarsh (1983):

1. Break the loop with the exchanger having the smallest heat load.
2. Remove the smallest heat load from the loop.
3. Restore ΔT_{min} by path relaxation. Normally, ΔT_{min} is violated when a loop crossing the pinch is eliminated.

The above heuristics are demonstrated with the help of an example below.

Example 4.1 Elimination of Loops in a Network
Break all the loops that cross the pinch in the MER network in Figure 3.1 for Case Study 4S1, and restore ΔT_{min} to get a revised network.

Solution. First, the number of independent loops needs to calculated; then, the loops must be identified and eliminated. If a ΔT_{min} violation occurs, heat loads must be shifted along a path.

Step 1. Calculation of Number of Loops
There is one independent loop that crosses the pinch in the MER network as per Equation 1.29. Note that the number of units in Figure 3.1 is six (four exchangers, one heater, and one cooler) while the minimum number as per Equation 1.26 is five.

Step 2. Identification of Loops
As the network in Figure 3.1 is simple, the loop can be identified by inspection. For complex networks, methods based on graph theory and incidence matrix (see next subsection) may be useful. The loop in terms of the exchangers involved and their corresponding heat loads is shown in Table 4.1.

Table 4.1 Representation of a Loop

Units Involved	1	4	3	2	1
Heat Loads	500	520	560	1840	500
Shifts in Heat Loads	−500	+500	−500	+500	−500
Revised Heat Loads	0	1020	60	2340	0

Units: heat load in kW.

Unit 1 has the smallest heat load (500 kW), so the first two design heuristics of Linnhoff et al. (1982) suggest that it is advisable to break the loop by removing unit 1.

Step 3. Removal of the Smallest Heat Load

To remove unit 1, its heat load must be made zero by subtracting 500 kW. This 500 kW must be alternately added to and subtracted from the other units that constitute the loop. This operation is shown in Table 4.1 along with the revised heat loads.

Step 4. Check for ΔT_{min} Violation on Revised Network

The revised network on the removal of unit 1 and on calculating the new values of the intermediate temperatures is shown in Figure 4.1. It is noticed that the ΔT at the hot end of unit 2 is $-12°C$, indicating an infeasibility. There is also a ΔT_{min} violation on the cold end of unit 2, which is clear on representing the split as:

	Heat Capacity Flow Rate	Temperature at Cold End
Branch H2a	$MC_{p,a}$	$125 - 2340/MC_{p,a}$
Branch H2b	$40 - MC_{p,a}$	$125 - 60/(40 - MC_{p,a})$

where $20 \le MC_{p,a} \le 25$.

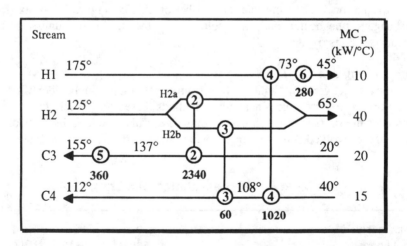

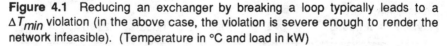

Figure 4.1 Reducing an exchanger by breaking a loop typically leads to a ΔT_{min} violation (in the above case, the violation is severe enough to render the network infeasible). (Temperature in °C and load in kW)

Step 5. Path Relaxation

The minimum approach temperature of 13°C needs to be restored by shifting heat loads along a path. A path is identified from a heater to a cooler that involves the "violating temperature" of 137°C. Such a path in terms of the units involved and their corresponding heat loads is shown in Table 4.2. Let the heat loads be shifted by an amount Q_s. To restore ΔT_{min} to 13°C, the outlet temperature on the hot end of exchanger 2 must be reset from 137° to 112°C. Therefore, a simple heat balance gives $Q_s = 500$ kW since $360 + Q_s = 20 (155 - 112)$. A further check may be performed at the cold end of unit 2 using $125 - (2340 - Q_s)/MC_{p,a} - 20 = 13$. This yields $Q_s = 500$ kW, as before. As per Table 4.2 (with $Q_s = 500$ kW), the new network involving five exchangers is shown in Figure 4.2 with revised heat loads and temperatures.

Table 4.2 Representation of Path for Energy Relaxation

Units Involved	5	2	3	4	6
Heat Loads	360	2340	60	1020	280
Shifts in Heat Loads	$+Q_s$	$-Q_s$	$+Q_s$	$-Q_s$	$+Q_s$
Revised Heat Loads	$360 + Q_s$	$2340 - Q_s$	$60 + Q_s$	$1020 - Q_s$	$280 + Q_s$

Units: heat load in kW.

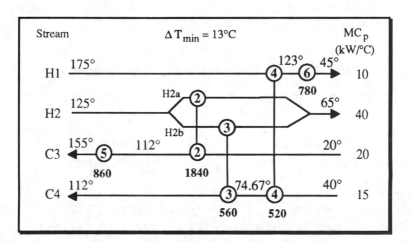

Figure 4.2 The infeasibility in the network in Figure 4.1 may be removed by energy relaxation along a path; however, this incurs an energy penalty on both hot and cold utility consumptions. (Temperature in °C and load in kW)

Compared to the MER network (whose hot utility requirement was 360 kW and cold utility requirement was 280 kW), an energy penalty of 500 kW is incurred for both hot as well as cold utility in this evolved network (Figure 4.2) featuring one less heat exchanger. Figure 4.2 has no more loops.

Step 6. Post-Analysis

It may be noted that Figure 4.2 may be obtained directly from Figure 3.1 by energy relaxation of 500 kW along the path involving units 5, 1, and 6. The number of exchangers can be further reduced to obtain the absolute minimum of four units by energy relaxation of 520 kW (to eliminate unit 4) using the path involving units 5, 2, 3, 4, and 6. At first glance, a network featuring only four units may appear odd since $N_{u,min} = 5$ as per Equation 1.26; however, this is consistent with Euler's Network Theorem (Equation 1.25) because there are two independent problems on eliminating unit 4 (one subsystem consists of streams H2, C3, and C4 and another of stream H1 only).

From the discussion so far, it is clear that a unit on a path may be eliminated on increasing both the heater and cooler duties by the load of the eliminated unit. However, a smaller penalty is incurred by loop breaking if the load of the eliminated unit can be redistributed to the other units forming the loop. This did not happen in Figures 4.1 and 4.2.

Can a loop be broken by removing a load other than the smallest heat load? Consider the same loop as in Table 4.1 but the removal of unit 4 (with a load of 520 kW) instead of unit 1 (with the smallest heat load). The procedure used above (Steps 3 through 5) is now repeated. As seen in Table 4.3, the revised heat loads are obtained by adding and subtracting 520 kW alternately from the various units that constitute the loop.

Table 4.3 Loop Breaking and Energy Relaxation

Units Involved in Loop	1	4	3	2	1
Heat Loads	500	520	560	1840	500
Shifts in Heat Loads	+520	−520	+520	−520	+520
Revised Heat Loads	1020	0	1080	1320	1020
Units Involved in Path	5	1	6		
Heat Loads	360	1020	280		
Shifts in Heat Loads	$+Q_s$	$-Q_s$	$+Q_s$	where $Q_s = 260$	
Revised Heat Loads	$360 + Q_s$	$1020 - Q_s$	$280 + Q_s$		

Units: heat load in kW.

It is observed that the revised network (Figure 4.3) on removal of unit 4 has a ΔT of $-13°C$ at the cold end of unit 1. To restore ΔT_{min} to $13°C$, the outlet temperature on the cold end of exchanger 1 needs to be reset from $73°$ to $99°C$. A heat balance over unit 6 gives $Q_s = 260$ kW since $280 + Q_s = 10\ (99 - 45)$.

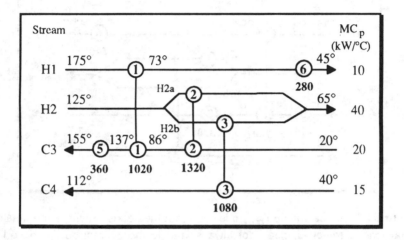

Figure 4.3 An exchanger with a load larger than the smallest heat load may be removed. The ΔT_{min} violation in the above case makes the network infeasible. (Temperature in °C and load in kW)

With revised loads as per Table 4.3, the final network (Figure 4.4) involves five units and incurs an energy penalty of 260 kW (compared to Figure 3.1) for both hot as well as cold utility.

On comparing with the MER network in Figure 3.1, the design in Figure 4.2 eliminates a unit of 500 kW, incurring an energy penalty of the same magnitude. On the other hand, the design in Figure 4.4 removes a unit of 520 kW at a significantly lower energy penalty (only 260 kW). Thus, in the above example, the heuristic recommending the removal of the smallest heat load does not lead to an evolved design optimal in terms of utility consumption. Such a conclusion was also made by Trivedi et al. (1990a), who proposed a systematic energy relaxation approach (discussed in detail in the next section) to account for loop-network interactions ignored by the simple "smallest heat load removal" heuristic.

Now, consider the removal of unit 3 (with a load of 560 kW) from Figure 3.1, using the same loop as in Table 4.1. Table 4.4 shows the revised heat loads obtained by adding and subtracting 560 kW alternately around the loop.

Interestingly, the load on unit 1 goes negative (–60 kW). This negative load may be removed by shifting 60 kW to unit 1 from the heater and cooler using the path involving units 5, 1, and 6.

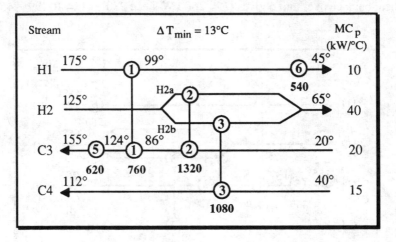

Figure 4.4 On restoring ΔT_{min}, it is seen that the heuristic (where the smallest heat load is removed) does not always lead to the optimal design based on energy consumption. Here, a unit of 520 kW has been eliminated by incurring an energy penalty of only 260 kW. (Temperature in °C and load in kW)

Table 4.4 Loop Breaking and Path Relaxations

Units Involved in Loop	1	4	3	2	1
Heat Loads	500	520	560	1840	500
Shifts in Heat Loads	–560	+560	–560	+560	–560
Revised Heat Loads	–60	1080	0	2400	–60
Units Involved in Path	5	1	6		
Heat Loads	360	–60	280		
Shifts in Heat Loads	$-Q_s$	$+Q_s$	$-Q_s$	where $Q_s = 60$	
Revised Heat Loads	300	0	220		
Units Involved in Path	5	2	D		
Heat Loads	300	2400	0		
Shifts in Heat Loads	$+Q_s$	$-Q_s$	$+Q_s$	where $Q_s = 560$	
Revised Heat Loads	$300 + Q_s$	$2400 - Q_s$	$0 + Q_s$		

Units: heat load in kW.

The resulting network (Figure 4.5) on removal of units 1 and 3 has a ΔT of $-15°C$ at the hot end of unit 2. To restore ΔT_{min} to 13°C, the outlet temperature on the hot end of exchanger 2 needs to be reset from 140° to 112°C. However, there is no path available for this purpose. A path may be created by seeding the network with a dummy exchanger (Grimes et al., 1982) denoted by D in Figure 4.5.

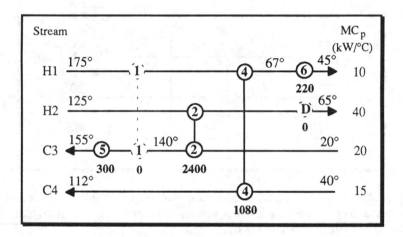

Figure 4.5 When a unit with a load larger than the smallest is removed during loop breaking, the loads on other units may become negative. The "negative load" exchanger (unit 1 above) may be removed if its load is reset to zero. (Temperature in °C and load in kW)

On carrying out conventional energy relaxation along a path involving units 5, 2, and D gives $Q_s = 560$ kW (from a heat balance over unit 2, i.e., $2400 - Q_s = 20$ [112 − 20]). With revised loads as per Table 4.4, the final network (Figure 4.6) involves five units and incurs a net energy penalty of 500 kW (compared to Figure 3.1) for both the hot as well as the cold utility.

Thus, two exchangers (unit 1 of 500 kW and unit 3 of 560 kW) and a stream split have been eliminated from the MER network in Figure 3.1. However, a new cooler (unit D of 560 kW) has been added in Figure 4.6.

In summary, it may be noted that the removal of a load larger than the smallest may lead to a lower utility penalty at times. However, it may result in negative loads on some of the remaining exchangers. The negative load may be reset to zero (and a second unit removed) by shifting loads around another loop or path involving the "negative load" exchanger. The seeding of a network with a dummy unit (with an initial heat load of zero) often proves beneficial, as it provides the necessary freedom for removing infeasibilities and restoring ΔT_{min} through additional paths and loops.

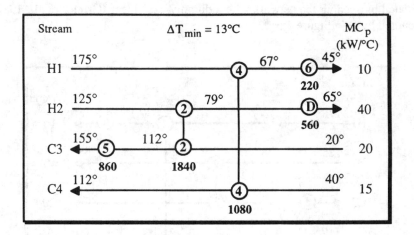

Figure 4.6 The introduction of a dummy exchanger (in this case, cooler D) creates a path to remove the infeasibility from the network in Figure 4.5. (Temperature in °C and load in kW)

Finally, it must be mentioned that restoration of ΔT_{min} after loop breaking need not always be done. In Figures 4.1, 4.3, and 4.5, the networks after loop breaking were thermodynamically infeasible because of a negative ΔT_{min}. If the network is feasible and there is no major violation in the ΔT_{min} given in the initial problem specification, it may be advisable at times not to restore ΔT_{min} nor incur an energy penalty. Note that the ΔT_{min} specification primarily ensures that the exchanger area does not become too high.

The following example illustrates a problem where restoration of ΔT_{min} is not necessary.

Example 4.2   Elimination of Loops Crossing Utility Pinches in Multiple Utility Designs

Break the loops that cross the utility pinches in the multiple utility design (ΔT_{min} = 20°C) given in Figures 3.4a and 3.4b for Case Study 4S1. It is not necessary to restore ΔT_{min} if no major violations occur.

Solution. The existence of multiple pinches in multiple utility designs leads to the final network having too many units. Some simplification is desirable to make the design practical.

Step 1. Simplification of the Design above the Process Pinch

If the utility pinch at 125°C is eliminated in Figure 3.4a, then only four units are necessary above the process pinch. The loop involving units 1, 7, 9, and 8 may be broken to reduce one unit (Table 4.5). It is best to remove unit 9 as it has the smallest load (5 kW). The network (Figure 4.7) on eliminating unit 5 has a small violation at the hot end of unit 7, where the ΔT is 19.75°C (instead of the required ΔT_{min} of 20°C).

Table 4.5 Shifting Loads around a Loop

Units Involved	1	7	9	8	1
Heat Loads	400	200	5	100	400
Shifts in Heat Loads	−5	+5	−5	+5	−5
Revised Heat Loads	395	205	0	105	395

Units: heat load in kW.

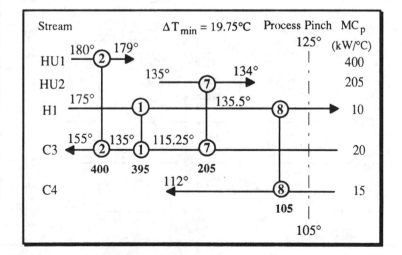

Figure 4.7 The ΔT_{min} need not be restored after loop breaking in cases where the violation is not serious. Thus, no utility penalty is incurred. (Temperature in °C and load in kW)

If the ΔT_{min} is not restored, then a ΔT_{min} penalty has been incurred while reducing the number of units. If the ΔT_{min} is restored, then a utility penalty has been incurred in reducing units (energy-capital tradeoff). Therefore, the number of units may be reduced at the sacrifice of utility (keeping ΔT_{min} fixed) or at the sacrifice of ΔT_{min} (keeping utility usage fixed).

Note that the ΔT_{min} in Figure 4.7 may be restored to 20°C by simply shifting 5 kW from HU2 to HU1. In general, simplification by reducing units in such designs requires sacrificing ΔT_{min} or load/level of intermediate utility.

Step 2. Simplification of the Design below the Process Pinch

If the utility pinch at 110°C is eliminated from Figure 3.4b, then only five units are required below the process pinch, so there is scope to eliminate three units from Figure 3.4b by breaking the three first-level loops. This is done simply by adding the load of one unit to the other. Thus, 50 kW from unit 13 is added to unit 5. Next, 50 kW from unit 12 is added to unit 4. Finally, 75 kW from unit 11 is added to unit 3. Figure 4.8 shows the resulting network. The ΔT_{min} violation of 1.875°C may be restored by shifting 75 kW from CU2 to CU1; however, VLP steam-raising opportunity is then lost. The small ΔT_{min} violation may be accepted, considering the design gains in terms of reduction of three units, removal of a stream split, and steam-raising ability.

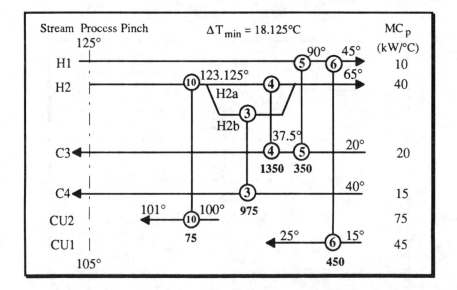

Figure 4.8 The above network is obtained on simplification from Figure 3.4b by breaking three first-level loops. A ΔT_{min} violation of 1.875°C must be tolerated to allow for VLP steam-raising. (Temperature in °C and load in kW)

4.1.3 Identification of Loops and Paths

Identification of all the loops by mere visual inspection is not an easy task in complicated networks with many streams and stream splits. A loop identification algorithm for HENs was given by Su and Motard (1984) based on a similar algorithm proposed by Forder and Hutchison (1969) for flowsheets. Pethe et al. (1989) proposed a technique for the purpose by representing the network in terms of a simple incidence matrix. Algorithms based on graph theory (Paton, 1969; Trivedi et al., 1990b) may also be used and may be most appropriate.

Example 4.3 Identification of Loops Using Incidence Matrix
Develop an MER network for Case Study 4S1 at ΔT_{min} = 20°C, and identify all the loops crossing the pinch in it. Use an incidence matrix representation for the purpose.

Solution. Figure 4.9 shows a possible MER design for Case Study 4S1 at ΔT_{min} = 20°C. It may be developed by the pinch design method as in Example 3.1. For the stream split, it may be noted that $20 \leq MC_{p,a} \leq 25$ and isothermal mixing is not possible. The MER network is expected to contain two independent loops (based on the unit targeting results in Table 1.11) and a maximum of three loops (as per Equation 4.1).

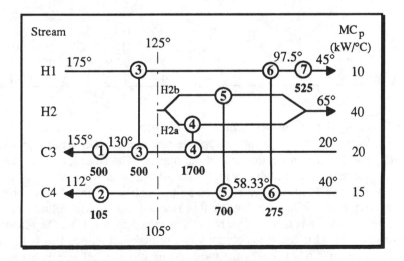

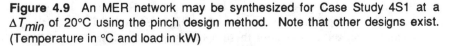

Figure 4.9 An MER network may be synthesized for Case Study 4S1 at a ΔT_{min} of 20°C using the pinch design method. Note that other designs exist. (Temperature in °C and load in kW)

The network may be represented by an incidence matrix as suggested by Pethe et al. (1989). After row transformations (i.e., normalizing the diagonal), a basis of column vectors is obtained. The loops may then be identified as linear combinations of the basis vectors that form the remaining columns.

Step 1. Generation of the Incidence Matrix

A network having N_s streams and N_u units is represented as an N_s x N_u incidence matrix. The rows represent the streams (both process and utility) while the columns represent the units (process stream exchangers as well as heaters/coolers). As each unit removes heat from some stream and supplies it to another, the matrix may be filled columnwise by placing +1 in the row from which heat is removed and −1 where it is supplied. Table 4.6 shows the incidence matrix for the MER network in Figure 4.9. (The purpose of column P will be explained later.) Table 4.6 assumes a single hot utility and a single cold utility. Specifically, the heaters are assumed to share the same hot utility source. Clearly, in the case of multiple utilities, every additional utility will result in an additional row in the matrix.

Table 4.6 Initial Incidence Matrix

Streams/Units	1	2	3	4	5	6	7	P
H1	0	0	+1	0	0	+1	+1	0
H2	0	0	0	+1	+1	0	0	0
C3	−1	0	−1	−1	0	0	0	0
C4	0	−1	0	0	−1	−1	0	0
HU	+1	+1	0	0	0	0	0	+1
CU	0	0	0	0	0	0	−1	−1

Step 2. Calculation of Number of Independent Loops

The next step is to perform appropriate row transformations to normalize the diagonal and determine the rank of the matrix. The basic idea is to detect the first nonzero entry in each row and make all entries below it zero.

The row operations to be performed on the matrix in Table 4.6 are briefly summarized below. On scanning the first row, +1 is found in column 3. To make all the entries below it zero, row H1 is added to row C3. On scanning the second row, +1 is found in column 4; so, to make all entries below it zero, row H2 is added to row C3. For the third and fourth rows, all entries below −1 in columns 1 and 2 may be made zero by adding rows C3 and C4 to row HU. At this stage, +1 appears in column 7 for row HU, so row HU is added to row CU. The normalized matrix after these operations is shown in Table 4.7, and its rank R_m is 5. Note that the columns in Table 4.7 may be re-ordered to yield a upper triangular matrix (not shown).

The number of independent loops is given by (Pethe et al., 1989)

$$N_l = N_u - R_m \qquad (4.2)$$

In this example, $N_l = 7 - 5 = 2$.

Step 3. Identification of the Loops

From Table 4.7, it is observed that $\{1, 2, 3, 4, 7\}$ represents a set of basis vectors in terms of the unit labels. One way to determine the basis vectors is to locate the columns corresponding to the first nonzero entry in each row. The remaining N_l columns can be expressed as linear combinations of these basis columns. This yields the two loops in the network as shown in the top half of Table 4.8. They can be verified by visual inspection of Figure 4.9.

Table 4.7 Final Incidence Matrix

Streams/Units	1	2	3	4	5	6	7	P
H1	0	0	+1	0	0	+1	+1	0
H2	0	0	0	+1	+1	0	0	0
C3	−1	0	0	0	+1	+1	+1	0
C4	0	−1	0	0	−1	−1	0	0
HU	0	0	0	0	0	0	+1	+1
CU	0	0	0	0	0	0	0	0

Table 4.8 Identification of Loops through Basis Vectors

Basis Columns $\{1, 2, 3, 4, 7\}$

Other Columns (as linear combinations)	*Loops*
Column 5 = Column 4 − Column 1 + Column 2	5 4 1 2 5
Column 6 = Column 3 − Column 1 + Column 2	6 3 1 2 6

Basis Columns $\{1, 3, 4, 5, 7\}$

Other Columns (as linear combinations)	*Loops*
Column 2 = Column 5 − Column 4 + Column 1	2 5 4 1 2
Column 6 = Column 5 − Column 4 + Column 3	6 5 4 3 6

Numbers in the above table refer to labels of units in Figure 4.9.

As the set of basis vectors is in general not unique, dependent loops may be identified. Suppose 2 is replaced by 5 in the basis columns (see bottom half of Table 4.8); then, an additional loop is identified. In summary, there are three loops [5 4 1 2 5], [6 3 1 2 6], and [6 5 4 3 6], of which two are independent.

Step 4. Identification of Paths

A path may be treated as a loop, provided a "pseudo link" is visualized between a heater and a cooler (Patel, 1993). The pseudo link is represented by the column vector P in Table 4.6. Then, the task of identifying paths reduces to finding loops (as done in the previous steps) that include column P. Table 4.9 shows four paths that exist in Figure 4.9.

Table 4.9 Identification of Paths through Incidence Matrix

Column P (as linear combination)	Paths
Col P = Col 7 − Col 3 + Col 1	7 3 1
Col P = Col 7 − Col 3 + Col 4 − Col 5 + Col 2	7 3 4 5 2
Col P = Col 7 − Col 6 + Col 2	7 6 2
Col P = Col 7 − Col 6 + Col 5 − Col 4 + Col 1	7 6 5 4 1

Numbers in the above table refer to labels of units in Figure 4.9.

Example 4.4 Identification of Loops Using Graph Theory

Identify all the cross-pinch loops in Figure 4.9 using graph theory.

Solution. A graph, where the nodes represent exchangers, is generated (see Figure 4.10) by examining the connections for each exchanger. This graph may then be used for identifying loops (Gibbs, 1969; Paton, 1969; Patel, 1993). It is convenient to define the following four sets: N is the set of nodes (exchangers), E is the set of edges (connections), X is the set of unexamined nodes, and G is the set of nodes in the graph. At the start, the set X is equal to set N, and set G is empty.

Step 1. Definition of the Initial Sets

In the context of Figure 4.9,

$N = \{1, 2, 3, 4, 5, 6, 7\}$; and
$E = \{1\text{-}2, 1\text{-}3, 1\text{-}4, 2\text{-}5, 2\text{-}6, 3\text{-}4, 3\text{-}6, 3\text{-}7, 4\text{-}5, 5\text{-}6, 6\text{-}7\}$.

This information appears in the first row of Table 4.10, which shows how the sets dynamically change as the graph is generated.

Step 2. Selection of a Node

A node (say, z) is to be selected for examination. How is it selected? Obviously, an unexamined node in the graph is a candidate. The last element of set G may be selected, provided it also belongs to set X. If not, the last but one element is chosen. The process is repeated till no z is available in set G. In this case, node z is chosen from set X. When no z is available in set X too, the graph generation is complete. On selecting node z, it is removed from set X and added to the graph.

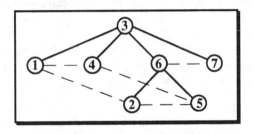

Figure 4.10 A graph may be generated with the exchangers forming the nodes; then, graph theory allows identification of loops. Note that edges 4-5 and 2-6 do not actually cross. The diagram should be viewed in three dimensions.

Table 4.10 Sets Used in Identification of Loops by Graph Theory

X	E	G
{1,2,3,4,5,6,7}	{1-2,1-3,1-4,2-5,2-6,3-4,3-6,3-7,4-5,5-6,6-7}	{ }
{1,2,4,5,6,7}	{1-2,1-4,2-5,2-6,4-5,5-6,6-7}	{3,1,4,6,7}
{1,2,4,5,6}	{1-2,1-4,2-5,2-6,4-5,5-6} (GL)	{3,1,4,6,7}
{1,2,4,5}	{1-2,1-4,2-5,4-5}	{3,1,4,6,7,2,5}
{1,2,4}	{1-2,1-4} (GL twice)	{3,1,4,6,7,2,5}
{1,4}	{1-4} (GL)	{3,1,4,6,7,2,5}
{1}	{ } (GL)	{3,1,4,6,7,2,5}
{ }	{ }	{3,1,4,6,7,2,5}

Numbers in the above table refer to labels of units in Figure 4.9. The node examined at each stage is underlined. (GL) indicates the "get_loop" operation performed at that stage.

As set G is empty at the start (the graph is yet to be generated), the node must be selected from set X. It is preferable to choose a node (exchanger) with many connections for the root to enable rapid generation of the graph.

If unit 3 is chosen as the root node, then $X = \{1, 2, 4, 5, 6, 7\}$ and $G = \{3\}$.

Step 3. Examination of a Node

All the edges for a node are identified. As each edge $(z\text{-}w)$ is examined, it is removed from set E. If $w \notin G$, then node w is added to set G. If $w \in G$, then the "get_loop" operation (described below) is carried out.

For node 3, the edges are 3-1, 3-4, 3-6, and 3-7. On removing these edges from set E, the result is $E = \{1\text{-}2, 1\text{-}4, 2\text{-}5, 2\text{-}6, 4\text{-}5, 5\text{-}6, 6\text{-}7\}$. The nodes 1, 4, 6, and 7 are included in set G (as they do not already belong to it). Therefore, $G = \{3, 1, 4, 6, 7\}$. The second row of Table 4.10 contains this information.

Step 4. Examination of Next Node

As node 7 belongs to both sets G and X, it may be chosen to be the next node for examination. Removal of edge 7-6 gives $E = \{1\text{-}2, 1\text{-}4, 2\text{-}5, 2\text{-}6, 4\text{-}5, 5\text{-}6\}$. As node 6 already belongs to set G, the "get_loop" operation is performed.

Step 5. The "Get_Loop" Operation

At this stage, an edge $z\text{-}w$ has been identified such that $w \in G$. A connection $w\text{-}r$ is next established, where r is a node on a branch of the graph. Finally, a set of connections is traced between r and z (say, $r\text{-} \ldots\ldots \text{-}z$). The candidate loop is obtained by adding the above three connections, i.e., candidate loop $= (z\text{-}w) + (w\text{-}r) + (r\text{-} \ldots\ldots \text{-}z) = z\text{-}w\text{-}r\text{-} \ldots\ldots \text{-}z$. If the number of units that constitute the candidate loop is even, it is accepted as a true loop. Otherwise, it is rejected since all the units lie on the same stream.

For our example, candidate loop $= (7\text{-}6) + (6\text{-}3) + (3\text{-}7) = 7\text{-}6\text{-}3\text{-}7$. As the number of units is odd (three), it is rejected because units 3, 6, and 7 all lie on the same stream (H1).

Step 6. Examination of Remaining Nodes

The remaining nodes are systematically examined in the same manner. Node 6 belongs to both sets G and X, so it may examined next. Removal of edges 6-2 and 6-5 gives $E = \{1\text{-}2, 1\text{-}4, 2\text{-}5, 4\text{-}5\}$. The nodes 2 and 5 are included in set G (as they do not already belong to it), so $G = \{3, 1, 4, 6, 7, 2, 5\}$ as shown in the third row of Table 4.10.

Node 5 is examined next, and edges 5-2 and 5-4 are removed from set E to give $E = \{1\text{-}2, 1\text{-}4\}$. As nodes 2 and 4 already belong to set G, the "get_loop" operation is performed. When edge 5-2 is considered, candidate loop $= (5\text{-}2) +$

(2-6) + (6-5) = 5-2-6-5. As the number of units is odd (three), it is rejected when it is noted that units 5, 2, and 6 all lie on the same stream (C4).   When edge 5-4 is considered, candidate loop = (5-4) + (4-3) + (3-6-5) = 5-4-3-6-5. As an even number of units are involved, loop [5 4 3 6 5] is the first loop detected.

Node 2 is examined next, and edge 2-1 is removed from set E. As node 1 already belongs to set G, the "get_loop" operation is performed to obtain candidate loop = (2-1) + (1-3) + (3-6-2) = 2-1-3-6-2. Loop [2 1 3 6 2] is another loop detected.

As nodes 7 and 6 have already been examined, node 4 is examined next. When edge 4-1 is considered, the "get_loop" operation is performed to obtain candidate loop = (4-1) + (1-3) + (3-4) = 4-1-3-4. As the number of units is odd, it is rejected because units 4, 1, and 3 all lie on the same stream (C3).

Lastly, node 1 is selected. As there are no edges associated with it, there is no examination necessary. As set X is empty, the graph generation is complete. The two loops detected in Figure 4.9 are [5 4 3 6 5] and [2 1 3 6 2] as seen on the final graph in Figure 4.10.

The above algorithm is suitable for computer implementation. It may be used to detect all loops, both dependent and independent, by repeated application of the procedure for all elements of N as the root nodes. Duplicate and invalid loops may be rejected, and only true loops stored. In Figure 4.9, these loops are [5 4 3 6 5], [2 1 3 6 2] and [2 1 4 5 2].

Now that sound methods for loop identification have been established, systematic procedures for energy relaxation that minimize the energy penalty during loop breaking are discussed next.

4.2 SYSTEMATIC ENERGY RELAXATION APPROACH

While a network is being evolved, a reasonable objective is to incur the least energy penalty. This is not always achieved if the simple heuristic of removing the smallest heat load is strictly adhered to (see Example 4.1). Trivedi et al. (1990a) have proposed a systematic energy relaxation method which looks into the loop-network interaction to minimize the energy penalty.

4.2.1 Classifying Exchangers Using Loop-Network Interaction

Loops cannot be treated in isolation, but must be considered in the context of the overall network. The analysis begins by identifying those exchangers that cause a hindrance to energy-optimum loop breaking and those that are favorable. Exchangers may, on this basis, be classified into three types (Trivedi et al., 1990a):

Type A

A Type A unit has the temperature difference at one end equal to ΔT_{min}, and the temperature of one its supply streams is fixed. If the other stream enters the unit on the other side, then its temperature must not be affected by loop breaking. The heat load of Type A units can only decrease. They often lie next to the pinch. In Figure 4.9, unit 4 is Type A.

Type B

A Type B unit is necessary for one of the streams passing through it to achieve its target temperature; hence, a Type B unit cannot be totally eliminated although its heat load could increase or decrease during the relaxation process. In Figure 4.9, units 4 and 5 together are Type B because they help stream H2 achieve its target temperature.

Type C

A Type C unit is capable of partly absorbing the heat load that is transferred whenever some (other) unit is eliminated. The remaining heat load is usually transferred along a path and defines the lower bound on the energy penalty to be incurred. In Figure 4.9, unit 6 is Type C because it is a "slack" exchanger in terms of the ΔT_{min} constraint at both ends. It can absorb a maximum of 375 kW, i.e., 10 (97.5 − 60).

After the exchangers are classified into their respective types, an attempt may be made to eliminate some of them.

4.2.2 Loop-Network Interaction and Load Transfer Analysis

LOop Network Interaction and load Transfer Analysis (abbreviated LONITA) was proposed by Trivedi et al. (1990a) to help identify units which on removal incur the least energy penalty. The order in which loops must be considered is not obvious in complicated problems. Some criteria for loop selection are enumerated below:

1. Break all lower level loops first (Su and Motard, 1984).
2. Break all process-process loops (PPLs) at all the levels, at this stage. (Note that PPLs pass through process streams only.)
3. Break process-utility loops (PULs) next. (Note that PULs pass through both process and utility streams.)
4. Break locked loops (LLs) finally, as they necessitate path relaxation. (Note that LLs have two Type A units placed adjacently in the loop. A decrease in the load of one unit leads to an undesirable increase in the load of the other unit when the loop is broken.)

Just as the conventional heuristic of removal of the smallest heat load results at times in designs that are suboptimal in energy consumption (as in

Example 4.1), the Su and Motard heuristic of giving priority to lower-level loops during loop breaking does not always yield the minimum energy penalty. In that sense, these are mere heuristics to guide the loop breaking process and must be judiciously used. For example, the rationale for giving priority to PPLs over PULs is to avoid early elimination of useful paths during loop breaking. Note that paths not only allow restoration of ΔT_{min}, but help positively in the controllability, flexibility, and resiliency of networks. To gain a proper understanding of LONITA, an example is considered next.

Example 4.5 Systematic Energy Relaxation Using LONITA
For the MER network in Figure 4.9, reduce the number of exchangers in the network by applying LONITA and selecting appropriate loops for breaking.

Solution. First, the loops in the network are identified and classified using LONITA. They are then broken using the criteria listed above.

Step 1. Classification of Units
Based on the discussion in Section 4.2.1, the following types of units are identified in Figure 4.9:
   Type A: unit 4 (load can only decrease)
   Type B: units 4, 5 (both units together cannot be eliminated)
   Type C: unit 6 (slack exchanger that can absorb a maximum of 375 kW)

Step 2. Identification and Classification of Loops
Based on the loop identification algorithms (see Section 4.1.3), a total of three loops are detected in Figure 4.9. These are classified below:
   Loop [5 4 3 6 5]   Level 2   Type PPL
   Loop [2 1 3 6 2]   Level 2   Type PUL
   Loop [2 1 4 5 2]   Level 2   Type PUL

Step 3. Evolution of MER Network to MNU Network
All loops are of the same level, so the PPL [5 4 3 6 5] is broken first, gaining precedence over the PULs. A positive (+) sign is attached to units whose loads increase during the loop breaking, and a negative (−) sign to units whose loads decrease. The signs using this convention may be assigned based on Type A units. As unit 4 is of Type A, it is assigned a (−) sign. In a loop, the (+) and (−) signs alternate to obey the heat balances. Therefore, LONITA suggests the following signs for the loop: [5(+) 4(−) 3(+) 6(−) 5(+)]. It suggests units with a (−) sign as potentially good candidates for elimination to avoid large energy penalties. Thus, LONITA has pruned the set of units for possible elimination to half the size.

On eliminating unit 6 (of load 275 kW), a violation occurs at the cold end of unit 3 (where $\Delta T = 6.25°C$). The path [1 3 7] may be used to restore ΔT_{min} to 20°C by incurring an energy penalty of 137.5 kW. The design with six units is shown in Figure 4.11.

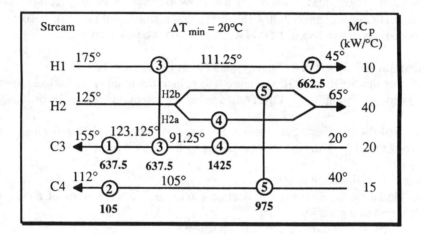

Figure 4.11 The evolved design with six units is obtained by eliminating unit 6 from the MER network in Figure 4.9. An energy penalty of 137.5 kW is incurred. (Temperature in °C and load in kW)

Note that unit 5 in the remaining loop [2 1 4 5 2] is Type A. Furthermore, careful examination of unit 4 shows that its load can only decrease. This is because the temperatures of both its supply streams are fixed, and there is no freedom on the temperature of 91.25°C since the driving force at the cold end of unit 3 is ΔT_{min}. Also, units 1 and 2 are Type B and cannot be eliminated. As the loads on both units 4 and 5 can only decrease, [2 1 4 5 2] is a locked loop. Therefore, it is not easy to evolve further in a meaningful manner. Elimination of any unit from Figure 4.11 is possible only at a high energy penalty. Consider the removal of unit 4 from the loop [2 1 4 5 2]. Unit 2 ends up with a negative load of −1320 kW (Figure 4.12). A dummy cooler (D) needs to be introduced on stream H2 to allow relaxation along the path [2 5 8]. It is not sufficient to reset the load on unit 2 to zero. Because it is a Type B unit that cannot be eliminated, its load must be reset to 105 kW by shifting 1425 kW along the path [2 5 8]. Also, unit 7 may be eliminated, as unit 3 is capable of absorbing more load. This is done by shifting −662.5 kW along the path [1 3 7]. The final design (Figure 4.13) for the minimum number of units

(MNU) has no stream split and saves two units compared to the MER network. However, it incurs a large energy penalty of 900 kW.

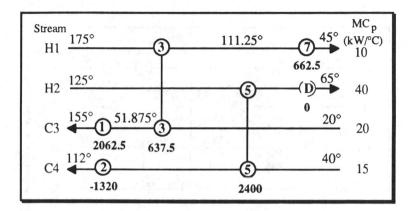

Figure 4.12 The negative heat load on unit 2 (Type B) must be reset to 105 kW (not zero) by introducing a dummy cooler. Also, unit 3 is a slack exchanger that can absorb more load. (Temperature in °C and load in kW)

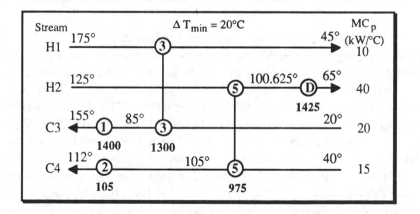

Figure 4.13 An MNU design is shown with five units and no stream splits. It incurs a penalty of 900 kW, compared to the MER network in Figure 4.9. (Temperature in °C and load in kW)

To investigate the possibility of a lesser energy penalty, one option is to restart the evolution by breaking the loop [2 1 3 6 2]. As it does not contain any Type A units, the conventional relaxation heuristic of removal of the smallest unit may be used. Removal of unit 2 (of load 105 kW) from the MER network (Figure 4.9) yields Figure 4.14.

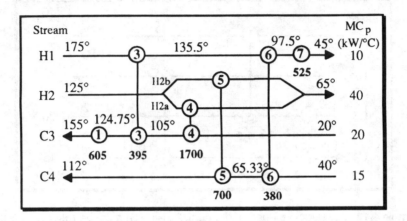

Figure 4.14 A violation occurs at the hot end of unit 5 on eliminating unit 2 from Figure 4.9. To restore ΔT_{min}, unit 5 itself must be removed. (Temperature in °C and load in kW)

A violation occurs at the hot end of unit 5 (where $\Delta T_{min} = 13°C$). There is no obvious way to restore ΔT_{min}, except by elimination of unit 5 itself. On shifting heat loads along the loop [5 4 3 6 5] to remove unit 5, the stream split disappears, but a negative load (of −305 kW) appears on unit 3. The negative load is set to zero by shifting −305 kW along the path [1 3 7]. Unit 3 is now eliminated, but there is a violation at the hot end of unit 4 of −15°C (see Figure 4.15). This is to be expected, as unit 4 is Type A in Figure 4.14 and its load cannot be increased from 1700 kW. To restore ΔT_{min}, the load of unit 4 must be reduced from 2400 kW to 1700 kW. The only hope is to seed the network with a dummy cooler on stream H2. Then, on shifting 700 kW along the path [1 4 D], the network in Figure 4.16 is obtained.

The final MNU design has no stream split and saves two units compared to the MER network. It incurs an energy penalty of only 395 kW (compared to the 900 kW in the MNU network previously obtained in Figure 4.13). This penalty obviously corresponds to the heat transferred across the pinch.

The MER network (Figure 4.9) has directly evolved into an MNU network (Figure 4.16). Is an intermediate network of six units possible? A solution may be obtained through Figure 4.15. The violation on the hot end of unit 4 may be reduced by shifting some of the heat load from unit 4 to unit 3, but only 70 kW can be transferred to unit 3 due to the temperature constraint on the cold end of unit 6. After shifting 70 kW along the loop [3 7 D 4 3], another 630 kW must be shifted along the path [1 4 D] to completely remove the violation. The resulting network (Figure 4.17), with six units and no stream splits, incurs an energy penalty of 325 kW.

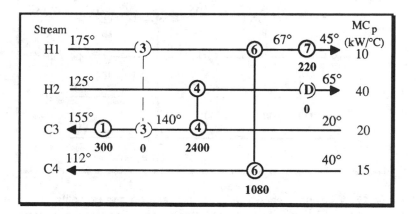

Figure 4.15 The negative heat load on unit 3 is reset to zero, and a dummy cooler is introduced to remove the ΔT_{min} violation at the hot end of unit 4. (Temperature in °C and load in kW)

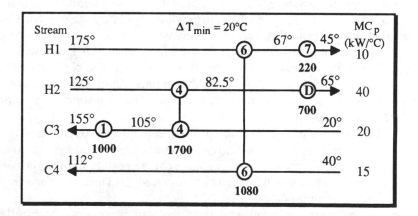

Figure 4.16 An alternative MNU design (with five units and no stream splits) incurs an energy penalty of only 395 kW compared to the design in Figure 4.13. (Temperature in °C and load in kW)

The above evolutions involved operations which are not very obvious and are, in that sense, tricky. Heuristics are available to guide the evolution process, but the lack of an algorithmic procedure makes it difficult to guarantee that the designs have the least energy penalties.

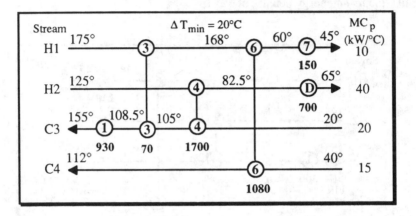

Figure 4.17 A design is shown with six units and no stream splits. It incurs a penalty of 325 kW, compared to the MER network in Figure 4.9. (Temperature in °C and load in kW)

Step 4. Generation of Search Tree for Evolution

All evolution possibilities need to be investigated in general, and a search tree may facilitate the task. An MER network is chosen for the root node and is expanded by identifying units for elimination. The associated energy penalty is calculated at each stage. Figure 4.18 shows such a tree with the MER network in Figure 4.9 forming the root node. All possible branches are explored, starting with units 2, 3, 4, 5, and 6 in turn. Units 1 and 7 cannot be removed and, hence, do not appear in the tree. The following observations may be readily made:

- Different branches often end up in the same design (represented by a single node in Figure 4.18 to avoid unnecessary repetition). In other words, the same design with a unique energy penalty results regardless of the order in which the units are removed, provided the same set of units is eliminated.
- As the maximum number of units that can be removed to yield an MNU design is known, the depth of the search tree is fixed (2 in Figure 4.18).

- The energy penalty does not necessarily increase monotonically with tree depth. For example, removal of unit 3 incurs an energy penalty of 500 kW; however, this energy penalty drops to 395 kW to yield a final MNU design with five units.
- A design with a low energy penalty initially does not necessarily evolve subsequently into a minimum energy penalty design. The most promising six-unit design (with an EP of 137.5 kW) evolves into a five-unit design (with an EP of 900 kW). On the other hand, the not-so-promising six-unit designs (with EPs of 325 and 500 kW) evolve into a five-unit design (with an EP of only 395 kW).

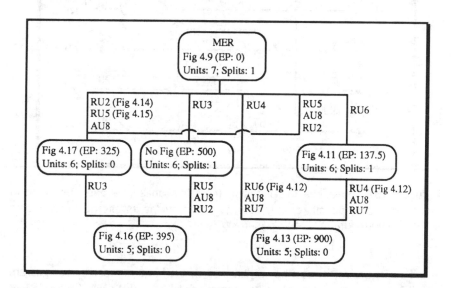

Figure 4.18 The search tree shows the evolutions possible starting with the MER network in Figure 4.9. The energy penalty (EP in kW) incurred at each stage after removing units (RU) and adding units (AU) is also shown. Unit 8 refers to the dummy cooler on stream H2.

Developing a complete search tree depicting the various evolutions is difficult and requires considerable effort. The problem is further complicated by the fact that the energy penalty in general depends on the MER network used to start with. Search trees similar to the one in Figure 4.18 need to be generated for the different initial MER networks. One must realize that different evolved designs with different energy penalties can emerge. For example, a different MNU design (although with the same energy penalty of

395 kW) is shown in Figure 4.19. It was evolved starting with an MER network (Figure 4.20), which is an alternative to the one in Figure 4.9.

In principle, the entire space of possible designs must be evolved starting with all feasible MER designs. The search space is thus frighteningly large, and the generation of the search trees is a Herculean task. A best-first search strategy (Trivedi et al., 1990b) may be used specially for large-sized problems to reduce the number of nodes to be investigated and expanded.

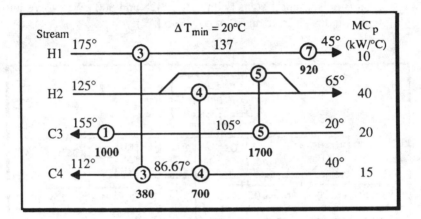

Figure 4.19 The above MNU design (with five units and one stream split) incurs an energy penalty of only 395 kW and provides an alternative to Figure 4.16. (Temperature in °C and load in kW)

To summarize, it must be emphasized that a large number of designs are possible for a given case study at a specified ΔT_{min}. For Case Study 4S1 at $\Delta T_{min} = 20°C$, there are four MER designs with seven units each (two designs appear in Figures 4.9 and 4.20), two designs with six units each (Figures 4.11 and 4.17), and three MNU designs with five units each (Figures 4.13, 4.16, and 4.19). These designs differ in terms of the number of stream splits and the utility consumptions. An MER design has minimum utility consumption but a higher number of units. On the other hand, an MNU design has the minimum number of units but a higher utility consumption. Can a network be designed that features minimum utility consumption as well as minimum number of units? Such a network which is both an MER design and an MNU design simultaneously is indeed possible in some cases.

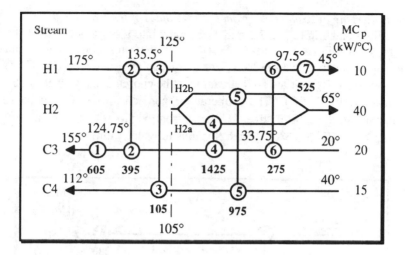

Figure 4.20 Alternative designs are possible for both the above- and below-pinch regions in Figure 4.9, so these designs can be combined in various ways to obtain four different MER networks. (Temperature in °C and load in kW)

4.3 ELIMINATING UNITS USING BYPASSES

An MER design usually contains more units than an MNU design because the pinch division forces streams crossing the pinch to be counted twice (see Equations 1.26 and 1.27). Wood et al. (1985) suggested a novel arrangement of stream splitting, mixing, and bypassing to reduce the units in an MER network by merging exchangers from the two sides of the pinch. An example is given below.

Example 4.6 Reducing the Number of Exchangers Using Bypasses

Consider the MER network in Figure 3.2 for Case Study 4S1 at $\Delta T_{min} =$ 13°C. Show that a unit can be eliminated to obtain an MNU design without incurring an energy penalty or violating the ΔT_{min} constraint. Use the bypassing scheme proposed by Wood et al. (1985).

Solution. The strategy of Wood et al. (1985) helps in the breaking of first-level loops and is guided by the following key observation. The MC_p criterion in the PDM may be satisfied above and below the pinch simultaneously by choosing the streams involved in the match to be of equal MC_p. Stream splitting is necessary to make the MC_p values identical. Furthermore, such a match would not incur an energy penalty in spite of crossing the pinch.

Step 1. Merging of Exchangers from the Two Sides of the Pinch

Units 1 and 4 in Figure 3.2 may be merged into a single unit (labeled S in Figure 4.21) with a heat load of 1020 kW, but this constitutes a match crossing the pinch (125°C/112°C). For the match to be valid, stream C3 must be split and the MC_p of the split stream must be chosen equal to that of stream H1 (namely, 10 kW/°C). The temperature approach in unit S must be 13°C throughout, so the split of stream C3 goes from 60° to 162°C whereas stream H1 goes from 175° to 73°C. Match S is thus fully determined.

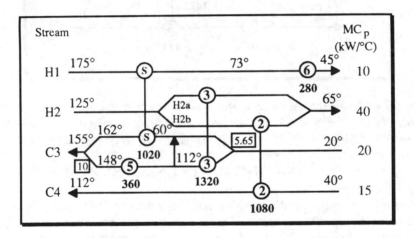

Figure 4.21 An MER-cum-MNU design for Case Study 4S1 at ΔT_{min} = 13°C may be synthesized with the help of stream splitting, mixing, and bypassing. (Temperature in °C and load in kW)

Step 2. Determining the Remaining Portion of the Split

Noting that stream H2 enters at the pinch temperature itself (125°C), the temperature approach at the hot end of unit 3 must be 13°C. Unit 3 will then lie completely below the pinch on the split of stream C3 having an MC_p value of 14.35 kW/°C (i.e., 1320/[112 − 20]). The network is completed as in Figure 4.21, using an additional split and mixing junction on stream C3 to ensure the necessary ΔT_{min} on the cold end of unit S. The split heat capacity flow rates shown in rectangular boxes in Figure 4.21 are unique. However, an alternative design (not shown) is possible by introducing a dummy cooler (D) on stream H2 and eliminating unit 6 by shifting 280 kW along the loop [6 S 3 D].

Step 3. Placement of Heater

The heater can be left in its original location in Figure 3.2, where the whole of stream C3 is heated from 137° to 155°C. However, it may be beneficial to relocate unit 5 on the split of stream C3 because a lower quality hot utility may suffice to heat this branch from 112° to 148°C.

Step 4. Post-Analysis

The network in Figure 4.21 satisfies the basic pinch design rules. The heater (unit 5) is completely above the pinch. Unit S satisfies the MC_p criterion because the streams involved in the match have equal MC_p. A bypass stream at the pinch temperature (112°C) mixes with a colder bypass stream (at 20°C) ensuring that no heat transfer or mixing of streams occurs across the pinch. However, it may be noted that both mixing junctions on stream C3 are nonisothermal and therefore may be thermodynamically highly irreversible. Although stream C3 is heated only marginally above its target temperature (162°C vs. 155°C) in this problem, the bypassing scheme could in general cause significant thermal degradation. When this occurs in an exchanger operating at ΔT_{min} throughout, the area requirement for the unit will typically be high. In essence, the major drawback of the bypassing scheme of Wood et al. (1985) is the relatively high area requirement for the network. Also, the match having equal MC_p values for the streams may require a large number of shells. Thus, Wood et al. (1985) concluded that designs like those in Figure 4.21 may prove more appropriate for systems having approximately parallel composite curves. Furthermore, they may appear promising for some retrofits, especially with their potential to use lower grade utilities. Jezowski (1990) has further generalized the Wood bypassing concept and shown it to be cost effective in some cases.

The demand on the area requirement for the various designs developed is an important consideration in determining their capital cost. In fact, the area targets established in Section 1.2 have not been fruitfully used thus far in network design. Methodologies are discussed in the next section to ensure that the area penalty is not unacceptably high at any stage of the network synthesis.

4.4 SOME DESIGN TOOLS TO ACHIEVE TARGETS

Six design tools are presented below that allow the network to approach the targets established in the first two chapters. These are the driving force plot, remaining problem analysis, hidden pinch phenomenon, diverse pinch concept, shrinking remaining problem analysis, and MC_p-ratio heuristic.

4.4.1 Driving Force Plot

The area and shell targets are based on an ideal model of vertical heat transfer and may be achieved only through a spaghetti-like design with an excessively large number of matches. Therefore, if the unit target is to be achieved, the area target will be only approximately met in practice. To design networks that approximate the minimum area predicted, the driving forces on the composite curve may be used as indicators.

The driving force plot (Linnhoff and Vredeveld, 1984) shows the vertical temperature difference, ΔT, between the HCC and CCC against the cold composite temperature, T_c. Instead of the cold composite temperature, T_c, for the abscissa, the temperature of the hot composite, T_h, or the average temperature of hot and cold composites, $[(T_h + T_c)/2]$, may be used. This information is readily available in a pinch tableau. For example, a driving force plot (DFP) may be generated (Figure 4.22) for Case Study 4S1 at a ΔT_{min} of 13°C using Table 2.8. The pinch corresponds to the point with the smallest vertical distance from the abscissa on such a plot.

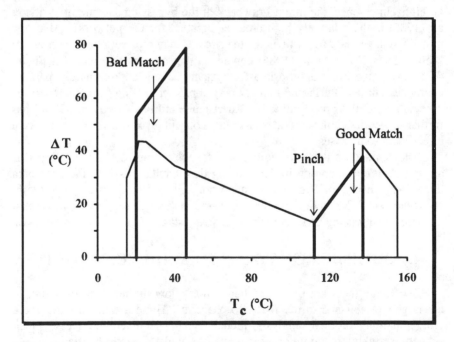

Figure 4.22 The driving force plot (DFP) helps the designer distinguish between good and bad matches. A good match must closely follow the vertical temperature differences on the composite curves.

On a DFP, a match may be represented as a linear segment (or a trapezium for emphasis). The trapeziums in Figure 4.22 correspond to matches 1 and 4 in Figure 3.2.

To obtain the area target, every match in the network must exactly follow the vertical temperature difference given by the DFP. This may imply too many matches (as pointed out earlier) and is rigorous only if the stream heat transfer coefficients are all equal. However, a guideline that may be useful for a designer in actual practice is to follow approximately (but closely) the temperature difference on the DFP for each exchanger. Note that the driving forces on any match should neither be higher nor lower than prescribed by the DFP. This is primarily because a higher ΔT on one match (lower area) will necessarily force a lower ΔT on some other match (higher area) resulting in a net increase in the overall area for the network. Specifically, Figure 4.22 shows that match 1 in Figure 3.2 exactly fits the DFP and is consequently a perfect match obeying vertical heat transfer. On the other hand, match 4 in Figure 3.2 provides higher ΔT than specified by the DFP in Figure 4.22; hence, it is a bad match from the viewpoint of minimum overall network area (although not energy). This is simply proven by actually calculating the areas of the various matches (see Section 4.4.2).

As the vertical heat transfer model forms the basis for both area and shell targeting, the DFP may be used for achieving near-minimum area as well as near-minimum shells in networks (Ahmad and Smith, 1989).

Though the DFP provides a useful picture of the appropriate driving forces in terms of the ideal ΔTs to be used for the matches during the network design, a more quantitative design tool would be preferable. Such a tool is provided by remaining problem analysis and is discussed next.

4.4.2 Remaining Problem Analysis

Ahmad and Smith (1989) discuss the use of remaining problem analysis (RPA) to estimate the penalty incurred during network synthesis with regard to the area and shell targets. The basic concept behind RPA is simple and powerful; however, it requires increased computational effort.

RPA can be used at any stage in the synthesis to predict the potential of the final network achieving the targets. Like the DFP, it allows the designer to distinguish between bad matches and good matches. At every stage of the design, the overall problem (comprising the complete stream data) may be decomposed into a match (the one under consideration) and a remaining problem (that excludes the portions of the streams involved in the match). For a perfect match (incurring no penalty), the sum of the targets for the match (M) and the remaining problem (RP) must equal the target for the overall problem (OP). This idea is used to define a penalty function for a match as follows:

$$Penalty = Target_M + Target_{RP} - Target_{OP} \qquad (4.3)$$

Note that $Target_M$ can accurately be obtained as it involves a single match (and, in that sense, is not really a target). Equation 4.3 has been used for area and shells by Ahmad and Smith (1989), but it is applicable and should prove very useful for energy and cost as well.

Example 4.7 Using Remaining Problem Analysis to Achieve Area Targets

Consider the MER network in Figure 3.2 for Case Study 4S1 at $\Delta T_{min} =$ 13°C. Use remaining problem analysis to evaluate the matches in terms of their area requirements. If some matches are not favorable in achieving the minimum network area, explore other match options in an attempt to develop a better MER design that approaches the area target. For the sake of simplicity, assume the mixing junction in Figure 3.2 to be isothermal.

Solution.   The PDM combined with RPA provides a powerful methodology for designing networks that meet the various targets (specifically, area, energy, and units in this example). RPA guides the designer in selecting good matches (when options exist) and in deciding match loads prudently (when the tick off heuristic leads to significant penalty). As the MER network in Figure 3.2 was designed with the PDM to obey the energy and unit targets, RPA is used below only for the area target.

Step 1. Target for the Overall Problem
    The countercurrent area target for the overall problem is determined using the procedure outlined in Example 1.4 to be 1640.06 m^2.

Step 2. Targets for Various Matches and Remaining Problems
    Consider the match labeled 1 in Figure 3.2. Its area is 214.53 m^2 (from $A = Q/[U\ LMTD] = 500/[0.1 \times 23.307]$). For the remaining problem, the portions of the streams involved in match 1 (namely, 175° to 125°C for stream H1 and 137° to 112°C for stream C3) are excluded. The remaining problem comprises the following streams: H1 (125° to 45°C), H2 (125° to 65°C), C3a (20° to 112°C), C3b (137° to 155°C), and C4 (40° to 112°C). The countercurrent area target for this remaining problem is 1425.53 m^2. So, the penalty as per Equation 4.3 is zero (i.e., 214.53 + 1425.53 − 1640.06), indicating that match 1 is an ideal match (as was also evidenced on the DFP in Figure 4.22).
    The penalties for the other three matches in Figure 3.2 are calculated in a similar fashion and are summarized in Table 4.11. Note that match 4 in Figure

3.2 incurs a high area penalty of 248.46 m^2 (about 15%) and is a bad match as was also indicated on the DFP in Figure 4.22.

Table 4.11 Area Penalties from Remaining Problem Analysis

Match	Load	Target$_M$	Target$_{RP}$	Area Penalty	% Penalty
1 (H1-C3)	500	214.53	1425.53	0	0
2 (H2b-C4)	1080	588.53	1089.61	38.09	2.3
3 (H2a-C3)	1320	834.88	927.69	122.51	7.5
4 (H1-C3)	520	79.83	1808.68	248.46	15.2

Units: load in kW (in Figure 3.2) and area (target and penalty) in m^2.

Step 3. Designing a Better Network to Approach the Area Target

Match 1 in Figure 3.2 is a perfect one with no penalty whatsoever. The area for the heater (labeled 5) is 109.86 m^2 (as $A = Q/[U\ LMTD] = 360/[0.1\ x\ 32.768]$). Thus, the total area for matches 1 and 5 above the pinch is 324.39 m^2, which exactly corresponds to the minimum area target (Table 2.8).

Below the pinch, all the matches in Figure 3.2 have area penalties (Table 4.11) associated with them. The area penalty for match 2 is 38.09 m^2 (2.3%), which may be acceptable. However, it would be desirable to reduce the area penalties of 7.5% and 15.2% associated with matches 3 and 4. One possibility that may be explored is to decrease the load of match 3. If the unit target is to be satisfied, then the load on match 4 must be increased to 800 kW so as to tick off stream H1. The load of match 3 is reduced to 1040 kW and the cooler of 280 kW is placed on stream H2a to obtain the network in Figure 4.23.

RPA is carried out on Figure 4.23 to generate the area penalties in Table 4.12. It is seen that the penalties on matches 3 and 4 have decreased to 4.9% and 11.4% respectively (compared to the earlier penalties of 7.5% and 15.2%). On considering the areas of the heater (109.86 m^2) and cooler (54.53 m^2), the total area of the network in Figure 4.23 is 1840.89 m^2. This is a reduction compared to the area of 1900.74 m^2 in Figure 3.2, although it is still higher than the target of 1640.06 m^2 by about 12%.

The DFP and RPA together help identify matches that are favorable or unacceptable in approaching the targets. They help determine cases where the tick off heuristic may not be appropriate. In such cases, units beyond the minimum target may be required.

Though the PDM suffices for achieving minimum energy consumption in a network, RPA is also suggested by Linnhoff and Hindmarsh (1983). It is indeed necessary in a few cases, as is demonstrated in the next section.

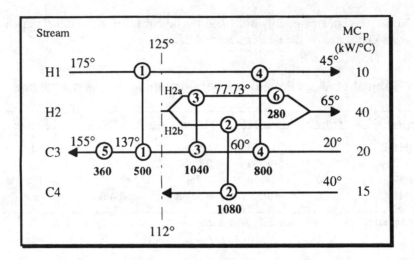

Figure 4.23 An alternative to the network in Figure 3.2 is developed by RPA; here, ticking off match 4 instead of match 3 helps approach the area target. (Temperature in °C and load in kW)

Table 4.12 Reduced Area Penalties from RPA

Match	Load	Target$_M$	Target$_{RP}$	Area Penalty	% Penalty
1 (H1-C3)	500	214.53	1425.53	0	0
2 (H2b-C4)	1080	588.53	1089.61	38.09	2.3
3 (H2a-C3)	1040	682.34	1037.86	80.14	4.9
4 (H1-C3)	800	191.10	1636.32	187.37	11.4

Units: load in kW (in Figure 4.23) and area (target and penalty) in m^2.

4.4.3 Hidden Pinch Phenomenon

Rev and Fonyo (1986) showed that the use of the tick off heuristic could sometimes cause violation of the minimum utility requirement during network synthesis. This is due to (what they termed) the hidden pinch, for which there is an allowable heat load on a match (that is less than the maximum possible). Determination of the hidden pinch and allowable heat load is easily done by RPA as demonstrated in the example below.

Example 4.8 Using Remaining Problem Analysis for Hidden Pinches
Consider the stream data in Case Study 4S1 with the following modification: the MC_p-value of stream H1 is 25 kW/°C (and not 10 kW/°C). Let this modified stream data be referred to as Case Study 4S1h. Design an MER network for Case Study 4S1h at ΔT_{min} = 20°C.

Solution. RPA was used in Example 4.7 to approach the area target; here, it is shown to be useful in achieving the energy target.

Step 1. Energy Targets Using PTA
The PTA gives the minimum utility requirements at ΔT_{min} = 20°C to be 1870 kW (for cold utility) and 0 kW (for hot utility). As no hot utility is required, this is a threshold problem.

Step 2. RPA for Detection of the Hidden Pinch
Use of the H/H heuristic for a threshold problem (see Section 3.4) or use of the PDM (considering the pinch at 175°C/155°C) suggests a match between streams H1 and C3. A maximum possible heat load of 2700 kW may be assigned to match 1 (Figure 4.24) by ticking off stream C3.

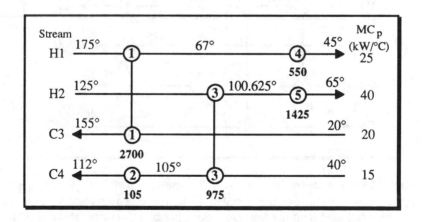

Figure 4.24 A hidden pinch prevents the above design from achieving the minimum energy consumption. Using the tick off heuristic for match 1 is inappropriate in this case as per RPA. (Temperature in °C and load in kW)

For the remaining problem, the portions of the streams involved in match 1 (namely, 175° to 67°C for stream H1 and the whole of stream C3) are excluded. Therefore, the remaining problem comprises the following streams:

H1 (67° to 45°C), H2 (125° to 65°C), and C4 (40° to 112°C). The energy targets for this remaining problem are 1975 kW (for cold utility) and 105 kW (for hot utility), which violate the minimum energy consumption. In fact, a pinch occurs for the remaining problem at 125°C/105°C, indicating that this is no longer a threshold problem. Using the conventional pinch design rules, the network is easily completed as in Figure 4.24. The design has an energy penalty of 105 kW and an extra unit (heater). Can the penalty be avoided?

Step 3. RPA for Determination of Allowable Heat Load

To assure minimum utility consumption, the match may be assigned only a limited heat load (and not the maximum possible). It is observed on Figure 4.24 that, without the heater, a match between streams H2 and C4 would cause a ΔT_{min} violation (125°C/112°C) at the hot end of unit 3. The only alternative is to match stream C4 with stream H1. To ensure a ΔT_{min} of 20°C for this match, the outlet temperature of stream H1 for match 1 must be at least 132°C (see Figure 4.25).

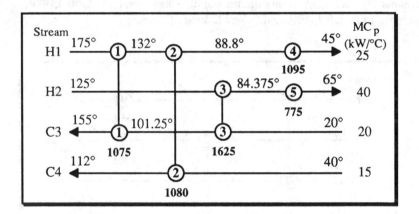

Figure 4.25 When there is a hidden pinch, the minimum energy consumption may be achieved by assigning matches (in this case, match 1) with only allowable heat load (and not the maximum possible). (Temperature in °C and load in kW)

The allowable heat load for match 1 cannot be greater than 1075 kW (i.e., 25[175 − 132]). As expected, RPA shows no energy penalty for the remaining problem now comprising streams H1 (132° to 45°C), H2 (125° to 65°C), C3 (20° to 101.25°C), and C4 (40° to 112°C).

The rest of the design may be easily completed using the fast matching algorithm (see Section 3.4) since the problem remains a threshold one throughout. Although the energy target is achieved, one additional unit above the minimum number (of four) is required as the tick off heuristic is inappropriate and hence not used.

Interestingly, there is a loop [5 3 1 4 5] in the network in Figure 4.25. Because the outlet temperature of stream C3 from unit 3 cannot exceed 105°C, it may be shown that 75 kW at best may be shifted from unit 5 to unit 4.

In summary, note that new pinches may be created during the process of network synthesis. Then, RPA may be used to assure minimum energy consumption. The tick off heuristic may be inappropriate, and matches should be assigned only allowable heat loads (and not the maximum possible). A combination of the fast algorithm and PDM may be useful in such cases.

4.4.4 Diverse Pinch Concept

When the stream heat transfer coefficients differ considerably, the concept of the diverse pinch may be profitably used (Rev and Fonyo, 1991). The energy and area targeting procedures for the case of the diverse pinch were established in Section 2.2. Here, the diverse pinch design methodology is discussed and compared with the PDM.

Example 4.9 Using Diverse Pinch to Achieve Minimum Area

Design a network for Case Study 4S1t for a minimum flux specification of Q''_{min} = 90/11 = 8.1818 kW/m^2 using the diverse pinch methodology. Compare the network with that from a conventional pinch design for the same utility consumption.

Solution. The stream film coefficients are ideally accounted for at the start of the design itself. This is the principle used in constructing the diverse-type heat cascade. The targets with the diverse pinch are then used in design as suggested by Rev and Fonyo (1991).

Step 1. Determination of the Minimum Utilities and Diverse Pinch

The diverse-type heat cascade (Example 2.4) gives the minimum utility requirements at Q''_{min} = 8.1818 kW/m^2 to be 1980 kW (for hot utility) and 400 kW (for cold utility). The pinch is at 44.09°C on the shifted temperature scale as per Table 2.13.

Step 2. Construction of Grid in Terms of Shifted Temperatures

The diverse pinch design is most conveniently done on a grid based on shifted temperatures (Figure 4.26). The individual stream shifts are given by

(Q''_{min}/h_j) as discussed in Section 2.2.1. For example, from Table 2.13, it is observed that stream H1 goes from 134.09° to 4.09°C in terms of shifted temperatures. The diverse pinch point can be represented as before by a vertical line on this "shifted-temperature" grid.

Step 3. Matching Using Conventional PDM Rules

The problem is separated into two regions by the diverse pinch division. The conventional pinch design rules apply to the above-pinch and below-pinch regions. Although the below-pinch design is trivial, consisting of only a cooler of 400 kW, many possibilities exist for the above-pinch design. Either hot stream can be matched with either cold stream. Figure 4.26 shows a network that gives preference to matching streams with similar film coefficients (i.e., streams H1 and C3 and streams H2 and C4). Matching streams with significant differences in heat transfer coefficients (i.e., streams H1 and C4 and streams H2 and C3) leads to a higher heat transfer area (which the reader should verify). Thus, streams H1 and H2 may be ticked off by matches 1 and 2, and the remaining heat loads on streams C3 and C4 may be satisfied by heaters.

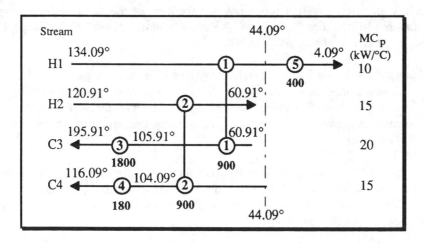

Figure 4.26 A diverse pinch design is most conveniently done on a grid based on shifted temperatures, which accounts for the heat transfer coefficients of the individual streams. (Temperature in °C and load in kW)

Step 4. Constructing the Temperature Approach Reserve Matrix

On the shifted-temperature grid in Figure 4.26, the ΔT at the cold end of match 1 is negative (namely, $\Delta T = 44.09 - 60.91 = -16.82$°C). Is this a valid match? The driving forces for the matches may be determined by constructing a temperature approach reserve matrix (TARM) as in Table 4.13. The elements of the TARM may be obtained by simply adding the individual temperature shifts of the hot and cold stream pair.

Table 4.13 Temperature Approach Reserve Matrix

Stream	C3 (40.91)	C4 (4.09)
H1 (40.91)	81.82	45
H2 (4.09)	45	8.18

Units: all values are temperatures in °C.

Note: values in parenthesis are individual stream shifts (Q''_{min}/h_j).

The actual ΔT for a match between hot stream jh and cold stream jc is then given by

$$\Delta T_{actual} \quad = \quad \Delta T_{shifted\ temp\ grid} + TARM(jh, jc) \qquad (4.4a)$$

$$\text{where} \quad TARM(jh, jc) \ = \ Q''_{min}/h_{jh} + Q''_{min}/h_{jc} \qquad (4.4b)$$

Using Equations 4.4a and 4.4b, the actual ΔT at the cold end of match 1 between streams H1 and C3 is 65°C (namely, −16.82 + 81.82). In addition to checking for sufficiency of driving forces for the various matches, the TARM is useful in general during loop breaking and energy relaxation. A high value for a TARM element implies greater scope for energy relaxation and more flexibility in shifting heat loads without causing unacceptable ΔTs. No evolution, however, is required for the network in Figure 4.26.

Step 5. Comparison with Conventional Pinch Design

For the same utility consumption (1980 kW for hot utility), the ΔT_{min} (equivalent to Q''_{min} = 8.1818 kW/m^2) is 49°C. This may be obtained by simple linear interpolation from Table 2.20. The pinch is at 89°C/40°C as per Table 2.11. Two possible designs above the pinch as well as below the pinch are shown in Figures 4.27 and 4.28. These may be obtained by the PDM. Note that the global ΔT_{min} of 49°C does not apply to the utility matches.

The below-pinch design involves an extra unit and consequently a loop [7 8 6 5 7]. Heat loads may be shifted along this loop, provided care is taken that ΔT_{min} is not violated. Stream C3 is split in the MC_p-ratio of 10:10 in

Figure 4.27 and in the ratio of 5:15 in Figure 4.28. Clearly, different intermediate designs between these two limits are possible.

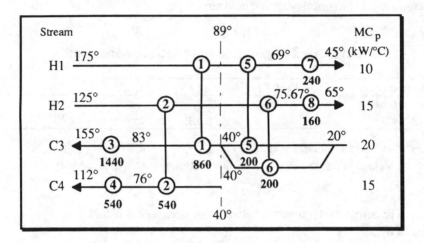

Figure 4.27 An MER design using the conventional PDM may be developed for comparison with the diverse pinch design in Figure 4.26. (Temperature in °C and load in kW)

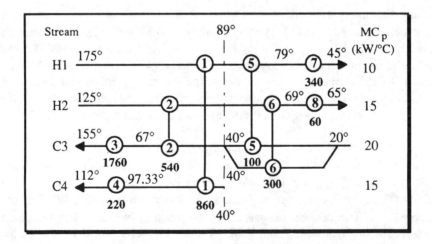

Figure 4.28 This conventional pinch design represents an alternative to the MER network in Figure 4.27. (Temperature in °C and load in kW)

Figure 4.27 is the preferred network as it has matches between streams with similar heat transfer coefficients, when compared to Figure 4.28. The areas for the networks in Figures 4.26 and 4.27 are computed in Tables 4.14 and 4.15. In Table 4.15, ΔT_{min} values of 25°C (for matches with hot utility) and 30°C (for matches with cold utility) are accepted.

Table 4.14 Area for Diverse Pinch Design

Match	Load	ΔT at ends	LMTD	U	Area
1 (H1-C3)	900	110 & 65	85.54	0.1	105.22
2 (H2-C4)	900	25 & 25	25	1	36.00
3 (HU-C3)	1800	25 & 114	58.66	0.19	161.11
4 (HU-C4)	180	68 & 79	73.36	1.33	1.84
5 (H1-CU)	400	60 & 30	43.28	0.18	50.83

Units: load in kW (in Figure 4.26), temperature in °C, overall heat transfer coefficient in kW/(m^2 °C), and area in m^2.

Table 4.15 Area for Conventional Pinch Design

Match	Load	ΔT at ends	LMTD	U	Area
1 (H1-C3)	860	92 & 49	68.26	0.1	126.00
2 (H2-C4)	540	49 & 49	49	1	11.02
3 (HU-C3)	1440	25 & 96	52.77	0.19	143.26
4 (HU-C4)	540	68 & 103	84.29	1.33	4.8
5 (H1-C3)	200	49 & 49	49	0.1	40.82
6 (H2-C3)	200	49 & 55.67	52.26	0.18	21.05
7 (H1-CU)	240	44 & 30	36.55	0.18	36.11
8 (H2-CU)	160	50.67 & 50	50.33	1	3.18

Units: load in kW (in Figure 4.27), temperature in °C, overall heat transfer coefficient in kW/(m^2 °C), and area in m^2.

The diverse pinch design (Figure 4.26) has a total network area of 355 m^2, which is very close to the NLP minimum area of 354.39 m^2 (see Table 2.25). On the other hand, the conventional pinch design (Figure 4.27) has a network area of 386.23 m^2. The basic difference between the two methodologies stems from the fact that the diverse pinch concept appropriately chooses a high

LMTD for a low *U* match and a low *LMTD* for a high *U* match. This is clearly observed in Table 4.14, where the H1-C3 match ($U = 0.1$ kW/m^2 °C) has an *LMTD* of 85.54°C and the H2-C4 match ($U = 1$ kW/m^2 °C) has an *LMTD* of 25°C. Furthermore, note that the conventional pinch design has a stream split and three extra units, which would certainly make it more costly than the diverse pinch design.

Closer examination of Figure 4.26 reveals that 360 kW is transferred in match 1 across the conventional pinch (89°C/40°C) from stream H1 (125° - 89°C) to stream C3 (22° - 40°C). At the same time, 360 kW is transferred in match 2 across the pinch from stream H2 (89° - 65°C) to stream C4 (40° - 64°C). As the pinch violations are equal in magnitude but opposite in direction, no energy penalty is incurred (see Chapter 5 also). Gundersen and Grossmann (1990) argued the benefits of such systematic crisscrossing through the pinch.

Linnhoff and Ahmad (1990) distinguished three effects that result from unequal heat transfer coefficients:

- lower heat transfer area is required for higher *h*-value and vice versa;
- lower total network area is possible by assigning higher ΔTs to streams with low *h*-values; and
- lower total network area is possible by matching streams with similar *h*-values as the thermal resistances operate in parallel.

Though the first effect is taken into account in the area targeting formula (Equation 1.7), the other two effects are not; typically, this leads to an error of less than 10% in the total area (Linnhoff and Ahmad, 1990). Importantly, the diverse pinch approach provides a methodology for this purpose in the case of unequal heat transfer coefficients.

4.4.5 Shrinking Remaining Problem Analysis

Remaining problem analysis provides a powerful tool that quantifies the approach to the various targets during synthesis. RPA as presented in Section 4.4.2 considers each match independently and does not recognize that a committed structure exists at every stage of the synthesis. Instead of excluding only the single match under consideration as in RPA, shrinking remaining problem analysis (SRPA) may be performed by excluding all matches committed to thus far. The remaining problem thus shrinks continuously during synthesis and comprises only those portions of the stream data where match options are to be yet exercised. The SRPA penalty is thus cumulative and continuously increases according to the following equation:

$$Penalty_{SRPA} = \Sigma \, Target_M + Target_{SRP} - Target_{OP} \qquad (4.5)$$

Note that $Target_M$ for each match is accurately known (and, in that sense, is not an estimated target). In other words, $\Sigma\ Target_M$ is the true value for the committed structure thus far, including the match under consideration.

Although Linnhoff and Ahmad (1990) briefly mention the idea of performing RPA on a cumulative basis, sufficient details are not provided. An example is solved below to illustrate the merits and disadvantages of the SRPA with regard to the classical RPA (of Section 4.4.2).

Example 4.10 Application of Shrinking Remaining Problem Analysis

Consider the MER network in Figure 3.1 for Case Study 4S1 at ΔT_{min} = 13°C. Use RPA as well as SRPA to evaluate the matches in terms of their area requirements. If some matches are not favorable in achieving the minimum network area, explore other match options in an attempt to develop a better MER design that approaches the area target. Assume stream H2 is split in the MC_p-ratio of 25:15.

Solution. The classical RPA as in Example 4.7 is first performed on Figure 3.1 to detect matches that are not favorable in achieving the area targets. The inadequacies with RPA may be then remedied through SRPA.

Step 1. Target for the Overall Problem and Optimum Split Ratio
The area target (without considering the stream split) is 1640.06 m^2 as per Table 1.15, but this does not appear appropriate for split-stream designs with nonisothermal mixing junctions. On performing area targeting on a five-stream data set with stream H2 split into two branches, H2a and H2b (Figure 3.1), with $20 \le MC_{p,a} \le 25$, the following results are obtained:

$MC_{p,a}$	20	21	22	23	24	25
Area (m^2)	1906.41	1835.85	1784.28	1745.12	1714.78	1691.22

The minimum countercurrent area target for the overall problem in Figure 3.1 is 1691.22 m^2, and the optimum split corresponds to $MC_{p,a} = 25$ kW/°C.

Step 2. Targets for Various Matches and Remaining Problems
The area for each match in Figure 3.1 and the area target for the corresponding remaining problems are determined. Then, penalties are calculated from Equation 4.3. The calculations are summarized in Table 4.16. The total area of the network in Figure 3.1 is 1836.78 m^2.

Step 3. Designing an Alternative Network
The design above the pinch incurs no area penalty (as in Example 4.7). Table 4.16 shows that all the matches in Figure 3.1 below the pinch have area

penalties associated with them. Can the area penalties of 3.7% and 7.2% associated with matches 3 and 4 be reduced? The strategy adopted in Example 4.7 may be used. The load on match 3 is decreased. For the unit target to be satisfied, the load on match 4 may be increased to 800 kW so as to tick off stream H1. However, this will cause a ΔT_{min} violation on the cold end of unit 4, so the cooler (unit 6) must remain with at least a load of 80 kW. Therefore, the load of match 3 is reduced to 360 kW, the load of match 4 is increased to 720 kW, and an additional cooler of 200 kW is placed on stream H2b to obtain the network in Figure 4.29.

Table 4.16 Area Penalties from RPA

Match	Load	Target$_M$	Target$_{RP}$	Area Penalty	% Penalty
1 (H1-C3)	500	214.53	1476.69	0	0
2 (H2a-C3)	1840	881.86	840.09	30.72	1.8
3 (H2b-C4)	560	430.77	1322.68	62.23	3.7
4 (H1-C4)	520	126.65	1685.45	120.87	7.2

Units: load in kW (in Figure 3.1) and area (target and penalty) in m^2.

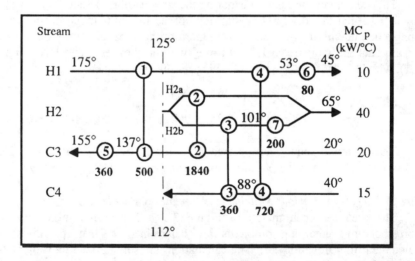

Figure 4.29 An alternative to the network in Figure 3.1 developed by RPA does not help approach the area target. This is better understood through SRPA. (Temperature in °C and load in kW)

RPA is performed on Figure 4.29 to generate the area penalties in Table 4.17. It is seen that the penalties on matches 3 and 4 have reduced to 1.4% and 5.4% respectively (compared to the earlier penalties of 3.7% and 7.2%). Surprisingly, the total area for matches 3 and 4 has increased from 557.42 m^2 (i.e., 430.77 + 126.65) to 590.71 m^2 (i.e., 276.92 + 313.79). On considering the areas of the heater (109.86 m^2) and two coolers (27.60 m^2 and 26.91 m^2), the total area of the network in Figure 4.29 is 1851.47 m^2. Although the penalties in Table 4.17 are lower than those in Table 4.16, there is an increase in the overall network area (on comparing Figures 3.1 and 4.29).

Table 4.17 Reduced Area Penalties from RPA

Match	Load	Target$_M$	Target$_{RP}$	Area Penalty	% Penalty
1 (H1-C3)	500	214.53	1476.69	0	0
2 (H2a-C3)	1840	881.86	840.09	30.72	1.8
3 (H2b-C4)	360	276.92	1437.80	23.50	1.4
4 (H1-C4)	720	313.79	1468.14	90.72	5.4

Units: load in kW (in Figure 4.29) and area (target and penalty) in m^2.

Step 4. Targets for Committed Structures and Shrinking Remaining Problems

Consider the match labeled 1 in Figure 3.1. Its area is 214.53 m^2 and the countercurrent area target for the remaining problem is 1476.69 m^2. Therefore, the penalty as per Equation 4.5 is zero (i.e., 214.53 + 1476.69 − 1691.22) indicating that match 1 is an ideal match (as was also seen earlier).

Next consider the heater labeled 5 in Figure 3.1. Its area is 109.86 m^2. For the remaining problem, the committed structure is excluded. In other words, portions of the streams involved in matches 1 and 5 (namely, 175° to 125°C for stream H1 and 155° to 112°C for stream C3) are excluded, so the shrinking remaining problem comprises the following streams: H1 (125° to 45°C), H2a (125° to 51.4°C), H2b (125° to 87.667°C), C3 (20° to 112°C), and C4 (40° to 112°C). The countercurrent area target for this SRP is 1366.83 m^2, so the SRPA penalty as per Equation 4.5 is zero (i.e., 214.53 + 109.86 + 1366.83 − 1691.22), indicating that the above-pinch design is good.

The SRPA penalties for the other matches (2, 3, 4, and 6 in that order) in Figure 3.1 are calculated in a similar fashion and are summarized in Table 4.18. An SRPA analysis is also performed on Figure 4.29, whose results are given in Table 4.19. It correctly predicts the cumulative area penalty to be higher for Figure 4.29 than for Figure 3.1.

Table 4.18 Area Penalties from SRPA

Match	Load	$\Sigma Target_M$	$Target_{SRP}$	SRPA Penalty	% Penalty
Start		0	1691.22	0	0
1 (H1-C3)	500	214.53	1476.69	0	0
5 (HU-C3)	360	324.39	1366.83	0	0
2 (H2a-C3)	1840	1206.25	515.70	30.73	1.8
3 (H2b-C4)	560	1637.02	199.76	145.56	8.6
4 (H1-C4)	520	1763.67	73.11	145.56	8.6
6 (H1-CU)	280	1836.78	0	145.56	8.6

Units: load in kW (in Figure 3.1) and area (target and SRPA penalty) in m^2.

Table 4.19 Area Penalties from SRPA

Match	Load	$\Sigma Target_M$	$Target_{SRP}$	SRPA Penalty	% Penalty
Start		0	1691.22	0	0
1 (H1-C3)	500	214.53	1476.69	0	0
5 (HU-C3)	360	324.39	1366.83	0	0
2 (H2a-C3)	1840	1206.25	515.70	30.73	1.8
3 (H2b-C4)	360	1483.17	287.90	79.85	4.7
4 (H1-C4)	720	1796.96	34.01	139.75	8.3
6 (H1-CU)	80	1824.56	26.91	160.25	9.5
7(H2b-CU)	200	1851.47	0	160.25	9.5

Units: load in kW (in Figure 4.29) and area (target and SRPA penalty) in m^2.

The benefit of the SRPA is that the true area (in terms of $\Sigma Target_M$) and the true penalty of the committed structure are available at every stage in the synthesis. The disadvantage is that the penalty associated with a match (being cumulative) is a function of the matches committed prior to it. An appropriate strategy may be to use both RPA and SRPA in cases where the penalties are unacceptably high. In essence, RPA and SRPA allow the designer the constant use of targets during design to evaluate and decide matches when options exist.

4.4.6 The MC_p-Ratio Heuristic

The MC_p-ratio heuristic is the simplest of the design tools proposed by Linnhoff and Ahmad (1990) for approaching the area target. It is not as sophisticated or powerful as the RPA/SRPA or DFP; however, it may be used sometimes as a rule of thumb for quick screening of options.

The MC_p criterion (Section 3.1.1) in the PDM not only guarantees that ΔT_{min} is maintained in pinch matches, but also helps approach the minimum area target in the most constrained region of the network. The MC_p criterion ensures that the temperature profiles of the individual pinch matches largely follow those of the composite curves (i.e., the profiles diverge away from the pinch). The following MC_p-ratio heuristic is a more rigorous statement of the design relation for achieving minimum area in the pinch region:

$$(MC_{p,h}/MC_{p,c})_{pinch\ match} \cong (\Sigma MC_{p,h}/\Sigma MC_{p,c})_{composite\ curves\ at\ pinch} \qquad (4.6)$$

Equation 4.6 states that true vertical heat transfer is possible if the MC_p-ratio of the pinch matches is identical to that of the composite curves.

Consider matches 1 and 2 in Figures 4.27 and 4.28. The MC_p-ratio of the composite curves just above the pinch is 0.714 (i.e., [10+15]/[20+15]). In Figure 4.27, the MC_p-ratio is 0.5 (match 1) and 1.0 (match 2). In Figure 4.28, the MC_p-ratio is 0.66 (match 1) and 0.75 (match 2). As per the MC_p-ratio heuristic, Figure 4.28 should yield a lower area than Figure 4.27 for these matches in the case of equal heat transfer coefficients. The heuristic does not appear to work well for this example. The areas for the two matches in the two networks do not show a significant difference.

4.5 EVOLUTION OF CONSTRAINED NETWORKS

For the case of constrained heat exchanger networks (CHENs), an initial design at the $QP_{p,min}$ condition (Figure 3.14) was generated in Section 3.5. Conventional loop breaking and path relaxation techniques may be used to evolve this network. However, in addition to these, there are two more degrees of freedom (the QP_p-QP_h tradeoff and the QP_p-QP_c tradeoff) that allow elimination as well as introduction of units to generate new network topologies (O'Young and Linnhoff, 1989) as illustrated below.

Example 4.11 Evolution of Constrained Heat Exchanger Networks

Evolve the $QP_{p,min}$ network (Figure 3.14) to a $QP_{p,max}$ network using the QP_p-QP_h and QP_p-QP_c tradeoffs. Evolve both the $QP_{p,min}$ and $QP_{p,max}$ networks further using conventional loop and path techniques. Note that Figure 3.14 is an MER network for Case Study 4S1c with $\Delta T_{min} = 20°C$ and the match between streams H2 and C3 forbidden.

Solution.   O'Young and Linnhoff (1989) developed the following important insights with regard to design and evolution of CHENs:
- the constrained PTA allows identification of streams that can and/or should bear the energy penalty;
- there are three constrained energy penalty components that interact with each other and may be exploited for flexibility and optimization in design; and
- a new type of evolution is possible by which heaters/coolers may be relocated for cost optimization, keeping the total energy consumption constant at its targeted value.

The first two aspects have been treated in detail in Section 3.5. It is the third aspect that is considered below, where utility level (operating cost) is traded off against capital cost at the same utility consumption level by changing the basic network structure.

Step 1. Examination of the QP_p-QP_h *Tradeoff*

The $QP_{p,min}$ network (Figure 3.14) corresponds to the $QP_{h,max}$ condition. The aim is to trade off the hot utility usage below the pinch against process-to-process heat exchange across the pinch. This is achieved through potential interactions between all the hot streams above the pinch and the cold imbalance below the pinch (MPRC). In general, the enthalpy matrix approach may be used after ticking off the essential $QP_{p,min}$ match at the start.

Figure 3.14 has $QP_{p,min} = 0$, $QP_{c,max} = 0$, and $QP_{h,max} = 900$ kW, so there is no $QP_{p,min}$ match. The enthalpy matrix is very simple, as the only hot stream above the pinch is H1 (500 kW) and the MPRC is stream C3 (900 kW). The only option is to maximize the heat recovery between streams H1 and C3 at 500 kW. This yields $QP_h = 400$ kW on stream C3, which corresponds to the $QP_{h,min}$ condition.

Starting with Figure 3.14, 500 kW from the below-pinch heater (unit 1) are replaced by cross-pinch process-to-process heat transfer between streams H1 and C3. To maintain a constant energy consumption level, the above-pinch process-to-process recovery is appropriately relaxed (Figure 4.30) by shifting 395 kW from unit 3 to unit 4 on stream C3 and by shifting 105 kW from unit 2 to a new heater on stream C4. Thus, evolution has been performed from one limiting condition ($QP_{h,max}$ in Figure 3.14) to the other ($QP_{h,min}$ in Figure 4.30). Clearly, various intermediate networks between these two limits exist.

Step 2. Examination of the QP_p-QP_c *Tradeoff*

The $QP_{p,min}$ network (Figure 3.14) also corresponds to the $QP_{c,max}$ condition, so the cold utility usage above the pinch may be traded off against process-to-process heat exchange across the pinch. This is achieved through possible interactions between all the cold streams below the pinch and the hot imbalance above the pinch (MPRH). In general, the enthalpy matrix approach may be used after ticking off the essential $QP_{p,min}$ match at the start.

Recall that Figure 3.14 has $QP_{p,min} = 0$ and $QP_{c,max} = 0$. Also, MPRH is zero. As there is no above-pinch cold utility in Figure 3.14, the question of evolving it to cross-pinch process heat transfer does not arise ($QP_{c,min} = QP_{c,max} = 0$) in this case study. As in the case of the QP_p-QP_h tradeoff, various intermediate networks are, in general, possible between the $QP_{c,min}$ and $QP_{c,max}$ limits.

It must be noted that the QP_p-QP_h and QP_p-QP_c tradeoffs are independent and, therefore, the order in which they are explored is unimportant. The two independently-evolved designs after exploring both tradeoffs may be combined to yield the final network. If both evolutions are to the $QP_{h,min}$ and $QP_{c,min}$ limits, then the final network is the $QP_{p,max}$ design (Figure 4.30).

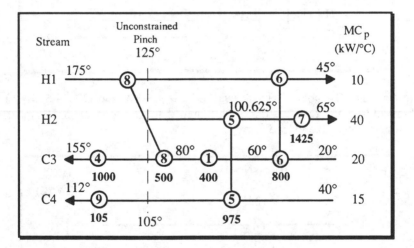

Figure 4.30 The $QP_{p,max}$ design above is obtained from the $QP_{p,min}$ design in Figure 3.14 by trading off hot utility below the pinch and cold utility above the pinch against process-to-process heat transfer across the pinch. (Temperature in °C and load in kW)

Step 3. Conventional Evolution for Relocating Heaters and Coolers

Conventional techniques of loop breaking and path relaxation may now be used to evolve the $QP_{p,min}$ network (Figure 3.14 with $QP_p = 0$) and the $QP_{p,max}$ network (Figure 4.30 with $QP_p = 500$). Thus, heaters and coolers may be relocated to the ends of the network. The first-order loop [3 6] in Figure 3.14 may be broken by shifting 395 kW from unit 3 to unit 6. Then, units 1 and 4 may be merged to yield a single heater of 1505 kW (Figure 4.31). Similarly, the first-order loop [8 6] in Figure 4.30 may be broken by shifting 500 kW from unit 8 to unit 6. Then, units 1 and 4 may be merged to yield a single heater of 1400 kW (Figure 4.32).

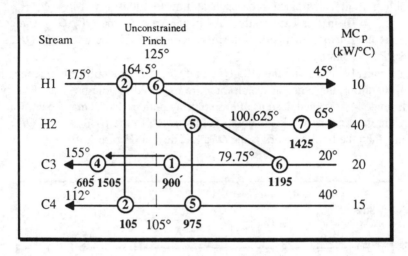

Figure 4.31 The above network (with QP_p = 395 kW and QP_h = 505 kW) is obtained from the $QP_{p,min}$ design in Figure 3.14 after conventional evolution. (Temperature in °C and load in kW)

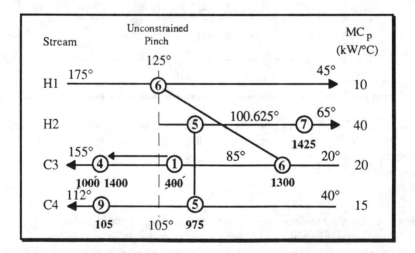

Figure 4.32 The above network (with QP_p = 500 kW and QP_h = 400 kW) is obtained from the $QP_{p,max}$ design in Figure 4.30 after conventional evolution. (Temperature in °C and load in kW)

In general, the $QP_{p,min}$ and $QP_{p,max}$ designs after evolution could have very different structures. Importantly, the above evolution methodology shows the entire gamut of design options available to the designer.

4.6 COST EVALUATION OF NETWORKS

A very large number of possible networks may be synthesized for a given stream data set, as demonstrated thus far. The selection of the best network among these is not straightforward. As practical considerations like safety, operability, controllability, flexibility, and layout are not easy to evaluate at this stage, the objective criterion typically used is total cost. A few cost models that vary in their level of detail are discussed below with the help of an example.

Example 4.12 Cost Estimation of Heat Exchanger Networks
Determine the total annualized cost (*TAC*) of the network in Figure 3.1 for the following cases:
   a) countercurrent exchangers;
   b) countercurrent exchangers with the maximum area per shell restricted to 350 m^2;
   c) 1-2 shell-and-tube exchangers; and
   d) 1-2 shell-and-tube exchangers accounting for piping and maintenance cost.

Given: Annual cost of unit duty ($/[kW.yr]): 120 (for HU) and 10 (for CU)
   Plant lifetime: 5 yr   Rate of interest: 10%
   Capital cost for carbon steel exchanger: $30000 + 750 A^{0.81}$ (*A* in m^2)
   Stream H2 is split in the MC_p-ratio of 25:15.

Solution. The operating cost is calculated using Equation 1.32 as $OC = (120)(360) + (10)(280) = 46 \times 10^3$ $/yr. The capital cost estimate depends on the model used.

a) For the case of countercurrent exchangers, the area of each unit in the network is calculated using $A = Q/(U\ LMTD)$. As the h-value for all the streams is 0.2 kW/(m^2 °C), the overall heat transfer coefficient, U, for each unit is 0.1 kW/(m^2 °C). The countercurrent areas are directly substituted into the equation for the exchanger capital cost (*CC*) to obtain the total capital cost for the network in Figure 3.1 as 618.24×10^3 $ (Table 4.20). Using the annualization factor, the total annual cost is given by $TAC = 46000 + (618.24 \times 10^3)(1 + 0.1)^5/5 = 245.14 \times 10^3$ $/yr. This compares well with the cost target of 240.42×10^3 $/yr in Table 1.15.

Table 4.20 Capital Cost Calculations for Countercurrent Units

Match	Load	ΔT at ends	LMTD	Area	CC
1 (H1-C3)	500	38 & 13	23.31	214.53	88.02
2 (H2a-C3)	1840	13 & 31.4	20.87	881.86	212.32
3 (H2b-C4)	560	13 & 13	13	430.77	132.05
4 (H1-C4)	520	50.33 & 33	41.06	126.65	67.86
5 (HU-C3)	360	25 & 42	32.77	109.86	63.74
6 (H1-CU)	280	48 & 30	38.30	73.11	54.26

Units: load in kW (in Fig 3.1), temperature in °C, area in m^2, and cost in 10^3 $.

b)   If there is the restriction that the maximum area per shell cannot exceed 350 m^2, then three shells of 293.95 m^2 may be used for match 2 and two shells of 215.39 m^2 may be used for match 3. Thus, from Equation 1.34 (keeping the distinction between shells and units in mind),

$$CC_{match\ 2} = 30000 + 750\ (3)\ (293.95)^{0.81} = 254.64 \times 10^3\ \$$$
$$CC_{match\ 3} = 30000 + 750\ (2)\ (215.39)^{0.81} = 146.41 \times 10^3\ \$$$

This gives the total annual cost as $TAC = 46000 + (674.93 \times 10^3)\ (1 + 0.1)^5/5$ = 263.40 x 10^3 $/yr.

c)   For the case of 1-2 shell-and-tube exchangers, the area of each unit in the network may be calculated using $A = Q/(U\ F\ LMTD)$ where F is the correction factor to the $LMTD$ to account for the noncountercurrent flow (see Chapter 6). An easy and equivalent way to obtain these areas is to use the targeting procedure in Example 1.5 with the single hot stream and single cold stream passing through the relevant exchanger. The number of shells for each 1-2 exchanger may also be obtained by the targeting method in Example 1.8. The areas and shells for the various matches in Figure 3.1 computed in this manner (using the software package HX for convenience) are given in Table 4.21. These are next substituted into the capital cost equation, e.g., $CC_{match\ 1}$ = 30000 + 750 (2) $(240.07/2)^{0.81}$ = 102.5 x 10^3 $. Then, TAC = 46000 + (796.6 x 10^3) $(1 + 0.1)^5/5$ = 302.6 x 10^3 $/yr.

d)   A cost model for piping and maintenance is discussed by Trivedi et al. (1989b) based on the suggestion originally made by Linnhoff (1979). It is possible to use the model only after the network structure is known and not at the targeting stage. The model is especially important when networks involve stream splits and, consequently, additional control valves, pipe-runs, and tees. The additional piping and maintenance cost (*PMC*) is then given by

$$PMC = [\beta N_{pipes}/(4\ N_{units}) + \beta (N_{nodes}/N_{u,min})^\gamma]\ (1/M + \phi)\ CC_s \qquad (4.7)$$

In the above equation, the term in square brackets is the correction for the additional piping. The term $\beta N_{pipes}/(4\ N_{units})$ is based on the simple observation (Linnhoff, 1979) that the number of pipe-runs is about four times the number of units in a simple network. The term $\beta(N_{nodes}/N_{u,min})^\gamma$, where γ is the pipework scale exponent, ensures that stream splitting becomes less attractive with increase in the number of units. The remaining term includes the amortization factor, M; the maintenance costs, ϕ; and the purchase cost of the shells, CC_s (excluding the installation cost).

For the network in Figure 3.1, the pipes required for streams H1, H2, C3, and C4 are 4, 6, 4, and 3 respectively, so $N_{pipes} = 17$. Since N_{nodes} represents the number of splitting and mixing junctions, N_{nodes} is 2 in Figure 3.1. The actual number of units (N_{units}) in Figure 3.1 is six whereas the Euler minimum unit target ($N_{u,min}$) is five. The purchase cost of the shells, CC_s, is 616.6 x 10^3 $ (i.e., 796600 − 6 x 30000). Using typical values for the other cost parameters ($\beta = 0.1$, $\gamma = 0.81$, $M = 5$, and $\phi = 0.2$) in Equation 4.7,

$$PMC = [0.1 \times 17/(4 \times 6) + 0.1(2/5)^{0.81}](1/5 + 0.2)(616.6 \times 10^3) = 29212 \ \$/yr.$$

Therefore, $TAC = 302.6 \times 10^3 + 29.21 \times 10^3 = 331.81 \times 10^3$ $/yr.

Table 4.21 Capital Cost Calculations for 1-2 Exchangers

Match	Load	Area	No. of Shells	CC
1 (H1-C3)	500	240.07	2	102.50
2 (H2a-C3)	1840	1090.94	4	311.88
3 (H2b-C4)	560	523.97	3	177.35
4 (H1-C4)	520	161.90	2	82.69
5 (HU-C3)	360	109.86	1	63.74
6 (H1-CU)	280	75.61	2	58.44

Units: load in kW (in Fig 3.1), area in m^2, and cost in 10^3 $.

The variety of network structures generated thus far for Case Study 4S1 is evaluated in Table 4.22 using the simplest model (countercurrent exchangers).

With the exception of Figures 3.1, 4.2, 4.9, 4.11, and 4.29 (where stream H2 is split in the MC_p-ratio of 25:15), the mixing junctions are chosen to be isothermal. It is observed that the designs at $\Delta T_{min} = 13°C$ yield near-optimal networks; thus, supertargeting is effective for pre-design optimization.

Table 4.22 Comparison of Different Designs

Network	ΔT_{min}	Utility (Hot)	Units	Splits	Area	TAC
Figure 3.1	13	360	6	1	1836.78	245.14
Figure 3.2	13	360	6	1	1900.74	248.24
Figure 3.3	13	360	6	2	1699.80	240.03
Figure 4.2	13	860	5	1	1707.17	287.98
Figure 4.4	13	620	5	1	1450.49	244.58
Figure 4.6	13	860	5	0	1230.51	259.33
Figure 4.9	20	605	7	1	1435.74	263.44
Figure 4.11	20	742.5	6	1	1335.58	264.29
Figure 4.13	20	1505	5	0	1035.96	331.25
Figure 4.16	20	1000	5	0	1123.56	271.84
Figure 4.17	20	930	6	0	1164.07	275.64
Figure 4.19	20	1000	5	1	1212.44	277.35
Figure 4.20	20	605	7	1	1450.57	264.86
Figure 4.21	13	360	5	2	2034.48	243.76
Figure 4.23	13	360	6	1	1840.84	246.74
Figure 4.29	13	360	7	1	1851.47	257.36

Units: temperature in °C, utility in kW, area in m^2, and cost in 10^3 $.

Any network close to the cost optimum may be chosen from Table 4.22 by the designer and evaluated further by considering intangibles like layout, operability, safety, flexibility, and controllability.

QUESTIONS FOR DISCUSSION

1. A consulting company develops a heat recovery system design for an inorganic bulk chemical process. You generate an alternative design through OPERA that appears cheaper in terms of energy consumption as well as capital expenditure. Is this possible? What about the classic energy-capital tradeoff?

2. Show that a spaghetti design with $(jh \times jc)$ matches for vertical heat transfer within an enthalpy interval can be evolved to a design with $(jh + jc - 1)$ matches by loop breaking. Specifically, evolve Figure 1.7b to Figure 1.8.

3. What is the valid range for the MC_p-ratio of the split on stream H2 in Figures 4.4, 4.8, 4.11, 4.19, and 4.21?

4. The allocation of multiple utilities in Section 1.7.3 minimizes operating cost but ignores capital cost considerations. Devise a supertargeting methodology for the case of multiple utilities using balanced composite curves that determines optimum ΔT_{min} values for process pinches as well as utility pinches. Specifically, how does one decide on the optimum distribution for the use of steam at different pressure levels? Are such tradeoffs important for threshold problems, too?

5. In case there is more than one loop in an MER network, is it a good strategy to break all the loops (ignoring ΔT_{min} violations in all intermediate networks) and to finally restore ΔT_{min} (by energy relaxation at the very end)?

6. Can loops not crossing the pinch exist in a network? Should such loops be broken during evolution? Does this lead to ΔT_{min} violations?

7. Does seeding networks with dummy exchangers allow topology traps to be overcome during evolution?

8. If an option existed during evolution, would it be preferable to eliminate a heater/cooler or a process-to-process exchanger?

9. Develop an alternative to the network in Figure 4.21 with the cooler on stream H2. Define all splits and mixing junctions clearly.

10. In Examples 4.7 and 4.10, RPA does not include heaters and coolers. Would it be meaningful to include them in the analysis and compute the associated area penalties?

11. Just before incorporating the cooler(s), Table 4.18 shows a higher penalty (8.6%) but a lower total area for the committed structure (1763.67 m^2) compared to Table 4.19 (penalty = 8.3% and area of committed structure = 1796.96 m^2). Comment on the above observation.

12. Why does the MC_p-ratio heuristic (see Section 4.4.6) not work well for matches 1 and 2 in Figures 4.27 and 4.28? Construct an example that shows the successful application of the MC_p-ratio heuristic.

13. In the capital cost expression (Equation 1.33), the exponent (c) for the heat exchange area is typically less than unity. Why?

14. A plant site can be divided into three different sections based on operability and layout considerations. What strategy may be adopted to decide whether it is worth integrating across these sections?

15. Comment briefly on the following aspects.

- Changing ΔT_{min} has a significant effect on the capital cost in the pinch region (rather than away from the pinch) due to the relatively large percent change that occurs in the driving forces in this region.
- A single utility can generate two utility pinches.
- In a hot oil circuit (an example of a variable temperature utility), it is meaningful to optimize the oil return temperature (considering tradeoffs) and maximize the oil supply temperature.
- RPA is a useful tool for determining the penalties that may be incurred due to preferred/compulsory matches.
- The DFP must be closely followed near the pinch and for the case of equal heat transfer coefficients.

PROBLEMS

4.A Loop Breaking and Path Relaxation

Break all the loops that cross the pinch in the MER networks (designed in Problem 3.A) for the following case studies. Restore ΔT_{min} using the concept of paths.

a) Case Study 4L2 at ΔT_{min} of 10°C
b) Case Study 4D1 at ΔT_{min} of 10°C
c) Case Study 4A2 (with temperatures reduced by 20°C) at ΔT_{min} of 20°C

Answers:

a) An evolved design appears in Figure 2.27 (Linnhoff et al., 1982, p. 45) with five units and an energy penalty of 7.5 kW.
    HU - C3 (27.5 kW); H1 - C4 (240 kW); H2 - C3 (112.5 kW);
    H1 - C3 (90 kW);   H2 - CU (67.5 kW)
b) An evolved design appears in Figure 8.6-6 (Douglas, 1988, p. 244) with five units and an energy penalty of 60 kW. However, the design below has a penalty of only 50 kW.
    HU - C4 (120 kW); H2 - C4 (240 kW); H1 - C3 (130 kW);
    H2 - C3 (50 kW);   H2 - CU (110 kW)
c) An evolved design appears in Figure 5 (Trivedi et al., 1990a, p. 605) with five units and an energy penalty of 200 kW.
    HU - C4 (4000 kW); H1 - C4 (4200 kW); H2 - C3 (5600 kW);
    H1 - CU (2400 kW); H2 - CU (1600 kW)

4.B Evolution of Split-Stream Design

Break all the loops that cross the pinch in the split-stream MER network (designed in Problem 3.B.b) for Case Study 4T1 at ΔT_{min} of 10°C. Restore ΔT_{min} using the concept of paths.

Answers:

For the two possible below-pinch designs, the two evolved designs are different and appear in Figures 6b & 6d (Trivedi et al., 1990a, p. 606). They feature five units and an energy penalty of 25 kW.

HU - C4 (145 kW)	HU - C4 (145 kW)
H1 - C4 (455 kW)	H1 - C4 (455 kW)
H2 - C3 (325 kW)	H2 - C3 (500 kW)
H1 - C3 (175 kW)	H1 - CU (175 kW)
H2 - CU (275 kW)	H2 - CU (100 kW)

4.C Elimination of Loops in Multiple Utility Designs

Break the loops that cross the utility pinches in the multiple utility design (ΔT_{min} = 10°C) in Problem 3.C for Case Study 4L3. It is desired to restore ΔT_{min} by sacrificing some of the VLP steam-raising at the expense of increased use of the cold utility at 15°C.

Answers:

An evolved design appears in Figure 2.39 (Linnhoff et al., 1982, p. 63) with seven units, no stream splits, and VLP steam-raising reduced from 460 kW to 300 kW.

Above-process Pinch (145° C)	*Below-process Pinch (145° C)*
HU - C3 (600 kW)	H2 - C3 (900 kW)
H1 - C3 (600 kW)	H2 - VLP (300 kW)
	H1 - C4 (2200 kW)
	H2 - C3 (1500 kW)
	H2 - CU (1700 kW)

4.D Elimination of Units Using Bypasses

Consider the MER network designed in Problem 3.A.a for Case Study 4L2 at ΔT_{min} = 10°C. Show that a unit can be eliminated to obtain an MNU design without incurring an energy penalty or violating the ΔT_{min} constraint. Use the bypassing scheme proposed by Wood et al. (1985).

Answers:

Two possible MER-cum-MNU designs appear in Figures 3b and 3c (Wood et al., 1985, p. 375).

HU - C3b (20 kW)	HU - C3b (20 kW)
H2 - C3a (120 kW)	H1 - C4 (240 kW)
H1 - C4 (240 kW)	H1 - C3b (30 kW)

H1 - C3b (90 kW)   H2 - C3a (180 kW)
H2 - CU (60 kW)   H1 - CU (60 kW)

4.E Remaining Problem Analysis to Achieve Area Targets

Consider the MER network designed in Problem 3.A.c for Case Study 4A2 at $\Delta T_{min} = 20°C$. Use RPA to evaluate the matches in terms of their 1-2 shell and tube area requirements. If some matches are not favorable in achieving the minimum area network, explore other match options to develop a better MER design that approaches the area target.

Answers:

A design using RPA appears in Figure 9 (Ahmad and Smith, 1989, p. 490) with an area of 2138.8 m^2, which is within 5.5% of the target area of 2027 m^2. There is a heater on stream C4 (4000 kW) and cooler on stream H2 (3800 kW).

Match	Load	Target$_M$	Target$_{RP}$	Area Penalty	% Penalty
H1 - C3	1600	359.95	1731.26	64.19	3.2
H2 - C4	1800	400.42	1626.60	0	0
H1 - C4	2600	258.25	1804.63	35.86	1.8
H2 - C3	1600	417.82	1620.25	11.05	0.6
H1 - C4	2400	266.94	1808.44	48.37	2.4

Units: load in kW and 1-2 area (target and penalty) in m^2.

4.F Diverse Pinch Concept to Design with ΔT Contributions

Ahmad et al. (1990) discussed the use of stream individual "ΔT contributions" based on the suggestions of Nishimura (1980) and Townsend (1989). According to this, the temperature difference between a hot stream and a cold stream may be expressed as the sum of two ΔT contributions where each $\Delta T_j = \kappa\, h_j^{-1/2}$. This is the diverse pinch concept as given by Equation 2.11, with the difference that the exponent is $-1/2$ rather than -1.

  a) Show that the exponent of $-1/2$ rather than -1 in the diversity relation does not affect the design in Figure 4.26.
  b) For Case Study 5A1, design an MER network at $\Delta T_{min} = 30°C$ using the conventional pinch design method.
  c) For Case Study 5A1, design an MER network for the same utility requirements as in a), using the diverse pinch concept and an exponent of $-1/2$.

Answers:

  a) Though $\kappa = 15.29$ and the pinch corresponds to a shifted temperature of 50.81°C (when the exponent is $-1/2$), the design is the same as in Figure 4.26.

b) A possible conventional pinch design appears in Figure 3b (Ahmad et al., 1990, p. 754) with an area of 351 m^2.

Above Pinch (144° C)

H2 - C5a (220.32 kW)
H3 - C5b (989.92 kW)
HU - C5 (1456.72 kW)

Below Pinch (144° C)

H1b - C5 (215.71 kW)
H1a - C4a (222.35 kW)
H2 - C4b (161.16 kW)
H3 - C4c (371.22 kW)
H1 - C4 (186.6 kW)
H1 - CU (1248.04 kW)

c) A possible diverse pinch design for κ = 13.56 (when the exponent is −1/2) appears in Figure 4b (Ahmad et al., 1990, p. 755) with an area of 265 m^2.

Above Pinch (145.4° C)

H2 - C5a (1027.58 kW)
H3 - C5b (204 kW)
HU - C4 (217.44 kW)
HU - C5 (1239.28 kW)

Below Pinch (145.4° C)

H1b - C5 (411.81 kW)
H1a - C4 (213.85 kW)
H2 - C4a (177.48 kW)
H3 - C4b (333.56 kW)
H1 - CU (1248.04 kW)

4.G Evolution of Constrained Heat Exchanger Network

Evolve the $QP_{p,min}$ networks (designed in Problem 3.G) to $QP_{p,max}$ networks using the QP_p-QP_h and QP_p-QP_c tradeoffs. Evolve both the $QP_{p,min}$ and $QP_{p,max}$ networks further using conventional loop and path techniques. Note that the starting MER networks correspond to Case Study 10O1 at ΔT_{min} of 20°C with the following forbidden matches: H1- C5, H1 - C9, H1 - C10, H2 - C6, H2 - C9, H2 - C10, H4 - C5, and H4 - C10.

Answers:

Evolved designs appear in Figures 26 and 27 (O'Young & Linnhoff, 1989)

$QP_{p,max}$ Design	Evolved $QP_{p,max}$	Evolved $QP_{p,min}$
HU - C5 (25)	HU - C5 (25)	HU - C5 (25)
HU - C9 (32)	HU - C9 (32)	HU - C9 (30.8)
H1 - C8 (1.8)	H1 - C8 (6.6)	H1 - C8 (1.8)
H1 - C6 (1.68)	H1 - C6 (9.03)	H1- C6 (9.03)
H2 - C5 (7)	H2 - C5 (15)	H2 - C5 (15)
H3 - C10 (1.2)	H3 - C10 (3.2)	H3 - C9 (1.2)
H1 - C8 (4.8)	H1 - C7 (1.07)	HU - C10 (13)
H1 - C7 (1.07)	HU - C10 (11.8)	H4 - C8 (4.8)
H1 - C6 (7.35)	H4 - C7 (6.805)	H4 - C7 (7.875)
H2 - C5 (8)	H1 - CU (13)	H3 - C10 (2.0)
HU - C10 (11.8)	H4 - CU (32.295)	H1 - CU (18.87)
H4 - C7 (6.805)		H4 - CU (26.425)
H3 - C10 (2)		

$QP_{p,max}$ *Design (Continued)*
H1 - CU (13)
H4 - CU (32.295)

4.H Designing at Topology Traps

A design in one topology region has a grossly different structure and does not evolve easily to a design in another topology region. This is primarily due to the pinch point changing location for two different ΔT_{min} values. This causes difficulties in locating a globally optimum structure. Can the difficulty be overcome by designing at the topology trap itself (where both pinches exist)? To answer this question, consider Case Study 4S1t, which has a topology trap at $\Delta T_{min} = 38.33°C$ (see Table 2.11).

a) Design an MER network for Case Study 4S1t at $\Delta T_{min} = 35°C$. Note that its structure is drastically different from the conventional pinch designs at $\Delta T_{min} = 49°C$ (given in Figures 4.27 and 4.28).

b) Now, design an MER network for Case Study 4S1t at $\Delta T_{min} = 38.33°C$. Discuss whether this multiple-pinched design provides a superstructure that includes the MER network in a) as well as those in Figures 4.27 and 4.28.

Answers:

a) *Above Pinch (37.5° C)*   *Below Pinch (37.5° C)*

Above Pinch (37.5° C)	Below Pinch (37.5° C)
H1 - C3 (200 kW)	H1 - CU (100 kW)
H2 - C3 (900 kW)	
H1 - C4 (1000 kW)	
HU - C3 (1600 kW)	
HU - C4 (80 kW)	

b) Two designs are possible for the above-pinch region.

Above Pinch (~59° C)	Above Pinch (~59° C)	Between Pinches
H1 - C3 (966.67 kW)	H1 - C4 (966.67 kW)	H1 - C3a (200 kW)
H2 - C4 (700 kW)	H2 - C3 (700 kW)	H2 - C3b (200 kW)
HU - C3 (1333.33 kW)	HU - C3 (1600 kW)	*Below Pinch (~39° C)*
HU - C4 (380 kW)	HU - C4 (113.33 kW)	H1 - CU (133.33 kW)

4.I Diverse Pinch Design

a) For Case Study 4S1h, determine the minimum utility requirements for a minimum flux specification of $Q''_{min} = 2.5 \text{ kW/m}^2$.

b) For Case Study 4S1h, design a network using the diverse pinch methodology for Q''_{min} of 2.5 kW/m^2.

c) Compare the diverse pinch design in b) with a conventional pinch design for the same utility consumption.

Answers:
   a) $Q_{hu,min} = 100$ kW; $Q_{cu,min} = 1970$ kW
   b) A possible diverse pinch design is given below.

Above Pinch (162.5° C)	*Below Pinch (162.5° C)*
HU - C3 (100 kW)	H1 - C3 (2600 kW)
	H2 - C4 (1080 kW)
	H1 - CU (650 kW)
	H2 - CU (1320 kW)

   c) Above the conventional pinch at 162.5°C, there is only a heater of
       100 kW on stream C3. However, three possible designs for the below-
       pinch region are given below.

Below Pinch	*Below Pinch*	*Below Pinch*
H1 - C3 (950 kW)	H1 - C3 (950 kW)	H1a - C3 (2600 kW)
H1 - C4 (237.5 kW)	H1 - C4 (180 kW)	H1b - C4 (180 kW)
H1 - C3 (1650 kW)	H2 - C4 (900 kW)	H2 - C4 (900 kW)
H2 - C3 (842.5 kW)	H1 - C3 (1650 kW)	H1b - CU (470 kW)
H1 - CU (412.5 kW)	H1 - CU (470 kW)	H2 - CU (1500 kW)
H2 - CU (1557.5 kW)	H2 - CU (1500 kW)	

CHAPTER 5

Methods Based on
Dual Approach Temperatures

If you are faced with a painful choice
between two opposing goals,
keep in mind that priorities can later be adjusted
and readjusted ad infinitum.

Alan Lakein

Within the category of systematic evolutionary methods for HENS, there exist the dual approach temperature methods (DATMs) in addition to the pinch design method (PDM). The concept of dual approach temperatures was used first by Challand et al. (1981) and later by Colbert (1982).

Using dual approach temperatures has the advantage of designing for fixed utility consumption, so these designs may be termed fixed energy recovery (FER) networks. Furthermore, the designer has more flexibility during network synthesis: the MC_p criterion used in the PDM is not as stringently observed in the DATMs, and so promising matches missed (not allowed) by the PDM are identified in the DATMs for some cases. Designs from DATMs usually have fewer numbers of shells and units and are therefore simpler and cheaper. As the DATMs greatly reduce the necessity for stream splitting, they typically lead to networks with better operability and controllability.

Methods based on dual approach temperatures have gained importance in recent years with the papers of Trivedi et al. (1989b), Colberg and Morari (1990), Gundersen and Grossmann (1990), Wood et al. (1991), Jezowski (1991a), and Suaysompol and Wood (1991). This chapter is based on these studies which suggest that the use of dual approach temperatures may yield superior and cost-effective solutions.

5.1 HRAT and EMAT

Dual approach temperature methods, as their name suggests, use two minimum temperature differences: HRAT and EMAT.

The heat recovery approach temperature (HRAT) sets the limit on the amount of process heat that can be recovered from the system. It is the minimum temperature difference between the composite curves and determines the minimum utility consumption. Thus, an increase in HRAT corresponds to the composite curves being further apart and leads to a decrease in total area and heat recovery.

The exchanger minimum approach temperature (EMAT) sets the lower limit on the temperature difference that must exist between the hot and cold streams of every exchanger in the network for reasonable sizing. A decrease in EMAT causes an increase in the number of shells but a decrease in the number of units and the complexity of the network.

It is generally unnecessary (and often inappropriate) to consider HRAT and EMAT to have the same value since the minimum temperature difference in an exchanger need not be determined by that between the composite curves. In general, EMAT ≤ HRAT; in that sense, the PDM is a special case in which EMAT = HRAT during MER network synthesis.

To gain a clear understanding of the two terms (HRAT and EMAT), consider the following examples. The MER network in Figure 3.1 has an HRAT of 13°C (based on the hot utility requirement of 360 kW) and an EMAT of 13°C (based on the minimum temperature driving forces for units 1, 2, and 3). This MER network was evolved in Example 4.1. The evolved network in Figure 4.6 has a hot utility requirement of 860 kW, and therefore HRAT is 27.286°C (see Table 2.6) based on the energy targeting from the composite curves. However, unit 2 has a minimum temperature approach of 13°C and, therefore, EMAT is 13°C for the network. Essentially, the HRAT is relaxed by the PDM during evolution whereas EMAT is held sacrosanct.

Importantly, the network in Figure 4.6 has two approach temperatures. Can this realization be exploited to develop a better method for synthesizing such a network? Recall that the steps involved in synthesizing Figure 4.6 were nontrivial. Also, the energy penalty during evolution (which directly relates to the HRAT relaxation) is not controlled by the designer. The energy penalty incurred after loop breaking and path relaxation cannot be predicted a priori as a target. Many MER networks are possible and each leads to different evolutions with different energy penalties. Therefore, many networks need to be generated and evaluated for the energy-capital tradeoff as seen in the previous two chapters. The DATMs are a response to these problems associated with the PDM.

Colbert (1982) introduced a variation of the temperature interval method of Linnhoff and Flower (1978) using dual temperatures. In Colbert's method, a large number of subnetworks (typically one less than twice the number of process streams) needs to be designed independently corresponding to the many temperature intervals defined by the EMAT. This design usually has too many units and loops; therefore, it needs extensive evolution as well as intra- and inter-subnetwork matching to obtain a practical network. Even for problems of a reasonable size, the effort is substantial and is best handled on a computer. As the method has other difficulties also (e.g., in identifying stream-split situations and in generating all possible network options because of topology traps), it is not worth pursuing further. Instead, the new method proposed by Trivedi et al. (1989b) is discussed in detail below since it draws on the desirable features of both the Colbert method and the PDM. It profitably uses the dual approach temperature concept from Colbert (1982) and divides the problem into only two independent subnetworks as in the PDM. The division is made at what Trivedi et al. (1989b) call the pseudo-pinch; hence, this synthesis procedure based on simple thermodynamic principles is appropriately termed the pseudo-pinch design method (PPDM). It does not call for a computer solution, and the resulting network typically requires minimal evolution (an advantage for industrial problems with many streams).

5.2 PSEUDO-PINCH DESIGN METHOD

Recall that the pinch must lie at a vertex on the composite curves. In fact, the pinch in any network necessarily lies at the point of entry of a hot or cold stream (Grimes et al., 1982). This holds good for the pseudo-pinch point also, which is defined in terms of the inlet temperature of this stream. Furthermore, as in the conventional pinch, there is a heat sink above the pseudo-pinch (where only hot utility is required) and a heat source below the pseudo-pinch (where only cold utility is required), but the pinch temperatures of all streams are fixed in the conventional PDM whereas they are not in the PPDM. Except for the stream that enters the network and which is used to define the pseudo-pinch, there is some flexibility in defining the pinch temperatures of the other streams. This flexibility may be exploited to generate simple and near-optimal designs.

Example 5.1 Network Generation Using Pseudo-pinch Design Method
Design a network using the PPDM for the stream data in Case Study 4S1. Given: HRAT = 20°C and EMAT = 13°C.

Solution. In this method, both the HRAT and EMAT must be specified ahead of design ensuring that HRAT $\geq$ EMAT. The method (Trivedi et al., 1989b; Wood et al., 1991) essentially requires determination of the heat to be passed across the pseudo-pinch and its appropriate distribution.

Step 1. Energy Targeting for HRAT and EMAT
The minimum utility requirements corresponding to the HRAT and EMAT need to be calculated. This is simply done, using the procedure (PTA)

employed in Example 1.1 by setting ΔT_{min} equal to HRAT and EMAT in turn. The results are summarized below:

ΔT_{min} (°C)	Hot Utility (kW)	Cold Utility (kW)
20 (= given HRAT)	605	525
13 (= given EMAT)	360	280

The utility consumption of the network is now fixed at 605 kW of hot utility and 525 kW of cold utility, corresponding to the HRAT. The difference between the utility requirements for the HRAT and EMAT is denoted by α_n. Thus, in this case,

$$\alpha_n = \text{Utility}_{HRAT} - \text{Utility}_{EMAT} = 605 - 360 = 525 - 280 = 245 \text{ kW}.$$

Since HRAT $\geq$ EMAT, α_n is never negative.

Now, the network is designed based on an EMAT of 13°C. With respect to the conventional pinch (125°C/112°C), there is excess energy equal to α_n being supplied to the system. This energy must be transferred across the conventional pinch as per the "more in, more out" principle. The excess energy is therefore allocated to the various streams that cross the pinch, which causes their pinch temperatures to change. The resulting pinch is termed the pseudo-pinch, as it allows "energy leakage" across the conventional pinch. Note that the pseudo-pinch is identical to the conventional pinch when HRAT = EMAT and $\alpha_n = 0$.

Step 2. Defining the Pseudo-Pinch

Defining the pseudo-pinch, in general, involves logically distributing α_n among streams crossing the conventional pinch. This leads to additional match options, primarily due to the flexibility afforded by EMAT being less than HRAT.

For Case Study 4S1, the conventional pinch (at an EMAT of 13°C) is identified by the entry of stream H2 (whose inlet temperature is 125°C), so the pseudo-pinch temperature of stream H2 is 125°C. Now, the pseudo-pinch temperatures for the cold streams are EMAT below this hot stream inlet temperature. In this case, the pseudo-pinch temperatures of streams C3 and C4 are 112°C (i.e., 125 – 13). If the conventional pinch (at EMAT) is identified by the entry of a cold stream, then the pseudo-pinch temperatures for the hot streams are EMAT above this cold stream inlet temperature.

Now, there is really no choice in this case study as there is only one other hot stream (H1) capable of carrying the 245 kW of heat across the conventional pinch. Therefore, α_n is completely allocated to stream H1, whose pseudo-pinch temperature becomes 149.5°C (i.e., 125 + 245/10).

The allocation of α_n in this case is easy, and the pseudo-pinch is unique. This is not the case in general, especially when the problem involves many streams. Various heuristics for the rational allocation of α_n in such cases will be enumerated later.

Step 3. Separate Matching for the Sink and Source Subproblems

The pseudo-pinch, like the conventional pinch, partitions the network in two: a sink subproblem and a source subproblem, which are in enthalpy balance on considering their respective hot and cold utility requirements (as dictated by the HRAT). Each subproblem may be independently designed (Figure 5.1), starting at the pseudo-pinch and moving away towards the hot or cold end. In that sense, the matching procedure is similar to that in the PDM. Of course, the maximum load for a match must be calculated subject to the EMAT constraint.

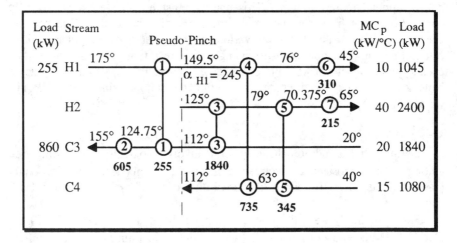

Figure 5.1 A network for fixed energy recovery (with HRAT = 20°C) synthesized using the pseudo-pinch design method (with EMAT = 13°C) involves no stream splits. (Temperature in °C and load in kW)

However, there is an important difference between the PDM and the PPDM: the feasibility criteria (the number criterion and the MC_p criterion discussed in Section 3.1.1) are not rigorously applicable at the pseudo-pinch, resulting in greater freedom during match selection. This occurs when the approach temperatures between the streams at the pseudo-pinch are greater than EMAT. The benefit is that stream splits required in the PDM are no longer necessary in the PPDM. This is evidenced in this example itself.

For Case Study 4S1, only streams H1 and C3 exist above the pseudo-pinch (Figure 5.1), so match 1 (with a heat load of 255 kW) allows stream H1 to be ticked off. The remaining heat of 605 kW on stream C3 is supplied by hot utility. This completes the design above the pseudo-pinch, which in this case is very simple.

Below the pseudo-pinch, stream C3 may be ticked off using match 3 (of 1840 kW) with stream H2. Now, the only possibility is to match streams H1 and C4, but it is not possible to tick off stream H1 due to the EMAT constraint. At best, the temperature difference at the cold end of the match can be equal to EMAT (13°C). If the outlet temperature of stream H1 for this match is T_{H1}, then an energy balance yields $10 (149.5 - T_{H1}) = 15 (112 - T_{H1} + 13)$ or $T_{H1} = 76$°C. Therefore, the maximum load for match 4 is 735 kW. The remaining load on stream C4 is satisfied by stream H2 through match 5 (of 345 kW). The design below the pseudo-pinch is completed using two coolers of 310 kW and 215 kW on streams H1 and H2 respectively. Note that match 4 would not be allowed by the PDM because it would violate the MC_p criterion at the conventional pinch. As a consequence, the number criterion in the PDM would force a stream split in the design below the conventional pinch.

An alternative to Figure 5.1 is possible for the design below the pseudo-pinch. However, it is more complex. Rather than the matches near the pseudo-pinch being H2-C3 and H1-C4 (Figure 5.1), consider the case where they are H1-C3 and H2-C4 (Figure 5.2).

From the EMAT constraint, the energy balance yields $10 (149.5 - T_{H1}) = 20 (112 - T_{H1} + 13)$ or $T_{H1} = 100.5$°C, so the maximum load for match 3 is 490 kW. If stream C4 is ticked off using stream H2, then $T_{H2} = 125 - 1080/40 = 98$°C. The design, however, cannot be completed since no further matches are possible due to EMAT constraints. The way to circumvent this difficulty is to assign match 4 only its allowable heat load and not the maximum possible (see the hidden pinch phenomenon in Section 4.4.3). For the design to proceed further, a match is needed between streams H2 and C3. This is possible if $T_{H2} = 87.5 + 13 = 100.5$°C and the load on match 4 is 980 kW. Now, the remaining load on stream C3 is satisfied by stream H2 through match 5 (of 1350 kW) and the remaining load on stream C4 by stream H1 through match 6 (of 100 kW). The design below the pseudo-pinch is completed using two coolers of 455 kW and 70 kW on streams H1 and H2 respectively.

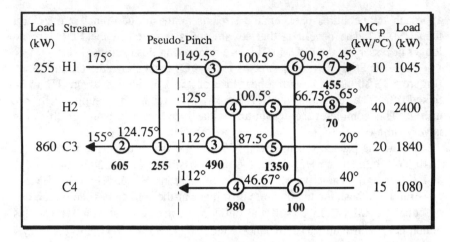

Figure 5.2 An alternative to the FER network (HRAT = 20°C and EMAT = 13°C) in Figure 5.1 is possible with a different cold-end design. (Temperature in °C and load in kW)

Step 4. Simplification of the Network through Loop Breaking

Figure 5.1 has two units more than the minimum. Of the three loops ([1 3 7 6 1], [4 5 7 6 4], and [1 3 5 4 1]), only two are independent. Figure 5.2 has three units more than the minimum; however, units 1 and 3 can obviously be combined into one. Then, of the three loops ([1 5 8 7 1], [4 6 7 8 4], and [1 5 4 6 1]), only two are independent.

Simplification of the networks may be attempted by loop breaking (i.e., by shifting heat loads). Recall that these are fixed energy recovery (FER) networks, and so no energy relaxation using paths is possible after loop breaking. Also, the EMAT constraint must be respected. It is observed that the elimination of units from Figure 5.1 and 5.2 is not straightforward. Systematic energy relaxation using LONITA (see Section 4.2) may be attempted (Trivedi et al., 1989b). Also, heat loads may be shifted around a loop without elimination of a unit in order to discover favorable changes towards the minimum network area or number of shells. An elegant network with only five units may be obtained from Figure 5.2, provided the EMAT is lowered to 11.5°C (from 13°C). However, this network is not developed here (see Problem 5.B.a) since it is generated more naturally by other methods (see next two sections).

Thus, it appears that the value of EMAT is critical. How is EMAT chosen? Wood et al. (1991) suggest that EMAT may be chosen to be about half of HRAT in value. An alternative way would be select EMAT as well as

HRAT based on network economics: Trivedi et al. (1989b) have shown that for a fixed HRAT, there is an optimum value of EMAT at which the capital cost is a minimum. For the same energy consumption, they argue that the dual approach temperatures lead to near-optimal designs whereas MER designs (with HRAT = EMAT) as well as MNU designs may not. Thus, HRAT may be varied in an outer loop and EMAT in an inner loop to locate the globally optimal design in terms of total annual cost. Of course, this approach, where both HRAT and EMAT are treated as optimization variables, involves considerable computational effort. Gundersen and Grossmann (1990) and Jezowski (1991a) have pointed out that EMAT is not an appropriate optimization variable in HENS. This is partly evident in the next two sections.

Note that the PPDM and PDM yield virtually identical results for very low approach temperatures. However, for approach temperatures greater than about 20°C, the PPDM has the potential of generating cost-effective designs not possible through the PDM.

Finally, consider the heat α_n to be passed across the pinch and its allocation among the streams crossing the pinch. Let α_j be the amount of heat allocated to the j-th stream. Then,

$$\Sigma \, \alpha_j = \Sigma \, \alpha_{jh} + \Sigma \, \alpha_{jc} = \alpha_n. \tag{5.1}$$

Equation 5.1 merely states that the total energy transfer across the conventional pinch point for the network is obtained by summing the individual contributions for all the streams (hot and cold).

If α_j is positive, then the pseudo-pinch temperature for a hot stream will be higher than its conventional pinch temperature by $(\alpha_{jh}/MC_{p,jh})$ and that for a cold stream will be lower by $(\alpha_{jc}/MC_{p,jc})$. For example, stream H1 in Figure 5.1 has a pseudo-pinch temperature of 149.5°C and a conventional pinch temperature of 125°C, corresponding to $\alpha_{H1} = \alpha_n = 245$ kW. Sometimes, α_j may need to be negative (see Section 5.3) for some streams in order to satisfy Equation 5.1. With respect to the conventional pinch, hot streams with a nonzero α_{jh} have their heat contents decreased by α_{jh} above the pinch and increased by α_{jh} below the pinch. On the other hand, cold streams have their heat contents increased by α_{jc} above the pinch and decreased by α_{jc} below the pinch.

The following heuristics are suggested (Trivedi et al., 1989b; Wood et al., 1991) for allocating the cross-pinch heat transfer, α_n, among the streams crossing the pinch.

a) Certain streams may be totally slipped. This implies that the entire enthalpy requirements of the stream are supplied either above or below the pseudo-pinch although the stream actually crosses the conventional pinch. Specifically, a hot stream may be completely satisfied below the pseudo-pinch or a cold stream completely satisfied above the pseudo-pinch. This is possible only if the heat excess of a hot stream above the conventional pinch or the heat shortfall of a cold stream below the pinch is not greater than α_n.

In the above case study, no stream could be completely slipped as α_n (245 kW) is not large enough to accommodate any hot stream below the pseudo-pinch or any cold stream above the pseudo-pinch. For stream H1 to be totally slipped, α_n would have to be 500 kW.

b) α_n may be allocated such that a part of a certain stream is slipped across, resulting in the complete matching of two streams (a subset equality). This is possible when the enthalpy requirements of a hot and cold stream on the same side of the pinch are similar. Also, EMAT for the match must be satisfied.

In Figure 5.1, it is seen that the enthalpy requirements would be identical (1080 kW) for streams H1 and C4 below the pinch if α_n could be set to 280 kW. Partial stream slipping of H1 would occur; however, a subset equality match would not be feasible due to EMAT at the cold end of the match being 5°C, which is less than the minimum allowable EMAT (13°C).

c) α_n may be allotted equally to all the streams crossing the pinch. For N streams, the temperature of each stream will change by $\alpha_n/(N\,MC_p)$.

d) The temperatures of all streams can be changed equally. The temperatures of hot streams will increase by $\alpha_n/(\Sigma\,MC_p)$ and the temperatures of cold streams will decrease by $\alpha_n/(\Sigma\,MC_p)$.

e) The amount transferred by each stream, α_j, may be determined by using weighting factors, say, inversely proportional to the stream heat transfer coefficients. This would lead to large approach temperatures for streams having small heat transfer coefficients and would be conceptually similar to the diverse pinch case.

The first two heuristics could normally be given preference (as in the flexible pinch design method discussed in the next section) because they lead to a reduction in the number of units required. If options exist during the application of the first two heuristics, then streams with a large heat capacity flow rate could be given preference to minimize the changes in the pinch temperatures. A combination of the above heuristics could be used for allocating α_n rather than any single heuristic. In fact, the allocation is solely at the discretion of the designer. Any strategy may be used so long as the network is partitioned into two subproblems in enthalpy balance. However, note that the PPDM as proposed by Trivedi et al. (1989b) only employed the simple heuristic of equal allocation of α_n among all the streams crossing the pinch. Wood et al. (1991) have demonstrated the use of the other heuristics within the context of the PPDM, but the flexible pinch design method (Suaysompol and Wood, 1991) appears to be a refined methodology for the purpose and is discussed in the next section.

Though the PPDM has advantages (e.g., reduced stream splitting) and works well for some examples (see Problem 5.B), it has some deficiencies as observed in Example 5.1. These are summarized below.

- EMAT must be specified prior to the design, and a poor choice of EMAT may lead to a suboptimal network. The use of EMAT as an optimization variable implies considerable computational effort and is inappropriate (Gundersen and Grossmann, 1990; Jezowski, 1991a).
- Although a variety of heuristics exist for the allocation of α_n providing flexibility, the lack of a definite procedure results in a certain arbitrariness and a dependence on the intuition of the designer.
- Although minimal evolution may typically be necessary, the difficulties in loop breaking encountered in the PDM would persist in those cases where network simplification is indeed required.

The methods outlined in the next two sections attempt to resolve these difficulties.

5.3 FLEXIBLE PINCH DESIGN METHOD

Introduced by Suaysompol and Wood (1991), this method is a variation of the PPDM. The flexible pinch design method (FPDM) attempts to eliminate the requirement for a variety of heuristics for the allocation of α_n; instead, it relies mainly on matching streams off completely. Also, the main relation used for the cross-pinch heat exchange is $\Sigma \alpha_{jh} + \Sigma \alpha_{jc} = 0$ (and not α_n as in Equation 5.1 for the PPDM).

The FPDM recognizes that the PDM is too inflexible if the ΔT_{min} constraint is rigidly followed. Therefore, it employs a variable approach temperature for the exchangers, with HRAT alone being specified ahead of design. Essentially, it recognizes that the approach temperatures for some streams at the flexible pinch need not be identical to the HRAT for the conventional pinch. As in the PPDM, the flexible pinch temperatures are defined through the cross-pinch heat transfer, but the objective here is to generate simple designs with few units. This automatically reduces the effort required for evolutionary improvement and network simplification.

As a matter of useful terminology, streams that lie entirely on one side of the conventional pinch may be called fixed streams. On the other hand, floating streams cross the conventional pinch and have exchangers on both sides of the pinch. Consequently, floating streams lead to additional matches beyond the Euler MNU target. The FPDM focuses attention on generating simple networks by reducing the number of matches between the floating streams and thus minimizing the task of loop breaking. A valuable strategy for generating networks that approach the MNU target would be the conversion of floating streams to fixed streams with respect to the flexible pinch boundary. This is possible by total stream slipping, i.e., by setting the flexible pinch temperatures of these streams to the appropriate supply or target temperatures.

In cases where total stream slipping causes excessive cross-pinch heat transfer, it may be advisable to attempt partial stream slipping so that a subset equality match is possible on one side of the flexible pinch.

In addition to total stream slipping and subset equality matches, mirror image matches provide a third strategy for designing simple networks that approach the MNU target. Mirror image matches are created by matching the same pair of floating streams on both sides of the flexible pinch and subsequently combining them into a single match (e.g., matches 1 and 3 in Figure 5.2). Such matches may sometimes yield high area and shell requirements because their temperature profiles converge on one side of the pinch (e.g., the temperature driving force for match 3 in Figure 5.2 reduces from 37.5° to 13°C). Caution must be specially exercised by minimizing the heat loads of tunnel matches (having long almost-parallel temperature profiles with tight driving forces) in cases where the heat capacity flow rate ratio, R, is close to unity.

The philosophy of the FPDM discussed above may be compacted into the following five heuristics (Suaysompol and Wood, 1991) for the development of simple FER designs:

- EMAT should be greater than HRAT/3 to ensure that matches do not have high area and shell requirements.
- Total slipping of a floating stream with a small heat content on one side of the pinch should be attempted to reduce the number of units required.
- Partial slipping of a floating stream should be attempted to form a subset equality match with another stream of similar heat content and result in a simpler network.
- For a tunnel match (which may be roughly identified when $0.8 \leq R \leq 1.25$ and $1/3 < $ EMAT/HRAT $< 2/3$), the heat load should be minimized by increasing the load of the adjacent units on both the hot and cold streams constituting this match; for other mirror image matches, the load should be maximized.
- The hot stream with the highest inlet temperature and the cold stream with the highest outlet temperature (as per the H/H heuristic in Section 3.4) should be matched in all other cases especially away from the flexible pinch to use the available driving forces effectively.

A subset equality match through partial stream slipping followed by combination with a mirror image match is beneficial and can be equivalent to total stream slipping (which may therefore be unnecessary).

Example 5.2 Network Generation Using Flexible Pinch Design Method

Design a network using the FPDM for the stream data in Case Study 4S1. Given: HRAT = 20°C.

Solution. In this method, only the HRAT needs to be specified ahead of design. The method essentially uses the principles outlined above, namely, stream slipping, subset equality matches, mirror matches, tunnel matches, and temperature matching by the H/H heuristic. Importantly, stream slipping determines the cross-pinch heat transfer in the FPDM. Furthermore, it must be balanced in accordance with $\Sigma\ \alpha_{jh} + \Sigma\ \alpha_{jc} = 0$, which in turn defines the flexible pinch without requiring the separate specification of EMAT.

For this case study, EMAT should be greater than 6.67°C as per the first heuristic. Based on the values of R, it is clear that there is no possibility of a tunnel match, so the loads for mirror image matches should be maximized.

Step 1. Energy Targeting for HRAT and Decomposition at Conventional Pinch

For the specified HRAT of 20°C, the minimum utility requirements using the PTA are 605 kW (hot utility) and 525 kW (cold utility). It may be noted that supertargeting (Section 1.6) may be used to obtain the optimum HRAT by investigating the capital-energy tradeoff.

The conventional pinch division at 125°C/105°C (for HRAT = 20°C) shows that the fixed stream is H2 and the floating streams are H1, C3, and C4. The heat contents for the various streams, on dividing the problem at the conventional pinch, are given below:

Stream	Above Pinch	Below Pinch
H1	500 kW	800 kW
H2	0 kW	2400 kW
C3	1000 kW	1700 kW
C4	105 kW	975 kW

Step 2. Stream Slipping and Definition of the Flexible Pinch

It is observed that 105 kW for stream C4 may be totally slipped from above the pinch, so $\alpha_{C4} = -105$ kW as shown in Figure 5.3. Next, the possibility of a subset equality match between streams H1 and C3 by partial stream slipping may be investigated above the pinch. However, as per the H/H heuristic, a heater may be appropriately located on stream C3 (which has the higher target temperature of 155°C). If this heater has the total target utility load of 605 kW, then its inlet temperature will be 124.75°C. This is greater than the outlet temperature of 112°C for stream C4, suggesting that a single heater design is suitable. Now, a nonsubset equality match is possible between stream H1 (of 500 kW) and the remaining portion of stream C3 (of 395 kW) after partial slipping of 105 kW for stream H1. This gives α_{H1} = +105 kW and satisfaction of the enthalpy balance in that $\alpha_{C4} + \alpha_{H1}$ = 0. Also, the flexible pinch is completely defined and the subnetwork design above the flexible pinch is completed.

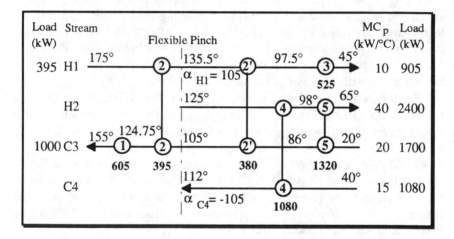

Figure 5.3 An elegant network for fixed energy recovery (with HRAT = 20°C) and featuring only five units may be synthesized using the flexible pinch design method. Units 2 and 2' are mirror image matches that can be combined. (Temperature in °C and load in kW)

Step 3. Mirror Image Matches and Design below the Flexible Pinch
  For the subnetwork design below the flexible pinch, mirror image matches need to be first considered. For such a match between streams H1 and C3, the load should be maximized since it does not produce a tunnel match. However, a check on EMAT must be performed as $MC_{p,H1} < MC_{p,C3}$. This shows that the ticking off of stream H1 is not feasible. Now, as per the H/H heuristic, a cooler may be appropriately located on stream H1. A single cooler design is attempted by assigning the total target utility load of 525 kW to this unit. The maximum possible heat load of 380 kW (from the remaining portion of stream H1) is assigned to the mirror image match 2', which yields an EMAT of 11.5°C at its cold end. As MC_p of stream H2 is larger than that of the cold streams,

the design may be easily completed with the help of the H/H heuristic. Stream C4 with the higher target temperature (112°C) is first ticked off using match 4 (of 1080 kW). Finally, the design below the flexible pinch is completed with match 5 of 1320 kW.

Step 4. Simplification and Final Design

The subnetworks above and below the flexible pinch are combined; then, the mirror image matches are consolidated as intended into a single exchanger with a heat load of 775 kW where stream H1 is cooled from 175° to 97.5°C and stream C3 is heated from 86° to 124.75°C. The final design features five units, which is the Euler MNU target. This is a reduction of two units compared with the networks in Figures 5.1 and 5.2 for the same energy consumption (HRAT of 20°C). Note that decreasing the number of process-process exchangers simplifies the network and reduces the associated pipework whereas decreasing the number of utility exchangers may imply some sacrifice on the control of target temperatures. Figure 5.3 has a lower EMAT (11.5°C at the cold end of unit 2') whereas Figures 5.1 and 5.2 were designed for an EMAT of 13°C. On the whole, the network from the FPDM appears more promising than those from the PPDM for this case study.

Of course, an important evaluation criterion is the area requirement for these networks. As EMAT is not specified in the FPDM, the driving forces for every match must be carefully checked. Also, the FPDM does rely to some extent on the experience and insight of the designer, especially in defining the flexible pinch temperatures. However, if done judiciously, the design from the FPDM will virtually feature no stream splitting and require no loop breaking.

The heuristics given above provide guidance in making appropriate design decisions. These decisions may be made more objectively and rationally by using a cost-based strategy to distinguish between suitable and unsuitable matches. However, it may not be convenient any longer to do the designs manually. Suaysompol and Wood (1991) discuss the development of a computer-aided design package for the application of the FPDM with the help of the A* heuristic search algorithm from artificial intelligence. Each match corresponds to a node in the search tree, and the node with the least value of the cost-based evaluation function is usually expanded. The computer-based strategy allows synthesis of networks that are cost effective, which cannot be guaranteed by the manual FPDM. Manually, it may be easy to synthesize only MNU designs as in Figure 5.3.

The next section describes a method based on key physical insights that does not require sophisticated computational models or involved heuristics.

5.4 COMPENSATION PRINCIPLE DESIGN METHOD

This method, proposed by Jezowski (1991a), is similar in some ways to the FPDM. However, it emphasizes a fundamental concept, which may be termed the compensation principle and may be simply stated as follows: cross-pinch heat transfer does not lead to an increase in the utilities beyond the minimum requirements, if the total heat transferred across the pinch in one direction is exactly compensated by an equal amount in the opposite direction. Thus, for every match with EMAT $\leq$ HRAT, the principle requires that a second match (or matches) exist such that the enthalpy balance equation ($\Sigma \ \alpha_{jh}$ + $\Sigma \ \alpha_{jc}$ = 0) is satisfied. This ensures an equal amount of heat passing the pinch in both directions and the net heat flow across the pinch being zero. Therefore, with dual approach temperatures, the utility requirements are identical to those in the case of MER networks. In fact, the division at the pinch is not necessary unless the single approach temperature is used. Such an observation was also made by Gundersen and Naess (1988).

Like the FPDM, the compensation principle design method (CPDM) requires only the HRAT as the specified parameter. It proposes simple general rules for generating cost-effective networks recognizing that the use of dual approach temperatures, cross-pinching, and crisscrossing leads to designs with lesser units and stream splits than by strictly observing the pinch division. However, the effect of the reduction in units and/or stream splits may be partially or totally nullified by the increase in the heat transfer area caused by crisscrossing. The tradeoff must be examined carefully for the particular case study to observe the net effect on the capital cost.

The major advantage of the method is that it requires no division of the problem at the pinch during actual implementation. This simplifies the procedure for placing matches and eliminates the need for complex heuristics. However, the method demands some intuition on the part of the designer and is easy to apply for small-sized problems.

Example 5.3 Network Using Compensation Principle Design Method

Design a network using the CPDM for the stream data in Case Study 4S1. Given: HRAT = 20°C.

Solution. The compensation principle is exploited in that the pinch division and the PDM rules are not followed. Instead, the synthesis method involves applying the Ponton and Donaldson (1974) fast matching algorithm, which is broadly consistent with the targeting principles. The decision to apply the fast matching algorithm from the hot end or the cold end of the network is problem dependent. On the basis of the H/H heuristic, the hot stream having the highest inlet temperature is matched with the cold stream having the highest outlet temperature. Importantly, the aim here is to achieve the Euler minimum unit target and to ensure that the area requirements are not high. If necessary, RPA may be used to approach the area and shell targets.

Step 1. Energy Targeting for HRAT

For the specified HRAT of 20°C, the minimum utility requirements are 605 kW (hot utility) and 525 kW (cold utility).

Step 2. Matching Using Hill Heuristic

A heater of 605 kW is placed on stream C3 (Figure 5.4) because it is the cold stream with the highest outlet temperature. Stream H1 (inlet temperature of 175°C) is matched with stream C3 (outlet temperature of 124.75°C). Ticking off stream H1 is not a feasible option, so the placement of the cooler is explored. A cooler of 525 kW is appropriately placed on stream H1. The remainder of stream H1 (775 kW) is then matched with stream C3. Next, stream H2 is matched with steam C4 (outlet temperature of 112°C). Stream C4 gets ticked off through match 4 (of load 1080 kW). Finally, the remaining portions of streams H2 and C3 are matched with unit 5 (of load 1320 kW).

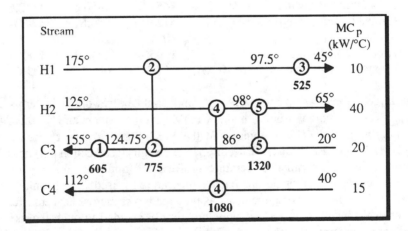

Figure 5.4 An elegant five-unit FER network may be synthesized using the compensation principle design method without any pinch decomposition. (Temperature in °C and load in kW)

Step 3. Verifying the Compensation Principle

For an HRAT of 20°C, the conventional pinch is at 125°C/105°C. The cross-pinch heat exchange is as follows:

Match 2: $10 (175 - 125) - 20 (124.75 - 105) = 105 \text{ kW}$
Match 4: $0 - 15 (112 - 105) = -105 \text{ kW}$

This is in accordance with the compensation principle.

Step 4. Evaluation of the Network

The network in Figure 5.4 from the CPDM is identical to the one generated by the FPDM in Figure 5.3 (provided the mirror image matches are merged). However, the CPDM generates the network with relatively little effort and, hence, appears superior to the FPDM. Also, systematic cross-pinching in accordance with the compensation principle has the potential to produce cost-effective solutions and may prove superior to the PDM. This is evident through the quantitative comparison of Figure 5.4 with Figure 4.20 (considering an isothermal mixing junction) given below:

	CPDM	PDM	Target
Hot Utility (kW)	605	605	605
Units	5	7	7 (MER)
Splits	0	1	--
Area (counterflow in m^2)	1451.8	1450.6	1312.6
Capital Cost (counterflow in 10^3 $)	512.9	580.6	574.2
Shells	10	16	10
Area (1-2 shell/tube in m^2)	1695	1646.2	1488.9
Capital Cost (1-2 shell/tube in 10^3 $)	617.3	699.1	641.6

The targets above are based on $N_{u,mer}$ and could be reworked in terms of $N_{u,min}$ for fair comparison with the CPDM results. Interestingly, there is virtually no area penalty in going from the MER (PDM) design to the MNU (CPDM) design. Recall that MER-cum-MNU designs (see Section 4.3) usually feature units with parallel temperature profiles and minimum driving forces across the pinch if a single approach temperature is used. Such units will typically have high area requirements as they lead to crisscross heat transfer in the critical pinch region. These units are simply avoided through the use of dual approach temperatures. In addition, the design in Figure 5.4 (with EMAT < HRAT) requires less units/shells and has a lower cost than the design in Figure 4.20 (with HRAT = EMAT). The decrease in the network capital cost is due to the fixed charge component in the cost law and the economy of scale and may occur even when the total area requirement increases.

From the above examples, it is seen that relaxing EMAT so as to have any value less than HRAT has the following benefits (Gundersen and Grossmann, 1990):

- a reduction in the number of units due to the increased flexibility and options during matching;
- no significant increase (and, at times, a decrease) in the total area requirements due to the improved approach to vertical heat transfer possible in the increased match options; and
- a reduction in the complexity of the network due to the elimination of splits, bypasses, and mixing junctions.

These gains are possible through the methods discussed in this chapter which employ dual approach temperatures during the network synthesis itself and not simply in the subsequent evolution. Gundersen and Grossmann (1990) have proposed a vertical MILP transshipment model allowing for EMAT relaxations. Their approach, being based on mathematical programming, is beyond the scope of the current chapter. However, some of their observations are noteworthy:

- The concept of a single approach temperature is valuable in several heat integration representations (e.g., composite curves) but may not be adequate for match selections during practical HENS.
- Although the original motivation for dual approach temperatures was the case of different heat transfer conditions, it is important even when streams have equal heat transfer coefficients (see Example 5.3).

In summary, it may be noted that it is exceedingly difficult to generate the globally optimal design on considering all factors (energy, area, units, shells, cost, operability, controllability, and safety) along with the associated tradeoffs. However, it is possible to produce a number of alternative designs that are near-optimal. Then, preliminary screening may be done by estimating the total annualized costs of these networks. Finally, detailed designs may be performed for the promising candidates and an analysis done to evaluate their operability, controllability, and safety characteristics.

The first five chapters may be regarded basic material on HENS and pinch technology. The remaining chapters may be better classified as advances and extensions of these principles.

QUESTIONS FOR DISCUSSION

1. Should HRAT be considered a parameter that specifies the desired heat recovery level (rather than a target)? Should HRAT be treated as a utility load specification (rather than a temperature specification)? If so, why not specify the utility load directly?

2. Comment briefly on the following aspects.

- There is a threshold value for the HRAT below which the minimum utility requirements do not change. Also, there is a threshold value for the EMAT below which the number of units does not change (as it reaches the minimum).

- The actual EMAT in a synthesized network can be higher than the specified EMAT.

3. Show through a qualitative sketch how the capital cost varies with EMAT for different values of HRAT. Explain why typically the capital cost vs. EMAT curve for a constant value of HRAT is roughly U-shaped.

4. Discuss the similarity between the zero HRAT calculation in HENS and the minimum reflux calculation in distillation. Does the zero HRAT specify the theoretical limit for the maximum level of energy recovery possible with an infinite heat exchange area?

5. What is the difference between a subset equality match and a nonsubset equality match? Why is it advantageous to force a subset equality match?

6. In Figure 5.2, is it possible to allocate the 245 kW (α_n) to stream C3 rather than to stream H1? Develop a network for this case.

7. What is the sign convention adopted for α_j? Specifically, in Figure 5.3, why is α_{H1} positive and α_{C4} negative? Is the sign convention consistent with that used during the calculation of the cross-pinch heat transfer in verifying the compensation principle (see step 3 in Example 5.3)?

8. Can the network in Figure 4.20 be evolved to the one in Figure 5.4 if ΔT_{min} is relaxed from 20° to 11.5°C?

9. Does it matter whether the application of the fast matching algorithm is started from the hot end or the cold end of the network? What is the corollary to the H/H heuristic when applying the fast matching algorithm from the cold end of the network?

10. The pinch division yields two independent threshold problems. Can these threshold problems be designed independently using the fast matching algorithm (rather than the PDM)?

11. Is it true that the DATMs attach a special importance to achieving the Euler minimum unit target? Is it therefore more appropriate to use $N_{u,min}$ rather than $N_{u,mer}$ in Equation 1.34 during supertargeting based on DATMs?

12. Does a solution always exist to the pinch design problem? Does a design satisfying the Euler minimum unit target always exist for every problem? Construct an example where the Euler target cannot be met.

13. In what situations can cross-pinching and crisscrossing be advantageous?

14. Supertargeting allows specifying an HRAT that brings the overall design into the region of the optimum. Allowing EMAT to be less than HRAT provides additional flexibility for optimization of the design. Comment.

15. Compare the FPDM and the CPDM. Does the CPDM yield good networks when the number of streams is large and/or the constraint imposed by the pinch is severe? Specifically, rework Example 5.2 (with the FPDM) and Example 5.3 (with the CPDM) for HRAT = 10°C.

PROBLEMS

5.A Determination of the Pseudo-Pinch
Determine the pseudo-pinch temperatures for Case Study 10L1 originally proposed by Linnhoff and Ahmad (1986) and later discussed by Trivedi et al. (1989b). The HRAT and EMAT are specified as 14.5°C and 10°C respectively. The cross-pinch heat transfer, α_n, may be allocated as per the following heuristic schemes.
   a) α_n may be allotted equally to all the streams crossing the pinch.
   b) Stream H3 may be totally slipped, followed by equal distribution of residual α_n between streams H1 and H2.
   c) α_js may be allocated in inverse proportion to the stream heat transfer coefficients (where $h_{H1} : h_{H2} : h_{H3} = 1 : 3.1 : 1.367$).

Answers:
   Conventional pinch determined by stream H4 (inlet temperature of 56°C), so pseudo-pinch temperatures are 56°C (for H4) and 46°C (for C10).
   α_n (2478 kW) to be allotted to streams H1, H2, and H3, which cross the pinch.
   a) 61.3°C (for H1); 72.5°C (for H2); and 90.6°C (for H3).
   b) 58.7°C (for H1); 64.3°C (for H2); and 125°C (for H3).
   c) 63.7°C (for H1); 63.8°C (for H2); and 92.9°C (for H3).

5.B Network Generation by Pseudo-pinch Design Method
Design networks using the PPDM for the following:
   a) Case Study 4S1. Given: HRAT = 20°C and EMAT = 11.5°C.

b) Case Study 4C2 (first discussed by Colbert [1982] and later by Trivedi et al. [1989b]). Given: HRAT = 16.67°C and EMAT = 3.33°C.

c) Case Study 4A2 (first discussed by Ahmad & Smith [1989] and later by Trivedi et al. [1989b]) with temperatures reduced by 20°C. Given: HRAT = 32°C and EMAT = 20°C. Simplify the designs wherever possible to obtain networks featuring only five units.

Answers:

a) Conventional pinch determined by stream H2 (inlet temperature of 125°C), so pseudo-pinch temperatures are 125°C (for H2) and 113.5°C (for C3 and C4).

α_n (275 kW) to be allotted to stream H1, whose pseudo-pinch is at 152.5°C.

A possible FER design appears below:

Above Pseudo-Pinch	Below Pseudo-Pinch
H1 - C3 (225 kW)	H1 - C3 (550 kW)
HU - C3 (605 kW)	H2 - C4 (1080 kW)
	H2 - C3 (1320 kW)
	H1 - CU (525 kW)

Network simplification is achieved by combining the 225 kW and 550 kW matches into a single exchanger of 775 kW.

b) Conventional pinch determined by stream H2 (inlet temperature of 150°C), so pseudo-pinch temperatures are 150°C (for H2) and 146.67°C (for C3).

α_n (40 kW) to be allotted to stream H1, whose pseudo-pinch is at 170°C.

A possible FER design appears in Figure 14 (Trivedi et al., 1989b, p. 677).

Above Pseudo-Pinch	Below Pseudo-Pinch
H1 - C3 (20 kW)	H2 - C3 (260 kW)
HU - C3 (80 kW)	H1 - C4 (260 kW)
	H2 - CU (180 kW)

c) Conventional pinch determined by stream C4 (inlet temperature of 120°C), so pseudo-pinch temperatures are 120°C (for C4) and 140°C (for H1 & H2).

α_n (900 kW) to be allotted to stream C3, whose pseudo-pinch is at 97.5°C.

Possible FER designs appear in Figures 15 & 16 (Trivedi et al., 1989b, p. 678)

Above Pseudo-Pinch	Below Pseudo-Pinch
H2 - C4 (1800 kW)	H1 - C3 (2400 kW)
H1 - C3 (2500 kW)	H2 - C3 (700 kW)

H1 - C4 (1700 kW)   H2 - CU (4700 kW)

HU - C4 (4900 kW)

Above Pseudo-Pinch   *Below Pseudo-Pinch*

H1 - C4 (4200 kW)   H2 - C3 (3100 kW)

H2 - C3 (1800 kW)   H1 - CU (2400 kW)

HU - C3 (700 kW)   H2 - CU (2300 kW)

HU - C4 (4200 kW)

Possible simplified designs with five units appear in Figures 5b & 17b (Trivedi et al., 1989b, p. 670 and 680)

HU - C4 (4900 kW)   HU - C4 (4900 kW)

H1 - C4 (3500 kW)   H2 - C3 (2500 kW)

H2 - C3 (5600 kW)   H1 - C4 (3500 kW)

H1 - CU (3100 kW)   H1 - C3 (3100 kW)

H2 - CU (1600 kW)   H2 - CU (4700 kW)

5.C Network Generation by Flexible Pinch Design Method

Design a network using the FPDM for the following:

a) Case Study 4A2. Given: HRAT = 32°C.

b) Case Study 9A2 (Aromatics Plant). Given: HRAT = 30°C.

Answers:

a) Conventional pinch determined by 152°C/120°C.

Flexible pinch temperatures: 163°C (for H1), 144°C (for H2), and 120°C (for C3 and C4), using $\alpha_{H1} + \alpha_{H2} = 340 - 340 = 0$.

A possible FER design appears in Figure 3 (Suaysompol & Wood, 1991, p. 461)

Above Flexible Pinch   *Below Flexible Pinch*

H2 - C3 (1600 kW)   H2 - C3 (900 kW)

H1 - C4 (3500 kW)   H1 - C3 (3100 kW)

HU - C4 (4900 kW)   H2 - CU (4700 kW)

The H2-C3 matches form a narrow tunnel, and the load should be minimized.

The mirror image matches of 1600 kW and 900 kW may be combined to form a single exchanger of 2500 kW. The network has an EMAT of 20°C.

The same design also appears in Figure 2 (Wood et al., 1991, p. 41).

b) Conventional pinch determined by 130°C/100°C.

Flexible pinch temperatures: 130°C (for H1 and H3), 127°C (for H4), 100°C (for C5 and C7), 98°C (for C6), and 80°C (for C8),

using $\alpha_{H4} + \alpha_{C6} + \alpha_{C8} = -1.3 + 0.1 + 1.2 = 0$.

A possible FER design appears in Figure 7 (Suaysompol & Wood, 1991, p. 463)

Above Flexible Pinch	Below Flexible Pinch
H4 - C7 (13.3 MW)	H4 - C7 (5.3 MW)
H3 - C8 (5.4 MW)	H3 - C8 (1.2 MW)
H1 - C6 (4.6 MW)	H1 - C6 (4.4 MW)
H2 - C5 (9.6 MW)	H1 - CU (4.6 MW)
H1 - C9 (15.1 MW)	H3 - CU (3.0 MW)
HU - C5 (10.4 MW)	H4 - CU (21.4 MW)
HU - C9 (16.9 MW)	

H4 - C7 (13.3 MW) and H3 - C8 (5.4 MW) are both subset equality matches. Note that the subset equality match between streams H1 and C5, though possible, is not selected as it requires too many shells.

Consolidating the mirror image matches yields H4 - C7 (18.6 MW), H3 - C8 (6.6 MW), and H1 - C6 (9.0 MW). The network has an EMAT of 12°C.

The final FPDM network has ten matches (five process-process exchangers, two heaters, and three coolers) and no stream splits. This may be compared with the PDM design given by Ahmad and Linnhoff (1989) that has eighteen units (twelve process-process exchangers, three heaters, and three coolers) and one stream split.

Note that a PPDM design (with EMAT = 20°C) having twelve matches (seven process-process exchangers, two heaters and three coolers) and no stream splits appears in Figure 4 (Wood et al., 1991, p. 42).

5.D Network Generation by Compensation Principle Design Method

Design a network using the CPDM for the following:
a) Case Study 4A2. Given: HRAT = 32°C.
b) Case Study 4C2. Given: HRAT = 16.67°C.
c) Case Study 4G1. Given: HRAT = 20°C.
d) Case Study 4L4. Given: HRAT = 30°C.
e) Case Study 4S1 (an alternative to Figure 5.4). Given: HRAT = 20°C.

Answers:
a) The MNU designs (with five units) reported as answers to Problem 5.B.c may be simply generated by the CPDM and also appear in Figure 2 (Jezowski, 1991a, p. 307). Note that 340 kW are transferred across the pinch in opposite directions in accordance with the compensation principle.

b) The design (with five units) reported as the answer to Problem 5.B.b may be simply generated by the CPDM by applying the fast matching algorithm starting form the cold end of the network. It appears in Figure 4 (Jezowski, 1991a, p. 309). Note that 300 kW are transferred across the pinch in opposite directions.

c) Three possible designs with five units appear in Figure 3 (Jezowski, 1991a, p. 308).

HU - C3 (1075 kW)	HU - C4 (1075 kW)	HU - C3 (1075 kW)
H1 - C4 (1800 kW)	H1 - C3 (1800 kW)	H1 - C4 (1400 kW)
H2 - C3 (1550 kW)	H2 - C4 (1175 kW)	H2 - C3 (1550 kW)
H2 - C4 (450 kW)	H2 - C3 (825 kW)	H2 - C4 (850 kW)
H2 - CU (400 kW)	H2 - CU (400 kW)	H1 - CU (400 kW)

d) A possible design by applying the fast matching algorithm starting from the cold end of the network appears in Figure 5 (Jezowski, 1991a, p. 310).

HU - C4 (162 MW)
H1 - C3 (86 MW)
H2 - C3 (144 MW)
H1 - C4 (118 MW)
H1 - CU (171 MW)

e) A possible alternative to Figure 5.4 may be generated by heating stream C4 first.

HU - C4 (605 kW)
H1 - C3 (775 kW)
H2 - C3 (1925 kW)
H2 - C4 (475 kW)
H1 - CU (525 kW)

5.E Compensation Principle Design with Multiple Utilities

Consider the stream data for Case Study 4S1h with the following utilities:

Utility	T_{in} (°C)	T_{out} (°C)	Cost ($/(kW.yr))
HU	180	179	120
CU1	15	25 (max)	10
CU2	112	113	5

The purpose of utility CU2 is specifically to raise VLP steam at 112°C from feed water at the same temperature.

a) Determine the sensitivity threshold for the above data.

b) Determine the usage levels (in kW) of the three utilities for minimum operating cost. Assume a ΔT_{min} (HRAT) of 25°C.

c) Design a network for maximum energy recovery for the multiple utility - multiple pinch problem under consideration if the H2-C3 match is forbidden. Use the PDM with HRAT = EMAT = 25°C.

d) Develop an alternative method using the CPDM with EMAT < HRAT. What is the actual (not net) cross-pinch heat transfer (in kW) across the process and utility pinches?

Answers:
   a) $\Delta T_{threshold} = 20°C$.
   b) 100 kW of HU, 1900 kW (MC_p = 190 kW/°C) of CU1 (15° to 25°C),
       70 kW (MC_p = 70 kW/°C) of CU2 (112° to 113°C)
       Process pinch is at 162.5°C. Utility pinch is at 112.5°C.
   c) A possible MER design with a two-way split on stream H1 is given
       below:

   Above Pinch (162.5° C) Below Pinch (112.5° C) Between Pinches
   HU - C3 (100 kW)   H1 - C3 (1600 kW)   H1a - C3 (1000 kW)
                           H2 - C4 (900 kW)   H1b - CU2 (70 kW)
                           H1 - CU1 (400 kW)   H1b - C4 (180 kW)
                           H2 - CU1 (1500 kW)

   d) A possible FER design using a balanced grid is given below:
   Match 1: HU - C3 (100 kW)   Match 4: H2 - C4 (1080 kW)
   Match 2: H1 - C3 (2600 kW)   Match 5: H2 - CU1 (1250 kW)
   Match 3: H2 - CU2 (70 kW)   Match 6: H1 - CU1 (650 kW)
   No cross-pinch heat transfer across process pinch. 250 kW of cross-
   pinch heat transfer in two opposing directions across utility pinch.

CHAPTER 6

Interfacing Network Synthesis with Detailed Exchanger Design

Design is concerned with how things ought to be,
with devising artifacts to attain goals.

Herbert Simon

The heat transfer equipment most commonly used is the shell and tube heat exchanger. Sizing and rating of such an exchanger based on the Kern method and the Bell-Delaware method are described in the first half of this chapter. Only the design procedures are given, and no attempt is made to present the derivations or detailed basis of the various design equations as these are available elsewhere (Kern, 1950; Bell, 1963, 1983; Peters and Timmerhaus, 1981; Taborek, 1983; Raman, 1985; Sinnott, 1989). Sizing involves determining the heat transfer area (or UA) for a new exchanger when all the end temperatures are specified. On the other hand, rating (or performance prediction) involves computing the outlet temperatures for an existing exchanger when UA and two inlet temperatures are known. The second half of the chapter discusses how the detailed design may be correlated with network synthesis through heat transfer coefficients and pressure drops.

6.1 BASIC THERMAL DESIGN

The assumptions typically made (Bowman et al., 1940) during the design of multipass shell and tube exchangers include constant heat capacity flow rates, negligible heat losses, and no phase change over a portion of the exchanger. The design procedure given below is applicable when isothermal condensation or vaporization occurs throughout the exchanger, provided appropriate correlations are used for the heat transfer coefficient on the condensing/vaporizing side. Furthermore, the shellside fluid temperature in any shellside pass is assumed uniform across a crosssection and the heat transfer areas in each pass are equal.

Four kinds of inputs (listed in Tables 6.1 and 6.2 along with their corresponding symbols) are required for the thermal design: process data, physical properties data, exchanger data, and limiting constraints. The value of a physical property should be available at any temperature of interest from a thermodynamic database.

Example 6.1 Basic Thermal Design of a Shell and Tube Exchanger

Perform the basic thermal design for the shell and tube exchanger whose input data appear in Table 6.1. Specifically, an exchanger is to be designed to cool 53650 kg/hr of process fluid from 98° to 65°C. Cooling water is available at 15°C, which may be heated to 25°C. It is desired to use 1-2 exchanger(s) with water on the tubeside. Determine the heat exchanger area required and the mean temperature difference assuming U to be 100 W/m^2 °C.

Table 6.1 Inputs for Thermal Design of an Exchanger

Quantity	Symbol	Units	Value
Process Data			
Temperature (inlet - hot stream)	T_{hi}	°C	98
Temperature (inlet - cold stream)	T_{ci}	°C	15
Temperature (outlet - hot stream)	T_{ho}	°C	65
Temperature (outlet - cold stream)	T_{co}	°C	25
Flow rate (hot stream)	M_h or M_s	kg/hr	53650
Flow rate (cold stream)	M_c or M_t	kg/hr	(Eq 6.2)
Physical Properties Data			
Specific heat (hot fluid)	C_{ph} or C_{ps}	J/(kg °C)	2684
Specific heat (cold fluid)	C_{pc} or C_{pt}	J/(kg °C)	4180
Density (hot fluid)	ρ_h or ρ_s	kg/m^3	777
Density (cold fluid)	ρ_c or ρ_t	kg/m^3	998
Viscosity (hot fluid)	μ_h or μ_s	kg/(m.s)	0.23×10^{-3}
Viscosity (cold fluid)	μ_c or μ_t	kg/(m.s)	1.00×10^{-3}
Thermal conductivity (hot fluid)	k_h or k_s	W/(m.°C)	0.11
Thermal conductivity (cold fluid)	k_c or k_t	W/(m.°C)	0.60

Physical properties at 81.5 °C for hot fluid (shell) and 20 °C for cold fluid (tube).

Solution. This is, in fact, the design of the cooler (unit 2) in Figure 1.2. First, an overall heat balance is performed; then, the mean temperature difference is calculated.

Step 1. Determination of Caloric or Average Fluid Temperatures

When viscosity effects are significant, the arithmetic mean temperature cannot be used as the temperature at which physical properties are required. In such cases, particularly for oils, the caloric temperatures based on the linear variation of U with temperature are more appropriate (see Problem 6.A for derivation). For the hot and cold fluid streams, the caloric temperatures may be calculated from

$$T_{hc} = T_{ho} + F_c (T_{hi} - T_{ho}) \qquad (6.1a)$$
$$T_{cc} = T_{ci} + F_c (T_{co} - T_{ci}) \qquad (6.1b)$$
where $F_c = [1/k_c + r/(r-1)]/[1 + \ln(1 + k_c)/\ln r] - 1/k_c \qquad (6.1c)$
and $\quad r = (T_{ho} - T_{ci})/(T_{hi} - T_{co}) \qquad (6.1d)$

Here, k_c is obtained from charts (Kern, 1950) corresponding to known values of the °API of the oil and its temperature range [i.e., $(T_{hi} - T_{ho})$ or $(T_{co} - T_{ci})$]. Note that a single value of F_c applies to both streams and is determined based on the controlling stream. The stream with the higher value of k_c is the controlling one (in the case where both streams are oils).

For nonviscous fluids, the arithmetic mean temperatures are appropriate. Thus, the physical properties in Table 6.1 are determined at 81.5°C and 20°C for the shellside fluid and tubeside fluid respectively.

Step 2. Computation for the Overall Heat Balance
The enthalpy lost by the hot stream equals that gained by the cold stream. Thus,

$$\text{Heat duty } Q = M_h C_{ph} (T_{hi} - T_{ho}) = M_c C_{pc} (T_{co} - T_{ci}) \qquad (6.2)$$

The above equation may be used as a check when all six quantities (the two flow rates as well as four end temperatures) are specified. Alternatively, when five of the quantities are known, then the sixth one may be determined. For example, substituting the values from Table 6.1 gives $Q = (53650/3600) (2684)$ $(98 - 65) = M_c (4180) (25 - 15) = 1319.97$ kW or $M_c = 31.58$ kg/s.

Step 3. Calculation of the Mean Temperature Difference
The logarithmic mean temperature difference for counterflow is

$$LMTD = [(T_{hi} - T_{co}) - (T_{ho} - T_{ci})]/\ln[(T_{hi} - T_{co})/(T_{ho} - T_{ci})] \qquad (6.3)$$

As the flow in shell and tube exchangers (except of the 1-1 type) is a mixture of cocurrent, countercurrent, and cross flow, the usual practice is to estimate the true mean temperature difference (*MTD*) from the *LMTD* by applying a correction factor to account for the departure from strict counterflow. Thus,

$$MTD = F\,(LMTD) \qquad \text{where } 0 < F \le 1. \tag{6.4}$$

The F correction factor (Bowman et al., 1940) depends on three dimensionless parameters:

$$P = (T_{co} - T_{ci})/(T_{hi} - T_{ci}) \tag{6.5a}$$
$$R = M_c C_{pc}/M_h C_{ph} = (T_{hi} - T_{ho})/(T_{co} - T_{ci}) \tag{6.5b}$$
$$N = UA/M_c C_{pc} \tag{6.5c}$$

It may be noted that P is the cold fluid temperature effectiveness, R is the heat capacity flow rate ratio (or the hot-fluid-to-cold-fluid-temperature-change ratio), and N is the number of transfer units. The alternative (but equivalent) definitions for P, R, and N (in terms of the hot fluid temperature effectiveness P) are

$$P = (T_{hi} - T_{ho})/(T_{hi} - T_{ci}) \tag{6.5'a}$$
$$R = M_h C_{ph}/M_c C_{pc} = (T_{co} - T_{ci})/(T_{hi} - T_{ho}) \tag{6.5'b}$$
$$N = UA/M_h C_{ph} \tag{6.5'c}$$

Now, F can be expressed conveniently in terms of the above definitions for P, R, and N by combining the overall heat balance in Equation 6.2 with the basic exchanger design equation (namely, $Q = UA\,F\,[LMTD]$). Thus,

$$F_{R \ne 1} = \ln[(1 - P)/(1 - RP)] / [N\,(R - 1)] \tag{6.6a}$$

For $R = 1$, the above equation becomes indeterminate, and l'Hopital's rule may be applied to obtain

$$F_{R = 1} = P / [N\,(1 - P)]. \tag{6.6b}$$

It is possible to derive a relationship between P, R, and N for various exchanger types. For a 1-2 shell and tube exchanger, the relation is (see Problem 6.B for derivation)

$$N = \frac{1}{\sqrt{R^2 + 1}} \ln\left[\frac{2 - P(R + 1 - \sqrt{R^2 + 1})}{2 - P(R + 1 + \sqrt{R^2 + 1})}\right]. \tag{6.7}$$

The popular NS - $2\,NS$ shell and tube exchanger (namely, exchangers of the type 2-4, 3-6, 4-8, etc.) merely refers to a series circuit of NS heat exchangers of the 1-2 shell and tube type. Since exchangers with multiple shell passes may be considered as trains of identical 1-2 exchangers connected in series with the fluids in overall counterflow (Bowman, 1936; Domingos, 1969), the following useful circuit relations (see Problem 6.C for derivation) are obtained:

$$(1 - R'P')/(1 - P') = [(1 - RP)/(1 - P)]^{NS} \qquad \text{for } R \neq 1 \qquad (6.8a)$$
$$P'/(1 - P') = NS \ P/(1 - P) \qquad \text{for } R = 1 \qquad (6.8b)$$

Here, primed quantities refer to the overall circuit. The expressions relate the overall P (denoted by P') for an exchanger train of NS identical shells in series to the individual P for each shell. Because the mass flow rates of the two streams do not change as they pass through the series circuit, $R' = R$. Furthermore, $N' = (NS) N$.

As the four terminal temperatures are known for this example, $R' = R = (98 - 65)/(25 - 15) = 3.3$ and $P' = (25 - 15)/(98 - 15) = 0.1205$ as per the definitions in Equation 6.5.

The minimum number of shell passes NS which results in a value of F greater than 0.75 - 0.8 must be used. Equations 6.8a (or equivalently Equation 1.21), 6.7, and 6.6a may be written in terms of the unknown number of shells NS as

$$P = [1 - (0.6849)^{1/NS}]/[3.3 - (0.6849)^{1/NS}]$$
$$N' = (NS/3.4482) \ln [(2 - 0.8518 \ P)/(2 - 7.7482 \ P)]$$
$$F = 0.1645/N'$$

On using the above equations with $NS = 1$, it is found that $P = 0.1205$, $N' = 0.1671$, and $F = 0.9848$. As F is very close to unity, one shell is sufficient; otherwise, the calculations would need to be repeated for $NS = 2, 3$, etc. until the $F > 0.75$ criterion is satisfied. Finally, from Equations 6.3, 6.4 and 6.5c,

$$LMTD = [(98 - 25) - (65 - 15)]/\ln[73/50] = 60.78.$$
$$MTD = 0.9848 (60.78) = 59.85°C; \text{ and}$$
$$A = 0.1671 (31.58) (4180)/100 = 220.53 \text{ m}^2.$$

6.1.1 A New Design Criterion Based on F-Slopes

The F correction factor (Bowman et al., 1940), used for multipass shell and tube exchanger design and defined in Equation 6.6, is examined in greater detail in this section. For the commonly used 1-2 exchanger, the following analytical expression for $F(P, R)$ is obtained by combining Equations 6.6 and 6.7:

$$F = \frac{\ln\left[(1-P)/(1-RP)\right]}{R-1} \frac{\sqrt{R^2+1}}{\ln\left[(2-\alpha P)/(2-\beta P)\right]} \tag{6.9}$$

where $\alpha \equiv R + 1 - \sqrt{(R^2 + 1)}$ and $\beta \equiv R + 1 + \sqrt{(R^2 + 1)}$. For $R = 1$, the term $\ln[(1 - P)/(1 - RP)]/(R - 1)$ in Equation 6.9 must be replaced by $P/(1 - P)$ as per l'Hopital's rule. Equation 6.9 forms the basis of the popular $F(P, R)$ chart used in heat exchanger design (Kern, 1950).

As pointed out in Example 6.1, common design practice requires $F > 0.75$ (Kern, 1950); however, this is merely a rule of thumb and must be judiciously used to avoid poor designs. In recent years, it has been emphasized (Taborek, 1979, 1983; Bell, 1983; Liu et al., 1985; Ahmad et al., 1988) that the slope $(\partial F/\partial P)_R$ is as important as the actual value of F. Regions of steep slope in the F chart must be avoided to ensure that a slight shift in the operating point (due to uncertainties or inaccuracies) does not cause a precipitous drop in exchanger performance or lead to an infeasible exchanger.

Ahmad et al. (1988) developed a constant slope criterion and presented it in a graphical form for $(\partial F/\partial P)_R = -2.8$. Their line of constant slope divides the F chart into two regions: a region of preferred designs above the line and another of unacceptable designs below it. Although the constant slope approach provides an effective criterion for design, Ahmad et al. (1988) acknowledged that its evaluation and usage were rather complex. Hence, they proposed a simpler alternative criterion that requires practical designs to be limited to some fraction (X_p) of the maximum asymptotic value of P (say, P_{max}) for a specified R. Thus,

$$P = X_p P_{max} \tag{6.10}$$

where $P_{max} = 2/\beta$ (Taborek, 1983) and $0 < X_p < 1$. They chose X_p to be 0.9 to guarantee that $F > 0.75$. The line $X_p = 0.9$ and the line $(\partial F/\partial P)_R = -2.8$ do not have identical profiles (as may be observed on Figure 3 in the paper by Ahmad et al., 1988). The constant X_p approach accepts designs that are sensitive (unacceptable) according to the constant slope approach for $R > 1$. On the other hand, it rejects designs that are good as per the constant slope criterion for $R < 1$. It is possible to eliminate this mismatch (Shenoy et al., 1993) to obtain a criterion based on the rigorous constant slope approach and which, at the same time, retains the advantages and simplicity of the constant X_p approach.

In other words, the constant slope criterion presented by Ahmad et al. (1988) in graphical form may be transformed into an analytical, easy-to-use form as follows. Equation 6.9 may be differentiated and then combined with Equation 6.10 to obtain

$$
\left(\frac{\partial F}{\partial P}\right)_R = \frac{\sqrt{R^2+1}}{\ln\left(\frac{1-X_p\alpha/\beta}{1-X_p}\right)}\left[\frac{1}{(1-2X_P/\beta)(1-2RX_P/\beta)} - \frac{\sqrt{R^2+1}}{(R-1)}\frac{\ln\left(\frac{\beta-2X_p}{\beta-2RX_p}\right)}{\ln\left(\frac{1-X_p\alpha/\beta}{1-X_p}\right)}\frac{1}{(1-X_p\alpha/\beta)(1-X_p)}\right].
$$

$$(6.11)$$

Equation 6.11 may be used for $R = 1$ provided the term $\ln[(\beta - 2X_p)/(\beta - 2RX_p)]/(R - 1)$ is replaced by $2X_p/(\beta - 2X_p)$. The difficulty with the use of Equation 6.11 as a design criterion lies in its complexity, which prevents X_p from being explicitly expressed as a function of R for a constant $(\partial F/\partial P)_R$.

To be consistent with Ahmad et al. (1988), $(\partial F/\partial P)_R$ is chosen to be -2.8 (which is the slope at $F = 0.75$ and $R = 1$). Then, the nonlinear Equation 6.11 is solved with $(\partial F/\partial P)_R = -2.8$ to obtain X_p (denoted by X_{pp}) for different values of R. On plotting this $X_{pp}(R)$ data, it is clear that the choice of $X_p = 0.9$ by Ahmad et al. (1988) is not appropriate and can be improved. It may be noted that X_{pp} varies from 1 (as $R \to 0$) to 0.777 (as $R \to \infty$). More importantly, the constant X_p approach is not consistent with the constant slope approach: the criterion of Ahmad et al. (1988) with $X_p = 0.9$ leads to the slope $(\partial F/\partial P)_R$ varying between -0.86 and -8.6 (for $0.1 \leq R \leq 10$).

Shenoy et al. (1993) showed that the $X_{pp}(R)$ data may be curve-fitted using the general correlational approach proposed by Churchill and Usagi (1972) to obtain the following equation:

$$
X_{pp} = 1 - 0.223/[1 + (0.223/(0.033 + 0.103\,R))^{1.4}]^{1/1.4}
$$

$$(6.12)$$

Equation 6.12 provides an excellent approximation to Equation 6.11 and maintains the slope $(\partial F/\partial P)_R$ close to -2.8 (with the actual slope varying between -2.75 and -2.93 primarily due to the sensitivity of derivative evaluation).

However, it is important to note that a criterion in terms of constant $(\partial F/\partial P)_R$ is fundamentally inappropriate for the following reason. Note that the value of F is independent of the convention used in Equations 6.5 and 6.5' for defining R and P. This is easily verified for the balanced 1-2 arrangement through Equation 6.9 by replacing R by $1/R$ and P by RP, which are the appropriate transformations for switching between the two conventions. This does not hold good for $(\partial F/\partial P)_R$ as is clear from Equation 6.11. In other words, a certain symmetry is required on the F-chart for equivalent points in the two conventions. Such a symmetry exists in terms of F but not in $(\partial F/\partial P)_R$. A simple way to circumvent this difficulty is to employ a chart of $F(X_p, R)$ as in Figure 6.1 rather than the conventional $F(P, R)$ chart.

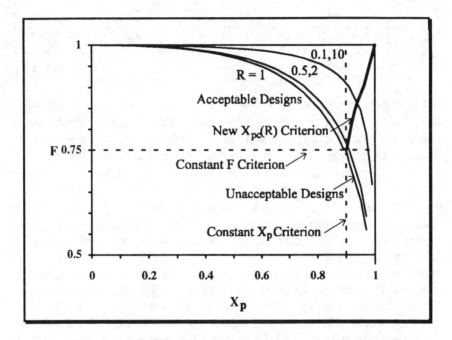

Figure 6.1 The F-curves for R and $1/R$ are identical in terms of X_p (not P). The above F-chart (for 1 shell pass-2 tube passes) allows development of a new criterion (Equation 6.13) rigorously based on constant slope $((\partial F/\partial X_p)_R \approx -1.64)$ and simplifies the procedure for shell estimation.

The criteria based on constant F and constant X_p correspond to horizontal and vertical lines on Figure 6.1. Both criteria are rules of thumb and lack a sufficiently strong theoretical basis. In what follows, a rigorous constant slope criterion is developed and presented in an analytical, easy-to-use form.

The appropriate slope criterion is based on constant $(\partial F/\partial X_p)_R$. Now, $(\partial F/\partial X_p)_R$ is chosen to be -1.64 (which is the slope at $F = 0.75$ and $R = 1$). This choice simply maintains consistency with previously used criteria (Ahmad et al., 1988) based on constant F ($= 0.75$) and X_p ($= 0.9$). From Equation 6.10,

$(\partial F/\partial X_p)_R = (2/\beta) (\partial F/\partial P)_R$, so this nonlinear equation is solved with $(\partial F/\partial X_p)_R = -1.64$ and the expression for the derivative $(\partial F/\partial P)_R$ from Equation 6.11 to obtain X_p (denoted by X_{pc}) for different values of R. It is observed that X_{pc} varies from 1 (as $R \to 0$ and $R \to \infty$) to 0.9 (at $R = 1$). The $X_{pc}(R)$ data is indeed symmetric, in that the values of X_{pc} are identical for R and $1/R$. This data is curve-fitted to obtain the following equation for the critical value of X_p (above which the F slope with respect to X_p is unacceptable):

$$X_{pc} = 1 - 0.1 \exp[-0.5 (\log R)^2]. \qquad (6.13)$$

Equation 6.13 provides an excellent approximation to the numerically-computed $(\partial F/\partial X_p)_R = -1.64$ criterion and is indistinguishable from it for most practical applications. Besides, it is much simpler and gives X_p explicitly in terms of R. Importantly, Equation 6.13 provides the required criterion for a 1-2 exchanger (single shell): a design is acceptable if $P \le P_c$, where P_c is defined as follows:

$$P_c = 2X_{pc}/[R + 1 + \sqrt{(R^2 + 1)}]. \qquad (6.14)$$

The criterion is now stated in terms of P (rather than F), and this proves to be an added advantage in multishell arrangements as pointed out by Ahmad et al. (1988) and explained next.

When the value of F is unacceptably low, designs with multiple 1-2 shells in series are explored. The determination of the number of shells typically demands a trial-and-error approach as discussed in Example 6.1. Starting with a single shell and systematically increasing their number, the least number of shells that yields an acceptable F value is obtained. Now, it is possible to devise an analytical method not involving trial-and-error, graphical construction, or charts that accounts for *both* an acceptable F value and an acceptable F slope. Substituting $P = P_c$ in Equations 6.8 and rearranging gives

$$NS_{min} = \ln[(1 - RP')/(1 - P')]/\ln[(1 - RP_c)/(1 - P_c)] \text{ for } R \neq 1 \quad (6.15a)$$
$$NS_{min} = [(1 - P_c)/P_c] [P'/(1 - P')] = 0.9 [P'/(1 - P')] \text{ for } R = 1. \quad (6.15b)$$

Equations 6.13, 6.14, and 6.15 allow explicit and straightforward evaluation of the minimum number of 1-2 shells for specified R and P'. The fractional value of NS_{min} is such that every shell in the series has $P = P_c$. The actual number of shells is obtained by rounding-off NS_{min} to the next largest integer, causing a marginal decrease in the individual P from P_c. The additive property for number of shells (Ahmad et al., 1988) is valid, on using Equations 6.15 in terms of fractional shells. This property assumes importance in problems with variable heat capacity flow rates and heat transfer coefficients as well as in synthesis of HENs involving 1-2 shells during area and shell targeting (Ahmad and Smith, 1989). Specifically, Equation 6.13 may be used for X_p in Equation 1.19 instead of a constant value of 0.9 as used by Ahmad and Smith (1989).

Example 6.1 may be reworked using the $X_{pc}(R)$ approach (Equation 6.13) as shown below:

$$X_{pc} = 1 - 0.1 \exp[-0.5 (\log 3.3)^2] = 0.9126$$
$$P_{max} = 2/[R + 1 + \sqrt{(R^2 + 1)}] = 2/(4.3 + \sqrt{11.89}) = 0.2581$$
$$P_c = 0.9126 \times 0.2581 = 0.2355$$
$$NS_{min} = \ln[(1-3.3\times0.1205)/(1-0.1205)]/\ln[(1-3.3\times0.2355)/(1-0.2355)]$$
$$= 0.3$$

Here, the result for the minimum number of 1-2 shells ($NS_{min} = 1$ on rounding off) agrees with that from the $F > 0.75$ criterion because of the high value of F (0.9848) in this case. In essence, Equations 6.13 and 6.14 provide a true constant slope criterion in an easy-to-use, interpretable form. It is seen on Figure 6.1 that the new criterion lies between the constant F criterion and the constant X_p criterion.

6.2 KERN'S METHOD

Kern (1950) proposed a "bulk flow" method for shell and tube exchanger design where the correlations are based on the total stream flow. While calculating the shellside mass velocity, this method assumes the flow area to be the maximum flow area that corresponds to the center of the shell. In practice, no tubes typically exist at the center of the shell and, instead, two equal maximum rows are placed on either side of it with fewer tubes than computed at the center. Such deviations, as well as the fact that the flow area is continuously changing on the shellside, are ignored in Kern's simplified approach. It also neglects the effect of bypass and leakage streams, all of which will be accounted for in the next section by the Bell-Delaware method.

However, Kern's method is simple to apply and accurate for preliminary designs where uncertainty in various design parameters is large enough not to justify the use of more rigorous methods.

Example 6.2 Design of a Shell and Tube Exchanger by Kern's Method

Design a shell and tube heat exchanger by Kern's method for the input data given in Tables 6.1 and 6.2. Note that pressure drop allowance is about 10 psi for the tubeside and 2 psi for the shellside (1 psi = 6.8948 kPa). A combined dirt factor of 3×10^{-4} m^2 °C/W should be provided.

It is desired to use an exchanger with steel (k = 36 W/m.°C) tubes of 19.1 mm o.d., 15.4 mm i.d., and 4.88 m length. The tubes may be laid out in 25.4 mm square pitch. The thickness of each tube sheet may be assumed to be 25 mm. It is preferable to use a split-ring floating head and to let the water flow in the tubes to avoid corrosion of the shell. The number of tube passes may be selected to obtain a tubeside design velocity of about 1.5 m/s. Calculate the heat transfer coefficients (based on average temperatures) and the pressure drops for the tubeside and shellside. Verify that the design satisfies the constraints on the pressure drop and dirt factor.

Table 6.2 Inputs for Exchanger Design

Quantity	Symbol	Units	Value
Exchanger Data			
Shell diameter	D_s	mm	890 (step 5)
Tube diameter (outer)	D_t	mm	19.1
Tube diameter (inner)	D	mm	15.4
Tube pitch	L_{tp}	mm	25.4
Pitch angle	A_{tp}	degree	90.0
Tube length (effective)	L	m	4.83
Number of tubes per shell	N_t		761 (step 1)
Number of tube passes	NT		6 (step 2)
Number of shell passes	NS		1 (Example 6.1)
Outer tube limit diameter	D_{otl}	mm	842.1 (step 4)
Baffle cut	B_c	%	25
Baffle spacing (central)	L_{bc}	mm	267 (step 6)
Baffle spacing (outlet)	L_{bo}	mm	= L_{bc}
Baffle spacing (inlet)	L_{bi}	mm	= L_{bc}
Number of sealing strips	N_{ss}		0 (Example 6.3)
Wall conductivity	k	W/(m °C)	36
Constraints			
Maximum pressure drop (shellside)	$\Delta P_{s,max}$	kPa	14
Maximum pressure drop (tubeside)	$\Delta P_{t,max}$	kPa	69
Minimum dirt resistance factor	$R_{d.min}$	(m^2°C)/W	3 x 10^{-4}

Solution. The detailed thermal design is performed first to calculate the heat transfer coefficients. Then, the hydrodynamic design is done to check for the pressure drops.

Step 1. Calculation of Number of Tubes

Note that steel tubes of 19.1 mm o.d., 15.4 mm i.d., and 4.88 m length (i.e., 3/4 in. x 16 ft.) are chosen. Allowing 50 mm for the thickness of the two tube sheets, the effective tube length, L, is found to be 4.83 m. Based on the initial trial value of 100 W/m^2 °C for U, the area, A, was calculated to be 220.53 m^2 in Example 6.1. So, the number of tubes required is

$$N_t = A/(\pi D_t L NS) \qquad (6.16)$$
$$= 220.53/(\pi \times 0.0191 \times 4.83 \times 1) = 761 \text{ (rounded off).}$$

Step 2. Calculation of Tubeside Reynolds Number

The tubeside mass velocity is calculated using

$$G_t = \rho_t v_t = M_t/[(\pi D^2/4)(N_t/NT)] \qquad (6.17)$$
$$= 31.58/[(\pi \times 0.0154^2/4)(761/NT)].$$

Note that the F equation for a 1-2 exchanger (Equation 6.9) may be applied to any exchanger with an even number of tube passes, so the minimum number of even tube passes is sought that gives an acceptable tubeside velocity. For $NT = 6$, $v_t = 1.34$ m/s and $G_t = 1336.7$ kg/(s.m^2).

The tubeside Reynolds number is then given by

$$Re_t = D G_t/\mu_t$$
$$= 0.0154 \times 1336.7/0.001 = 20585.$$

Step 3. Calculation of Tubeside Heat Transfer Coefficient

Correlations for the Nusselt number ($Nu_t = h_t D/k_t$) used for this purpose are given below:

$$h_t D/k_t = 1.86 (Re_t Pr_t D/L)^{1/3} (\mu_t/\mu_{tw})^{0.14} \quad \text{for } Re_t \leq 2100 \qquad (6.18a)$$
$$= 0.116 (Re_t^{2/3} - 125) Pr_t^{1/3} [1 + (D/L)^{2/3}] (\mu_t/\mu_{tw})^{0.14}$$
$$\text{for } 2100 < Re_t \leq 10000 \qquad (6.18b)$$
$$= C_t Re_t^{0.8} Pr_t^{1/3} (\mu_t/\mu_{tw})^{0.14} \quad \text{for } Re_t > 10000 \qquad (6.18c)$$

where $C_t = 0.021$ (for gases), 0.023 (for nonviscous liquids), and 0.027 (for viscous liquids).

Using the "turbulent regime" expression with $C_t = 0.023$ and Prandtl number $Pr_t = C_{pt} \mu_t/k_t = 4180 \times 0.001/0.6 = 6.97$ gives

$$h_t/(\mu_t/\mu_{tw})^{0.14} = 0.023 (k_t/D) Re_t^{0.8} Pr_t^{1/3}$$
$$= 0.023(0.6/0.0154)(20585)^{0.8}(6.97)^{1/3} = 4832.8 \text{ W/m}^2 \text{ °C.}$$

For water, the viscosity correction is not required. On setting $(\mu_t/\mu_{tw})^{0.14} = 1$, $h_t = 4832.8$ W/m^2 °C.

Step 4. Calculation of Outer-tube-limit Diameter

An empirical equation useful for the calculation of the diameter of the circle through the centers of the outermost tubes of the tube bundle is

$$D_{ctl} = D_t (N_t/K_1)^{1/n_1}.$$

K_1 and n_1 depend on the number of tube passes and the pitch. For six tube passes and a square pitch, it is found by table-lookup (Sinnott, 1989) that $K_1 = 0.0402$ and $n_1 = 2.617$. Thus, noting that the outer-tube-limit diameter (D_{otl}) corresponds to the tube-bundle-circumscribed circle,

$$D_{ctl} = 19.1 (761/0.0402)^{1/2.617} = 823 \text{ mm},$$
and $D_{otl} = D_{ctl} + D_t = 823 + 19.1 = 842.1$ mm.

Step 5. Calculation of Shell Inside Diameter

First, the bundle diametrical clearance must be established. Based on Sinnott (1989, Figure 12.10), the following equations are proposed (all dimensions in mm):

$$
\begin{aligned}
D_s - D_{ctl} &= 0.01 D_{ctl} + 8 && \text{for fixed and U-tube} \\
&= 0.0075 D_{ctl} + 87 && \text{for pull-through floating head} \\
&= 0.028 D_{ctl} + 44.4 && \text{for split-ring floating head} \\
&= 38 && \text{for outside packed head}
\end{aligned}
$$

For the split-ring floating head, $D_s - D_{ctl} = 0.028 (823) + 44.4 = 67$ mm.

This directly gives the shell inside diameter as $D_s = 823 + 67 = 890$ mm. In practice, the nearest standard pipe size (35") would be used. Calculations are continued below with the value of 890 mm for D_s.

Step 6. Choice of Baffle Spacing

The baffle spacing ranges from 0.2 to 1.0 times the shell diameter. Let

$$L_{bc} = 0.3 D_s = 0.3 \times 890 = 267 \text{ mm}.$$

Step 7. Calculation of Shellside Reynolds Number

The shellside mass velocity is based on crossflow at the shell centerline and is calculated using

$$G_s = \rho_s v_s = M_s/[L_{bc} (L_{tp} - D_t) (D_s/L_{tp})] \tag{6.19}$$
$$= (53650/3600)/[267(25.4 - 19.1)(890/25.4)10^{-6}] = 252.8 \text{ kg/(s.m}^2).$$

Next, the equivalent diameter is calculated from

$$D_e = 4 [\sqrt{3}/4 L_{tp}^2 - \pi/8 D_t^2]/(\pi/2 D_t) \qquad \text{for triangular pitch}$$
$$D_e = 4 [L_{tp}^2 - \pi/4 D_t^2]/(\pi D_t) \qquad \text{for square pitch}$$

For our case of a square pitch,

$$D_e = 4 [25.4^2 - \pi/4 \times 19.1^2]/(\pi \times 19.1) = 23.9 \text{ mm}.$$

The shellside Reynolds number is then given by

$$Re_s = D_e G_s/\mu_s$$
$$= 0.0239 \times 252.8/0.00023 = 26282.$$

Step 8. Calculation of Shellside Heat Transfer Coefficient
The correlation for the Nusselt number ($Nu_s = h_s D_e/k_s$) is

$$h_s D_e/k_s = 0.36 Re_s^{0.55} Pr_s^{1/3} (\mu_s/\mu_{sw})^{0.14}. \tag{6.20}$$

Using Prandtl number $Pr_s = C_{ps} \mu_s/k_s = 2684 \times 0.00023/0.11 = 5.61$ gives

$$h_s/(\mu_s/\mu_{sw})^{0.14} = 0.36(0.11/0.0239)(26282)^{0.55}(5.61)^{1/3} = 794 \text{ W/m}^2 \text{ °C}.$$

The procedure for viscosity correction is demonstrated below. Because the hot fluid, metal, and water offer thermal resistances in series, an expression for the wall temperature on the shellside (hot fluid side) may be obtained as

$$T_{sw} = T_{hc} - (U/h_s) (T_{hc} - T_{cc}) = 81.5 - (100/794) (81.5 - 20) = 73.8 \text{ °C}.$$

The caloric (in this case, average) temperatures for hot and cold fluids calculated in Example 6.1 as $T_{hc} = 81.5°C$ and $T_{cc} = 20°C$ are used here. As the viscosity of the hot fluid at 73.8°C is 0.25 x 10^{-3} kg/(m.s), it is seen that $(\mu_t/\mu_{tw})^{0.14} = (0.23/0.25)^{0.14} = 0.988$. Thus, the correction is negligible for a low-viscosity fluid.

Step 9. Calculation of Dirt Factor
The overall heat transfer coefficient under clean conditions is given by

$$U_c = [(1/h_s) + (1/h_t) (D_t/D) + D_t/(2k) \ln(D_t/D)]^{-1} \qquad (6.21a)$$
$$= [(1/794) + (1/4832.8)(19.1/15.4) + 0.0191/(2 \times 36) \ln(19.1/15.4)]^{-1}$$
$$= 635.6 \text{ W/m}^2 \text{ }^\circ\text{C.}$$

The overall dirt resistance factor is then found from

$$R_d = (1/U) - (1/U_c) \qquad (6.21b)$$
$$= (1/100) - (1/635.6) = 8.4 \times 10^{-3} \text{ m}^2 \text{ }^\circ\text{C/W.}$$

As the calculated dirt factor is higher than the minimum specified (3×10^{-4} m^2 °C/W), the constraint on the fouling allowance is satisfied and the exchanger will operate satisfactorily till the scheduled maintenance shutdown.

Step 10. Calculation of Tubeside Pressure Drop
Correlations for the friction factor in the case of smooth pipes are

$$f_t = 16/Re_t \qquad\qquad\qquad \text{for } Re_t \le 2100 \quad (6.22a)$$
$$= 0.046 \, Re_t^{-0.2} \qquad\qquad \text{for } Re_t > 2100 \quad (6.22b)$$

The tubeside pressure drop due to friction is then given by

$$\Delta P_{tf} = 2 f_t G_t^2 L \, NT \, NS/[D \, \rho_t \, (\mu_t/\mu_{tw})^{ac}] \qquad (6.23)$$

where $ac = 0.35$ (for $Re_t \le 2100$) and 0.14 (for $Re_t > 2100$).
Substituting appropriate values,

$$f_t = 0.046 \, (20585)^{-0.2} = 6.31 \times 10^{-3}$$
$$\Delta P_{tf} = 2(6.31 \times 10^{-3}) \, 1336.7^2 \times 4.83 \times 6 \times 1/[0.0154 \times 998] = 42.52 \text{ kPa.}$$

The return losses are given by

$$\Delta P_{tr} = 1.25 \, G_t^2 \, NT \, NS/\rho_t$$
$$= 1.25 \times 1336.7^2 \times 6 \times 1/998 = 13.43 \text{ kPa.}$$

Therefore, the total tubeside pressure drop is

$$\Delta P_t = \Delta P_{tf} + \Delta P_{tr} = 42.52 + 13.43 = 55.95 \text{ kPa.}$$

Step 11. Calculation of Shellside Pressure Drop
Various correlations for the shellside friction factor are available:

$$f_s = 0.4475 \, Re_s{}^{-0.19} \tag{6.24a}$$
$$f_s = 0.324 \; Re_s{}^{-0.16} \tag{6.24b}$$

The shellside pressure drop due to friction is then given by

$$\Delta P_s = 2 f_s \, G_s{}^2 \, D_s \, (N_b + 1) \, NS/[D_e \, \rho_s \, (\mu_s/\mu_{sw})^{0.14}] \tag{6.25}$$

where N_b is the number of baffles required. It may be approximately calculated from

$$N_b = L/L_{bc} - 1 = 4.83/0.267 - 1 = 17 \text{ (rounded off to nearest integer).}$$

Substituting appropriate values,

$$f_s = 0.4475 \, (26282)^{-0.19} = 0.0647$$

or $f_s = 0.324 \, (26282)^{-0.16} = 0.0636$. Both correlations give similar results.
$\Delta P_s = 2 \, (0.0647) \, 252.8^2 \times 890 \times 18 \times 1/[23.9 \times 777] = 7.14$ kPa.

The pressure drop in the shell nozzles is neglected here and will usually be significant only for gases. The maximum pressure drops represent the limiting abilities of the external pumps to transport the fluids along the exchanger; in this case, the calculated pressure drops on both the tubeside and the shellside are lower than their specified limits. Thus, the design is acceptable in terms of both fouling allowance and pressure drop constraints.

6.3 BELL-DELAWARE METHOD

Kern's method as described in the previous section provides a good model for tubeside predictions. However, it oversimplifies the description of the shellside flow and, consequently, does not typically provide satisfactory predictions for the shellside design. The Bell-Delaware method (Bell, 1963, 1981, 1983) captures a more realistic picture of the heat exchanger geometry and is therefore more complex than Kern's method. It is presumably the best method available in the open literature. Better predictions are possible with commercially available software packages, but the sophisticated models used by them are proprietary.

In the Bell-Delaware method, the shellside heat transfer coefficient and pressure drop are estimated starting with correlations for flow over ideal tube banks. The effects of leakage, bypassing, and flow in the window zone are then incorporated through suitable correction factors. The method can be used to study the effects of constructional tolerances and the use of sealing strips.

Example 6.3 Shellside Design by Bell-Delaware Method
Rework Example 6.2 using the Bell-Delaware method. Specifically, calculate the heat transfer coefficient and the pressure drop for the shellside.

Solution. The basis for the Bell-Delaware method is the model for flow across the tube bundle of a baffled shell and tube exchanger proposed by Tinker (1951) and modified by Palen and Taborek (1969). The assumption of the entire shell fluid flowing across the tube bundle is not valid; instead, it is suggested that there exist a variety of flow paths, of which the five main ones (labeled A through F, with no D stream) are described below.

Stream A is the tube-to-baffle leakage that flows through the clearance between the tube outer diameter and the tube hole in the baffle.
Stream B is the main stream that flows across the bundle.
Stream C is the bundle-to-shell bypass flowing between the tube bundle diameter and the shell inner diameter. This bypass will be significant in the case of pull through bundles, and may be reduced through the use of sealing strips.
Stream E is the baffle-to-shell leakage that flows through the clearance between the edge of a baffle and the shell wall.
Stream F is the pass-partition stream, which corresponds to the fluid that flows through the gap in the tube arrangement due to the pass-partition plates. It provides a low pressure drop path when the gap is vertical and does not contain dummy tubes. However, the effect of these tube partition bypass lanes is usually negligible.

Whereas streams C, E, and F cause the effective heat transfer to decrease, stream A leads to the effective pressure drop to increase. In general, stream A has a relatively small effect whereas streams C and E have a larger effect.
A simplified version of the method recommended by Taborek (1983) appears below. The shellside heat transfer coefficient (h_s) is obtained by applying five correction factors (J_c, J_l, J_b, J_r, and J_s) to the ideal crossflow coefficient (h_{ic}). The shellside pressure drop (ΔP_s) is obtained by summing the individual pressure drops from three distinct zones: baffled crossflow zones (ΔP_c), window zones (ΔP_w), and end zones (ΔP_e).

Step 1. Calculation of Crossflow Area at Shell Centerline
The crossflow area, S_m, at the shell centerline within one baffle spacing (L_{bc}) includes the bundle-to-shell bypass channel and the crossflow area between the tubes. Thus,

$$S_m = L_{bc} \{D_s - D_{otl} + (L_{tp} - D_t) [(D_{otl} - D_t)/L_{tpe}]\}$$

where $L_{tpe} = L_{tp}$ for 30° and 90° layouts
and $L_{tpe} = L_{tp}/\sqrt{2}$ for 45° layout.

Thus, $S_m = 267 \{890 - 842.1 + (25.4 - 19.1) [(842.1 - 19.1)/25.4]\} \times 10^{-6}$
         $= 6.73 \times 10^{-2}$ m^2.

Step 2. Calculation of Shellside Reynolds Number
The shellside crossflow mass velocity is defined as

$$G_s = M_s/S_m$$
$$= (53650/3600)/0.0673 = 221.5 \text{ kg/(s.m}^2)$$

In principle, the calculation of G_s is identical to that in Kern's method, but it is more accurately performed accounting for the bundle-to-shell bypass.

However, instead of the Reynolds number being based on the equivalent diameter, it is now based on the outside tube diameter. Thus, the shellside Reynolds number for the Bell-Delaware method is given by

$$Re_S = D_t G_s/\mu_s$$
$$= 0.0191 \times 221.5/0.00023 = 18391.$$

Step 3. Calculation of Heat Transfer Coefficient for Ideal Crossflow (h_{ic})
The heat transfer coefficient for pure crossflow over an ideal tube bank may be found from the following:

$$h_{ic} = j_h (k_s/D_t) Re_S Pr_s^{1/3} (\mu_s/\mu_{sw})^{0.14} \qquad (6.26)$$
where $j_h = a_1 (Re_S)^{a_2} (1.33 D_t/L_{tp})^a$
and $a = a_3/[1 + 0.14 (Re_S)^{a_4}]$

The viscosity correction $(\mu_s/\mu_{sw})^{0.14}$ is valid only for liquids. The coefficients a_1, a_2, a_3, and a_4 depend on the tube arrangement and Reynolds number. For a square pitch and $10^4 < Re_S \leq 10^5$, the table provided by Taborek (1983) yields $a_1 = 0.37$, $a_2 = -0.395$, $a_3 = 1.187$, and $a_4 = 0.37$.

Thus, $a = 1.187/[1 + 0.14 (18391)^{0.37}] = 0.189$
         $j_h = 0.37(18391)^{-0.395} (1.33 \times 19.1/25.4)^{0.189} = 7.65 \times 10^{-3}$
         $h_{ic} = 7.65 \times 10^{-3} (0.11/0.0191) \times 18391 \times (5.61)^{1/3} = 1440$ W/m^2 °C.

Step 4. Calculation of Segmental Baffle Window Correction Factor (J_c)
The heat transfer coefficient is based on ideal crossflow and needs to be corrected for the effects of baffle window flow. The window flow velocity is

higher than the crossflow velocity at the shell centerline and increases with decreasing baffle cut (B_c). It also depends on the fraction of number of tubes in the window (F_w). The correction factor, J_c, is approximately linear for B_c between 15 and 45% and is given by

$$J_c = 0.55 + 0.72 \, (1 - 2F_w)$$
where $F_w = \theta_{ctl}/360 - \sin \theta_{ctl} /(2\pi)$
and $\theta_{ctl} = 2 \cos^{-1}[(D_s/D_{ctl}) \, (1 - 2 \, B_c/100)]$.

Substituting appropriate values,

$$\theta_{ctl} = 2 \cos^{-1}[(890/823) \, (1 - 2 \times 25/100)] = 114.5°$$
$$F_w = 114.5/360 - \sin 114.5 /(2\pi) = 0.173$$
$$J_c = 0.55 + 0.72 \, (1 - 2 \times 0.173) = 1.02$$

Note that J_c varies between 0.65 (for large baffle cuts) and 1.15 (for small baffle cuts). J_c should be near unity for a well designed exchanger.

Step 5. Calculation of Baffle Leakage Correction Factor (J_l)
The baffle leakage correction factor (J_l) accounts for the effects of stream A flowing through the tube-to-baffle clearance and stream E flowing through the baffle-to-shell clearance. If L_{tb} is the diametrical clearance between the tube and the baffle hole (usually 0.8 mm for easy assembly of the tube bundle), then the tube-to-baffle hole leakage area per baffle is given by

$$S_{tb} = (\pi/4) \, [(D_t + L_{tb})^2 - D_t^2] \, N_t \, (1 - F_w)$$

The correlation for the shell-to-baffle diametrical clearance is (Taborek, 1983)

$$L_{sb} = 3.1 + 0.004 \, D_s \qquad \text{(in mm)}$$

Then, the shell-to-baffle leakage area per baffle is given by

$$S_{sb} = (\pi/2) \, D_s \, L_{sb} \, [1 - \theta_{ds}/360]$$
where $\theta_{ds} = 2 \cos^{-1}(1 - 2 \, B_c/100)$.

The correlation for the baffle leakage correction factor (J_l) is (Taborek, 1983)

$$J_l = 0.44 \, (1 - r_s) + [1 - 0.44 \, (1 - r_s)] \exp (-2.2 \, r_{lm})$$
where $r_s = S_{sb}/(S_{sb} + S_{tb})$
and $r_{lm} = (S_{sb} + S_{tb})/S_m$.

Now, the calculations are carried out in the following manner:

$$S_{tb} = (\pi/4)\,[(19.1 + 0.8)^2 - 19.1^2](761)(1 - 0.173) \times 10^{-6} = 1.54 \times 10^{-2}\ m^2$$
$$L_{sb} = 3.1 + 0.004 \times 890 = 6.66\ mm$$
$$\theta_{ds} = 2\cos^{-1}(1 - 2 \times 25/100) = 120°$$
$$S_{sb} = (\pi/2) \times 890 \times 6.66\,[1 - 120/360] \times 10^{-6} = 6.21 \times 10^{-3}\ m^2$$
$$r_s = 6.21 \times 10^{-3}/(6.21 \times 10^{-3} + 1.54 \times 10^{-2}) = 0.287$$
$$r_{lm} = (6.21 \times 10^{-3} + 1.54 \times 10^{-2})/\,6.73 \times 10^{-2} = 0.321$$
$$J_l = 0.44(1 - 0.287) + [1 - 0.44\,(1 - 0.287)]\exp(-2.2 \times 0.321) = 0.652$$

This is the most significant of the correction factors. The shell-to-baffle leakage escapes contact with the tubes and, hence, has a strong adverse effect on the heat transfer. J_l gives more weightage to the shell-to-baffle leakage compared to the tube-to-baffle leakage. J_l must be greater than 0.6 (preferably 0.7 to 0.9) for a good design. The value of J_l (if too low) may be increased by increasing baffle spacing and/or tube pitch.

Step 6. Calculation of Bundle Bypass Correction Factor (J_b)

The bypass correction factor (J_b) accounts for the effect of stream C flowing around the tube bundle (where the flow resistance is notably less than through the tube field), so J_b depends on the shell-to-bundle clearance area (S_b). It typically varies from 0.7 (for large clearances as in pull-through bundles) to 0.9 (for small clearances as in fixed-tubesheet and U-tube bundles). To obtain higher local velocities and heat transfer coefficients, sealing strips may be used to force the bypass (C) stream systematically back into the main (B) stream. To start with, no sealing strips are used. If the value of J_b is too low (less than 0.7), then J_b may be increased by increasing the number of sealing strips (N_{ss}).

The bundle-to-shell bypass area within one baffle is given by

$$S_b = L_{bc}\,[(D_s - D_{otl}) + L_{pl}]$$

where L_{pl} is the appropriate tube lane partition half-width and accounts for the effect of the additional bypass through the tube partition pass lanes.

The number of effective tube rows in the crossflow region (between baffle tips in one baffle section) is

$$N_{tcc} = (D_s/L_{pp})\,(1 - 2\,B_c/100)$$

where L_{pp} is the tube pitch parallel to the flow. L_{pp} is equal to $\sqrt{3}/2\,L_{tp}$ (for $A_{tp} = 30°$), $L_{tp}/\sqrt{2}$ (for $A_{tp} = 45°$), and $L_{pp} = L_{tp}$ (for $A_{tp} = 90°$).

Now the correction factor, J_b, may be estimated from

$$J_b = \exp\{-C_{bh}(S_b/S_m)[1-(2N_{ss}/N_{tcc})^{1/3}]\} \quad \text{for } N_{ss}/N_{tcc} < 0.5$$
$$= 1 \qquad\qquad\qquad\qquad\qquad\qquad \text{for } N_{ss}/N_{tcc} \geq 0.5$$

where $C_{bh} = 1.35$ for $Re_S \leq 100$ (laminar flow),
and $C_{bh} = 1.25$ for $Re_S > 100$ (transition and turbulent flow).

Now, the calculations are performed as follows (assuming $L_{pl} = 0$):

$$S_b = 267(890-842.1) \times 10^{-6} = 1.28 \times 10^{-2}\ m^2$$
$$N_{tcc} = (890/25.4)(1-2 \times 25/100) = 17.52$$
$$J_b = \exp\{-1.25(1.28 \times 10^{-2}/6.73 \times 10^{-2})[1-(2 \times 0/17.52)^{1/3}]\} = 0.789$$

Step 7. Calculation of Adverse Temperature Gradient Buildup Correction Factor at Low Reynolds Number (J_r)

The correction factor (J_r) accounts for the effect of adverse temperature gradient development through the boundary layer in laminar flow and is given by the following formulae:

$$J_r = 1 \qquad\qquad\qquad\qquad\qquad\quad \text{for } Re_S > 100$$
$$= (J_r)_r + [(20-Re_S)/80][(J_r)_r - 1] \quad \text{for } 20 \leq Re_S \leq 100$$
$$= (J_r)_r \qquad\qquad\qquad\qquad\qquad \text{for } Re_S < 20 \text{ (deep laminar)}$$

where $(J_r)_r = (10/N_c)^{0.18}$.

N_c is the total number of tube rows crossed in the entire exchanger. Thus,

$$N_c = (N_{tcc} + N_{tcw})(N_b + 1) \text{ and } N_{tcw} = (0.8/L_{pp})[D_s B_c/100 - (D_s - D_{ctl})/2]$$

For this case study, $J_r = 1$.

Step 8. Calculation of Unequal Baffle Spacing Correction Factor (J_s)

The correction factor (J_s) accounts for the effect of the increased baffle spacing often employed at inlet and outlet sections. This increase is required to accommodate nozzles that are located away from the tubesheets. J_s is calculated from the following correlation:

$$J_s = [N_b - 1 + (L_{bi}/L_{bc})^{1-n} + (L_{bo}/L_{bc})^{1-n}]/[N_b - 1 + L_{bi}/L_{bc} + L_{bo}/L_{bc}]$$

where $n = 1/3$ for $Re_S \leq 100$ and $n = 0.6$ for $Re_S > 100$.

In this case study, $L_{bi} = L_{bo} = L_{bc}$. Thus, $J_s = 1$. Usually, it is between 0.85 and 1.0.

Step 9. Calculation of Shellside Heat Transfer Coefficient (h_s)
The shellside heat transfer coefficient (h_s) can now be obtained as

$$h_s = h_{ic} J_c J_l J_b J_r J_s \qquad (6.27)$$
$$= 1440 \times 1.02 \times 0.652 \times 0.789 \times 1 \times 1$$
$$= 1440(0.525) = 755.6 \ W/m^2 \ °C.$$

The net effect (product) of the five correction factors should be typically greater than 0.5 for a reasonably good design and never below 0.4.

Step 10. Calculation of Pressure Drop for Ideal Crossflow Section (ΔP_{ic})
The pressure drop for crossflow over an ideal tube bank (between baffle tips in one baffle compartment) may be found from the following:

$$\Delta P_{ic} = 2 f_i G_s^2 N_{tcc}/[\rho_s (\mu_s/\mu_{sw})^{0.14}] \qquad (6.28)$$

where $f_i = b_1 (Re_S)^{b2} (1.33 D_t/L_{tp})^b$
and $\quad b = b_3/[1 + 0.14 (Re_S)^{b4}]$

The coefficients b_1, b_2, b_3, and b_4 depend on the tube arrangement and Reynolds number. For a square pitch and $10^4 < Re_S \leq 10^5$, the table provided by Taborek (1983) yields $b_1 = 0.391$, $b_2 = -0.148$, $b_3 = 6.3$, and $b_4 = 0.378$.

Thus, $\quad b = 6.3/[1 + 0.14 (18391)^{0.378}] = 0.936$
$\quad\quad\quad f_i = 0.391(18391)^{-0.148} (1.33 \times 19.1/25.4)^{0.936} = 9.14 \times 10^{-2}$
$\quad\quad\quad \Delta P_{ic} = 2 \times 9.14 \times 10^{-2} \times 221.5^2 \times (17.52)/[777 \times 1] = 202.3 \ Pa.$

Step 11. Calculation of Correction Factors for Baffle Leakage (R_l) and Bundle Bypass (R_b)
The baffle leakage correction factor (R_l) and the bundle bypass correction factor (R_b) are the analogs of J_l (step 5) and J_b (step 6) respectively, but for pressure drop. The correlations for R_l and R_b are (Taborek, 1983)

$$R_l = \exp [-1.33 (1 + r_s) (r_{lm})^p]$$

where $p = -0.15 (1 + r_s) + 0.8$

$$R_b = \exp \{-C_{bp} (S_b/S_m) [1 - (2 N_{ss}/N_{tcc})^{1/3}]\} \text{ for } N_{ss}/N_{tcc} < 0.5$$
$$= 1 \qquad\qquad\qquad\qquad \text{ for } N_{ss}/N_{tcc} \geq 0.5$$

where $C_{bp} = 4.5$ for $Re_S \leq 100$ (laminar flow),
and $C_{bp} = 3.7$ for $Re_S > 100$ (transition and turbulent flow).

Substituting appropriate values,

$$p = -0.15 (1 + 0.287) + 0.8 = 0.6$$
$$R_l = \exp [-1.33 (1 + 0.287) (0.321)^{0.6}] = 0.424$$
$$R_b = \exp \{-3.7 (1.28 \times 10^{-2}/6.73 \times 10^{-2}) [1 - (2 \times 0/17.52)^{1/3}]\} = 0.495$$

R_l is usually between 0.4 and 0.5. Low values of R_l are obtained when the baffles are closely spaced. R_b is usually between 0.5 (for a pull-through floating head with one or two pairs of sealing strips) and 0.8 (for a good fixed tube sheet design).

Step 12. Calculation of Pressure Drop in the Crossflow (ΔP_c)
The baffled crossflow pressure drop is given by

$$\Delta P_c = (N_b - 1) \Delta P_{ic} R_l R_b \qquad\qquad\qquad (6.29)$$
$$= (17 - 1) 202.3 \times 0.424 \times 0.495 = 678.4 \text{ Pa.}$$

Step 13. Calculation of Pressure Drop in the Baffle Window (ΔP_w)
The total pressure drop in all the windows crossed (N_b in number) may be estimated from the following correlations, which use the mass velocity, G_w, based on the geometric mean of S_m (the crossflow area) and S_w (the net window area).

$$\Delta P_w = N_b [(2 + 0.6 N_{tcw}) G_w^2/(2\rho_s)] R_l \qquad \text{for } Re_S > 100 \qquad (6.30)$$

where $G_w = M_s/\sqrt{(S_m S_w)}$
and $S_w = (\pi D_s^2/4) [\theta_{ds}/360 - \sin \theta_{ds}/(2\pi)] - N_t F_w (\pi D_t^2/4)$

On substituting appropriate values,

$$S_w = \{(\pi 890^2/4)[120/360 - \sin 120/(2\pi)] - 761 \times 0.173 (\pi 19.1^2/4)\} \times 10^{-6}$$
$$= 8.38 \times 10^{-2} \text{ m}^2$$
$$G_w = (53650/3600)/\sqrt{(0.0673 \times 0.0838)} = 198.4 \text{ kg/(s.m}^2)$$

$N_{tcw} = (0.8/25.4)[890 \times 25/100 - (890 - 823)/2] = 5.95$

$\Delta P_w = 17\,[(2 + 0.6 \times 5.95)\,198.4^2/(2 \times 777)]\,0.424 = 1017.7$ Pa.

Step 14. Calculation of Pressure Drop in the End Zones (ΔP_e)

The crossflow pressure drop in the end zones (i.e., the inlet and outlet baffle spacings) may be estimated from the following:

$$\Delta P_e = \Delta P_{ic}\,(1 + N_{tcw}/N_{tcc})\,R_b\,R_s \qquad (6.31)$$

where $R_s = (L_{bc}/L_{bi})^{2-n} + (L_{bc}/L_{bo})^{2-n}$

and $n = 1$ for $Re_S \leq 100$ and $n = 0.2$ for $Re_S > 100$.

In this case study, $L_{bi} = L_{bo} = L_{bc}$. Thus, $R_s = 2$ and $\Delta P_e = 202.3\,(1 + 5.95/17.52)\,0.495 \times 2 = 268.3$ Pa.

Step 15. Calculation of Shellside Pressure Drop (ΔP_s)

The total nozzle-to-nozzle shellside pressure drop (ΔP_s) can now be obtained as

$$\Delta P_s = \Delta P_c + \Delta P_w + \Delta P_e \qquad (6.32)$$
$$= 678.4 + 1017.7 + 268.3 = 1.96 \text{ kPa.}$$

The above calculation does not include the pressure drop due to nozzles and impingement devices. Note that Kern's method overpredicts the shellside pressure drop (as 7.14 kPa) because it neglects leakage and bypassing effects that could cause a significant reduction (as high as 70 to 80%) in the pressure drop for a typical exchanger.

Step 16. Checking against Constraints on Pressure Drop and Dirt Factor

The shellside pressure drop is within the maximum specified limit (14 kPa). The overall heat transfer coefficient under clean conditions is given by (see step 9 in Example 6.2)

$$U_c = [(1/755.6) + (1/4832.8)(19.1/15.4) + 0.0191/(2 \times 36)\ln(19.1/15.4)]^{-1}$$
$$= 610.8 \text{ W/m}^2\,^\circ\text{C.}$$

The overall dirt resistance factor is then found from

$$R_d = (1/100) - (1/610.8) = 8.36 \times 10^{-3} \text{ m}^2\,^\circ\text{C/W.}$$

The calculated dirt factor is higher than 3×10^{-4} m^2 °C/W (the minimum specified); thus, the constraint on the fouling allowance is satisfied. There is only a small difference between the overall heat transfer coefficients predicted by Kern's method and the Bell-Delaware method for this case study.

The issue of design modification is an important one and is now addressed. What if the set of exchanger geometrical parameters selected does not lead to a design that satisfies the constraints? If the rating shows the dirt factor to be lower than the minimum specified or the pressure drop(s) to be higher than the maximum allowable, then a different (usually bigger) exchanger is selected and re-rated. If the pressure drop(s) turn out to be substantially less than available, then parameters may be re-chosen to utilize more of the allowable pressure drop(s) in order to obtain a smaller and cheaper exchanger. Some qualitative rules (Bell, 1981) for logical modification of exchanger configurations are given below.

- If the amount of heat transfer is to be increased,
  then the area of the exchanger may be increased
  or the heat transfer coefficients may be increased
  or the mean temperature difference may be increased.
- If the area of the exchanger is to be increased,
  then the length of the tubes may be increased
  or the diameter of the tubes may be increased
  or the use of multiple shells in series/parallel may be considered.
- If the tubeside coefficient is to be increased,
  then the number of tube passes may be increased.
- If the shellside coefficient is to be increased,
  then the baffle spacing may be decreased
  or the baffle cut may be decreased.
- If the mean temperature difference is to be increased,
  then the use of multiple shells in series may be considered
  or the use of a countercurrent arrangement may be considered.
- If the tubeside pressure drop is to be decreased,
  then the diameter of the tube may be increased
  or the length of the tube may be decreased
  or the number of tube passes may be decreased.
- If the shellside pressure drop is to be decreased,
  then the baffle spacing may be increased
  or the baffle cut may be increased
  or the tube pitch may be increased
  or the use of double/triple segmental baffles may be considered.

The possible modifications are many, so choosing the appropriate ones to obtain satisfactory operation with a proper consideration of the tradeoffs and costs is a complex task. A simple, and novel, alternative has been recently proposed by Polley et al. (1991) and is presented in the next section.

6.4 RAPID DESIGN ALGORITHM

Both Kern's method and the Bell-Delaware method are presented as rating methods in the previous two sections. In other words, the geometry of the exchanger is almost fully specified by the designer in a subjective manner, and rating involves subsequent calculation of the heat transfer coefficients and the pressure drops for both the tubeside and shellside. As the computational effort is significant, a design that satisfies the dirt factor and pressure drop constraints is often accepted without thoroughly investigating other options that may prove to be more promising.

Polley et al. (1991) have presented design algorithms for the rapid sizing of shell and tube and compact heat exchangers on a more objective basis. These algorithms may be used in conjunction with the detailed rating methods presented in the earlier sections as well as programs for mechanical design and tube bundle vibration analyses to arrive at an optimal design. The philosophy is to utilize completely the maximum allowable pressure drops on both the hot and cold streams rather than use these specifications as mere constraints. Full use of both pressure drops ensures that the exchanger is designed for the highest possible velocities (and consequently heat transfer coefficients). Thus, the exchanger is the smallest possible for a given service (and presumably the most economical). For this purpose, it is necessary to have simple relationships (Peters and Timmerhaus, 1981; Polley et al., 1990) between the pressure drop and the heat transfer coefficient involving the fewest geometrical parameters for the exchanger. Such equations for the turbulent regime are derived below.

Relation between ΔP_t and h_t for Tubeside
On neglecting return losses, Equations 6.22b and 6.23 give

$$\Delta P_t = 2 \, (0.046) \, (D \, v_t \, \rho_t/\mu_t)^{-0.2} \, \rho_t \, v_t^2 \, L \, NT \, NS/[D \, (\mu_t/\mu_{tw})^{0.14}]$$

Combining Equations 6.16 and 6.17 to eliminate N_t gives

$$\rho_t \, v_t = M_t \, /[(A \, D^2)/(4 \, D_t \, L \, NS \, NT)]$$

Substituting in the tubeside pressure drop expression gives

$$\Delta P_t = 2(0.046)(D\rho_t/\mu_t)^{-0.2}\rho_t/[D(\mu_t/\mu_{tw})^{0.14}]\, v_t^{1.8}\,(\rho_t\, v_t\, A\, D^2)/(4\, D_t\, M_t)$$
$$= 0.023\,(D\,\rho_t/\mu_t)^{-0.2}\,(\rho_t^2\, D)/[D_t\, M_t\,(\mu_t/\mu_{tw})^{0.14}]\, A\, v_t^{2.8}$$

The Dittus-Boelter equation may be written as $h_t = 0.023\,(k_t/D)\,(D\,v_t\,\rho_t/\mu_t)^{0.8}$ $(C_{pt}\mu_t/k_t)^{1/3}\,(\mu_t/\mu_{tw})^{0.14}$ (see Equation 6.18c). Then, it may be used to eliminate v_t and get the desired relation given below:

$$\Delta P_t = K_t\, A\, h_t^{3.5} \tag{6.33}$$

where $K_t = 1/(0.023)^{2.5}\, D^{1/2}\,\mu_t^{11/6}/(M_t\rho_t k_t^{7/3}\, C_{pt}^{7/6})(D/D_t)[(\mu_t/\mu_{tw})^{-0.14}]^{4.5}$.

Note that K_t depends on the tube diameter, the physical properties, and the flow rate of the tubeside fluid.

Relation between ΔP_s and h_s for Shellside: Kern's Method
    The correlations presented by Kern (1950) for the shellside give reasonable predictions when the baffle cut is 25% and the ratio of baffle spacing to shell diameter is unity. Equations 6.24a and 6.25 give

$$\Delta P_s = 2(0.4475)(D_e\, v_s\,\rho_s/\mu_s)^{-0.19}\rho_s v_s^2\, D_s(N_b + 1)NS/[D_e\,(\mu_s/\mu_{sw})^{0.14}]$$

Now, the number of tubes is approximately given by $N_t = \pi\, D_s^2/(4\, L_{tp}^2)$. Assuming equispaced baffles, $L = (N_b + 1)L_{bc}$ (or else, $L = [N_b - 1]L_{bc} + L_{bi} + L_{bo}$ if the entrance and/or exit baffle spacings are different from the central baffle spacing). Combining these expressions with Equations 6.16 and 6.19 gives

$$\rho_s\, v_s = M_s\,(N_b + 1)\,(\pi\, D_t\, NS\,)\,\pi\, D_s^2/(4\, L_{tp}^2)/[A\,(L_{tp} - D_t)\,(D_s/L_{tp})]$$

Substituting for the term $[D_s(N_b + 1)NS]$ in the shellside pressure drop expression gives

$$\Delta P_s = 2(0.4475)(D_e\rho_s/\mu_s)^{-0.19}\rho_s/[D_e(\mu_s/\mu_{sw})^{0.14}]v_s^{1.81}$$
$$\rho_s v_s L_{tp} 4A(L_{tp}-D_t)/(\pi^2 D_t\, M_s)$$
$$= 3.58/\pi^2\,(D_e\rho_s/\mu_s)^{-0.19}(\rho_s^2 L_{tp}(L_{tp}-D_t))/[D_e D_t M_s(\mu_s/\mu_{sw})^{0.14}]\, A\, v_s^{2.81}$$

Equation 6.20 may be written as $h_s = 0.36\,(k_s/D_e)\,(D_e\, v_s\,\rho_s/\mu_s)^{0.55}$ $(C_{ps}\mu_s/k_s)^{1/3}\,(\mu_s/\mu_{sw})^{0.14}$ and used to eliminate v_s. This yields

$$\Delta P_s = K_s\, A\, h_s^{5.1} \tag{6.34}$$

where $K_s = 67 L_{tp}(L_{tp}-D_t) D_e^{1.1} \mu_s^{1.3}/(D_t M_s \rho_s k_s^{3.4} C_{ps}^{1.7})[(\mu_s/\mu_{sw})^{-0.14}6.1}$.

Note that K_s depends on the bundle equivalent diameter (along with the diameter and pitch of the tubes), the physical properties, and the flow rate of the shellside fluid. For consistency with rating calculations, the accurate value of (2.81/0.55) for the exponent may need to be used in place of 5.1.

Relation between ΔP_s and h_s for Shellside: Bell-Delaware Method
From Equations 6.29 through 6.32, the total shellside pressure drop is given by

$$\Delta P_s = (N_b - 1) \Delta P_{ic} R_l R_b + N_b [(2 + 0.6 N_{tcw}) M_s^2/(S_m S_w)/(2\rho_s)] R_l$$
$$+ \Delta P_{ic} (1 + N_{tcw}/N_{tcc}) R_b R_s.$$

Substituting $\Delta P_{ic} = 2 f_i \rho_s v_s^2 N_{tcc}/(\mu_s/\mu_{sw})^{0.14}$ and assuming $S_w = S_m$ gives

$$\Delta P_s = [(N_b - 1) R_l R_b + (1 + N_{tcw}/N_{tcc}) R_b R_s] 2 f_i \rho_s v_s^2 N_{tcc} (\mu_s/\mu_{sw})^{-0.14}$$
$$+ N_b [(1 + 0.3 N_{tcw}) \rho_s v_s^2] R_l.$$

Now, N_b may be eliminated using Equation 6.16 in the form $N_t = A/(\pi D_t [N_b + 1] L_{bc} NS)$ to give

$$\Delta P_s = [(1+0.3 N_{tcw})\rho_s R_l + 2 f_i \rho_s N_{tcc}(\mu_s/\mu_{sw})^{-0.14} R_l R_b] [A/(\pi D_t N_t L_{bc} NS)-1] v_s^2$$
$$+ 2 f_i \rho_s N_{tcc} (\mu_s/\mu_{sw})^{-0.14} R_b [(1 + N_{tcw}/N_{tcc}) R_s - R_l] v_s^2.$$

Equations 6.26 and 6.27 together yield $h_s = j_h k_s (v_s \rho_s/\mu_s) (C_{ps}\mu_s/k_s)^{1/3} (\mu_s/\mu_{sw})^{0.14} J_c J_l J_b J_r J_s$ and are used to eliminate v_s. This yields the equation form originally proposed by Panjeh Shahi (1991):

$$\Delta P_s = (K_{s1} A + K_{s2}) h_s^2 \tag{6.35}$$

where $K_{s1} = [(1 + 0.3 N_{tcw})\rho_s R_l + 2 f_i \rho_s N_{tcc}(\mu_s/\mu_{sw})^{-0.14} R_l R_b]/(\pi D_t N_t L_{bc} NS)$
$[j_h k_s \rho_s/\mu_s (C_{ps}\mu_s/k_s)^{1/3} (\mu_s/\mu_{sw})^{0.14} J_c J_l J_b J_r J_s]^{-2}$
and $K_{s2} = [2 f_i \rho_s N_{tcc} (\mu_s/\mu_{sw})^{-0.14}(1 + N_{tcw}/N_{tcc})R_b R_s - (1 + 0.3 N_{tcw})\rho_s R_l$
$- 4 f_i \rho_s N_{tcc} (\mu_s/\mu_{sw})^{-0.14} R_l R_b]$
$[j_h k_s \rho_s/\mu_s (C_{ps}\mu_s/k_s)^{1/3} (\mu_s/\mu_{sw})^{0.14} J_c J_l J_b J_r J_s]^{-2}$

Note that K_{s1} and K_{s2} depend in a very complicated manner on the ideal heat transfer factor (j_h), the ideal friction factor (f_i), the correction factors ($J_c, J_l, J_b, J_r, J_s, R_l, R_b, R_s$), the geometry ($N_{tcw}, N_{tcc}, D_t, N_t, L_{bc}, NS$), and the fluid physical properties. Furthermore, the baffles must be equally spaced, and the baffle cut must be chosen so that the ratio of crossflow area to window area is unity. This choice may be appropriate (Polley et al., 1991) as the pressure drop will be used to overcome friction and enhance heat transfer (rather than be wasted in alternate acceleration and deceleration of the fluid).

For most practical applications in exchanger optimization and network synthesis, Equation 6.35 is too complex whereas Equation 6.34 is not sufficiently accurate. It is desirable to have a shellside relation in the simple form of Equation 6.34 that accounts for bypassing and leakage effects. Polley et al. (1991) proposed such an equation ($\Delta P_s = K'_s A h_s^{4.412}$) for exchangers with geometric similarity having crossflow area equal to window area. However, the expression for K'_s is not readily available. In what follows, an expression is derived based on a simple rule of thumb: the combined effect of all the correction factors for a reasonably good design is usually about 60% and may be incorporated through a safety factor.

Relation between ΔP_s and h_s for Shellside: Simple Bypass-Leakage Model

The correlations used below are given in Peters and Timmerhaus (1981). The shellside pressure drop is given by

$$\Delta P_s = 2 b_o (D_t v_s \rho_s/\mu_s)^{-0.15} \rho_s v_s^2 (N_b + 1) N_r NS$$

where b_o is a coefficient that depends on tube size and arrangement and N_r is the number of rows of tubes across which the shell fluid flows.

If N_c is the number of clearances between tubes at the shell equator, then $\rho_s v_s = M_s (N_b + 1)/[L (L_{tp} - D_t) N_c]$. Combining this expression with Equation 6.16 gives

$$\rho_s v_s = M_s (N_b + 1) (\pi D_t N_t NS)/[A (L_{tp} - D_t) N_c].$$

Substituting for the term $[(N_b + 1)NS]$ in the shellside pressure drop expression gives

$$\Delta P_s = 2b_o(D_t\rho_s/\mu_s)^{-0.15} \rho_s v_s^{1.85} N_r \rho_s v_s A (L_{tp} - D_t) N_c/(\pi D_t N_t M_s)$$
$$= 2b_o/\pi N_r N_c (D_t\rho_s/\mu_s)^{-0.15} \rho_s^2 (L_{tp} - D_t)/(D_t N_t M_s) A v_s^{2.85}.$$

The shellside heat transfer coefficient is given by $h_s = (a_o/F_s) (k_s/D_t) (D_t v_s \rho_s/\mu_s)^{0.6} (C_{ps}\mu_s/k_s)^{1/3}$. Here, a_o is 0.26 for an in-line tube bank and 0.33 for a staggered tube bank. F_s is a safety factor (with a recommended value of 1.6) to account for bypassing between outermost tubes and shell, leakage between baffles and shell, and leakage between tubes and tube holes in baffles. Eliminating v_s yields the final equation given below:

$$\Delta P_s = K_{ss} A h_s^{4.75} \tag{6.36}$$

where $K_{ss} = 2b_o/[\pi(a_o/F_s)^{19/4}] (N_r N_c/N_t) (L_{tp}-D_t) D_t^{3/4}\mu_s^{17/12}/(M_s\rho_s k_s^{19/6} C_{ps}^{19/12})$. As per Peters and Timmerhaus (1981),

$(N_r N_c/N_t)$ can be taken as 1;
$2b_o/[\pi(a_o/F_s)^{19/4}]$
$$= 3567.2 [0.044 + 0.08(L_{pp}/D_t)/(L_{pn}/D_t - 1)^{0.43+1.13D_t/L_{pp}}]$$
for in-line tubes; and
$2b_o/[\pi(a_o/F_s)^{19/4}] = 1149.5[0.23 + 0.11/(L_{pn}/D_t - 1)^{1.08}]$
for staggered tubes.

Here, L_{pn} is the pitch transverse to flow, and L_{pp} is the pitch parallel to flow. Note that K_{ss} depends only on the physical properties and the flow rate of the shellside fluid (along with the diameter and pitch of the tubes).

Rapid Design Equations

A system of three equations in three unknowns (h_t, h_s, and A) is formed by combining the above equations relating ΔP, h, and A (one equation for the tubeside and one equation for the shellside) with the basic exchanger design equation:

$$Q = UA F (LMTD) \tag{6.37}$$

where $U = [(1/h_s) + R_{ds} + (1/h_t + R_{dt}) (D_t/D) + D_t/(2k) \ln(D_t/D)]^{-1}$, and R_{dt} and R_{ds} are the dirt resistance factors on the tubeside and shellside respectively. It is more convenient to solve the system of three equations by reduction (Duvedi, 1993) to a single equation in a single unknown (say, A).

For Kern's method, Equations 6.33, 6.34, and 6.37 may be combined to obtain

$$k_1 A^{1/5.1} + k_2 A^{1/3.5} + k_3 A + k_4 = 0 \tag{6.38}$$

where $k_1 = (K_s/\Delta P_s)^{1/5.1}$,
$\quad k_2 = (K_t/\Delta P_t)^{1/3.5} (D_t/D)$,
$\quad k_3 = -F (LMTD)/Q$,
and $\quad k_4 = R_{ds} + R_{dt} (D_t/D) + D_t/(2k) \ln(D_t/D)$.

For the Bell-Delaware method, Equations 6.33, 6.35, and 6.37 yield

$$d_1 A^2 + d_2 A^{9/7} + d_3 A + d_4 A^{4/7} + d_5 A^{2/7} + d_6 = 0 \qquad (6.39)$$

where $d_1 = [F\ (LMTD)/Q]^2$,

$d_2 = -2F\ (LMTD)/Q\ (D_t/D)\ (K_t/\Delta P_t)^{2/7}$,

$d_3 = -2F\ (LMTD)/Q\ [R_{ds} + R_{dt}\ (D_t/D) + D_t/(2k)\ \ln(D_t/D)] - (K_{s1}/\Delta P_s)$,

$d_4 = (D_t/D)^2\ (K_t/\Delta P_t)^{4/7}$,

$d_5 = 2(D_t/D)\ (K_t/\Delta P_t)^{2/7}\ [R_{ds} + R_{dt}\ (D_t/D) + D_t/(2k)\ \ln(D_t/D)]$,

and $d_6 = [R_{ds} + R_{dt}\ (D_t/D) + D_t/(2k)\ \ln(D_t/D)]^2 - (K_{s2}/\Delta P_s)$.

For the simple bypass-leakage model, Equation 6.38 is applicable, with the exponent 5.1 replaced by 4.75 and the term K_s replaced by K_{ss}. The rapid design algorithm (RDA) is illustrated next for Kern's method below.

Example 6.4 Rapid Design of Shell and Tube Exchanger -- Kern's Method

Design a 1-6 shell and tube heat exchanger by Kern's method for the input data given in Table 6.1. Note that the pressure drop allowed is 42 kPa on the tube side and 7 kPa on the shellside.   A dirt resistance factor of 1.5×10^{-4} m^2 °C/W should be provided on each side.

It is desired to use an exchanger with tubes of 19.1 mm o.d. and 15.4 mm i.d., laid out in 25.4 mm square pitch. The design objective is to use both the stream pressure drops fully, in accordance with the RDA.

Solution.   The required exchanger area is determined first using Equation 6.38; then, the film heat transfer coefficients and detailed exchanger geometry are obtained. Finally, the performance is confirmed using rating procedures. Note that the metal wall resistance and viscosity corrections are neglected throughout. In this example, the number of shell passes and tube passes is specified. In general, the shell passes must be determined based on the F-criterion and the tube passes based on Equations 6.16, 6.17, and 6.23.

Step 1. Calculation of Heat Transfer Area

The coefficients in Equation 6.38 are calculated after determining K_t and K_s using the expressions given in Equations 6.33 and 6.34:

$K_t = 1/(0.023)^{2.5}(0.0154)^{1/2}(0.001)^{11/6}/(31.58 \times 998 \times 0.6^{7/3} \times 4180^{7/6})(15.4/19.1)$

    $= 2.46 \times 10^{-11}$

$K_s = 67(0.0254)(0.0254-0.0191)(0.0239)^{1.1}(0.00023)^{1.3}/$

$\qquad\qquad\qquad\qquad (0.0191 \times 14.9 \times 777 \times 0.11^{3.4} \times 2684^{1.7})$

    $= 3.93 \times 10^{-14}$

$k_1 = (3.93 \times 10^{-14}/7000)^{1/5.1} = 4.2 \times 10^{-4}$
$k_2 = (2.46 \times 10^{-11}/42000)^{1/3.5} (19.1/15.4) = 5.5 \times 10^{-5}$
$k_3 = -0.9848 (60.78)/1319969 = -4.5 \times 10^{-5}$
$k_4 = 1.5 \times 10^{-4} + 1.5 \times 10^{-4}(19.1/15.4) = 3.4 \times 10^{-4}.$

Then, Equation 6.38 is solved to obtain $A = 28.4$ m^2.

Step 2. Calculation of Heat Transfer Coefficients and Velocities
    The heat transfer coefficients are calculated from Equations 6.33 and 6.34, and the velocities from Equations 6.18 and 6.20.

$h_t = [42000/(2.46 \times 10^{-11} \times 28.4)]^{1/3.5} = 8649.5$ W/m^2 °C.
$h_s = [7000/(3.93 \times 10^{-14} \times 28.4)]^{1/5.1} = 1235.8$ W/m^2 °C.
$v_t = [0.001/(0.0154 \times 998)] [8649.5 \times 0.0154/0.6/0.023/(6.97)^{1/3}]^{1/0.8}$
$\quad = 2.77$ m/s.
$v_s = [0.00023/(0.0239 \times 777)] [1235.8 \times 0.0239/0.11/0.36/(5.61)^{1/3}]^{1/0.55}$
$\quad = 0.73$ m/s.

Step 3. Calculation of Tubecount and Tube Length
    The tubecount and tube length are obtained from Equations 6.17 and 6.16.

$N_t = 31.58/[(\pi \times 0.0154^2/4) (1/6) (998 \times 2.77)] = 367.98$ (rounded off to 368).
$L = 28.4/(\pi \times 0.0191 \times 368 \times 1) = 1.286$ m.

Step 4. Calculation of Shell Diameter and Number of Baffles
    The approximation, $N_t = \pi D_s^2/(4 L_{tp}^2)$, used in deriving Equation 6.34, is now utilized to determine the shell diameter. Equation 6.19 then gives the baffle spacing, L_{bc}, which is subsequently used to obtain the number of baffles, N_b, from $L = (N_b + 1)L_{bc}$.

$D_s = (4 \times 368 \times 25.4^2/\pi)^{1/2} = 549.8$ mm.
$L_{bc} = (53650/3600)/[777 \times 0.73 \times (25.4 - 19.1)(549.8/25.4)10^{-6}] = 192.7$ mm.
$N_b = 1286/192.7 - 1 = 5.67$ (rounded off to 6).

Step 5. Checking Performance Using Rating Procedures
    The final step is to verify the "rapid" designs using the rating procedures given in Sections 6.2 and 6.3. Following the steps used in Examples 6.2 and 6.3, the heat transfer coefficients and pressure drops reported in Table 6.3 are obtained by both the Kern and Bell-Delaware methods. The value of D_{otl} in mm is estimated using $D_s - D_{otl} = 25 + 0.017 D_s$ (obtained by fitting the

diametrical shell-to-bundle clearance curve on Figure 14 in Section 3.3.5 [Taborek, 1983]).

Table 6.3 Checking RDA Using Kern and Bell-Delaware Ratings

Quantity	RDA	Kern	Bell-Delaware
Exchanger area (A in m^2)	28.4	28.4	28.4
Shell diameter (D_s in mm)	549.8	549.8	549.8
Tube length (L in m)	1.286	1.286	1.286
Number of tubes per shell (N_t)	368	368	368
Outer tube limit diameter (D_{otl} in mm)			515.5
Baffle spacing (L_{bc} in mm)	192.7	183.7	183.7
Tubeside Reynolds number (Re_t)	42573	42568	
Shellside Reynolds number (Re_s)	58941	61832	42789
Tubeside coefficient (h_t in W/m^2 °C)	8649.5	8642.3	
Shellside coefficient (h_s in W/m^2 °C)	1235.8	1270.7	1168.5
Overall coefficient (U_c in W/m^2 °C)	1049.7	1074.7	1000.7
Tubeside pressure drop (ΔP_t in kPa)	42.0	41.9	
Shellside pressure drop (ΔP_s in kPa)	7.0	8.1	2.9

Based on Kern's method, the "rapid" design in Table 6.3 requires an area of 28.4 m^2 to achieve an overall heat transfer coefficient of 1074.7 W/m^2 °C. This is a significantly better design compared to the one in Example 6.2, which required an area of 220.5 m^2 to achieve an overall heat transfer coefficient of 635.6 W/m^2 °C. Both designs have similar pressure drop considerations, and so oversizing forces the operational velocities to be lower and the additional area to be inefficient. Note that the "rapid" design has velocities of 2.77 m/s (tubeside) and 0.73 m/s (shellside) whereas the corresponding values for the design in Example 6.2 are 1.34 m/s and 0.325 m/s.

Table 6.3 shows good agreement between the RDA and the Kern rating results. The differences observed are due to the rounding off necessary on the number of tubes and baffles. The design overall heat transfer coefficient (dirty) is 776 W/m^2 °C. This yields a dirt resistance factor of 3.6 x 10^{-4} m^2 °C/W (for Kern's method) and 2.9 x 10^{-4} m^2 °C/W (for the Bell-Delaware method), which is close to the desired allowance of 3 x 10^{-4} m^2 °C/W. The Bell-Delaware design in Table 6.3 uses no sealing strips and has $J_c J_l J_b J_r J_s = 0.49$ (which is a bit low). Use of one sealing strip improves this product of the five correction factors to 0.57; then, the shellside heat transfer coefficient increases to 1364.5 W/m^2 °C, the shellside pressure drop increases to 3.6 kPa, and the dirt resistance factor increases to 4.1 x 10^{-4} m^2 °C/W.

As expected, the pressure drop from the Bell-Delaware method is lower than that from the RDA; on the other hand, the shellside pressure drop from Kern's method is higher (because of the decrease in baffle spacing required for making the baffles equispaced). Note that the pressure drops in the RDA exclude those in the headers and nozzles, which must be accounted for separately.

Polley et al. (1991) determined the exchanger geometry for desired performance by solving Equation 6.39 based on the Bell-Delaware method rather than Equation 6.38 based on Kern's method. However, the RDA for the Bell-Delaware method involves an iterative procedure due to the complicated form of the K_{s1} and K_{s2} expressions in Equation 6.35. Initial guesses and updates are needed for the ideal heat transfer factor (j_h), the ideal friction factor (f_i), the baffle cut (B_c), and the shell diameter (D_s).

As the RDA for the Bell-Delaware method is relatively complex, the methodology adopted in Example 6.4 is recommended: use the RDA based on Kern's method (Equation 6.38), and then perform detailed rating of the exchanger by the full Bell-Delaware method. To obtain better results, the fact that Kern's method usually overpredicts the shellside pressure drop could be accounted for while specifying ΔP_s in Equation 6.38. This strategy has numerous benefits: the inputs (in terms of detailed exchanger geometry) to be specified a priori are minimal; iterations are not required; the search space of possible exchanger geometries is greatly reduced; the pressure drop on the tubeside is fully utilized; and the actual shellside pressure drop, being lower than the allowable, takes care of the inaccurate (unsafe) pressure drops sometimes predicted by the Bell-Delaware method. Be aware that in the RDA the velocities are kept as high as the pressure drop limits permit; however, the vibration limits need to be checked to avoid severe tube vibration problems.

6.5 AREA TARGETING BASED ON PRESSURE DROPS

The relations between the frictional pressure drops and the film heat transfer coefficients derived in the form $\Delta P = K A\, h^{(3 - mf)/mh}$ (where mf and mh are the exponents for Re in the frictional factor and heat transfer correlations respectively; and K is a constant dependent on the fluid physical properties, flow rate, and some key exchanger dimensions) allow development of network area targets for specified stream pressure drops. These provide better area estimates than those from arbitrarily assumed stream heat transfer coefficients and form the basis of the interfacing methodology (Polley and Panjeh Shahi, 1991) described in the next section.

Example 6.5 Calculation of Target Area for Given Pressure Drops

For Case Study 4S1, determine the heat transfer area (for $\Delta T_{min} = 13°C$) for a network employing 1-2 shell and tube exchangers with stream pressure drops as specified in Table 6.4. The area estimate must be consistent with detailed designs (to be performed later using RDA) by Kern's method according to the additional data given in Table 6.4.

It is desired to use exchangers with tubes of 19.1 mm o.d. and 15.4 mm i.d., laid out in 25.4 mm square pitch. The hot streams are to be placed on the shellside, and the cold streams on the tubeside. Also, the heat transfer coefficients for utilities may be assumed to be constant at 5000 W/(m^2 °C) for hot utility and 2500 W/(m^2 °C) for cold utility.

Table 6.4 Inputs for Pressure-drop-based Area Targeting

Stream	H1	H2	C3	C4
Pressure drop ΔP in kPa	30	25	10	10
Dirt factor R_d in (m^2°C)/W	0.00015	0.00015	0.00015	0.00015
Spec. heat C_p in J/(kg °C)	1658	2684	2456	2270
Density ρ in kg/m^3	716	777	700	680
Viscosity μ in kg/(m.s)	0.24×10^{-3}	0.23×10^{-3}	0.23×10^{-3}	0.23×10^{-3}
Conductivity k in W/(m.°C)	0.11	0.11	0.12	0.13

Solution. The area targeting is essentially based on the algorithm suggested by Polley and Panjeh Shahi (1991) although the actual implementation is different. Polley and Panjeh Shahi have provided a flow chart for the calculations that involves updating the stream coefficients until the area target has converged.

The procedure below sets up a system of simultaneous equations (equal in number to the streams) to be solved for the individual stream heat transfer coefficients. Once the heat transfer coefficients are known, standard area targeting procedures based on the BATH formula (see Section 1.2) may be used.

Step 1. Calculation of Matchwise UA Distribution Matrix

It is convenient to determine a distribution matrix whose elements correspond to the UA-values of the individual matches that constitute the spaghetti design. Using Equation 1.39b as the starting point, the following formula may be written down based on hot streams:

$$(UA)_{mh} = (MC_p)_{jh} \left(\frac{dT_{jh}}{F \, LMTD} \right)_i \frac{(MC_p)_{jc}}{\sum_{jc} (MC_p)_{jc}} \tag{6.40}$$

The necessary calculations for Equation 6.40 may be done with the extended pinch tableau (see Table 2.8) given below. An extra row of F correction factors is added to account for the 1-2 shell and tube exchangers.

Interval	0	1	2	3	4	5	6	7	8
T_h	45	59	65	68.6	74.6	125	175	179	180
T_c	15	20	21.25	25	40	112	137	137	155
LMTD	--	34.30	41.33	43.67	38.93	22.07	23.31	39.97	32.77
F	--	.9900	.9993	.9988	.9900	.8411	.8936	1.00	.9972

Substitution of the values from the above tableau, along with appropriate values for MC_p depending on the stream population in each interval, yields the matchwise UA distribution matrix (Table 6.5).

Table 6.5 Matchwise UA Distribution Matrix

Interval	Match							
	H1-C3	H1-C4	H1-CU	H2-C3	H2-C4	H2-CU	HU-C3	HU-C4
1			3707.1					
2	600.4		760.5					
3	340.4		430.5	1371.6		1732.1		
4	1555.6			6232.3				
5	15515	11640		62059	46558			
6	24004							
7								
8								11031
$(UA)_m$	42016	11640	4898	69663	46558	1732	11031	

Units: UA in W/°C.

Sample calculations for the matrix elements in enthalpy interval 5 (containing H1, H2, C3, and C4), using Equation 6.40 are given below:

$(UA)_{H1\text{-}C3}$ = 10000 (125 − 74.6)/(0.8411 x 22.07) x 20/35 = 15515 W/°C
$(UA)_{H1\text{-}C4}$ = 10000 (125 − 74.6)/(0.8411 x 22.07) x 15/35 = 11640 W/°C
$(UA)_{H2\text{-}C3}$ = 40000 (125 − 74.6)/(0.8411 x 22.07) x 20/35 = 62059 W/°C
$(UA)_{H1\text{-}C4}$ = 40000 (125 − 74.6)/(0.8411 x 22.07) x 15/35 = 46558 W/°C

Summing each column (over all enthalpy intervals) of Table 6.5, the $(UA)_m$ for each hot stream - cold steam pair is obtained.

Step 2. Calculation of K_t *and* K_s

The expressions in Equations 6.33 and 6.34 are used to determine K_t and K_s.

$$K_s = 67(0.0254)(0.0254 - 0.0191)(0.0239)^{1.1}(0.00024)^{1.3}/$$
$$(0.0191 \times 6.03 \times 716 \times 0.11^{3.4} \times 1658^{1.7})$$
$$= 2.53 \times 10^{-13} \text{ (for stream H1)}.$$

$$K_s = 67(0.0254)(0.0254 - 0.0191)(0.0239)^{1.1}(0.00023)^{1.3}/$$
$$(0.0191 \times 14.9 \times 777 \times 0.11^{3.4} \times 2684^{1.7})$$
$$= 3.93 \times 10^{-14} \text{ (for stream H2)}.$$

$$K_t = 1/(0.023)^{2.5}(0.0154)^{1/2}(0.00023)^{11/6}/$$
$$(8.14 \times 700 \times 0.12^{7/3} \times 2456^{7/6})(15.4/19.1)$$
$$= 7.30 \times 10^{-10} \text{ (for stream C3)}.$$

$$K_t = 1/(0.023)^{2.5}(0.0154)^{1/2}(0.00023)^{11/6}/$$
$$(6.61 \times 680 \times 0.13^{7/3} \times 2270^{7/6})(15.4/19.1)$$
$$= 8.42 \times 10^{-10} \text{ (for stream C4)}.$$

Step 3. Calculation of Stream Heat Transfer Coefficients

The values of $(UA)_m$ determined in step 1 for each hot stream - cold steam pair may be used to obtain the total stream contact area by summing over all the matches in which the particular stream is involved. This contact area may then be substituted in Equation 6.33 or Equation 6.34, depending on whether the stream is located on the shellside or tubeside. Thus,

$$
\begin{aligned}
A_{m,j} &= \Sigma \, (1/h_{jh} + 1/h_{jc}) \, (UA)_m &&(6.41)\\
&= \Delta P_t / \{ K_t \, [1/h_j \, (D/D_t) - R_{dt}]^{-3.5} \} &&\text{if stream } j \text{ is on tubeside}\\
&= \Delta P_s / [K_s \, (1/h_j - R_{ds})^{-5.1}] &&\text{if stream } j \text{ is on shellside}
\end{aligned}
$$

Application of the above equation for Case Study 4S1 yields

$(1/h_{H1} + 1/h_{C3})\, 42016 + (1/h_{H1} + 1/h_{C4})\, 11640 + (1/h_{H1} + 1/h_{CU})\, 4898$
$= 30000/[2.53 \times 10^{-13}\, (1/h_{H1} - 0.00015)^{-5.1}]$ (for stream H1).

$(1/h_{H2} + 1/h_{C3})\, 69663 + (1/h_{H2} + 1/h_{C4})\, 46558 + (1/h_{H2} + 1/h_{CU})\, 1732$
$= 25000/[3.93 \times 10^{-14}\, (1/h_{H2} - 0.00015)^{-5.1}]$ (for stream H2).

$(1/h_{C3} + 1/h_{H1})\, 42016 + (1/h_{C3} + 1/h_{H2})\, 69663 + (1/h_{C3} + 1/h_{HU})\, 11031$
$= 10000/[7.30 \times 10^{-10}\, (1/h_{C3}\, (15.4/19.1) - 0.00015)^{-3.5}]$
(for stream C3).

$(1/h_{C4} + 1/h_{H1})\, 11640 + (1/h_{C4} + 1/h_{H2})\, 46558 + (1/h_{C4} + 1/h_{HU})\, 0$
$= 10000/[8.42 \times 10^{-10}\, (1/h_{C4}\, (15.4/19.1) - 0.00015)^{-3.5}]$
(for stream C4).

The above system of four equations in four unknowns [given $h_{HU} = 5000$ W/(m^2°C) and $h_{CU} = 2500$ W/(m^2°C)] is simultaneously solved to obtain

$h_{H1} = 741.7$ W/(m^2 °C), $h_{H2} = 891.4$ W/(m^2 °C),
$h_{C3} = 769.8$ W/(m^2 °C),   and $h_{C4} = 903.6$ W/(m^2 °C).

Note that these heat transfer coefficients include the fouling resistance.

Step 4. Calculation of Area Target
   With the heat transfer coefficients determined in step 3, the uniform BATH formula is used as in Section 1.2 to obtain the total area target for 1-2 shell and tube exchangers as 441.25 m^2.

6.6 THE INTERFACING METHODOLOGY

   The network synthesis procedures discussed so far require the specification of stream heat transfer coefficients. In a practical scenario, these would not be known a priori; consequently, prior to detailed exchanger design, these would need to be guessed. A poor guess for the heat transfer coefficients would lead to a large discrepancy between the heat exchanger areas predicted from the network synthesis and those actually obtained through the detailed design. Polley and Panjeh Shahi (1991) have developed an interfacing methodology to accomplish consistency between the synthesis results and the detailed designs by performing both the operations using allowable stream pressure drops as the common basis.
   Network synthesis based on assumed heat transfer coefficients and without employment of the interfacing methodology has the inherent danger of the final design implemented being suboptimal. An inappropriate set of stream

heat transfer coefficients will result in an incorrect estimate of the optimal ΔT_{min} through supertargeting and, consequently, a wrong initialization of the network topology. For a fixed energy recovery level, it will lead to an erroneous capital cost optimization due to inaccurate network area predictions.

Example 6.6   Interfacing Network Synthesis with Exchanger Design for Consistent Area Predictions

For Case Study 4S1, determine the heat transfer area (for ΔT_{min} = 13°C) for a network employing 1-2 shell and tube exchangers

a) when heat transfer coefficients are all assumed to be 200 W/(m² °C); and b) when stream pressure drops are specified as in Table 6.4.

For both the assumed h case and the specified ΔP case, compare the area predictions at three stages: the targeting stage, the network synthesis stage, and the design stage.

The heat transfer coefficients for utilities may be assumed to be constant at 5000 W/(m² °C) for hot utility and 2500 W/(m² °C) for cold utility.

Solution. The network areas are predicted for the assumed h case first, and then for the specified ΔP case. Note that at ΔT_{min} = 13°C, the utility requirements are 360 kW and 280 kW for hot and cold utility respectively.

Step 1. Area Prediction at Targeting Stage for Assumed h Case

Using the method described in Example 1.5, the target area for a network of 1-2 shell and tube exchangers is determined to be 1796.18 m².

Step 2. Area Prediction at Synthesis Stage for Assumed h Case

Figure 3.1 shows a possible MER network synthesized using the pinch design method. The area for each of the six units in Figure 3.1 is determined using $A = Q/(U \, F \, LMTD)$ and is given below:

Unit	1	2	3	4	5	6
Load (kW)	500	1840	560	520	360	280
Area (m²)	240.07	1090.94	523.97	161.90	57.29	40.83

The MC_p-ratio of the split on stream H2 is taken to be 25:15. Thus, the "synthesis" area is 2115 m².

Step 3. Area Prediction at Design Stage for Assumed h Case

The available pressure drop on a stream is fully used during the design stage as per the RDA. As the true pressure drop is directly proportional to the actual area (as per Equations 6.33 and 6.34), it may be assumed (Polley and

Panjeh Shahi, 1991) that the available pressure drop is linearly distributed according to the predicted area for the case where a stream has more than one exchanger on it. So, using the "synthesis" area from step 2, the pressure drops across each unit are found. The pressure drop distribution among the different exchangers is given in Table 6.6.

Table 6.6a Exchanger Designs by RDA (Kern)- Assumed h

Quantity	Unit 1	Unit 2	Unit 3
Expected area from synthesis (m^2)	240.07	1090.94	523.98
Tubeside pressure drop (ΔP_t in kPa)	1.73	7.86	7.64
Shellside pressure drop (ΔP_s in kPa)	16.27	25.00	25.00
Actual exchanger area (A in m^2)	64.39	283.39	121.72
Number of shell passes (NS)	2	4	3
Number of tube passes (NT)	4	8	6
Shell diameter (D_s in mm)	599.8	843.9	593.6
Tube length (L in m)	1.225	1.362	1.576
Number of tubes per shell (N_t)	438	867	429
Number of baffles (N_b)	11	9	15
Tubeside coefficient - dirty (W/m^2 °C)	728.3	733.7	864.5
Shellside coefficient - dirty (W/m^2 °C)	763.7	809.9	857.6
Overall coefficient - dirty (W/m^2 °C)	372.8	385.0	430.5

Table 6.6b Exchanger Designs by RDA (Kern)- Assumed h

Quantity	Unit 4	Unit 5	Unit 6
Expected area from synthesis (m^2)	161.89	57.29	40.83
Tubeside pressure drop (ΔP_t in kPa)	2.36	0.41	--
Shellside pressure drop (ΔP_s in kPa)	10.97	--	2.77
Actual exchanger area (A in m^2)	39.89	17.93	13.25
Number of shell passes (NS)	1	1	1
Number of tube passes (NT)	2	2	2
Shell diameter (D_s in mm)	346.3		187.9
Tube length (L in m)	4.554	2.577	5.133
Number of tubes per shell (N_t)	146	116	43
Number of baffles (N_b)	26		14
Tubeside coefficient - dirty (W/m^2 °C)	852.3	700.6	2500.0
Shellside coefficient - dirty (W/m^2 °C)	775.0	5000.0	738.9
Overall coefficient - dirty (W/m^2 °C)	405.9	614.5	570.3

The design of each unit is undertaken by the RDA as in Example 6.4. All exchangers use tubes of 19.1 mm o.d. and 15.4 mm i.d., laid out in 25.4 mm square pitch. The hot streams are placed on the shellside, and the cold streams on the tubeside. The baffle cut is 25%. Table 6.6 shows the results based on Kern's method.

The actual pressure drops from rating procedures will be slightly different from the specified ones due to the rounding off required for the number of tubes and the equispaced baffles. The actual areas are significantly lower than those expected from the synthesis for all the six units. The total "design" area is 540.57 m^2, about 75% lower than the "synthesis" area. Can better consistency be achieved using pressure drop specifications to determine the heat transfer coefficients?

Step 4. Area Prediction at Targeting Stage for Specified ΔP Case

From Example 6.5, the target area for a network of 1-2 shell and tube exchangers for the pressure drops specified in Table 6.4 is 441.25 m^2.

Step 5. Area Prediction at Synthesis Stage for Specified ΔP Case

Using the stream heat transfer coefficients from Example 6.5, the overall heat transfer coefficients are recalculated. Then, the area for each of the six units in Figure 3.1 is determined using $A = Q/(U\,F\,LMTD)$ as in step 2. Now, the "synthesis" area is 513.93 m^2, being distributed among the units as given below:

Unit	1	2	3	4	5	6
Load (kW)	500	1840	560	520	360	280
Area (m^2)	63.55	264.10	116.77	39.77	16.52	13.22

Step 6. Area Prediction at Design Stage for Specified ΔP Case

Now, using the "synthesis" area from step 5, the available pressure drop is linearly distributed as in step 3 (i.e., according to the predicted areas when a stream has more than one unit on it). Thus, the pressure drops across each unit on both the tubeside and shellside are found as given in Table 6.7.

The design of each unit is performed by the RDA using the full pressure drop available as in step 3. Table 6.7 shows the results based on Kern's method. The actual areas are very close to those expected from the synthesis for all the units. The small difference in the case of units 2 and 3 is due to the stream split. The total "design" area is 539.98 m^2, slightly higher (about 5%) than the "synthesis" area of 513.93 m^2.

Table 6.7a Exchanger Designs by RDA (Kern) - Given ΔP

Quantity	Unit 1	Unit 2	Unit 3
Expected area from synthesis (m^2)	63.55	264.10	116.77
Tubeside pressure drop (ΔP_t in kPa)	1.85	7.67	7.45
Shellside pressure drop (ΔP_s in kPa)	16.35	25.00	25.00
Actual exchanger area (A in m^2)	63.73	284.36	122.09
Number of shell passes (NS)	2	4	3
Number of tube passes (NT)	4	8	6
Shell diameter (D_s in mm)	592.2	848.3	596.4
Tube length (L in m)	1.244	1.352	1.566
Number of tubes per shell (N_t)	427	876	433
Number of baffles (N_b)	11	9	14
Tubeside coefficient - dirty (W/m^2 °C)	741.8	728.8	858.9
Shellside coefficient - dirty (W/m^2 °C)	765.8	809.4	857.2
Overall coefficient - dirty (W/m^2 °C)	376.8	383.5	429.0

Table 6.7b Exchanger Designs by RDA (Kern) - Given ΔP

Quantity	Unit 4	Unit 5	Unit 6
Expected area from synthesis (m^2)	39.77	16.52	13.22
Tubeside pressure drop (ΔP_t in kPa)	2.54	0.48	--
Shellside pressure drop (ΔP_s in kPa)	10.23	--	3.42
Actual exchanger area (A in m^2)	39.80	17.19	12.81
Number of shell passes (NS)	1	1	1
Number of tube passes (NT)	2	2	2
Shell diameter (D_s in mm)	341.5		187.9
Tube length (L in m)	4.671	2.653	4.966
Number of tubes per shell (N_t)	142	108	43
Number of baffles (N_b)	25		15
Tubeside coefficient - dirty (W/m^2 °C)	868.0	735.0	2500.0
Shellside coefficient - dirty (W/m^2 °C)	766.1	5000.0	770.9
Overall coefficient - dirty (W/m^2 °C)	406.9	640.8	589.2

The design methodology based on pressure drops has been shown to achieve better consistency in the network synthesis and design tasks. The "synthesis" area is higher than the target area because the synthesized network has fewer units than a spaghetti design and therefore has nonvertical heat transfer to some extent. Polley and Panjeh Shahi (1991) have also stressed that nonvertical stream matching could lead to larger differences between the "synthesis" area and the "design" area. This is because the area predictions based on pressure drop are still based on the BATH formula, which assumes a vertical heat transfer model. In fact, there is a double penalty: an initial area penalty due to inefficient use of driving forces on the composite curves and an additional area penalty due to the decrease in heat transfer coefficients (caused by the initial area increase, given that the pressure drops are fixed). This suggests the importance of careful utilization of available pressure drops.

It is clear that the original guess of 200 W/(m^2 °C) for the heat transfer coefficients of all the process streams is an inferior one (too conservative), and a value of 800 W/(m^2 °C) would be more reasonable (of course, how does one make such a reasonable guess?). Certainly, a better estimate of the average heat transfer coefficient would improve matters. However, it would not be a sound way of resolving the problem, especially when the actual heat transfer coefficients vary over a wide range. Also, an iterative method to determine good values for the heat transfer coefficients is not practical as the activities of process synthesis and detailed engineering are usually disconnected in the real world.

Various refinements of the methodology may be desirable in actual practice as illustrated by Polley and Panjeh Shahi (1991). The Bell-Delaware method may be used for detailed design as well as for the RDA. Optimization of ΔT_{min} may be performed based on pressure drop considerations prior to design: the supertargeting procedure from Section 1.6 holds, but the pressure drops specified need to be converted to heat transfer coefficients as in Example 6.5 for each value of ΔT_{min}, and then the energy-capital tradeoff examined.

It is seen that pressure drops are critical in exchanger design. How are these specified? Unless a definite procedure for the same exists, their specification would be as arbitrary as the assumption of the heat transfer coefficients. Of course, the available head is known for gravity flows and industrial standards exist for allowable pressure drops for forced flows. A more objective procedure (Jegede and Polley, 1992a) for the determination of optimum pressure drops is discussed next based on the tradeoff between the fluid flow power consumptions and the exchanger area capital cost.

6.7 STREAM PRESSURE DROP OP   IIZATION

The rapid design algorithm (Section 6.4) pro   ᵤₛ a straightforward procedure for the full use of the allowable pressure drops when they are specified a priori. This ensures an exchanger of the smallest size with near-minimal capital cost. However, the power   sts and the capital costs associated with pumping liquids and/or compressing gases through an exchanger often constitute a significant fraction (15% to 35%) of the total annual costs. Hence, it is meaningful to determine the optimum pressure drops for the streams taking the economics and tradeoffs into account (Peters and Timmerhaus, 1981; Jegede and Polley, 1992a). Considering the economics for example, the design may use large pressure drops if power is cheap relative to capital. Considering the tradeoffs, for example, the heat exchanger area will be less (due to higher heat transfer coefficients) but the power consumption will be more (due to higher pressure drops) if the stream velocities are increased.

The optimization presented below by way of example for liquids pumped through a single shell and tube exchanger is a valuable extension of the RDA. It has the potential of being adapted to compressed gases and other exchanger types (e.g., air coolers, plate-and-frame exchangers, and spiral plate exchangers). Extension to heat exchanger networks is possible, allowing a three-way tradeoff between energy, area, and power consumption (Jegede, 1990).

Example 6.7 Optimum Design of Shell and Tube Exchanger

Determine the optimum pressure drops for a shell and tube heat exchanger based on Kern's method for the input data (temperatures, flow rates, and physical properties of the two liquid streams) given in Table 6.1. The following additional cost data are also provided:

Capital cost of exchanger (\$): $a + b\,A^c$	$= 30000 + 750\,A^{0.81}$
Capital cost of pump (\$): $e + f\,(M\,\Delta P/\rho)^g$	$= 2000 + 5\,(M\,\Delta P/\rho)^{0.68}$
Cost of power (\$/W hr): C_{pow}	$= 0.00005$
Pump efficiency : η	$= 70\%$
Plant operation (hrs/yr): H	$= 8000$
Annualization factor (/yr): A_f	$= 0.322$

Solution. The pressure drops, heat transfer coefficients, and exchanger area are simultaneously optimized in this formulation (Jegede and Polley, 1992a). The three equations (6.33, 6.34, and 6.37) that constitute the RDA for Kern's method are given below and continue to be used in this development:

$$\Delta P_t = K_t\,A\,h_t^{3.5}$$
$$\Delta P_s = K_s\,A\,h_s^{5.1}$$
$$Q = [1/h_s + 1/h_t\,(D_t/D) + k_4]^{-1}A\,F\,(LMTD)$$

Note that k_4 (see Equation 6.38) includes the dirt and metal wall resistances. However, the pressure drops are no longer specified; in fact, they form part of the set of optimization variables (ΔP_t, ΔP_s, h_t, h_s, and A). Two more equations are needed, and these may be obtained by minimizing the total annual cost.

The total cost consists of five components: the capital cost of the exchanger, the capital cost for two pumps (one for the tubeside and another for the shellside), and the operating (power) costs of these two pumps. Thus,

$$TAC = A_f \{a + b\, A^c + e + f(M_t \Delta P_t/\rho_t)^g + e + f(M_s \Delta P_s/\rho_s)^g\}$$
$$+ C_{pow}\, H/\eta\, \{M_t \Delta P_t/\rho_t + M_s \Delta P_s/\rho_s\}$$

In the above equation, there is a three-way tradeoff (between ΔP_t, ΔP_s, and A), but it is clear from Equation 6.38 that there are only two degrees of freedom. So, for the optimum, the derivatives of TAC with respect to any two of the variables must be zero. On eliminating ΔP_t, ΔP_s, and h_s, it is possible to rewrite the expression completely in terms of A and h_t as follows:

$$TAC = A_f \{a + b\, A^c + e + f(M_t K_t\, A\, h_t^{3.5}/\rho_t)^g$$
$$+ e + f(M_s K_s\, A\, [A\, F\, LMTD/Q - 1/h_t\, (D_t/D) - k_4]^{-5.1}/\rho_s)^g\}$$
$$+ C_{pow}\, H/\eta\, \{M_t K_t\, A\, h_t^{3.5}/\rho_t$$
$$+ M_s K_s\, A\, [A\, F\, LMTD/Q - 1/h_t\, (D_t/D) - k_4]^{-5.1}/\rho_s\} \quad (6.42)$$

Setting $\partial(TAC)/\partial A = 0$ and $\partial(TAC)/\partial h_t = 0$ gives the final system of two equations (Duvedi, 1993) that needs to be solved simultaneously for the two unknowns (A and h_t).

$$p_1\, A^{c-1} + p_2\, A^{g-1}\, h_t^{3.5g} + p_3\, h_t^{3.5}$$
$$+ p_4\, A \quad [A\, F\, LMTD/Q - 1/h_t\, (D_t/D) - k_4]^{-6.1}$$
$$+ p_5 \quad\ \ [A\, F\, LMTD/Q - 1/h_t\, (D_t/D) - k_4]^{-5.1}$$
$$+ p_6\, A^g \quad [A\, F\, LMTD/Q - 1/h_t\, (D_t/D) - k_4]^{-5.1g-1}$$
$$+ p_7\, A^{g-1}\, [A\, F\, LMTD/Q - 1/h_t\, (D_t/D) - k_4]^{-5.1g} \qquad = 0 \qquad (6.43a)$$

$$p_8\, A^g\, h_t^{3.5g-1} + p_9\, A\, h_t^{2.5}$$
$$+ p_{10}\ \ A/h_t^2\, [A\, F\, LMTD/Q - 1/h_t\, (D_t/D) - k_4]^{-6.1}$$
$$+ p_{11}\, A^g/h_t^2\, [A\, F\, LMTD/Q - 1/h_t\, (D_t/D) - k_4]^{-5.1g-1} = 0 \qquad (6.43b)$$

where
$$p_1 = A_f\, bc,$$
$$p_2 = A_f\, fg\, (M_t K_t/\rho_t)^g,$$
$$p_3 = C_{pow}\, H/\eta\, (M_t K_t/\rho_t),$$
$$p_4 = -5.1\, C_{pow}\, H/\eta\, (M_s K_s/\rho_s)\, F\, LMTD/Q,$$
$$p_5 = C_{pow}\, H/\eta\, (M_s K_s/\rho_s),$$
$$p_6 = -5.1\, A_f\, fg\, (M_s K_s/\rho_s)^g\, F\, LMTD/Q,$$

$$p_7 = A_f fg \, (M_s K_s/\rho_s)^g,$$
$$p_8 = 3.5 \, A_f fg \, (M_t K_t/\rho_t)^g,$$
$$p_9 = 3.5 \, C_{pow} \, H/\eta \, (M_t K_t/\rho_t),$$
$$p_{10} = -5.1 \, C_{pow} \, H/\eta \, (M_s K_s/\rho_s) \, (D_t/D),$$

and $\quad p_{11} = -5.1 \, A_f fg \, (M_s K_s/\rho_s)^g \, (D_t/D).$

Substituting the cost data and other known values for the example problem gives

$$195.7 \, A^{-0.19} + 6.39 \times 10^{-9} \, A^{-0.32} \, h_t^{2.38} + 4.45 \times 10^{-13} \, h_t^{3.5}$$
$$- 9.96 \times 10^{-20} \, A \quad [4.5 \times 10^{-5} \, A - 1.24/h_t - 3.4 \times 10^{-4}]^{-6.1}$$
$$+ 4.31 \times 10^{-16} \quad [4.5 \times 10^{-5} \, A - 1.24/h_t - 3.4 \times 10^{-4}]^{-5.1}$$
$$- 1.32 \times 10^{-14} \, A^{0.68} \, [4.5 \times 10^{-5} \, A - 1.24/h_t - 3.4 \times 10^{-4}]^{-4.468}$$
$$+ 5.7 \times 10^{-11} \, A^{-0.32} \, [4.5 \times 10^{-5} \, A - 1.24/h_t - 3.4 \times 10^{-4}]^{-3.468} \quad = 0$$

$$2.24 \times 10^{-8} \, A^{0.68} \, h_t^{1.38} + 1.56 \times 10^{-12} \, A \, h_t^{2.5}$$
$$- 2.72 \times 10^{-15} \quad A/h_t^2 \quad [4.5 \times 10^{-5} \, A - 1.24/h_t - 3.4 \times 10^{-4}]^{-6.1}$$
$$- 3.6 \times 10^{-10} \, A^{0.68}/h_t^2 \, [4.5 \times 10^{-5} \, A - 1.24/h_t - 3.4 \times 10^{-4}]^{-4.468} = 0$$

The above two equations are solved simultaneously to obtain $A = 26.18$ m^2 and $h_t = 5407.2$ W/(m^2 °C). Then, ΔP_t, ΔP_s, h_s, and TAC corresponding to the optimum exchanger design may be determined to be 7.49 kPa, 23.97 kPa, 1619.1 W/(m^2 °C), and 14920 \$/yr respectively. The optimization may also be performed by nonlinear programming (see Chapter 10).

This chapter has discussed methodologies for bridging the gap between the design acts performed during the systems engineering and the detailed engineering phases. This is important not only for grassroots designs, but also for retrofitting, as demonstrated in the next chapter.

QUESTIONS FOR DISCUSSION

1. Derive Equations 6.6 that define the F correction factor.

2. Give an example where the new $X_p(R)$ criterion in Equation 6.13 and the $X_p = 0.9$ criterion yield identical results for 1-2 area and shell targeting. Specifically, consider Case Study 4S1 as discussed in Examples 1.5, 1.7, and 1.8. Give an example where the two criteria yield different results.

3. Give the basis (derivation) for the following equations:

$$\theta_{ctl} = 2\cos^{-1}[(D_s/D_{ctl})\,(1 - 2\,B_c/100)]$$
$$\theta_{ds} = 2\cos^{-1}(1 - 2\,B_c/100)$$
$$S_{tb} = (\pi/4)\,[(D_t + L_{tb})^2 - D_t^2]\,N_t\,(1 - F_w)$$
$$S_{sb} = (\pi/2)\,D_s\,L_{sb}\,[1 - \theta_{ds}/360]$$

State any approximations made. Draw a sketch showing the relevant geometrical parameters.

4. Explain briefly why ΔP_t involves R_b, R_l, and $(N_b - 1)$; why ΔP_w involves R_l and N_b; and why ΔP_t involves R_b and R_s.

5. What are the various types of heat exchangers commonly used in the chemical process industries? For a specific application, how is the type selected (based on cost, temperature, pressure, size, and fluid specifications)?

6. Is it meaningful to have a capital cost equation for an exchanger in terms of UA rather than A (see Equation 1.33)? Justify your answer.

7. What are the advantages and disadvantages of the various types of shell and tube heat exchangers (e.g., fixed tubesheet, U-tube, and floating head)? What factors (e.g., thermal expansion stresses, fouling, hazardous leaks) dictate the choice between the various types?

8. What is the "3-letter" nomenclature suggested by the Tubular Exchanger Manufacturers Association (TEMA) for the description of various exchanger configurations? Specifically, what is a BEU type exchanger? Do the three letters designate the front-end head type (e.g., A, B, C, D, N, Y), the shell type (E, F, G, H, J, K, X), and the rear-end head type (L, M, N, P, S, T, U, W) respectively?

9. What are compact exchangers? In compact exchangers, is it more meaningful to use exchanger volume/weight (rather than surface area) as the design parameter? Discuss the extension of the rapid design algorithm to compact exchangers and two-phase flows.

10. How can area targets be established when the stream heat transfer coefficients vary (linearly) with temperature? What about stream pressure drop specifications in this case?

PROBLEMS

6.A Caloric Temperature Concept for Variable Heat Transfer Coefficient

Consider a countercurrent heat exchanger where the overall heat transfer coefficient, U, varies linearly with the hot stream temperature, T_h, and the cold stream temperature, T_c. Let $\Delta T = T_h - T_c$, and subscripts 1 and 2 denote the two terminals of the exchanger.

a) Show that $(U - U_1)/(U_2 - U_1) = (\Delta T - \Delta T_1)/(\Delta T_2 - \Delta T_1)$.

b) Show that $Q = A (U_1\Delta T_2 - U_2\Delta T_1)/\ln[(U_1\Delta T_2)/(U_2\Delta T_1)]$. Start with a macroscopic energy balance and obtain a first-order differential equation for ΔT as a function of distance along the exchanger. Then, substitute the result in a) and perform the integration over the entire exchanger length.

c) The expression in b) was first derived by Colburn (1933). Show that this expression leads to the caloric temperature concept defined in Equations 6.1. Note that $Q = U_c A \, LMTD$, $F_c = (U_c - U_2)/(U_1 - U_2)$, and $k_c = (U_1 - U_2)/U_2$.

6.B Performance Equation for 1-2 Shell and Tube Exchanger

Derive an expression for the correction factor, F, in a 1-2 shell and tube exchanger operating under steady-state conditions (with constant U and MC_p s) by the following procedure. Use the definitions of R, P, and N in Equations 6.5.

a) Perform an overall energy balance over a portion of the exchanger.

b) Perform a differential energy balance over an area, dA, to obtain three ordinary first-order differential equations for the temperature distributions of the shellside fluid, of the tubeside fluid in parallel flow, and of the tubeside fluid in counter flow.

c) Show that the three differential equations in b) may be rewritten to obtain a single differential equation for the temperature distribution of the shellside fluid. Solve this ordinary second-order differential equation imposing appropriate boundary conditions.

d) Use the result in c) to obtain $N(P, R)$ as in Equation 6.7. Rearrange the $N(P, R)$ expression to obtain the exchanger performance relation in the explicit form given below:

$$P = \cfrac{2}{1+R+\sqrt{R^2+1}\left[\cfrac{1+\exp(-N\sqrt{R^2+1})}{1-\exp(-N\sqrt{R^2+1})}\right]}$$

6.C Exchanger Trains in Series-parallel Circuits

a) Consider a train of *NS* exchangers in a series circuit with overall counter flow (see figure below).

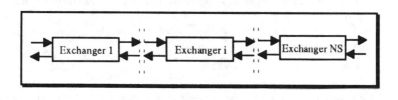

Starting with the definitions of *R*, *P*, and *N* in Equations 6.5, show that

$(1 - R'P')/(1 - P') = \prod [(1 - R_iP_i)/(1 - P_i)]$ for $R \neq 1$
$P'/(1 - P')\qquad = \sum [P_i/(1 - P_i)]$ for $R = 1$
$R' = R_i$ and $N' = \sum N_i$

where primed quantities refer to the overall train and subscript *i* denotes an individual exchanger. Note that $\prod$ and $\sum$ respectively stand for product and summation of terms (running from 1 to *NS*).
If the *NS* exchangers are identical (each characterized by *R*, *P*, *N*), show that (see Equations 6.8)

$(1 - R'P')/(1 - P') = [(1 - RP)/(1 - P)]^{NS}$ for $R \neq 1$
$P'/(1 - P')\qquad = NS\ P/(1 - P)$ for $R = 1$
$R' = R$ and $N' = (NS)\,N$

b) Consider a train of *NS* exchangers in a series circuit with overall parallel flow (see figure below).

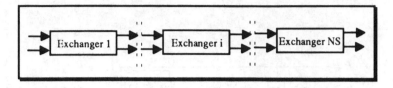

Show that

$$(1 - R'P' - P') = \prod [1 - R_iP_i - P_i]$$
$$R' = R_i \quad \text{and} \quad N' = \sum N_i$$

If the NS exchangers are identical (each characterized by R, P, N), show that

$$(1 - R'P' - P') = [1 - RP - P]^{NS}$$
$$R' = R \quad \text{and} \quad N' = (NS)\,N$$

c) Consider a train of NS exchangers in a series-parallel circuit (see figure below).

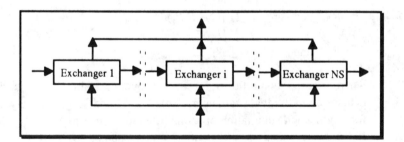

For the case of equally-divided hot fluid in parallel, show that

$$(1 - P') = \prod [1 - P_i]$$
$$R' = R/NS \quad \text{and} \quad N' = \sum N_i$$

Further, if the NS exchangers are identical (each characterized by R, P, N), then show that

$$(1 - P') = [1 - P]^{NS}$$
$$R' = R/NS \quad \text{and} \quad N' = (NS)\,N$$

For the case of equally-divided cold fluid in parallel, show that

$$(1 - R'P') = \prod [1 - RP_i]$$
$$R' = (NS) R \qquad \text{and} \qquad N' = (1/NS) \sum N_i$$

Further, if the NS exchangers are identical (each characterized by R, P, N), then show that

$$(1 - R'P') = [1 - RP]^{NS}$$
$$R' = (NS) R \qquad \text{and} \qquad N' = N$$

6.D Useful Equations for Counter and Parallel Flow Exchangers

It is convenient to have design equations in the form $N(P, R)$ and performance prediction equations in the form $P(N, R)$.

a) For true countercurrent flow, show that

$$N_{R \neq 1} = \ln[(1 - P)/(1 - RP)] / [R - 1]$$
$$N_{R = 1} = P/(1 - P)$$

$$P_{R \neq 1} = \{1 - \exp[N(R - 1)]\} / \{1 - R \exp[N(R - 1)]\}$$
$$P_{R = 1} = N/(1 + N)$$

b) For true parallel flow, show that

$$N \quad = -\ln[1 - RP - P] / [R + 1]$$
$$P \quad = \{1 - \exp[-N(R + 1)]\} / [R + 1]$$

6.E Exchanger Rating by Bell-Delaware Method

Predict the performance of the 1-2 shell and tube exchanger whose data appear in the table below. The metal wall resistance is 0.00003 m^2 °C/W and the dirt resistance on both sides may be neglected. It is desired to use an exchanger with tubes of 16 mm o.d. and 13.5 mm i.d., laid out in 20.8 mm triangular pitch. The clearances to be used for the shellside rating by the Bell-Delaware method are 5.5 mm (baffle-to-shell), 0.5 mm (tube-to-baffle), and 10 mm (bundle-to-shell).

Some other geometrical data are as follows:

Tube length = 2.276 m   Shell diameter = 591 mm
Number of tubes = 652   Number of baffles = 8
Baffle cut = 24 %   Baffle spacing = 250 mm

Quantity	Units	Shellside	Tubeside
Temperature (inlet)	°C	100	7.1
Temperature (outlet)	°C	42.5	16.6
Flow rate	kg/s	22.4	77.96
Specific heat	J/(kg °C)	2407	4187
Density	kg/m^3	740	1000
Viscosity	kg/(m.s)	0.494 x 10^{-3}	1.0 x 10^{-3}
Thermal conductivity (hot fluid)	W/(m.°C)	0.105	0.61

Answers:

This example is discussed by Polley et al. (1991) and is adapted (with some modifications) from Taborek (1983).

$h_t = 6000$ W/(m^2 °C) $h_s = 1241$ W/(m^2 °C)

$\Delta P_t = 11.66$ kPa $\Delta P_s = 13.7$ kPa

6.F Area Targeting Based on Pressure Drops

For Case Study 9P1 (the aromatics plant) discussed by Polley and Panjeh Shahi (1991), determine the heat transfer area (for $\Delta T_{min} = 25$°C) for a network employing 1-2 shell and tube exchangers

a) when heat transfer coefficients are all assumed to be 500 W/(m^2 °C); and

b) when stream pressure drops are specified as in the tables below. Use the procedure employed in Example 6.5 (based on Kern's method).

It is desired to use exchangers with tubes of 19.1 mm o.d. and 15.4 mm i.d., laid out in 25.4 mm square pitch. The hot streams are to be placed on the shellside and the cold streams on the tubeside. Also, the heat transfer coefficients for utilities may be assumed to be constant at 1000 W/(m^2 °C) for hot utility and 2500 W/(m^2 °C) for cold utility.

Hot Stream	H1	H2	H3	H4
Pressure drop ΔP in kPa	120	80	90	60
Dirt factor R_d in (m^2°C)/W	0.00018	0.00014	0.00018	0.00018
Spec. heat C_p in J/(kg °C)	2000	2192	1877	5480
Density ρ in kg/m^3	500	55	676	697
Viscosity μ in cPs	0.25	0.01	0.28	0.31
Conductivity k in W/(m.°C)	0.11	0.026	0.11	0.11

Cold Stream	C5	C6	C7	C8	C9
Pressure drop ΔP in kPa	20	20	30	15	80
Dirt factor R_d in (m²°C)/W	0.00018	0.00018	0.00014	0.00018	0.00018
Spec. heat C_p in J/(kg °C)	2000	1590	10000	1580	2740
Density ρ in kg/m³	464	570	1.0	685	667
Viscosity μ in cPs	0.16	0.30	0.01	0.27	0.21
Conductivity k in W/(m.°C)	0.11	0.11	0.17	0.11	0.11

Answers:

Hot utility = 24480 kW

a) Area Target (assumed h) = 12888.95 m².

b) Area Target (specified ΔP) = 7315.42 m². The heat transfer coefficients in this case are

h_{H1} = 817.0 W/(m² °C), h_{H2} = 593.8 W/(m² °C),

h_{H3} = 881.1 W/(m² °C), h_{H4} = 997.5 W/(m² °C),

h_{C5} = 846.0 W/(m² °C), h_{C6} = 739.6 W/(m² °C),

h_{C7} = 1411.7 W/(m² °C), h_{C8} = 770.1 W/(m² °C),

and h_{C9} = 1241.8 W/(m² °C).

CHAPTER 7

Retrofitting

*Awareness is the first step
to correction and improvement in any skill.*

Edward de Bono

Modification of existing plants has been achieved by various retrofitting techniques that may be classified into four broad categories: computer search, mathematical programming, inspection, and pinch technology.

Given an existing network, a computer search technique (Jones et al., 1986) can be used to choose from a number of simulated MER networks a new network with the most favorable economics and with minimal change. The method may not prove efficient in many cases due to three reasons: the element of chance in hitting or missing the best network, the large amount of computational effort involved in simulating many networks, and the difficulty in retrofitting to identify a design having a structure reasonably close to the existing one and simultaneously transferring zero heat across the pinch.

Numerical techniques formulate the network modification into various mathematical programming problems (Ciric and Floudas, 1989, 1990b, 1990c; Yee and Grossmann, 1991). For example, a mixed integer linear programming (MILP) model has been proposed (Ciric and Floudas, 1990b) for retrofit at a level of matches based on a classification of the possible structural modifications. Such formulations are beyond the scope of this chapter.

In this chapter, a retrofit example problem is first solved by inspection and then more systematically by pinch technology.

7.1 RETROFIT BY INSPECTION

Retrofit by inspection (Tjoe and Linnhoff, 1986) has been tried as an intuitive method to achieve energy savings in a plant. As early as in 1983, a modification of an ICI plant was reported (Boland, 1983) to have been achieved by matching of certain streams that seemed likely candidates.

Example 7.1 Modification by Inspection
For Case Study 4S1, modify the network in Figure 1.2 by inspection to obtain an energy savings. Assume $U = 0.1$ kW/(m^2 $^\circ$C).

Given: Payback period = 2 years (approximate)
          Annual cost of unit duty in \$/(kW.yr): 120 (for HU) and 10 (for CU)
          Capital cost of exchanger (\$) = $30000 + 750A^{0.81}$ (A is area in m^2)

Solution.   The approach to modify a network for energy savings by inspection involves the following steps.

Step 1. Calculation of Total Network Area
    Assuming counterflow exchangers, the area of each exchanger is simply found from $A = Q/(U\ LMTD)$.   In Figure 1.2, exchanger 2 has an area of 358.92 m^2 (namely, 1080/[0.1 x 30.09]) and exchanger 3 has an area of 256.19 m^2 (namely, 1300/[0.1 x 50.74]).   Thus, the total exchanger area for the existing network is 615.11 m^2.   The utility area is not considered during the area calculations employed for retrofitting because their duties will be reduced as the modifications are implemented (Ahmad and Polley, 1990).

Step 2. Selection of Positions for New Exchangers
    Inspection suggests contacting streams H2 and C3 (at the cold end of the network) as this would decrease the heat duties of both the heater and cooler. In this case study, there is no other obvious option in terms of the streams to match.   When options exist, it may be preferable to choose streams with large heat capacity flow rates.
    The introduction of the new exchanger (labeled N in Figure 7.1) would affect the temperatures of the downstream exchangers (namely, unit 3) and consequently lead to additional area.
    The heat load of the new match (denoted by Q_N) must be chosen using some constraints on the EMAT (exchanger minimum approach temperature) to ensure efficient use of the match, but what is the maximum range for Q_N?   If the use of the hot utility must be completely eliminated, then the maximum value of Q_N is 1320 kW.   This is infeasible due to EMAT constraints.   Since unit 3 must have a positive driving force at its cold terminal, $45 - (20 + Q_N/20) > 0$.   This implies that the feasible range is $0 \le Q_N < 500$.

Step 3. Calculation of Energy Savings
    The total energy required for the network in Figure 7.1 is $(1400 - Q_N)$ kW (hot utility) and $(1320 - Q_N)$ kW (cold utility), so the energy savings is Q_N kW (hot utility) and Q_N kW (cold utility).

Step 4. Calculation of Change in Area
    The total area for the network in Figure 7.1 is calculated using a methodology similar to that in step 1.   The area of unit 2 does not change in the retrofitting process.

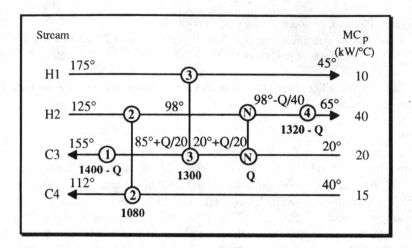

Figure 7.1 The network retrofitted by inspection indicates the need for a new exchanger and recalculation of duties of downstream exchangers. (Temperature in °C and load in kW)

However, exchanger 3 (after retrofitting) has an area of $200 \ln[(90 - Q_N/20)/(25 - Q_N/20)]$ and the new exchanger requires an area of $400 \ln[(78 - Q_N/40)/(78 - Q_N/20)]$. Details of calculations for area (in m^2) appear below:

Unit	Area (before)	Area (after retrofit)
2	358.92	358.92
3	256.19	$1300/\{0.1 \times 65/\ln[(90 - Q_N/20)/(25 - Q_N/20)]\}$
N	0	$Q_N/\{0.1 \times (Q_N/40)/\ln[(78 - Q_N/40)/(78 - Q_N/20)]\}$

Thus, change in area = required new area − existing area
$$= 200 \ln[(90 - Q_N/20)/(25 - Q_N/20)]$$
$$+ 400 \ln[(78 - Q_N/40)/(78 - Q_N/20)] - 256.19.$$

Step 5. Cost Analysis
The yearly savings are obtained simply as follows:

Savings = 120 (hot utility saved in kW) + 10 (cold utility saved in kW)
$$= 130 \, Q_N \ \$/yr.$$

The investment for the new area on exchanger (N) is

$$30000 + 750 \{400 \ln[(78 - Q_N/40)/(78 - Q_N/20)]\}^{0.81}.$$

Further, the investment for the extra area on unit 3 is

$$750 \, N_s \, \{(200/N_s) \ln[(90 - Q_N/20)/(25 - Q_N/20)] - 256.19/N_s\}^{0.81},$$

where $N_s = 1$ for $0 < Q_N \leq 418.5$, $N_s = 2$ for $418.5 < Q_N \leq 483.5$, etc. Here, N_s is the number of new shells required and emerges from the constraint on the maximum area per shell of 310 m^2. Note that the existing network has an average area per match (and per shell) of about 310 m^2.

The payback period (obtained by dividing the investment by the savings) may be computed for various levels of energy recovery (Q_N) as shown below:

Q_N (kW)	New Area (unit N in m^2)	Extra Area (unit 3 in m^2)	Investment (10^3 $)	Savings (10^3 $/yr)	Payback (yr)
120	16.33	41.09	52.42	15.6	3.36
240	34.80	102.17	75.11	31.2	2.41
280	41.52	130.38	84.10	36.4	2.31
320	48.54	165.18	94.36	41.6	2.27
360	55.90	209.96	106.54	46.8	2.28
400	63.63	271.62	121.92	52.0	2.34
480	80.27	581.74	204.65	62.4	3.28

Thus, the project economics shows that the network in Figure 7.1 with $Q_N = 360$ kW provides an energy savings of 26.5% (given by 720/2720) for a payback period of 2.28 years and an investment of 106.54 x 10^3 $. In addition to the payback, the project scope is often restricted by a ceiling on investment. In general, a higher savings requires a higher investment and often a longer payback period. However, the high payback periods for low values of Q_N in this case study emphasize that it is not meaningful to place a new exchanger (with an installation cost of 30,000 $) to recover small amounts of energy. Though other locations are possible for the new exchanger, N, contacting streams H2 and C3 at the cold end of the network is favorable in this case. A similar analysis performed by placing exchanger N at the hot end of the process shows the payback period to be 3.58 years for a energy recovery level of 360 kW.

7.2 RETROFIT-FIXED HEAT TRANSFER COEFFICIENTS

The methods of retrofitting by inspection and computer search carry the potential risk of the network not being the optimum (Tjoe and Linnhoff, 1987) while the use of mathematical programming makes strong computational demands. Retrofitting by pinch technology provides a promising alternative. The method has been successfully used in industry, e.g., in Union Carbide (Linnhoff and Vredeveld, 1984), for an ethylene plant (Linnhoff and Witherell, 1986), and in a refinery (Lee et al., 1989; Farhanieh and Sunden, 1990; Fraser and Gillespie, 1992). The basic methodology (Tjoe and Linnhoff, 1986) looks at the economics of plant operation and of modification to provide targets. The next step involves network modification to achieve the set targets.

7.2.1 Targeting Based on Constant h-values

The targeting procedure is based on energy and area targets (discussed earlier in Chapter 1) as well as on the concept of area efficiency.   An investment vs. savings plot is used to obtain a target for retrofit design (Tjoe and Linnhoff, 1986).

Example 7.2 Establishing Retrofit Targets for Constant h-values
For the stream data in Case Study 4S1 and the network in Figure 1.2, determine a target ΔT_{min} for retrofit design. The cost data are the same as given earlier in Example 7.1.

Solution. First, the area efficiency of the existing network is calculated and a path for retrofitting is established on an area-energy plot. Next, the area-energy plot is converted to an investment-savings plot and the payback period line used to determine the target ΔT_{min}. In retrofits, these plots only consider heat recovery area, i.e., process-to-process exchangers. In other words, the area for heat transfer with utilities is not considered.

Step 1. Calculation of Area Efficiency of Existing Network
The area efficiency is defined as

$$\alpha = A_{ideal}/A_{existing} \tag{7.1}$$

where A_{ideal} is the ideal target area based on the composite curves corresponding to the existing utility levels and $A_{existing}$ is the actual area of the existing network.
If the method employed in Example 1.4 is used to determine A_{ideal}, then ΔT_{min} must be known a priori. This would involve use of the PTA to obtain utility requirements for various ΔT_{min} and a trial-and-error procedure to ascertain the ΔT_{min} for the existing utility level.

A better alternative is to shift the CCC horizontally so that the cold utility requirement corresponds to that in the existing network. This is simply done by generating preliminary composite curve data as in Table 2.1, the only difference being that Q_f is set equal to the current cold utility usage (1320 kW in this case) for the preliminary CCC. This information is converted into the classical tableau representation (Table 7.1) as discussed in Chapter 2. The tableau is extended by two additional rows to facilitate calculations for LMTD and area in each interval within the region of overlap of the composites. Summing the areas for the individual intervals gives the ideal target area corresponding to the existing utility level.

Table 7.1 Pinch Tableau Extended for Area Calculation

Vertices	A	B	a	b	C	D	c	d
Q	0	200	1320	1720	3200	3700	4240	5100
T_h	45	65	87.4	95.4	125	175	--	--
T_c	--	--	20	40	82.29	96.57	112	155
ΔT	--	--	67.4	55.4	42.71	78.43	--	--
$LMTD_i$	--	--	--	61.20	48.78	58.77	--	--
A_i	--	--	--	65.36	303.39	85.07	--	--

ΔT_{min}= 42.71°C, $Q_{hu,min}$= 1400 kW, $Q_{cu,min}$= 1320 kW, and T_p= 103.643°C.

Units: temperature in °C, enthalpy in kW, and area in m².

Thus, A_{ideal} = 65.36 + 303.39 + 85.07 = 453.82 m². $A_{existing}$ has already been obtained in step 1 of Example 7.1 to be 615.11 m². Finally, substituting in Equation 7.1 gives α = 453.82/615.11 = 0.7378.

Step 2. Calculation of Area Targets for Various Energy Levels
The area and energy targets (for various positions of the CCC) are established by constructing tableaux as in Table 7.1. The tableau may be revised for a given shift based on the procedure in Chapter 2. Alternatively, the targets for various ΔT_{min} may be calculated as in Chapter 1 to yield the results in Table 7.2 (see first four columns).

Step 3. Calculation of the Retrofit Curve
One of the aims of retrofitting is to improve the use of area; hence, the efficiency should not decrease and may be chosen to be $\alpha_{existing}$ (namely, 0.7378) to provide a most conservative estimate for further calculations.

However, when α is very low (i.e., $\alpha < 0.9$), the usage of an incremental value of $\Delta\alpha = 1$ is recommended (Silangwa, 1986; Ahmad and Polley, 1990). Thus, the maximum area to be used in designing the new network may be obtained as

$$A_{max,retr} = A_{ideal}/\alpha_{existing} \qquad\qquad \text{for } \alpha \geq 0.9 \qquad (7.2a)$$
$$A_{max,retr} = (A_{ideal} - A_{ideal1}) + A_{existing} \qquad \text{for } \alpha < 0.9 \qquad (7.2b)$$

where A_{ideal1} is the value of A_{ideal} obtained in step 1. The above equations may be rewritten in a unified, general form as

$$A_{max,retr} = (A_{ideal} - A_{ideal1})/\Delta\alpha + A_{existing} \qquad\qquad (7.2c)$$

with $\Delta\alpha = 1$ for $\alpha < 0.9$ and $\Delta\alpha = \alpha_{existing}$ for $\alpha \geq 0.9$. Equation 7.2c reduces to Equation 7.2a on using $\alpha_{existing} = A_{ideal1}/A_{existing}$ (see definition in Equation 7.1).

Table 7.2 Retrofit Curve on Area-Energy Plot

Col. A ΔT_{min}	Col. B $Q_{cu,min}$	Col. C $Q_{hu,min}$	Col. D A_{ideal}	Col. E $A_{max,retr}$ $\Delta\alpha = 1$	Col. F $A_{max,retr}$ $\Delta\alpha = .7378$
5	120	200	2050.61	2211.91	2779.45
10	220	300	1621.91	1783.20	2198.37
13	280	360	1455.83	1617.12	1973.22
15	350	430	1303.54	1464.84	1766.85
20	525	605	1028.96	1190.26	1394.68
25	700	780	839.18	1000.48	1137.45
30	875	955	697.60	858.89	945.54
35	1050	1130	586.29	747.58	794.67

Units: temperature in °C, enthalpy in kW, and area in m^2.

On using Equation 7.2c in the form $A_{max,retr} = (A_{ideal} - 453.82)/1 + 615.11 = A_{ideal} + 161.29$, column E of Table 7.2 for $A_{max,retr}$ may be obtained from column D. Plotting A_{ideal} vs. energy (i.e., column D vs. column C of Table 7.2 for an area-energy plot based on hot utility) gives the ideal target curve for grassroots design while $A_{max,retr}$ vs. energy (i.e., column E vs. column C) gives the retrofit curve for $\Delta\alpha = 1$. Figure 7.2 shows the target curve and the retrofit curves for $\Delta\alpha = 1$ as well as $\Delta\alpha = \alpha_{existing} = 0.7378$ (column F vs. column C of Table 7.2).

Step 4. Calculation of Energy Savings and Extra Area Required
   This step is similar to steps 3 and 4 in Example 7.1. The formulae are

energy savings = current utility usage − target utility required
$$\text{(based on hot or cold)} \qquad (7.3a)$$
extra area = required new area − existing area $\qquad (7.3b)$

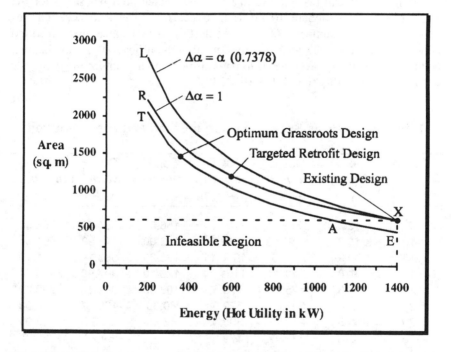

Figure 7.2 The above area-energy plot allows the designer to choose a path for retrofitting the existing network (marked X). Two likely retrofit paths are curve XL ($\Delta\alpha = \alpha_{existing} = 0.7378$) and curve XR ($\Delta\alpha = 1$). Curve EAT is the ideal target, below which is the region of infeasible designs.

   Subtracting the entries in columns B and C of Table 7.2 from the current utility usages (namely, 1320 kW for cold, and 1400 kW for hot) gives the energy savings in Table 7.3. Similarly, subtracting the existing area of 615.11 from columns E and F of Table 7.2 gives the extra area required in Table 7.3.

Step 5. Economic Analysis of Investment vs. Savings

The cost-saving analysis is similar to step 5 in Example 7.1. Columns E, F, and G are obtained by substituting the values in columns B, C, and D in the following formulae:

Savings in $/yr   = 130 (energy saved in kW)
Investment in $ = $30000 N_m + 750 N_s (A/N_s)^{0.81}$
                  where $N_m = [A/310]$ and $N_s = [A/310]$.

Here, N_m is the number of matches and N_s is the number of shells after rounding off. Recall that the average area per match (and per shell) for the existing network is about 310 m^2 (i.e., 615.11/2). The area cost requires making these assumptions (Ahmad and Polley, 1990) since the retrofitted network (and hence the area distribution after modification) is not known. Note also that energy saved needs to be multiplied by the sum of the unit costs for hot and cold utility (i.e., 120 + 10 = 130).

Table 7.3 Analysis Based on Energy Saved and Area Required

Col. A ΔT_{min}	Col. B Energy Savings	Col. C Extra Area $\Delta\alpha = 1$	Col. D Extra Area $\Delta\alpha=.7378$	Col. E Savings	Col. F Investment $\Delta\alpha= 1$	Col. G Investment $\Delta\alpha=.7378$
5	1200	1596.80	2164.34	156.00	594.52	756.06
10	1100	1168.09	1583.26	143.00	417.93	591.67
13	1040	1002.01	1358.11	135.20	383.12	501.19
15	970	849.73	1151.74	126.10	307.99	414.54
20	795	575.15	779.57	103.35	207.12	293.29
25	620	385.37	522.34	80.60	166.37	196.08
30	445	243.78	330.43	57.85	94.35	153.91
35	270	132.47	179.56	35.1	69.26	80.23

Units: temperature in °C, energy in kW, area in m^2, savings in 10^3 $/yr, and investment in 10^3 $.

Step 6. Identification of Target ΔT_{min}

Based on the specified payback period (of two years), the required target is the point where the investment is twice the savings. Note that initially (ΔT_{min} < 20°C) the investment is greater than twice the savings, but the inequality essentially gets reversed later (ΔT_{min} > 20°C). Thus, Table 7.3 yields a target

ΔT_{min} of 20°C (for $\Delta\alpha = 1$), at which the savings is 103.35 x 10^3 \$/yr (an energy savings of 795 kW) and the investment required is 207.12 x 10^3 \$ (an extra area of 575.15 m^2). Figure 7.3 shows the investment-savings plot along with the payback line.

From Table 7.3, it may be deduced that a payback period of two years cannot be achieved for $\Delta\alpha = \alpha_{existing} = 0.7378$. This is also observed on Figure 7.3 where there is no intersection between the two-year payback line and the curve XL (for $\Delta\alpha = \alpha_{existing}$). If the payback period is increased (by rotating the payback line about the origin in Figure 7.3), then a reasonable target ΔT_{min} is 25°C (for $\Delta\alpha = 0.7378$). For this case, the annual savings is 80.6 x 10^3 \$/yr (an energy savings of 620 kW) and the investment required is 196.08 x 10^3 \$ (an extra area of 522.34 m^2), giving a payback period of 2.43 years.

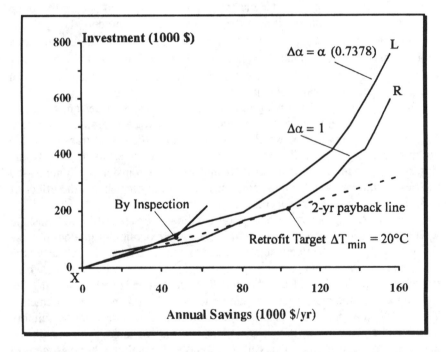

Figure 7.3 The above investment-savings plot allows the designer to choose a target ΔT_{min} for retrofitting, using the (dotted) straight line for the specified payback. The retrofit target by pinch technology is seen to be a significant improvement over the retrofit by inspection.

Note that the area-energy plot is continuous, smooth, and differentiable only between two consecutive significant shifts (see Table 2.9). However, accurate determination of the target ΔT_{min} by using the golden section search technique (see step 6 in Example 2.2) between two significant shifts has a further complication due to the nonsmooth nature of the investment-savings plot arising out of the average area per match/shell assumption. Moreover, for most industrial applications, determination of the target ΔT_{min} within an accuracy of 1°C should suffice.

7.2.2 Understanding the Retrofit Targeting Philosophy

To understand retrofit targeting, the two important representations (Tjoe and Linnhoff, 1986) on which it is based must be understood. These are the area-energy plot (Figure 7.2) and the investment-savings plot (Figure 7.3).

On the area-energy plot, the curve EAT represents ideal networks having different ΔT_{min} but satisfying both the energy and area targets. The point T signifies a small ΔT_{min} (with low energy consumption and high area requirement), and the point E signifies a large ΔT_{min} (with high energy consumption and low area requirement). In general, the goal is to optimize on energy consumption using the required area in the most effective way.

Retrofitting involves moving from the point X (existing design) to a new point having favorable retrofit economics. The movement should be towards the ideal target curve EAT, which corresponds to vertical heat transfer on the composite curves. Recall that the vertical heat transfer model yields the rigorous minimum area for the case of uniform heat transfer coefficients and the near-minimum area for most other cases. On the other hand, the effect of crisscross heat transfer is an increase in the overall area.

For an existing design to be a good candidate for retrofitting, it must lie well above the ideal target curve. In this case, the existing design does not use the installed area effectively due to crisscross heat exchange. Clearly, the minimum energy consumption for the existing area is given by point A, which therefore is the ideal point to move to. This corresponds to reorganizing the existing area in the most effective manner to optimize on energy requirements. This is often not practically possible for the installed area may not be suitably distributed among the existing exchangers and the rearrangements within the network may be restricted. This calls for some additional capital investment to use the existing exchangers more effectively and to provide the necessary new area. Many retrofit paths (like XL and XR on Figure 7.2) with varied cost effectiveness are possible. Path XR is lower than path XL on the plot and calls for less investment to achieve the same savings. It is difficult to define a best path in practice, as it depends on process considerations and plant layout. The region within which the best path must lie may however be defined as shown next.

The following four regions may be distinguished on Figure 7.2:

- the region below the curve EAT corresponds to infeasible designs (for retrofit as well as grassroots design purposes). This is simply because the curve EAT is the ideal target curve and corresponds to the best possible design. Note that, for accurate determination of the curve EAT, the areas must be obtained by the NLP formulation (Section 10.2) when the heat transfer coefficients are unequal.
- the region above the curve XL corresponds to designs that do not offer promising economics. This is because the curve XL (for constant α) provides an upper bound for a retrofit project as it corresponds to the existing area efficiency level. Thus, the minimum expectation is that the retrofitted network will use the heat transfer area as efficiently as (if not better than) the existing design.
- the region bounded by XA, AE, and EX again corresponds to designs that do not typically promise favorable economics. This is because the line segment XA shows the heat exchange area that already exists, and it is not sensible to discard such area for which the investment has already been made.
- the remaining region bounded by LX, XA, and AT thus signifies retrofit projects with promising economics.

Does retrofitting involve transformation of the existing design to the optimum grassroots design? This is often not the right approach since features in the existing design may provide unique opportunities or place unique constraints. If the optimum grassroots design requires less area than the existing process (i.e., it lies on the portion of the curve EA in Figure 7.2), then it does not provide a reasonable target for it calls for rejection of already installed area. On the other hand, if it requires more area than the existing process (i.e., it lies on the portion of the curve AT as in Figure 7.2), then the optimum grassroots design does provide a reasonable target, provided it meets the scope of the retrofit project in terms of the desired savings, the ceiling on investment, and an acceptable payback. Note that the slope of a typical retrofit path is such that the payback period increases with investment level and energy savings. Specifically, the optimum grassroots design for our case study is at a ΔT_{min} of 13°C (see Section 1.6). For this retrofit target, Table 7.3 shows the annual savings to be 135.2 x 10^3 \$/yr (an energy savings of 1040 kW) and the investment required to be 383.12 x 10^3 \$ (an extra area of 1002 m^2) giving a payback period of 2.83 years (which is greater than specified). The three economic criteria (savings, investment, and payback) should be typically conceived as constraints and not objectives for the retrofit project. In fact, if our target within the imposed constraints is conservative (e.g., based on constant α), then the retrofit design will perform better than anticipated in terms of these economic criteria.

Although the area-energy plot provides the key physical basis for establishing retrofit targets, the final representation to be used for assessment is the investment-savings plot that captures the current economics as well as the appropriate tradeoff between capital and energy costs. Figure 7.3 shows that pinch technology allows better retrofit targeting than inspection. Through pinch technology, lower payback periods may be achieved because higher annual savings are possible for a given investment level or lower investments are required for a specified energy savings. In the case study under consideration, retrofit by inspection yielded an energy savings of 26.5% (given by 720/2720) for a payback period of 2.28 years and an investment of 106.54×10^3 \$. In fact, the maximum energy savings by inspection was limited to 36.8%. Pinch technology indicates that an energy savings of 58.5% (given by 1590/2720) is possible with a payback period of two years and an investment of 207.12×10^3 \$. Is a design methodology available that synthesizes the retrofitted network and guarantees satisfaction of the above targets? Such a design technique does exist and is presented in Section 7.2.4.

7.2.3 Understanding the Crisscross Principle

Understanding the principles of crisscrossing and cross-pinching (Linnhoff and Ahmad, 1990) helps in retrofit design.

From the previous section, it is clear that networks with near-vertical heat transfer will lie close to the ideal target curve (EAT in Figure 7.2) on the area-energy plot. As the crisscross heat transfer in a network increases, the design will be located further away from the ideal target curve. This crisscrossing effect is quantitatively captured by the area efficiency parameter, α, with a value of unity signifying true vertical heat transfer. A low value of α signifies severe crisscross heat transfer and ineffective use of heat exchange area. Therefore, the important principle is that there should be no crisscross heat transfer in order to achieve the area target predicted by the uniform BATH formula.

From Section 1.1.4, it is clear that networks with maximum energy recovery must not transfer any heat across the pinch, or else a double energy penalty is incurred. Thus, the important principle is that there should be no cross-pinch heat transfer in order to achieve the energy target predicted by the problem table algorithm.

These two principles are in agreement in the pinch region since vertical heat transfer (no crisscrossing) is essential for no heat transfer to occur across the pinch (no cross-pinching). In other words, satisfying the energy target is a necessary, though not sufficient, condition for satisfying the area target.

Next, consider the point X on Figure 7.2 (the existing design with crisscrossing). Compared to point E, the network at point X shows no energy penalty and consequently must have no net cross-pinch heat transfer. To move from point X to point E, alterations are required to make the matches near-vertical. When compared with other points on the curve EA, the network at point X shows an energy penalty and consequently has cross-pinch heat transfer. To move from point X to these points, alterations are required in the cross-pinch exchangers. Network X is not optimal because of crisscrossing (with or without cross-pinching). Therefore, the principle given earlier in Section 1.1.4 to eliminate cross-pinching during retrofitting is not sufficient. More generally, it may be stated that crisscrossing must be eliminated to approach the energy target and area target simultaneously.

7.2.4 Retrofit Design for Constant *h*-values

The retrofit targets established in Section 7.2.1 are crucial as they provide valuable insight into designing energy saving retrofits. Specifically, they yield an appropriate beginning point for design and an estimate of the project investment and the expected payback period.

The retrofit design procedure presented below was originally proposed by Tjoe and Linnhoff (1987). It exploits the interactions among the heat exchangers in a network so that the investment made in new equipment yields greater savings than expected from its size and its individual heat recovery level. This is possible if the added equipment leads to more effective use of the heat transfer area in the existing exchangers.

Example 7.3 Revamp of an Existing Network for Constant *h*-values
For Case Study 4S1, develop an improved network (energy saving retrofit) starting with the existing network in Figure 1.2. Attempt to make the minimal changes to the existing network and to reuse/reposition existing equipment as far as possible in a cost-effective manner. Install new exchangers where necessary. The cost data are the same as given earlier in Example 7.1.

Solution. The retrofit design, which focuses on eliminating crisscrossing and not merely cross-pinching, is accomplished in the following steps. The methodology begins with an assessment of the existing network to determine exchangers that are already well-positioned and hence should not be disturbed. Next, exchangers that are poorly placed are detected. The attempt is to improve stream alignment (or reduce crisscrossing) in as many exchangers as possible by making changes in as few exchangers as possible. Note that every change propagates through the downstream paths (Linnhoff and Kotjabasakis, 1986) in the network, and the effects may be felt in several units. The installation of new area is finally considered where necessary. As the area efficiency parameter, α, is an effective measure of the crisscrossing, it is used throughout the design process. The effort is to improve it continually and not to allow it to drop below the value for the existing plant.

Step 1. Establishing Retrofit Targets

The retrofit targeting in Example 7.2 suggested the network be designed at a target ΔT_{min} of 20°C for $\Delta \alpha = 1$ and for a payback period of two years. This would yield operating cost savings of 103.35 x 10^3 \$/yr (an energy savings of 795 kW) and require an investment of 207.12 x 10^3 \$ (an extra area of 575.15 m^2). The hot and cold utility required will be 605 kW and 525 kW respectively. The minimum area theoretically required is 1028.97 m^2. The procedure is now directed towards approaching this ideal target in the retrofit.

Step 2. Using Remaining Problem Analysis for Existing Network

The performance of each exchanger in the overall context of the existing network (rather than in isolation) may be assessed using the concepts of RPA (see Section 4.4.2). The extended pinch tableau (see representation in Table 7.1) is useful in determining the ΔT_{min} and the area of the remaining problem for the given utility consumption. On accepting exchanger 2 in Figure 1.2, the remaining problem comprises the following streams: H1 (175° to 45°C); H2 (98° to 65°C); and C3 (20° to 155°C). Table 7.4 shows the extended pinch tableau for this stream data (corresponding to the remaining problem of unit 2) for the targeted hot utility consumption of 605 kW. The countercurrent area is seen to be 783.21 m^2 and the $\Delta T_{min,rp}$ is 11.75°C for the remaining problem.

For each exchanger, an area efficiency, α_e, is defined (Tjoe and Linnhoff, 1987) below to indicate the performance of the unit with respect to the entire network. If α_e denotes the maximum area efficiency possible for the overall network after accepting the exchanger, e, then

$$\alpha_e = A_{ideal}/(A_e + A_{rp,e}) \tag{7.4}$$

where A_{ideal} is the ideal target area (1028.97 m^2 as per Table 7.2 at the target ΔT_{min} of 20°C), A_e is the existing area of the exchanger (358.92 m^2 for unit 2), and $A_{rp,e}$ is the area of the remaining problem (783.21 m^2 as per Table 7.4).

For unit 2, this yields $\alpha_e = 1028.97/(358.92 + 783.21) = 0.9$ and $\Delta T_{min,rp} = 11.75$°C. A similar analysis for unit 3 in Figure 1.2 yields $\alpha_e = 1028.97/(256.19 + 2784.90) = 0.34$ and $\Delta T_{min,rp} = 0.25$°C.

Table 7.4 Extended Pinch Tableau for RP of Unit 2

Q	0	200	525	1850	2620	3225
T_h	45	65	<u>71.5</u>	98	175	--
T_c	--	--	20	<u>86.25</u>	<u>124.75</u>	155
ΔT	--	--	51.5	11.75	50.25	--
$LMTD_i$	--	--	--	26.90	26.49	--
A_i	--	--	--	492.58	290.63	--

$\Delta T_{min,rp} = 11.75°C, Q_{hu,min}= 605$ kW, $Q_{cu,min}= 525$ kW, $A_{rp,e} = 783.21$ m^2.

Units: temperature in °C, enthalpy in kW, and area in m^2.

Step 3. Accepting Good Exchangers

Recall that a value of α_e close to unity implies near-vertical heat transfer and an exchanger that may be accepted in its present location. A low value of α_e implies an inefficient exchanger that has the potential to be better positioned. An indeterminate value of α_e (i.e., $A_{rp,e}$) implies an exchanger that is terribly inefficient and must be repositioned to operate the network at the given utility level.

Even if α_e for an exchanger is high, the unit may not be accepted if $\Delta T_{min,rp}$ is very low because difficulty will be experienced during design on account of small temperature driving forces. Note that $\Delta T_{min,rp}$ is the temperature difference for the composite curves of the remaining problem to operate the network at the given utility level.

In essence, an exchanger may be accepted only if both α_e and $\Delta T_{min,rp}$ are high (Tjoe and Linnhoff, 1987). Acceptance of good exchangers implies that the portions of the streams associated with them are excluded from further analysis. This guarantees that a good exchanger is unaffected and retained in its original location.

For the present example, a reasonable guideline is to accept an exchanger if α_e is greater than the target value of 0.86 (i.e., 1028.97/1190.19) and $\Delta T_{min,rp}$ is greater than the target of 20°C. Based on this criterion, neither of the exchangers (units 2 and 3 in Figure 1.2) can be accepted. Although unit 2 has a moderately high value of the area efficiency, its $\Delta T_{min,rp}$ is not high enough and accepting it may result in tight temperature driving forces for the remaining design. Thus, units 2 and 3 may both be considered inefficient and improvements attempted in terms of their stream alignment by altering their loads and by considering network interactions. This is most conveniently done on the driving force plot.

Step 4. Generating the Driving Force Plot

The concept of the driving force plot (DFP) was introduced in Section 4.4.1. It allows the designer to distinguish exchangers that are well aligned (those which closely follow the vertical temperature differences on the composite curves) from those that are poorly aligned.

Tjoe and Linnhoff (1987) recommend the use of a DFP in the form of T_h vs. T_c for retrofit studies rather than in the form of ΔT vs. T_c (Figure 4.22). The DFP in Figure 7.4 is generated using the T_h and T_c rows of the pinch tableau (Table 7.5), after accepting the good exchangers (none in our case study). The plot is piecewise linear, with the slope equal to the heat capacity flow rate ratio of the CCC to the HCC in the given enthalpy interval. If the plot is close to the 45° line ($T_h = T_c$), then the problem involves tight temperature driving forces. The minimum approach of the DFP to the 45° line corresponds to the pinch. For an exchanger not to be a cross-pinch match, it must lie fully in the first quadrant (be above pinch) or third quadrant (be below pinch) when the origin in Figure 7.4 is made to coincide with the pinch.

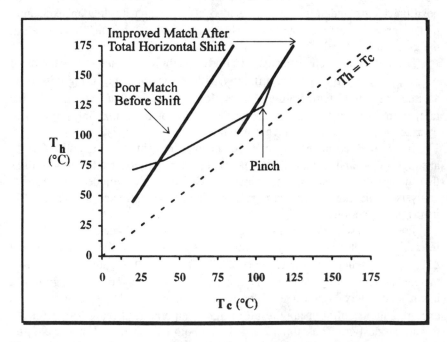

Figure 7.4 The driving force plot (DFP) helps the designer recognize a poorly aligned match and shift the associated line segment so as better to follow the vertical temperature differences on the composite curves.

Table 7.5 Pinch Tableau for Generating DFP

Q	0	200	525	925	3200	3445	3700	4305
T_h	45	65	71.5	79.5	125	149.5	175	--
T_c	--	--	20	40	105	112	124.75	155
ΔT	--	--	51.5	39.5	20	37.5	50.25	--
$LMTD_i$	--	--	--	45.24	28.65	27.84	43.56	--
A_i	--	--	--	88.43	794.00	88.01	58.53	--

$\Delta T_{min} = 20°C$, $Q_{hu,min} = 605$ kW, $Q_{cu,min} = 525$ kW, $A = 1028.97$ m^2.

Units: temperature in °C, enthalpy in kW, and area in m^2.

Every exchanger in a network may be represented by overlaying a line segment on the DFP, whose lower end-point corresponds to the cold terminal of the exchanger with coordinates $(T_{c,in}, T_{h,out})$ and whose upper end-point corresponds to the hot terminal of the exchanger with coordinates $(T_{c,out}, T_{h,in})$. When this line segment exactly coincides with the DFP, the exchanger is a perfect match, obeying vertical heat transfer.

Step 5. Improving Inefficient Units by Repositioning on DFP

Exchanger 3 in Figure 1.2 is represented on Figure 7.4 by a line segment with end-points having coordinates of (20, 45) and (85, 175). It shows a very poor fit to the DFP (an inefficient unit as was evidenced earlier in step 2 with $\alpha_e = 0.34$ and $\Delta T_{min,rp} = 0.25°C$).

The alignment of the exchanger may be improved by changing the end-temperatures or the heat capacity flow rate ratio. In the case of exchanger 3, consider shifting the hot terminal of the line segment horizontally to the right on Figure 7.4. Specifically, the hot stream inlet temperature $(T_{h,in})$ is kept constant at 175°C while the cold stream outlet temperature $(T_{c,out})$ is increased from 85° to 124.75°C (as per Table 7.5). The temperatures for the other end-point are calculated as follows (see Chapter 6 for details of equations used).

Using the overall heat transfer coefficient ($U = 0.1$ kW/m^2 °C) and the existing area of unit 3 (256.19 m^2), the number of transfer units is obtained by $N = UA/MC_{p,cold} = 0.1 \times 256.19/20 = 1.28$. As the heat capacity flow rate ratio is 2 ($R = MC_{p,cold}/MC_{p,hot} = 20/10$), the performance of the countercurrent exchanger is given by (see Example 6.D):

$$P = \{1 - \exp[N(R-1)]\}/\{1 - R\exp[N(R-1)]\} = 0.4194.$$

From the definitions of P and R, $(124.75 - T_{c,in})/(175 - T_{c,in}) = 0.4194$ and $(175 - T_{h,out})/(124.75 - T_{c,in}) = 2$. The required temperatures for the other terminal are therefore $T_{c,in} = 88.47°C$ and $T_{h,out} = 102.44°C$.

Though the shifted exchanger continues to be cross-pinched, it shows a significant improvement in terms of alignment with the DFP. This improvement may be quantified in terms of α_e and $\Delta T_{min,rp}$ (see step 2). Unit 3 after shifting yields $\alpha_e = 1028.97/(256.19 + 795.56) = 0.98$ and $\Delta T_{min,rp} = 13°C$. Further improvement of $\Delta T_{min,rp}$ from 13° to about 20°C may be considered by more changes in local conditions (e.g., by decrease in the exchanger load). Here, unit 3 is accepted at this stage.

Figure 7.5 shows a similar analysis for unit 2. Various possibilities exist for the shifting of unit 2. Table 7.6 shows the improvements in α_e and $\Delta T_{min,rp}$ based on shifting the cold terminal of the line segment vertically downwards or horizontally to the right.

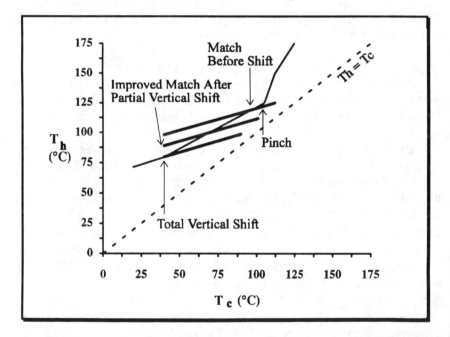

Figure 7.5 The match alignment is improved on the driving force plot (DFP) by shifting the cold terminal of the exchanger vertically downwards. After the shift, it is observed that the unit has a lower load and is no longer cross-pinched.

Table 7.6 Improvements Possible through Shifting Unit 2

Shift	$T_{h,in}$	$T_{h,out}$	$T_{c,in}$	$T_{c,out}$	α_e	$\Delta T_{min,rp}$
Total Vertical	98.62	80	40	89.66	0.93	20
Partial Vertical	111.8	89	40	100.8	0.96	20
Total Horizontal	112.7	98	66.43	105.62	0.93	20.47
Partial Horizontal	118.85	98	53.22	108.82	0.96	21.66

Units: temperature in °C.

The partial vertical shift of unit 2 yields α_e = 1028.97/(358.92 + 711.45) = 0.96 and $\Delta T_{min,rp}$ = 20°C, which are improvements from the corresponding values (before the shift) of 0.9 and 11.75°C. The partial vertical shift is preferred to the partial horizontal shift because it takes care of a portion of stream C4 starting from one end. Unit 2 is accepted at this stage since the shifted exchanger is no longer cross-pinched and its alignment to the DFP can be further improved only by changing the slope of the line segment (i.e., the heat capacity flow rate ratio of the exchanger).

In this case study, the shifting of units 2 and 3 is independent as is clear from Figure 1.2. In general, the sequence in which the shifts are effected may be of consequence, being critical in the most constrained regions (i.e., near the pinch).

Essentially, the match alignment may be improved on the DFP by adjusting hot and/or cold temperatures at one terminal of the exchanger and sometimes by introducing a stream split.

Step 6. Matching the Remaining Streams

The adjustments in the terminal temperatures of the exchangers were decided in the previous step without any concern about how they would be effected. In fact, they may be brought about without physical changes to the units themselves. Hence, the portions of the streams associated with these existing process exchangers may be excluded in the remainder of the synthesis to ensure that these well-placed matches are not disturbed. Figure 7.6 shows the remaining problem after excluding the shifted unit 2 (i.e., stream H2 from 111.8° to 89°C and stream C4 from 40° to 100.8°C) and the shifted unit 3 (i.e., stream H1 from 175° to 102.5°C and stream C3 from 88.5° to 124.75°C). The pinch tableau for the remaining problem at a utility consumption of 605 kW (for hot utility) and 525 kW (for cold utility) shows a $\Delta T_{min,rp}$ of 13°C and a pinch at 125°C/112°C.

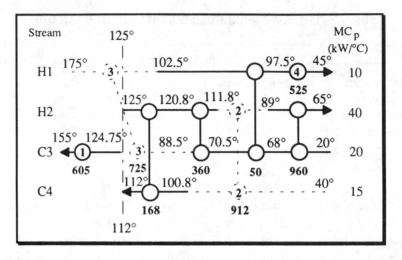

Figure 7.6 The design for the remaining problem after excluding the shifted exchangers (shown in the dashed sections) is completed by the conventional pinch design method. (Temperature in °C and load in kW)

Recognizing the pinch division, the design is completed by the conventional PDM. During this design activity, concepts like DFP, RPA, and compatibility with the existing structure may be exploited. The design in Figure 7.6 is not unique. An alternative to Figure 7.6 may involve placing the cooler of 525 kW on stream H2; in this case, the load on the H1-C3 match will be 575 kW (rather than 50 kW) and that on the H2-C3 match will be 435 kW (rather than 960 kW). This finally leads to the network obtained in Figure 7.1 by inspection and is therefore not discussed here.

Step 7. Evolving to the Final Retrofit Design
Figure 7.6 shows the need for installing new area in four locations. The number of new units may be reduced through conventional evolution by shifting loads along loops and paths (Section 4.1). There are three first-level loops in Figure 7.6 that may be broken by simply adding the load of one unit to another. Thus, 50 kW may be added to unit 3 and 168 kW to unit 2. Also, the units with loads of 360 kW and 960 kW may be combined into a single unit. Clearly, verticality in heat transfer will be lost to some extent and the area requirement will increase. However, this tradeoff between units and area may result in reduced capital investment for the retrofit and must be explored.

The evolved design for the retrofit in Figure 7.7 requires only a single new unit (labeled N) of 528.7 m^2 (2 shells) and an increased area of 38.74 m^2 on unit 3. Thus,

Investment = $750(38.74)^{0.81} + 30000 + 2(750)(528.7/2)^{0.81} = 181.93 \times 10^3$ \$;
Savings = $130 \times 795 = 103.35 \times 10^3$ \$/yr; and payback = 1.76 years.

This is less than the targeted payback period of two years, and the area is used more efficiently ($\alpha = 0.87$, which is better than the earlier value of 0.74).

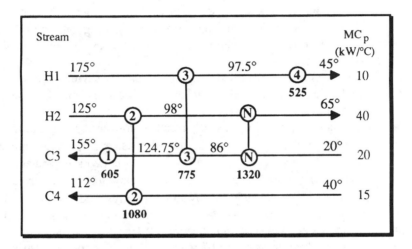

Figure 7.7 The final network for the retrofit is obtained by evolution and shows the payback on the project to be 1.76 years. (Temperature in °C and load in kW)

7.3 RETROFIT - SPECIFIED PRESSURE DROPS

Only the thermal considerations involved during process retrofitting were systematically handled in the previous section. Considering the effect of the retrofit on the flow systems only as an afterthought may not be appropriate. In reality, it may prove disastrous in situations where the cost of the changes required in pumps and/or compressors turns out to be exorbitant. In other cases, the additional heat transfer area required may be traded off against allowable pressure drop.

Given the methodology for the effective utilization of pressure drops in networks (see Chapter 6), flow considerations may be incorporated to form an integral part of both the retrofit targeting and design procedures as discussed by Polley et al. (1990) and demonstrated below.

7.3.1 Targeting Based on Specified ΔP-values

It may be useful to recapitulate at this stage some important features of retrofit targeting (from Section 7.2.1):

- Retrofitting by inspection to meet subjectively set economic criteria often yields a suboptimal solution.
- Predicting the project scope is possible ahead of network synthesis based on economic criteria (for investment ceiling and maximum payback) by relating the additional heat transfer area required to the energy savings.
- Setting the correct project scope in terms of energy savings in a retrofit is vital since the network structure is dependent on the energy consumption level.
- Increasing the project scope usually results in longer payback periods.
- Setting the project scope may be done by using a retrofit curve of constant α if the area efficiency of the existing network is greater than 0.9. Otherwise (for $\alpha < 0.9$), the area efficiency may be improved by adding the new area in an ideal way and this corresponds to $\Delta\alpha = 1$.

Furthermore, it is important to recognize the following deficiencies of the retrofit targeting procedure discussed in Section 7.2.1:

- The stream heat transfer coefficients are specified on the basis of existing exchanger performance. This works well when a single exchanger exists (before retrofitting) and will continue to exist (after retrofitting) on a stream. Difficulties arise in specifying a single heat transfer coefficient for a stream on which there exist many exchangers with varying heat transfer coefficients. Matters worsen if a new exchanger is to be installed on such a stream and is to be designed according to an assumed heat transfer coefficient.
- The stream heat transfer coefficients are assumed to be independent of the scope of the retrofit project. Frequently, stream pressure drop constraints come into play due to the existing flow system (pumps, compressors, and available gravitational head). On the area-energy plot, the area requirement increases as the project scope increases; for fixed pressure drops, this implies that the heat transfer coefficients decrease with increasing project scope (as per Equations 6.33 and 6.34). Thus, the retrofit targets based on fixed heat transfer coefficients will tend to be optimistic. Furthermore, pressure drop constraints may require placement of new exchangers in parallel (not in series) with existing units.

Except for these deficiencies that need to be taken care of by area targeting based on pressure drops (see Section 6.5), the rest of the retrofit targeting procedure is essentially the same as in Section 7.2.1.

Example 7.4 Establishing Retrofit Targets for Constant ΔP-values

For the stream data in Case Study 4S1 and the network in Figure 1.2, determine a target ΔT_{min} for retrofit design introducing pressure drop considerations. The cost data are the same as given earlier in Example 7.1. The physical properties for the four streams are given in Table 6.4. All exchangers use tubes of 19.1 mm o.d. and 15.4 mm i.d., laid out in 25.4 mm square pitch, and a baffle cut of 25%.

Solution. Two important aspects must be noted during retrofits subject to pressure drop constraints: a single appropriate value for the heat transfer coefficient of each stream must be determined and the fall in heat transfer coefficients must be considered in generating the ideal target curve on the area-energy plot.

Of course, careful considerations must be given to specifying the stream pressure drops available for heat transfer (after considering the head loss due to pipeflow, valves, fittings, and gravity along with ways of minimizing unnecessary losses). These useful pressure drops are then employed in the area targeting algorithm given in Section 6.5. Advantages of pump replacement may be investigated at the targeting stage itself by appropriately modifying stream pressure drops and studying the effects on the area-energy plot.

Step 1. Rating of Existing Exchanger Performance

For retrofitting purposes, heaters and coolers are not considered. The specifications and the ratings (by Kern's method) for the existing process-to-process exchangers in Figure 1.2 are given in Table 7.7. The existing 1-2 area is 703.7 m^2. The unusually high dirt factors indicate overdesigned exchangers.

Step 2. Conversion from Exchanger Basis to Stream Basis

The pressure drops and heat transfer coefficients are obtained on an exchanger basis from Step 1. As the flow path for a stream through a HEN is known, its pressure drop may be determined by appropriate addition of the pressure drops of the exchangers on it. Because there is only a single process exchanger on each stream in Figure 1.2, the exchanger pressure drop is the stream pressure drop in this case.

Heat transfer coefficients may be predicted on a clean basis using Equations 6.33 and 6.34, and then modified to coefficients on a dirty basis using the dirt resistances. Note that K_t and K_s values required for this purpose are already available in Example 6.5. Table 7.8 summarizes the results on a streamwise basis. They are easy to obtain because each stream has only one unit in Figure 1.2. The dirty h-values for the tubeside are based on the outside tube diameter.

Table 7.7 Existing Exchanger Specifications and Ratings

Quantity	Unit 2	Unit 3
Exchanger area (A in m^2)	398.8	304.9
Number of shell passes (NS)	2	2
Number of tube passes (NT)	6	4
Shell diameter (D_S in mm)	751.8	657.3
Tube length (L in m)	4.83	4.83
Number of tubes per shell (N_t)	688	526
Number of baffles (N_b)	15	36
Tubeside coefficient - clean (W/m^2 °C)	875.9	903.0
Shellside coefficient - clean (W/m^2 °C)	814.1	713.2
Tubeside dirt resistance (m^2 °C/W)	0.00363	0.00356
Shellside dirt resistance (m^2 °C/W)	0.00363	0.00356
Tubeside pressure drop (ΔP_t in kPa)	6.68	4.92
Shellside pressure drop (ΔP_S in kPa)	11.65	29.14

Table 7.8 Stream Pressure Drops and Heat Transfer Coefficients

Stream	H1	H2	C3	C4
Pressure drop ΔP in kPa	29.14	11.65	4.92	6.68
Dirt factor R_d in (m^2°C)/W	0.00356	0.00363	0.00356	0.00363
Clean coefft in W/(m^2 °C)	713.2	814.1	903.0	875.9
Dirty coefft in W/(m^2 °C)	201.5	205.8	172.8	169.0

Step 3. Calculation of Area Efficiency of Existing Network

The calculation is similar to step 1 in Example 7.2. The existing utility level (1400 kW of hot utility and 1320 kW of cold utility) corresponds to a ΔT_{min} of 42.71°C (see Table 7.1) and an ideal 1-2 area target of 459.21 m^2. Thus, from Equation 7.1, $\alpha = 459.21/703.7 = 0.6527$.

The area target of 459.21 m^2 is obtained using the method from Example 6.5 and corresponds to the following stream heat transfer coefficients:

$$h_{H1} = 209.27 \text{ W/(m}^2 \text{ °C)} \qquad h_{H2} = 209.26 \text{ W/(m}^2 \text{ °C)}$$
$$h_{C3} = 214.95 \text{ W/(m}^2 \text{ °C)} \quad \text{and} \quad h_{C4} = 221.08 \text{ W/(m}^2 \text{ °C)}.$$

Thus, the same utility consumption (1400 kW of hot utility at ΔT_{min} = 42.71°C) has three 1-2 areas associated with it: 703.7 m^2 is the actual existing area, 525.18 m^2 is the ideal target area based on the heat transfer coefficients (in Table 7.8), and 459.21 m^2 is the ideal target area based on the specified pressure drops (in Table 7.8). The difference between 703.7 m^2 and 525.18 m^2 may be attributed to crisscross heat transfer and the difference between 525.18 m^2 and 459.21 m^2 to inferior utilization of pressure drops.

Step 4. Calculation of Area Targets for Various Energy Levels
The area and energy targets for various ΔT_{min} may be calculated to yield the results in Table 7.9 (see first four columns).

Table 7.9 Retrofit Curve on Area-Energy Plot (Specified ΔP)

Col. A ΔT_{min}	Col. B $Q_{cu,min}$	Col. C $Q_{hu,min}$	Col. D A_{ideal}	Col. E $A_{max,retr}$ $\Delta\alpha = 1$
5	120	200	2402.7	2647.2
10	220	300	1812.6	2057.1
15	350	430	1468.7	1713.2
20	525	605	1083.0	1327.5
25	700	780	899.3	1143.8
26	735	815	856.6	1101.1
27	770	850	817.6	1062.1
28	805	885	780.6	1025.1
29	840	920	746.0	990.5
30	875	955	713.7	958.2
35	1050	1130	580.5	825.0
40	1225	1305	517.1	761.5
42.71	1320	1400	459.2	703.7

Units: temperature in °C, enthalpy in kW, and area in m^2.

The area targets in Table 7.9 for $25°C \leq \Delta T_{min} \leq 30°C$ are computed using the procedure discussed in Section 6.5 (with the pressure drops fixed at values specified in Table 7.8).

Step 5. Calculation of the Retrofit Curve

Since α is very low, an incremental value of $\Delta \alpha = 1$ is used. On using Equation 7.2c in the form $A_{max,retr} = (A_{ideal} - 459.21)/1 + 703.7 = A_{ideal} + 244.49$, column E of Table 7.9 for $A_{max,retr}$ is obtained. Figure 7.8 shows the ideal target curve (i.e., column D vs. column C of Table 7.9 for an area-energy plot based on hot utility) and the retrofit curve for $\Delta \alpha = 1$ based on fixed pressure drop (column E vs. column C of Table 7.9). The retrofit curve for $\Delta \alpha = 1$ is parallel to the ideal one, being displaced by 244.49 units vertically upwards in this case. In Figure 7.8, the ideal target and retrofit curves appear to converge because of the effect of the changing slope although in reality they do not.

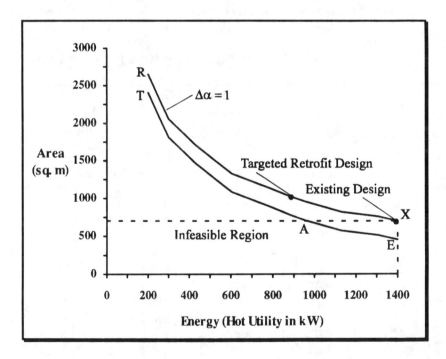

Figure 7.8 The above area-energy plot shows the likely path (curve XR for $\Delta \alpha$ = 1) for retrofitting the existing network (marked X). Curve EAT is the ideal target based on fixed stream pressure drops.

Step 6. Calculation of Energy Savings and Extra Area Required
    The calculation is based on Equations 7.3. Subtracting the entries in columns B and C of Table 7.9 from the current utility usages (namely, 1320 kW for cold and 1400 kW for hot) gives the energy savings in Table 7.10. Similarly, subtracting the existing area of 703.7 from column E of Table 7.9 gives the extra area required in Table 7.10.

Step 7. Economic Analysis of Investment vs. Savings
    Columns D and E are obtained by substituting the values in columns B and C in the following formulae:

Savings in $/yr   = 130 (energy saved in kW)
Investment in $ = $30000 N_m + 750 N_s (A/N_s)^{0.81}$
            where $N_m = [A/352]$ and $N_s = [A/176]$.

As before, N_m is the number of matches and N_s is the number of shells after rounding off. The average area per match for the existing network is about 352 m^2 (i.e., 703.7/2), and the average area per shell is about 176 m^2 (i.e., 703.7/4).

Table 7.10 Economic Analysis for Fixed Pressure Drop Retrofit

Col. A ΔT_{min}	Col. B Energy Savings	Col. C Extra Area $\Delta\alpha = 1$	Col. D Savings	Col. E Investment $\Delta\alpha = 1$	Col. F Payback
5	1200	1943.5	156.00	734.44	4.71
10	1100	1353.4	143.00	502.92	3.52
15	970	1009.5	126.10	375.90	2.98
20	795	623.8	103.35	239.24	2.31
25	620	440.1	80.60	187.93	2.33
26	585	397.4	76.05	177.79	2.34
27	550	358.4	71.50	168.32	2.35
28	515	321.4	66.95	121.83	1.82
29	480	286.8	62.40	113.72	1.82
30	445	254.5	57.85	106.02	1.83
35	270	121.3	35.10	66.56	1.90
40	95	57.8	12.35	50.07	4.05

Units: temperature in °C, energy in kW, area in m^2, savings in 10^3 $/yr, investment in 10^3 $, and payback in years.

Step 8. Identification of Target ΔT_{min}

Based on the specified payback period (of about two years), Table 7.10 suggests a target ΔT_{min} of 28°C, at which the retrofit project indicates a savings of 66.95 x 10^3 \$/yr and an investment of 121.83 x 10^3 \$. An energy savings of 36.8% (515 kW) defines the scope of the project. In accordance with the retrofit curve for $\Delta\alpha = 1$, the extra area of 321.4 m^2 is to be installed, ideally leading to an increased value of the area efficiency (namely, $\alpha = 780.6/1025.1 = 0.7615$, an improvement from 0.6527).

At the target ΔT_{min} of 28°C, the stream heat transfer coefficients are:

$$h_{H1} = 204.3 \text{ W/(m}^2 \text{ °C)} \qquad h_{H2} = 203.5 \text{ W/(m}^2 \text{ °C)}$$
$$h_{C3} = 207.4 \text{ W/(m}^2 \text{ °C)} \qquad \text{and} \qquad h_{C4} = 213.6 \text{ W/(m}^2 \text{ °C)}.$$

These coefficients are expectedly lower in value than those obtained at ΔT_{min} of 42.71°C in Step 3. The corresponding stream contact areas $A_{m,j}$ (as per the definition in Equation 6.41) are:

$$A_{m,H1} = 251.45 \text{ m}^2 \qquad A_{m,H2} = 529.16 \text{ m}^2$$
$$A_{m,C3} = 481.71 \text{ m}^2 \qquad \text{and} \qquad A_{m,C4} = 298.90 \text{ m}^2 .$$

With the scope of the retrofit project defined, the design is discussed next.

7.3.2 Retrofit Design for Constant ΔP-values

Both the effective utilization of pressure drops and temperature driving forces must be considered during network design. With the overall pressure drop on a stream being fixed, the distribution of pressure drops among the exchangers on a stream must be such that the pressure drop on a particular match is neither too high nor too low. This is primarily because a high ΔP on one exchanger (low area) will necessarily force a lower ΔP on some other exchanger (high area), resulting in a net increase in the overall area for the network. In a sense, this argument for ΔPs is similar to that for ΔTs put forth in Section 4.4.1.

A useful rule of thumb (Polley et al., 1990) is to use the same stream velocities and consequently the same heat transfer coefficients for exchangers on a stream in order to meet the minimum area target through effective pressure drop utilization. This involves determining the uniform film heat transfer coefficient consistent with the target, and then making modifications in existing exchangers to approach the desired conditions. Furthermore, the addition of new area may be skillfully done to achieve this objective. Of course, the rapid design algorithm provides the necessary powerful tool for optimum exchanger design through full utilization of pressure drop.

Example 7.5 Revamp of an Existing Network for Constant ΔP-values

For Case Study 4S1, develop an improved network (energy saving retrofit) starting with the existing network in Figure 1.2. Attempt to make the minimal changes to the existing network and to reuse/reposition existing equipment as far as possible in a cost-effective manner. Install new exchangers where necessary. The cost data are the same as given earlier in Example 7.1. Consider both effective utilization of temperature driving forces as well as pressure drops.

Solution. The steps are similar to those in the retrofit design procedure in Section 7.2.4: an analysis of the existing network to determine the acceptable exchangers, improvement of the poorly positioned exchangers, installation of necessary new area to meet the retrofit project target/scope, and network simplification to obtain the final design. The DFP and RPA prove to be powerful tools at each stage of the analysis.

For pressure drop based retrofits, two kinds of RPA (Polley and Panjeh Shahi, 1990) may be conveniently employed: RPA(h) where the area targets are based on constant heat transfer coefficients and RPA(P) where the area targets are based on fixed pressure drops. RPA(h) uses the procedure in Section 7.2.4 and yields $\alpha_{e,h}$, which is an indicator of the efficient utilization of temperature driving forces. Units with $\alpha_{e,h}$ near unity are thermally well positioned. Units with $\alpha_{e,h}$ less than the efficiency of the existing network need improvement through repositioning. RPA(P) is an analogous procedure that uses the area targeting from Section 6.5 and yields $\alpha_{e,P}$, which is an indicator of the efficient utilization of pressure drops (along with the temperature driving forces). The two RPAs, when used together, reveal whether the problem is with the poor use of pressure drops or temperature driving forces. For example, a unit with a high value of $\alpha_{e,h}$ and a low value of $\alpha_{e,P}$ uses temperature driving forces efficiently but pressure drops poorly.

Step 1. Establishing Retrofit Targets

The retrofit targeting in Example 7.4 suggested the network be designed at a target ΔT_{min} of 28°C for a project payback of 1.82 years. This would yield operating cost savings of 66.95 x 10^3 \$/yr (an energy savings of 515 kW) and require an investment of 121.83 x 10^3 \$ (an extra area of 321.4 m^2). The hot and cold utility required will be 885 kW and 805 kW respectively. The minimum area theoretically required is 1025.1 m^2.

Step 2. Using RPA (Based on h *and* ΔP) *for Existing Network*

The performance of each exchanger in the overall context of the existing network is assessed using the concepts of RPA as in step 2 of Example 7.3. The extended pinch tableau is useful in determining the ΔT_{min} and the area of the remaining problem for the given utility consumption. Table 7.11 shows the extended pinch tableau for the stream data corresponding to the remaining problem of unit 2 in Figure 1.2 for the targeted hot utility consumption of 885 kW. The countercurrent area is seen to be 436.0 m^2 and the $\Delta T_{min,rp}$ is 25.75°C for the remaining problem. Thus, in Equation 7.4, A_{ideal} = 725.6 m^2 (the ideal target area corresponding to the heat transfer coefficients at the target ΔT_{min} of 28°C in step 8 of Example 7.4), A_e = 1080(1/0.2035 + 1/0.2136)/30.09 = 344.5 m^2 (the existing countercurrent area of unit 2), and $A_{rp,e}$ = 436.0 m^2 (the area of the remaining problem as per Table 7.11). Note that RPA(h) uses countercurrent areas since its purpose is to track the temperature driving forces.

For unit 2, RPA(h) yields $\alpha_{e,h}$ = 725.6/(344.5 + 436.0) = 0.93 and $\Delta T_{min,rp}$ = 25.75°C. Similarly, RPA(h) for unit 3 in Figure 1.2 yields $\alpha_{e,h}$ = 725.6/(249.0 + 806.7) = 0.69 and $\Delta T_{min,rp}$ = 13.71°C.

Table 7.11 Extended Pinch Tableau for RP of Unit 2

Q	0	200	805	1850	2620	3505
T_h	45	65	<u>77.1</u>	98	175	--
T_c	--	--	20	<u>72.25</u>	<u>110.75</u>	155
ΔT	--	--	57.1	25.75	64.25	--
$LMTD_i$	--	--	--	39.37	42.11	--
A_i	--	--	--	258.34	177.68	--

$\Delta T_{min,rp}$ = 25.75°C, $Q_{hu,min}$= 885 kW, $Q_{cu,min}$= 805 kW, $A_{rp,e}$ = 436.02 m^2.

Units: temperature in °C, enthalpy in kW, and area in m^2.

RPA(P) is conceptually similar to RPA(h), but is more complicated since the fixed pressure drops (after excluding the pressure drops for the particular unit under analysis) must be converted into heat transfer coefficients using the algorithm in Section 6.5. Also, RPA(P) uses 1-2 areas because its purpose is to track the pressure drops. Application of RPA(P) yields $\alpha_{e,P}$ = 780.6/(398.8 + 458.1) = 0.91 for unit 2 and $\alpha_{e,P}$ = 780.6/(304.9 + 930.2) = 0.63 for unit 3.

Unit 3 certainly needs to be improved as it is inefficient overall. Unit 2 is relatively efficient but could also do with some improvement.

Polley and Panjeh Shahi (1990) argue that corrections for the temperature driving force must be made prior to those for the pressure drops because appropriate use of area is the first necessity (given the coupling between area and pressure drop). The driving force plot (DFP) allows for efficiency improvement through suitable thermal repositioning.

Step 3. Improving Inefficient Units by Repositioning on DFP

The pinch tableau (Table 7.12) is generated without accepting either of the exchangers. For a utility consumption of 885 kW, the pinch is at 125°C/97°C. The cross-pinch heat transfer by units 1, 2, and 3 in Figure 1.2 is 515 kW (i.e., 20 [97 − 85] − 15 [112 − 97] + 10 [175 − 125]). This expectedly equals the targeted energy savings.

The DFP in Figure 7.9 is generated using the T_h and T_c rows of the pinch tableau (Table 7.12).

Table 7.12 Pinch Tableau for Generating DFP

Q	0	200	805	1205	3200	3700	3725	4585
T_h	45	65	<u>77.1</u>	<u>85.1</u>	125	175	--	--
T_c	--	--	20	40	<u>97</u>	<u>111.29</u>	112	155
ΔT	--	--	57.1	45.1	28	63.71	--	--
$LMTD_i$	--	--	--	50.86	35.87	43.44	--	--
A_i	--	--	--	76.53	537.87	111.15	--	--

ΔT_{min} = 28 °C, $Q_{hu,min}$= 885 kW, $Q_{cu,min}$= 805 kW, A_{rp} = 725.55 m^2.

Units: temperature in °C, enthalpy in kW, and area in m^2.

In general, exchangers near the pinch with large violations may be corrected first. However, in this example, the order in which corrections are made is not important.

Exchanger 3 is represented on Figure 7.9 by a line segment with endpoints having coordinates of (20, 45) and (85, 175). It shows a very poor fit to the DFP (an inefficient unit with $\alpha_{e,h}$ = 0.69, $\alpha_{e,P}$ = 0.63, and $\Delta T_{min,rp}$ = 13.71°C). As in Example 7.3, consider shifting the hot terminal of the line segment horizontally to the right on Figure 7.9. Specifically, the hot stream inlet temperature ($T_{h,in}$) is kept constant at 175°C while the cold stream outlet temperature ($T_{c,out}$) is increased from 85° to 115°C. The temperatures for the other end-point are calculated using expressions from Problems 6.B and 6.C. The number of transfer units is $N = UA/MC_{p,cold} = 0.785$, the heat capacity flow rate ratio is $R = MC_{p,cold}/MC_{p,hot} = 2$, and the temperature effectiveness is calculated to be $P = 0.324$. From the definitions of P' and R, $(115 - T_{c,in})/(175 - T_{c,in}) = 0.422$ and $(175 - T_{h,out})/(115 - T_{c,in}) = 2$. The temperatures for the other terminal are therefore found to be $T_{c,in} = 71.28°C$ and $T_{h,out} = 87.56°C$.

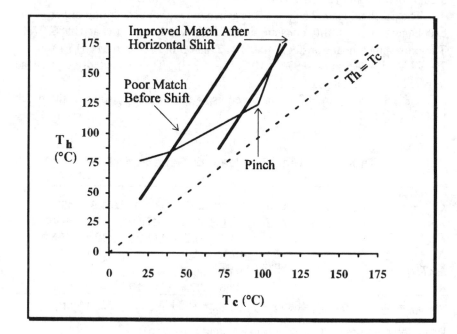

Figure 7.9 The driving force plot (DFP) shows a poorly aligned match, which may be improved by shifting the associated line segment for a better fit.

The improvement may be quantified in terms of $\alpha_{e,h}$, $\alpha_{e,P}$, and $\Delta T_{min,rp}$ (by repeating the calculations in step 2). Unit 3 after shifting yields $\alpha_{e,h} = 725.6/(253.5 + 504.4) = 0.96$, $\alpha_{e,P} = 780.6/(304.9 + 543.2) = 0.92$, and $\Delta T_{min,rp} = 18.67°C$. There is an overall improvement in unit 3, which is accepted at this stage.

Exchanger 2 is represented on Figure 7.10 by a line segment with end-points having coordinates of (40, 98) and (112, 125). Improving the alignment of the exchanger by changing the heat capacity flow rate ratio (slope of the line segment) involves a stream split and may be avoided.   As in Example 7.3, consider shifting the cold terminal of the line segment vertically downwards. Specifically, the cold stream inlet temperature $(T_{c,in})$ is kept constant at 40°C while the hot stream outlet temperature $(T_{h,out})$ is decreased from 98° to 91.55°C (say, a halfway vertical shift).   The temperatures for the other end-point are then calculated and found to be $T_{h,in}$ = 115.84°C and $T_{c,out}$ = 104.77°C.   The partial vertical shift of unit 2 yields $\Delta T_{min,rp}$ = 33.83°C (a reasonable improvement) and insignificant changes in α_e.   Unit 2 is accepted at this stage.

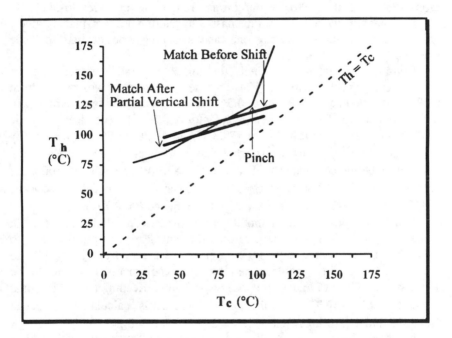

Figure 7.10 The match alignment may be slightly improved on the driving force plot (DFP) by shifting the cold terminal of the exchanger vertically downwards.

Step 4. Adjusting the Pressure Drops for Units

Polley and Panjeh Shahi (1990) recommend the uniform use of stream pressure drop.   The ideal pressure drop across a unit may be obtained from:

$$\Delta P_i = \alpha \, (A_{unit}/A_{m,j}) \, \Delta P_j \tag{7.5}$$

Equation 7.5 yields

$$\Delta P_i = 0.7615 \,(398.8/298.9) \,6.68 \quad = 6.79 \text{ kPa (tubeside for unit 2)},$$
$$\Delta P_i = 0.7615 \,(398.8/529.2) \,11.65 \;= 6.69 \text{ kPa (shellside for unit 2)},$$
$$\Delta P_i = 0.7615 \,(304.9/481.7) \,4.92 \quad = 2.37 \text{ kPa (tubeside for unit 3)},$$
$$\text{and} \quad \Delta P_i = 0.7615 \,(304.9/251.5) \,29.14 = 26.9 \text{ kPa (shellside for unit 3)}.$$

Comparison with the existing pressure drops in Table 7.7 shows that the shellside pressure drop for unit 2 and the tubeside pressure drop for unit 3 need to be decreased. A variety of options is available for modifying the pressure drops (Polley et al., 1990): reconfiguring from shells in series to shells in parallel; changing the number of tube passes by altering headers, placing a new unit in parallel with an existing one, and swapping tubeside and shellside streams.

Consider placing the two shells for unit 2 in parallel (rather than the present arrangement in series). The tubeside pressure drop will reduce from 6.68 kPa to 0.96 kPa (by a factor of $0.5^{2.8}$) and the shellside pressure drop will reduce from 11.65 kPa to 1.66 kPa (by a factor of $0.5^{2.81}$). The clean shellside heat transfer coefficient drops from 778.8 W/(m^2 °C) to 531.9 W/(m^2 °C) (by a factor of $0.5^{0.55}$), giving a dirty shellside coefficient of 181.5 W/(m^2 °C). The dirty tubeside coefficient turns out to be 183.1 W/(m^2 °C). Keeping the cold terminal of unit 2 at 91.55°C/40°C, the other terminal is calculated to be at 112.26°C/95.23°C (using $N = 2.423$, $R = 0.375$, and $P = 0.764$).

Further, consider a similar modification (shells in parallel rather than in series) for unit 3. The tubeside pressure drop will decrease from 4.92 kPa to 0.71 kPa and the shellside pressure drop will decrease from 29.14 kPa to 4.16 kPa. The dirty shellside and tubeside coefficients turn out to be 181.4 W/(m^2 °C) and 173.7 W/(m^2 °C) respectively. Keeping the hot terminal of unit 3 at 175°C/115°C, the other terminal is calculated to be at 105.72°C/80.36°C (using $N = 1.353$, $R = 2$, and $P = 0.366$).

For both units, the modifications lead to shorter line segments (Figure 7.11) and better fits to the DFP. These improvements may be confirmed through RPA.

Step 5. Matching the Remaining Streams
    The existing exchangers were modified in the previous steps for better thermal positioning and better use of pressure drops. To ensure that these well-placed matches are not disturbed, the portions of the streams associated with these exchangers may be excluded in the remainder of the synthesis. Figure 7.12 shows the remaining problem after excluding the modified unit 2 (i.e., stream H2 from 112.26° to 91.55°C and stream C4 from 40° to 95.23°C)

and the modified unit 3 (i.e., stream H1 from 175° to 105.72°C and stream C3 from 80.36° to 115°C).

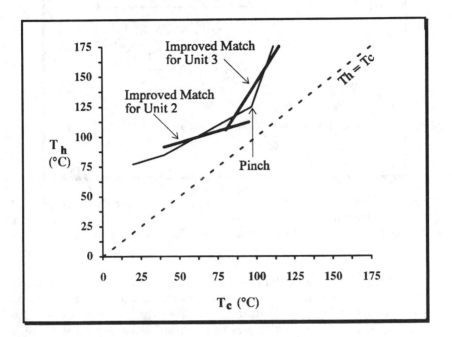

Figure 7.11 The better use of pressure drops leads to improvements in the alignment on the DFP for both units, caused by a reduction in their heat loads.

The pinch tableau for this remaining problem at a utility consumption of 885 kW (for hot utility) and 805 kW (for cold utility) shows a $\Delta T_{min,rp}$ of 18.67°C and a pinch at 125°C/106.33°C. With the identification of the pinch division, the design is completed by conventional design methods maintaining compatibility with the existing structure. A possible design appears in Figure 7.12.

Step 6. Evolving to the Final Retrofit Design
Figure 7.12 shows the need for a new heater (85 kW) and installation of new area in four locations. Network simplification may be achieved by reducing the number of new units through conventional evolution. There will be a sacrifice to some extent in terms of the area requirements.

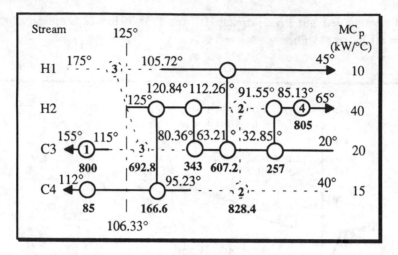

Figure 7.12 The design for the remaining problem after excluding the modified existing exchangers (shown in the dashed sections) is completed by conventional design methods. (Temperature in °C and load in kW)

The heater of 85 kW is eliminated by shifting the load to the other heater (unit 1). The match load of 166.6 kW will now increase to 251.6 kW whereas the match load of 343 will decrease to 258 kW. The two matches between streams H2 and C4 in Figure 7.12 may be combined into a single match with a load of 1080 kW. This will correspond to the existing unit 2 requiring additional shells. Similarly, the two matches between streams H2 and C3 may be consolidated into a single new unit with a load of 515 kW. However, the two matches between streams H1 and C3 cannot be merged due to insufficient temperature driving forces. At this stage, the existing exchangers in their new positions are rated to check for proper operation. Although unit 2 shows satisfactory performance (ΔP_t = 0.96 kPa, ΔP_s = 1.66 kPa, h_t (clean) = 503 W/(m^2 °C), h_s (clean) = 556 W/(m^2 °C), F = 0.84), the value of the F-correction factor for unit 3 is found to be less than 0.75. With its load decreased from 692.8 kW to 662 kW, F improves to 0.78. Correspondingly, the match load of 607.2 kW increases to 638 kW. Unit 3 now shows satisfactory performance (ΔP_t = 0.71 kPa, ΔP_s = 4.16 kPa, h_t (clean) = 518.6 W/(m^2 °C), h_s (clean) = 487.1 W/(m^2 °C), F = 0.78). The final retrofit design appears in Figure 7.13.

Step 7. Rapid Design of New Exchangers
The rearrangement of the shells from series to parallel for units 2 and 3 has made pressure drops available on every stream for the new exchangers.

These may be fully utilized by units N1, N2, and N3 in Figure 7.13 through the rapid design algorithm.

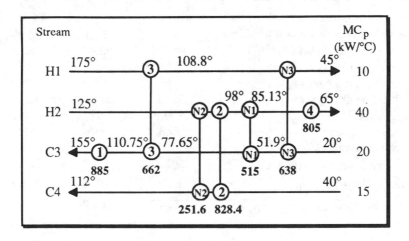

Figure 7.13 The final network for the retrofit obtained by evolution satisfies the pressure drop constraints as well as the economic constraints. (Temperature in °C and load in kW)

The design of each unit (new exchanger and additional shells) is undertaken by the RDA as in Example 6.4. All exchangers use tubes of 19.1 mm o.d. and 15.4 mm i.d., laid out in 25.4 mm square pitch. The hot streams are placed on the shellside and the cold streams on the tubeside. The baffle cut is 25%. The dirt factor allowances on each side of the new exchangers are chosen to be 0.0015 m^2 °C/W. Table 7.13 shows the results based on Kern's method.

Step 8. Analysis of Final Retrofit Design
The final design for the retrofit in Figure 7.13 saves 515 kW of utility and requires an additional area of 309 m^2. Thus, the retrofit project gives a savings of 66.95 x 10^3 $/yr and requires an investment of 130.29 x 10^3 $ for a payback of 1.95 years. Importantly, the pressure drop constraints are satisfied and, therefore, there are no hidden costs due to pump replacement. In contrast to Section 7.2, more realistic values for the heat transfer coefficients are used at all stages. Also, the designs of the new units are simultaneously obtained, bridging the gap between the systems engineering and the detailed engineering to a large extent.

Table 7.13 Designs of New Units by RDA (Kern)

Quantity	Unit N1	Unit N2	Unit N3
Tubeside pressure drop (ΔP_t in kPa)	1.88	5.72	2.33
Shellside pressure drop (ΔP_s in kPa)	6.71	3.28	24.99
Actual exchanger area (A in m^2)	125.78	82.42	100.83
Number of shell passes (NS)	1	1	2
Number of tube passes (NT)	4	4	2
Shell diameter (D_S in mm)	666.0	476.1	435.6
Tube length (L in m)	3.882	4.977	3.637
Number of tubes per shell (N_t)	540	276	231
Number of baffles (N_b)	13	11	25
Tubeside coefficient - dirty (W/m^2 °C)	306.4	356.7	322.6
Shellside coefficient - dirty (W/m^2 °C)	385.8	376.5	375.4
Overall coefficient - dirty (W/m^2 °C)	170.78	183.2	173.5

7.4 DEBOTTLENECKING

The discussion thus far has concentrated on energy saving retrofits. Plant debottlenecking is different in that its objective is increased throughput, subject to an investment ceiling (payback period is usually not an economic criterion here). Ahmad and Polley (1990) argue that an effective strategy is to avoid the purchase of extra capacity for expensive equipments (e.g., pumps, compressors, furnaces, refrigeration systems) by dexterous adjustments in the heat recovery system.

7.4.1 Targeting for Debottlenecking

Ahmad and Polley (1990) suggest a two-step, pressure-drop-based targeting procedure for plant debottlenecks: first, a process simulation at the desired increased throughput with additional utility usage (temporarily assumed) to obtain the required process temperatures, and then an improvement in the energy recovery using the area-energy plot considering the additional utility requirement (assumed earlier) as the scope of an energy saving retrofit.

Example 7.6 Establishing Debottlenecking Targets
For the network in Figure 1.2 (with the stream data in Case Study 4S1 corresponding to the current operation), it is desired to increase the throughput

by 5%. Determine a target ΔT_{min} for this debottlenecking study. For all streams, the heat transfer coefficients (including dirt factors) are 200 W/(m^2 °C) and the dirt factors are assumed constant at 0.00015 m^2 °C/W. The cost data are the same as given earlier in Example 7.1. The physical properties for the four streams are given in Table 6.4. The specifications of the existing exchangers are shown in Table 7.7.

Solution. In practice, the reactor and distillation column in Figure 1.1 would need to be simulated for the increased throughput to determine changes in stream temperatures, compositions, and flow rates. This would indicate possible bottlenecks in the column and reactor that need to be handled independently. Here, it is assumed that only flow rates in the network in Figure 1.2 are increased by 5% and temperatures remain unaffected. From Table 7.7, the process-to-process exchanger area currently existing is 703.7 m^2 (i.e., 398.8 + 304.9).

Step 1. Simulation of Existing Network for Increased Throughput
The increase in the velocity of each stream by 5% leads to an increase in the clean heat transfer coefficients by a factor of $(1.05)^{0.55}$ for the shellside and $(1.05)^{0.8}$ for the tubeside. Thus, the stream heat transfer coefficients (including fouling) change to 205.3 W/(m^2°C) (shellside) and 207.7 W/(m^2°C) (tubeside).

The increased heat transfer coefficients and flow rates may be used with the expressions from Problems 6.B and 6.C to compute the modified heat loads. For unit 3, the heat capacity flow rate ratio is $R = MC_{p,cold}/MC_{p,hot} = 2$ (unchanged), the number of transfer units is $N = UA/MC_{p,cold} = 0.75$, and the temperature effectiveness is calculated to be $P = 0.319$. From the definitions of P' and R, $(T_{c,out} - 20)/(175 - 20) = 0.418$ and $(175 - T_{h,out})/(T_{c,out} - 20) = 2$. This gives $T_{c,out} = 84.78$°C and $T_{h,out} = 45.44$°C. Thus, the heat load of unit 3 has increased from 1300 kW to 1360.4 kW. A similar analysis shows the heat load of unit 2 to have increased from 1080 kW to 1129.4. The revised network in Figure 7.14 assumes that 79.2 kW of additional utility is supplied to meet the target temperatures and account for the 5% increase in throughput.

Step 2. Generation of the Area-Energy Plot
The area targets are generated for different energy levels to plot the ideal area-energy target curve as in Figure 7.15. The ideal target curve for the increased throughput lies above the corresponding curve for the existing throughput. This is because the heat recovery area is inadequate to operate at the higher throughput and therefore must be compensated for through the use of additional utility. In other words, every point on the area-energy plot

(including the existing design) moves horizontally to the right when the flow rates are increased, due to the extra utility required for a given area.

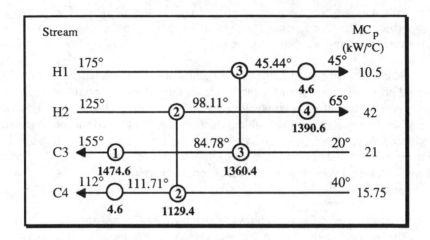

Figure 7.14 The existing network at the increased throughput level exhibits higher pressure drops and higher utility requirements. (Temperature in °C and load in kW)

Step 3. Area Requirement to Overcome Utility Load Limitations

If the operation for the increased throughput is to be conducted at the existing utility level (1400 kW of hot utility), then the energy savings to be targeted is 79.2 kW. The area efficiency of the existing network at the increased throughput (point I in Figure 7.15) may be determined using an extended pinch tableau at a hot utility level of 1479.2 kW. It yields $\Delta T_{min} = 42.96°C$ and $A_{ideal,1-2} = 492.7$ m^2 (point A). As $\alpha_e = 492.7/703.7 = 0.7$, the retrofit curve (IH in Figure 7.15) based on the $\Delta\alpha = 1$ criterion may be used. Once it is determined that $A_{ideal,1-2}$ is 537.8 m^2 at a hot utility level of 1400 kW (point E), the additional area (A_H) required for the increased heat recovery is targeted to be 45.16 m^2. This new area has an installed cost of about 46.4 x 10^3 \$, which may be compared with the cost of installing additional equipment to overcome the utility load limitations.

Step 4. Area Requirement to Overcome Pressure Drop Constraints

If pressure drop constraints exist for the operation at the increased throughput, then additional area may be installed in parallel with existing units. Equations 6.33 and 6.34 may be used to quantify these effects. Using the notation from Figure 7.15, let subscript I denote the operation of the stream with existing area and increased throughput. Also, let subscript P denote the operation of the stream with increased area and increased throughput.

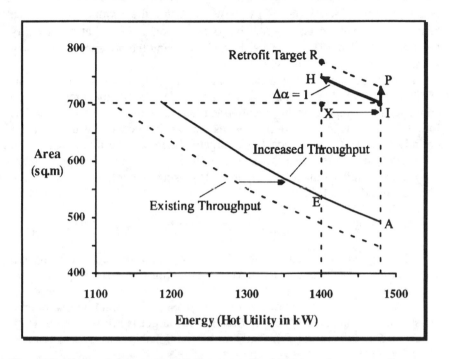

Figure 7.15 For the increased throughput, the ideal target curve on the area-energy plot moves to the right and the existing design X moves to point I. The increased area requirements to overcome utility load limitations (H) and pressure drop constraints (P) are added to obtain the retrofit target (R).

The reduction in the heat transfer coefficient (from h_I to h_P) caused by the area (A_H) added for increased heat recovery is given by (Ahmad and Polley, 1990)

$$(h_P/h_I)^m \, (A_{existing} + A_H)/A_{existing} = \Delta P_P/\Delta P_I \qquad (7.6)$$

where exponent m is 3.5 (for tubeside) and 5.1 (for shellside). Because pressure drop varies as the square of the throughput, the reduction in the heat transfer coefficient for this case study is given by

$$(h_P/h_I)^m = 703.7/(703.7 + 45.16) \, (1/1.05)^2 = 0.852.$$

The above calculation presumes that there is no pressure drop available in excess of the existing specification. Thus, the stream heat transfer coefficients (increased in step 1 for the increased throughput) now reduce to 199 W/(m^2°C) (shellside) and 198.4 W/(m^2 °C) (tubeside). Using these values (h_p) for the heat transfer coefficients, $A_{ideal,1\text{-}2}$ is found to be 511.99 m^2 at a hot utility level of 1479.2 kW. Thus, this area helps in reducing the pressure drop but not in increasing the heat recovery.

At first glance, it may appear that the existing area is sufficient (as 703.7 m^2 > 511.99 m^2) and may be used to overcome the pressure drop constraints by simple rearrangement, given that the existing area efficiency is low (0.7). This is usually not possible as the existing network structure is not entirely consistent with the ideal area target. Therefore, the more appropriate approach would be to target the additional area (A_P) required for the pressure drop compensation (line IP in Figure 7.15) using the following expression (Ahmad and Polley, 1990):

$$A_P = A_{ideal,h_p}/\alpha_e - A_{existing} = 511.99/0.7 - 703.7 = 27.71 \text{ m}^2 \quad (7.7)$$

This new area has a cost of about 11.1 x 10^3 \$, which may be compared with the total cost (capital and power consumption) of installing additional equipment (pumps/compressors) to overcome the pressure drop constraints.

The extra area requirements for utility limitations and pressure drop constraints are additive. Thus, the final retrofit target is given by point R on Figure 7.15 and totally requires 72.87 m^2 of additional area at an installed cost of 57.5 x 10^3 \$.

If the project is considered an energy saving retrofit, it is seen that the payback is 5.58 years (i.e., 57500/[79.2 x 130]). The high payback is due to the fixed cost coefficient being high (30000 \$). An attractive option is to expand the project scope for greater energy savings.

The above debottlenecking targeting procedure employs the suggestions of Ahmad and Polley (1990) and uses the area targeting based on constant heat transfer coefficients. The use of area targeting based on specified pressure drops (Section 6.5) may be explored to invent an improved procedure.

The debottlenecking targets are important since they allow the process engineer to assess alternatives prior to detailed retrofit design. Cost comparisons for additional area, installation of utility equipment, and installation of pumps/compressors may be quickly attempted. After utilization of the existing equipment (pumps, compressors, and utility exchangers) to their maximum limits, new installations (involving addition to existing equipment or replacement for increased throughput) could be considered. Thus, the optimum strategy may be determined, which could involve a combination of extra utility, new pumps/compressors, and additional area. Often, the attractive option is to install additional area because it is cost effective (in terms of both energy and capital).

7.4.2 Retrofit Design for Debottlenecking

The retrofit design procedure for debottlenecking is identical to that adopted in the earlier revamps. The example below provides the final retrofit design that meets both the pressure drop and utility constraints.

Example 7.7 Revamp of an Existing Network for Increased Throughput

For Case Study 4S1, develop an improved network (debottlenecking retrofit) starting with the existing network in Figure 1.2. The objective is to increase the throughput by 5% maintaining the pressure drops and utilities at their existing levels.

Solution. The steps are identical to those in Example 7.5; therefore, the detailed procedure is not given below.

The retrofit targeting in Example 7.6 suggests the network be designed at a target ΔT_{min} of 40.81°C for a hot utility of 1400 kW and a cold utility of 1316 kW. This would require an investment of 57.5 x 10^3 $ (an extra area of 72.87 m^2).

The existing exchangers may be assessed using RPA. For unit 2, RPA(h) yields $\alpha_{e,h}$ = 510.4/(359 + 236.8) = 0.86 and $\Delta T_{min,rp}$ = 4.8°C. Similarly, RPA(h) for unit 3 in Figure 1.2 yields $\alpha_{e,h}$ = 510.4/(256 + 359.7) = 0.83 and $\Delta T_{min,rp}$ = 26.5°C. Both units could be improved through suitable thermal repositioning using the DFP.

For a utility consumption of 1400 kW, the pinch is at 125°C/84.19°C. Both units 2 and 3 exhibit cross-pinch heat transfer, which is undesirable. Based on the insights from Example 7.5, adjustments may be made for the existing units to improve their use of temperature driving forces as well as pressure drops.

Consider placing the two shells for unit 2 in parallel (rather than in series). The pressure drops and heat transfer coefficients on both the shellside and tubeside will reduce. With the hot terminal of unit 2 kept at 125°C/112°C, the other terminal is calculated to be at 114.43°C/83.82°C (using N = 1.588, R = 0.375, and P = 0.684). Next, consider a similar modification (shells in parallel rather than in series) for unit 3. Keeping the hot terminal of unit 3 at 175°C/88.33°C, the other terminal is calculated to be at 86.28°C/43.97°C (using N = 0.91, R = 2, and P = 0.339).

With these adjustments in the existing exchangers, units 2 and 3 in their new positions are rated. They show satisfactory performance. For unit 2, ΔP_t

= 1.05 kPa, ΔP_s = 1.81 kPa, h_t (clean) = 523.1 W/(m^2 °C), h_s (clean) = 571.2 W/(m^2 °C), and F = 0.86. For unit 3, ΔP_t = 0.77 kPa, ΔP_s = 4.54 kPa, h_t (clean) = 539.2 W/(m^2 °C), h_s (clean) = 500.4 W/(m^2 °C), and F = 0.79. Then, a possible final design may be developed as in Figure 7.16.

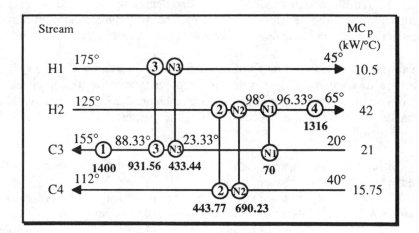

Figure 7.16   The final network for the increased throughput shows that additional area may be placed in a cost-effective manner to overcome the pressure drop constraints as well as the utility equipment limitations. (Temperature in °C and load in kW)

Figure 7.16 shows the need for a new match (N1) and additional shells (N2 and N3) on units 2 and 3. The rearrangement of the shells from series to parallel for units 2 and 3 has made pressure drops available for the new exchangers. Using the rapid design algorithm, these pressure drops may be fully utilized by units N1 (ΔP_t = 0.58 kPa and ΔP_s = 0.91 kPa), N2 (ΔP_t = 5.63 kPa and ΔP_s = 8.93 kPa), and N3 (ΔP_t = 3.57 kPa and ΔP_s = 24.6 kPa).

The area requirements for N1, N2, and N3 are 1.43 m^2, 31.66 m^2, and 35.61 m^2 respectively. Thus, the design requires 68.7 m^2 of total retrofit area at a cost of 56.9 x 10^3 $, which is in agreement with the targeted predictions. Note that unit N1 is too small to be practical during actual implementation; therefore, the scope of the retrofit project should be expanded in this case from merely debottlenecking to enhanced energy savings.

The last two chapters have stressed that pressure drop (and the associated power consumption) should be treated as a resource of great importance and that it is insufficient to concentrate on merely energy and area. In this context, retrofitting involves using existing resources (energy, area, and pressure drop) as far as possible in an efficient manner before adding new resources. Debottlenecking assumes special importance when certain resources (e.g., equipments like pumps and furnaces in oil refineries) are limiting. The cost-effective solution is sought, which requires use of the "right" resource (e.g., additional area rather than a new pump or furnace) or the optimal mix of a variety of resources.

QUESTIONS FOR DISCUSSION

1. What factors of importance are not included in the retrofitting analysis presented in this chapter? Specifically, discuss the effect of the following: nonuniform exchanger specifications (different materials of construction, pressure ratings, and types), widely different heat transfer coefficients, plant layout (space constraints and physical distances between streams), auxiliary equipment (pipes and valves), fouling, operability, controllability, maintenance costs, multiple utilities, and process modifications.

2. Does the ideal target curve on the area-energy plot effectively define the maximum number of units required to reach the area target? Does this maximum number correspond to $N_{u,vht}$ (see Equation 1.28)?

3. What is the nature of the area-energy plot for threshold problems? Is there a threshold value of energy below which the EAT curve in Figure 7.2 becomes vertical (and there is no area-energy tradeoff)? Does the vertical portion of the ideal target curve correspond to an area-unit tradeoff? What are its implications for retrofitting?

4. Describe how a plot of the theoretical limit to the minimum payback period may be generated as a function of ΔT_{min}. Would the extra retrofit area in this case be given by $(A_{AT} - A_{existing})$, where A_{AT} is the area at a point on the AT portion of the curve in Figure 7.2? Can this curve itself be used as a path for retrofitting or would it be too over-optimistic?

5. Is establishing retrofit targets vital? Why? Specifically, what would happen if the retrofit design in Example 7.3 was conducted at a target ΔT_{min} of 13°C (rather than 20°C)?

6. Is it possible to devise a continuous retrofit targeting algorithm (along the lines discussed in Chapter 2)? Would this have any advantages?

7. In some cases, retrofit targeting based on the "constant α" and "$\Delta\alpha = 1$" criteria may be conservative. Discuss alternative paths for

retrofitting that may result in better targets. Is it possible to consider the area distribution of the existing network and compare it with the "ideal" matchwise area distribution (see Table 1.17)? Note that the deficiency with α-based retrofit targeting methods stems from the fact that α is defined in terms of the total area and therefore does not capture the matchwise area distribution.

8. Can there be crisscrossing without cross-pinching? Can there be cross-pinching without crisscrossing? Is the compensation principle (see Section 5.4) compatible with the crisscross principle and the pinch principle?

9. The DFP provides a useful guide to the selection of matches, loads, and stream-split ratios. However, it is not an appropriate tool when the stream heat transfer coefficients differ considerably since smaller driving forces may be used for higher heat transfer coefficients and vice versa. Comment.

10. Comment on the following strategies that are alternatives to those discussed in this chapter and are sometimes used to develop a retrofit network design.
- Cross-pinch exchangers are first eliminated, and then compatibility with the existing network is sought using heat load loops and paths.
- An optimal grassroots design is generated, which is evolved to the existing design to obtain intermediate networks that provide alternative revamp schemes; then, heat load loops/paths are used to restore the *UA*-values of the existing units and the different retrofit options are evaluated.

11. List the different options available for modifying an existing HEN (e.g., adding more area (shells) on an existing exchanger, adding a new exchanger, repiping to allow different stream(s) to transfer heat in an existing exchanger, and relocating existing exchangers with no necessity for repiping). What are the costs associated with each of these modifications?

12. Can the retrofit design in Figure 7.7 be modified (say, by increasing the load on unit 3) for further energy savings? Note that the payback period of 1.76 years has the scope to be increased to the allowable two years.

PROBLEMS

7.A Retrofit Targeting Based on Constant α Criterion
For the Case Study 5T1 (Tjoe and Linnhoff, 1986), determine the target ΔT_{min} (to an accuracy of 1°C) for an energy saving retrofit following the procedure used in Example 7.2. The desired payback period is two years.

Given: Existing network uses 17597 kW (hot utility), 15510 kW (cold utility), and 2312.24 m^2 (area). Average area is 578 m^2 (per match) and 289 m^2 (per shell).

Answers:

$\alpha = 0.9025$; $\Delta T_{min} = 19°C$

Savings = 328.63 x 10^3 £/yr (for 5186.9 kW)

Investment = 652.77 x 10^3 £/yr (for 2342.71 m^2)

7.B Retrofit Targeting Based on $\Delta\alpha$ Criterion

For the Case Study 9T1 (Tjoe and Linnhoff, 1986; 1987), examine the scope for an energy saving retrofit following the procedure used in Example 7.2. The desired payback period is two years.

Given: Existing network uses 27100 kW (hot utility), 23495 kW (cold utility), and 4011 m^2 (area).

a) Determine the target ΔT_{min} based on the constant α criterion.

b) Determine the target ΔT_{min} based on the $\Delta\alpha = 1$ criterion.

c) What would be the payback period if a target ΔT_{min} of 10°C is used (along with the constant α criterion)?

Answers:

a) $\alpha = 0.8588$; $\Delta T_{min} = 26°C$

Savings = 207.36 x 10^3 £/yr (for 3600 kW)

Investment = 422.64 x 10^3 £/yr (for 1677.14 m^2)

b) $\Delta T_{min} = 22°C$

Savings = 281.09 x 10^3 £/yr (for 4880 kW)

Investment = 559.76 x 10^3 £/yr (for 2221.28 m^2)

c) Payback period = 4.3 years

Savings = 525.31 x 10^3 £/yr (for 9120 kW)

Investment = 2256.36 x 10^3 £/yr (for 8953.83 m^2)

7.C Energy Saving Retrofit Targeting of a Refinery

Farhanieh and Sunden (1990) have discussed the retrofitting of a heat exchanger network for a refinery in Goteborg, Sweden. For the Case Study 8F1, examine the scope for an energy saving retrofit following the procedure used in Example 7.2.

Note that the stream heat transfer coefficients are assumed to vary linearly between the supply and target temperatures. One approach to account for the variation of heat transfer coefficients with temperature is to segment streams into substreams with appropriate average h-values.

Given: Existing network uses 3062 kW (cold utility) and 2882 m^2 (area).
   a) Determine the ideal target curve on the area-energy plot.
   b) Determine the savings-investment plot for retrofitting based on the $\Delta\alpha$ = 1 criterion. Assume the capital cost of exchanger area (in SEK) = $12500A^{0.65}$ (A is area in m^2) with no maximum area for match/shell.
   c) How can the linear variation of the heat transfer coefficients be directly incorporated into the procedure discussed in Example 1.4 (without subdividing the streams into substreams)?

Answers:

ΔT_{min}	Energy Target	Area Target	Savings	Investment $\Delta\alpha= 1$
5	1799.18	7533.22	2.526	3.149
10	2001.73	5541.93	2.121	2.253
15	2208.98	4476.54	1.706	1.684
20	2435.03	3740.55	1.254	1.222
25	2661.08	3223.73	0.802	0.830
30	2887.13	2833.08	0.350	0.447

Units: temperature in °C, energy (cold utility) in kW, area in m^2, savings in 10^6 SEK/yr, and investment in 10^6 SEK.

7.D Debottlenecking Targeting of a Crude Distillation Unit

Ahmad and Polley (1990) have discussed the debottlenecking of a crude distillation unit operating under winter conditions in Europe. For the Case Study 8A1, examine the scope for a 10% increase in throughput following the procedure used in Example 7.6. Note that the MC_p values for Case Study 8A1 in Appendix B reflect the 10% increase. For the increased throughput, 106700 kW of hot utility (from furnace) are required. However, the furnace duty is limited to 100000 kW.

Given: Existing network uses 7445 m^2 (area). Average area is 930 m^2 (per match) and 400 m^2 (per shell).
   a) What is the additional area required to overcome the utility load limitations (or, equivalently, for an energy savings of 6700 kW)? Use the $\Delta\alpha$ = 1 criterion and ignore changes in heat transfer coefficients for simplicity.
   b) What is the additional area required to overcome the pressure drop constraints (or, equivalently, to compensate for the net decrease in heat transfer coefficients by a factor of $[1.1^{0.8} \times 0.88]$ for the cold streams)?

Use Equation 7.7 and ignore changes in heat transfer coefficients for hot streams.

c) What is the total installed cost of the retrofit for the predicted extra area?

Answers:

a) $\alpha = 0.8304$ (existing), $\Delta T_{min} = 38.69°C$, additional area = 1719.86 m^2

b) additional area = 196.56 m^2

c) Cost = 497.8×10^3 $/yr (for 1916.42 m^2)

7.E Pressure-Drop-Based Supertargeting and Use of DFP/RPA for Grassroots Design -- Crude Preheat Train Case Study

Consider the stream data given in Case Study 7S1 for a crude preheat train in an oil refinery.

a) Determine the optimum value of ΔT_{min} by supertargeting based on the limiting pressure drops specified. Also, determine other relevant targets.

Given:

Stream	LN	K	HN	HD	LD	B	C
ΔP (kPa)	13.614	19.154	14.615	8.059	15.154	19.403	124.58
$R_d \, 10^4 (\text{m}^2°C/W)$	11.375	11.265	10.625	6.96	7.98	4.83	8.884
C_p (J/[kg°C])	2390	2700	2608.3	3116.7	2885	3237	2865
ρ (kg/m^3)	747	792	772	866	833	921	825
μ (cP)	0.15	0.35	0.25	0.5	0.45	0.55	0.9
k (W/[m °C])	0.125	0.125	0.125	0.125	0.125	0.125	0.125

b) Design an MER network at the optimum ΔT_{min} obtained in a). What is its synthesis area? Discuss any possible improvements to the network. *Hint:* Use the DFP and RPA to improve exchanger placement and to decide on an appropriate split ratio.

Answers:

a) Optimum $\Delta T_{min} = 16.66°C$, $T_p = 38.33°C$, $N_{u,mer} = 11$,
$Q_{hu,min} = 84002$ kW, $Q_{cu,min} = 3693.8$ kW,
$A_{12} = 15702.88 \text{ m}^2$, $S_{min} = 29$, $TAC = 11299.85 \times 10^3$ $/yr.
Based on the specified pressure drops, the stream heat transfer coefficients are:

Stream	LN	K	HN	HD	LD	B	C
h (W/m^2 °C)	463.99	401.94	427.29	397.38	431.05	518.09	434.01

b) A possible MER design featuring a five-way split appears below:

Above Pinch	Below Pinch
K - Ca (9482.8 kW)	K - CU (1079.6 kW)
LD - Cb (24580.2 kW)	HD - CU (934.6 kW)
HD - Cc (16793.0 kW)	HN - CU (1042.9 kW)
HN - Cd (6970.0 kW)	LN - CU (636.7 kW)
LN - Ce (5194.9 kW)	
B - Ce (59257.2 kW)	

The crude stream is split optimally as follows: Ca (74.9 kW/°C), Cb (137.3 kW/°C), Cc (72.3 kW/°C), Cd (68.4 kW/°C), and Ce (220.3 kW/°C).

The synthesis area is then found to be 18758.76 m^2.

CHAPTER 8

From Combined Heat and Power to Total Site Integration

*It isn't the stuff,
but the power to make the stuff, that is important.*

Richard Feynman

The synthesis of optimal heat and power systems involves appropriately combining power-producing and power-consuming systems with heat recovery networks so as to minimize costs and maximize efficiency.

Appropriate placement of heat engines and heat pumps in process networks and the design procedure for practical systems are discussed first. Then, a case study on heat engine integration is solved using the pinch method (PM) and the operating line method (OLM). The economic aspects of heat pumping are also presented. Finally, the exergy concept and the procedures for three important targets (shaftwork, total sites, and emissions) are explained.

8.1 HEAT ENGINES AND HEAT PUMPS

A heat engine (see Figure 8.1a) is a device which accepts heat Q_1 from a source at temperature T_1, rejects heat Q_2 to a sink at a lower temperature T_2, and generates work W. From thermodynamics,

$$W = Q_1 - Q_2 \qquad \text{first law} \qquad (8.1a)$$
$$W/Q_1 \leq \eta_c \qquad \text{second law} \qquad (8.1b)$$
$$\text{and } \eta_c = 1 - T_2/T_1 \qquad \text{Carnot efficiency} \qquad (8.1c)$$

Because real heat engines are irreversible, an equation introducing a machine efficiency, η_e, for the heat engine may be written as

$$W = \eta_e \eta_c Q_1 \qquad 0 \leq \eta_e < 1. \qquad (8.2)$$

A heat pump (see Figure 8.2a) is a heat engine operating in reverse: it accepts heat Q_2 at temperature T_2, rejects heat Q_1 at a higher temperature T_1, and therefore consumes work W. From thermodynamics,

$$W = Q_1 - Q_2 \qquad \text{first law} \qquad (8.3a)$$
$$W/Q_1 \geq \eta_c \qquad \text{second law} \qquad (8.3b)$$
$$\text{and } \eta_c = 1 - T_2/T_1 \qquad \text{Carnot efficiency} \qquad (8.3c)$$

Equation 8.3b may be rewritten for real (irreversible) heat pumps as

$$W = \eta_c \, Q_1/\eta_e = Q_1/\text{COP} \quad 0 \leq \eta_e < 1 \qquad (8.4)$$

where COP denotes the coefficient of performance. For ideal (reversible) heat engines and pumps, the equalities in Equations 8.1b and 8.3b apply; therefore, $\eta_e = 1$.

8.1.1 Appropriate Placement Relative to the Pinch

Recall that the process acts as a heat sink above the pinch and as a heat source below the pinch. Appropriate placement of a heat engine (Townsend and Linnhoff, 1983) requires that the engine be placed either entirely above the pinch or below it, but not across it (Figure 8.1). Above the pinch (Figure 8.1b), the engine takes heat Q_1 and rejects heat $(Q_1 - W)$ into the sink. Thus, the integrated system uses W units of additional heat in excess of the process requirements. However, it produces an equal amount of work. In other words, heat is converted to shaftwork at a marginal efficiency of 100% because of the integration. Proper placement results in the integrated system generating work from fuel on a one-to-one basis (Townsend and Linnhoff, 1982). Strictly speaking, rather than a complete conversion of heat, the wastage of heat is totally minimized. Essentially, the heat Q_2 rejected by the engine substitutes an equal amount of hot utility. In fact, there is a double benefit (compared to the nonintegrated system in Figure 8.1a) because the same amount of cold utility is also saved. This could lead to a net capital savings due to smaller equipment inspite of the tighter integration (Linnhoff and Turner, 1980, 1981). Of course, this is all possible only if the engine exhaust heat is over the right temperature range. When the engine exhaust is colder than the pinch temperature and the engine is placed across the pinch (Figure 8.1c), the engine rejects heat $(Q_1 - W)$ into a process source. This heat, therefore, simply cascades through the below-pinch subsystem and requires cold utility for its satisfaction. Thus, improper placement yields no benefit from the integration since the performance is no superior to a stand-alone engine, i.e., the total heat requirement and efficiency for the integrated system are no different from those for the separate systems. Below the pinch (Figure 8.1d), the engine is placed such that it absorbs heat Q_1 from the process source and thereby reduces the cold utility demand; again, the engine is properly placed and has a marginal efficiency of 100% as it converts the excess process heat into work rather than waste heat.

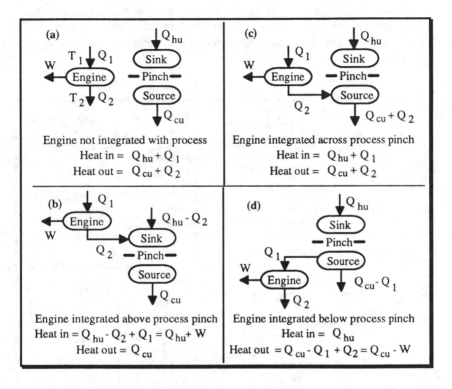

Figure 8.1 For a heat engine and a process to be properly integrated (i.e., lead to 100% efficient work generation), the engine must reject heat above the pinch and accept heat below the pinch.

There is a load limit in terms of the maximum amount of work that can be generated at 100% efficiency. Figure 8.1 shows that the exhaust heat Q_2 from an engine integrated above the pinch can at best replace all the hot utility requirement of the process ($Q_{hu,min}$), and the maximum heat Q_1 absorbed by an engine integrated below the pinch can equal the cold utility requirement ($Q_{cu,min}$). Generation of any additional work is possible only at the stand-alone engine efficiency since this would cause heat flow across the pinch and therefore consume utilities more than the minimum required.

In a few practical heat engines (e.g., open-cycle gas turbines and Diesels), some exhaust heat must be rejected at virtually ambient temperatures. In these cases, proper placement requires the engine to be placed below the pinch (which is at above-ambient temperature). If placed above the pinch, then it is not fully appropriate because the overall efficiency decreases depending on the amount of exhaust heat crossing the pinch and leaving at ambient conditions.

Appropriate placement of a heat pump (Townsend and Linnhoff, 1983) demands that the pump be across the pinch and not entirely above or below it (Figure 8.2).

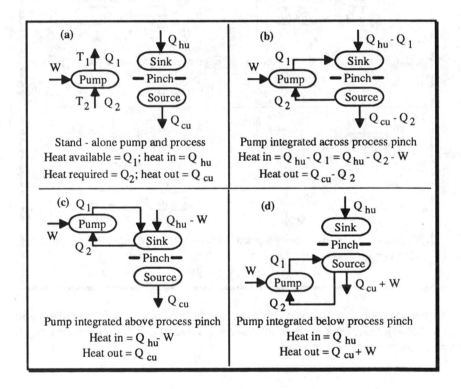

Figure 8.2 For proper integration of a heat pump and a process, the pump must be placed across the pinch, and not entirely on either side of it.

If placed across the pinch (Figure 8.2b), it takes heat Q_2 from the process source below the pinch and rejects heat $(Q_2 + W)$ to the process sink above the pinch. Thus, the work input, W, leads to a reduction of $(Q_2 + W)$ in the hot utility and of Q_2 in the cold utility. Therefore, the stand-alone efficiency is achieved through proper placement, subject to a maximum utility savings dictated by $Q_2 \leq Q_{cu,min}$ and $Q_1 \leq Q_{hu,min}$. Beyond this load limit, there will be a mere degradation of the work inputted to the pump. If placed above the pinch (Figure 8.2c), then the hot utility gets reduced by W units; however, W units of work are also added and so no net energy savings result. Basically, Q_2 units of heat are circulated in a loop and work is converted into heat (hot utility) on a one-to-one basis. This is not meaningful in most practical situations. When placed below the pinch (Figure 8.2d), Q_2 units of heat are circulated in a loop; also, work, W, is added to the heat source and so cold utilities are increased by W units, leading to improper placement and wastage.

8.2 OPTIMAL INTEGRATION OF HEAT ENGINES

All heat engines use a working fluid that absorbs the heat from a source, generates the work while passing through mechanical devices, and rejects the waste heat to a sink. Proper placement (see Figure 8.1) ensures that the work generating efficiency is 100%, but maximizing the amount of work that can be extracted at 100% efficiency requires minimizing the driving force losses between the engine working fluid and the process. This is conveniently done on a temperature-enthalpy diagram by matching the heat-absorbing and heat-rejecting profiles of the engine with the source and sink profiles of the process. The ideal matching possible, based on thermodynamics, is described first and then the practical designs are developed.

8.2.1 Matching Heat Profiles of Engine and Process

Just as there is a load limit, there is a level limit for the integration (Townsend and Linnhoff, 1982). The exhaust heat, Q_2, from an engine can be supplied over a range of temperature levels (rather than at the highest process level) as long as it is confined to the process sink above the pinch. Similarly, the heat, Q_1, can be absorbed by an engine over a range of temperature levels (rather than at the lowest process level) as long as it is confined to the process source below the pinch. However, this does not affect the work generating efficiency, which is always 100% when the engine is appropriately placed. What it does affect is the amount of work that can be generated. To maximize W, Equation 8.2 suggests that $\eta_e \eta_c$ be maximized. The engine cycle efficiency can be increased by decreasing the temperature, T_2, at which the heat is rejected or by increasing the temperature, T_1, at which the heat is accepted. In fact, η_e is maximized by reducing temperature differences between the process and the working fluid in the engine exchangers that reject and accept heat. However, the temperature level chosen for the integration should not cause an infeasibility in the heat cascade. What is the best achievable thermodynamically? Ideally, the exact quantity of heat for each temperature interval in the heat cascade (obtained through the PTA) must be supplied by the engine exhaust above the pinch or absorbed by the engine below the pinch. Then, after the engine is integrated, there are no residual heat flows between intervals and the engine exchangers operate throughout at the minimum allowable temperature difference. Simply speaking, the profiles for the engines must exactly coincide with the process GCC profiles in terms of shifted temperatures.

Consider the GCC for Case Study 4S1 at ΔT_{min} = 20°C (Figure 1.10), which shows a hot utility requirement of 605 kW. This utility requirement can be supplied by the exhaust from a heat engine placed entirely above the pinch. The thermodynamically best fit is given by an engine exhaust stream, whose T-Q profile is countercurrent to and lies completely above the process sink profile (for 100% work generating efficiency). Also, it should be capable of supplying all the necessary process heating (as per the load limit). If $\Delta T_{min,e}$ is the ΔT_{min} contribution assigned to the engine working fluid, then the engine exhaust profile must be everywhere exactly $\Delta T_{min,e}$ above the process sink profile (to maximize engine efficiency and, hence, work output). If $\Delta T_{min,e}$ is chosen to be 10°C (consistent with the global ΔT_{min}), then the exhaust stream fit for Figure 1.10 as per Table 1.1 would have a segment from 175° to 132°C (of 430 kW) and another segment from 132° to 125°C (of 175 kW) in terms of unshifted temperatures. Such a profile comprising parallel segments (shown dotted in Figure 8.3) maximizes work output, but each segment requires a separate power cycle to account for the change in exhaust flow rate. It also usually leads to a complex heat recovery network and, hence, simplification in terms of fewer number of power cycles is desirable in practice. With a small degradation in driving forces (and a consequent reduction in the work output at 100% efficiency), the two-cycle profile can be approximated by a single-cycle profile of 605 kW running straight from 175° to 125°C (Figure 8.3). The matching of an engine heat-absorption profile below the pinch to meet the cold utility demands of the process network follows an analogous procedure.

The best single-cycle or multiple-cycle fit to the process profile may be determined in general by inspection. Multiple cycles help minimize driving forces and typically give superior energy performance over single cycles, but the capital cost and complexity for multiple engine cycles are higher. Therefore, the tradeoff between the revenue gained due to marginal increase in work output and the increase in exchanger area/capital cost must be evaluated.

In cases where the process profile is reentrant (i.e., the GCC has a "pocket"), there exists a local heat source above the pinch or a local heat sink below it. In such cases (e.g., below-pinch region in Figure 1.10), the possibility exists for generating work at 100% efficiency by adding "interprocess" heat engines (Townsend and Linnhoff, 1983).

The local heat sink in the below-pinch region (nonmonotonic portion of Figure 1.10) is reversed in direction (for the convenience of profile matching) and shifted to a position where interprocess temperature differences are maximized (Figure 8.3). The thermodynamic best three-cycle fit for this portion may be approximated by a single-cycle fit on tolerating some loss in work output. Then, two engines may be operated: one between the process source and the reversed nonmonotonic portion and another between the process source and the utility sink. The two engines may accept heat over any temperature range of the process source. This determines their cycle efficiencies.

Now that the ideal thermodynamic fits for heat engines have been explained, attention is next focused on practical heat engines.

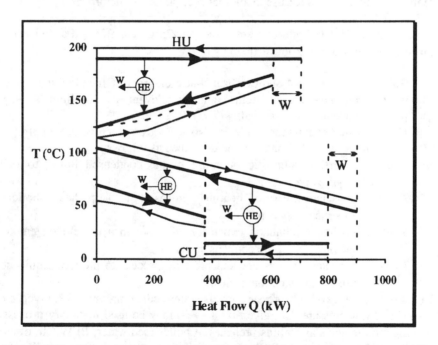

Figure 8.3 Above the pinch, the ideal two-cycle heat engine exhaust profile may be approximated by a single-cycle fit. Below the pinch, more work output is possible by exploiting the local heat sink and using two engines. For clarity, engine profiles are shown in heavy lines in terms of unshifted temperatures.

8.2.2 Practical Heat Engines

Some of the important heat engines are steam turbine Rankine cycles, open-cycle gas turbines, closed-cycle gas turbines, organic Rankine cycles, and Diesels. Of these, the first two are more popular in industry. Their design is given below based on methods discussed by Townsend and Linnhoff (1983). The various engine cycles have definite profiles that impose constraints on the minimum driving force achievable or the maximum heat recovery possible.

Example 8.1 Steam Rankine Cycle Design
    Design a combined heat and power (CHP) system using a steam turbine to obtain 4500 kW of shaft power and meet a process hot utility demand of 1580 kW (for Case Study 4S1t at ΔT_{min} = 20°C). In addition, 2248.4 kg/hr of 1718 kPa steam (at 205°C) and 4733 kg/hr of 414.1 kPa steam (at 145°C) are required to be produced for use in another section (X) of the process.

Given: Turbine inlet conditions are fixed at 475°C and 9750 kPa and each turbine stage has an isentropic efficiency of 0.85.

    Solution. Figure 8.4 shows a schematic diagram of the Rankine cycle (the most popular type of heat engine), and the relevant working fluid (steam) heat profile on a temperature-enthalpy (T-H) diagram.
    The ideal thermodynamic cycle is also shown on a temperature-entropy (T-S) diagram in Figure 8.4 and may be described as follows:

1 → 2:   reversible adiabatic pumping of the condensed liquid to the required pressure;

2 → 5:   constant pressure heating process in a preheater-boiler-superheater system;

5 → 10: reversible adiabatic expansion of vapor in a turbine system to the pressure of the condenser; and

10 → 1: constant temperature and pressure process to produce liquid at point 1 by rejection of heat.

Exhaust steam at two back-pressure levels (intermediate pressure, IP, given by 6 → 7 and low pressure, LP, given by 8 → 9) may be used to satisfy process heating requirements as well as preheat the boiler feed water, BFW. In other words, portions of 6 → 7 and 8 → 9 (excluding segments corresponding to internal-cycle heat transfer with 2 → 3) are steam Rankine cycle exhaust profiles that usually may be matched against process sink profiles.
    The objective in this example problem is first to match profiles and then to calculate the steam turbine cycle heat input/efficiency.

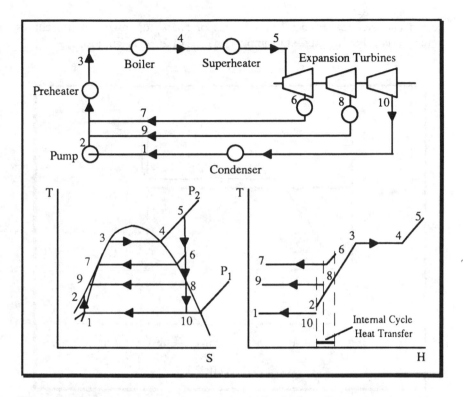

Figure 8.4 The thermodynamics of a steam Rankine cycle may be shown on a T-S diagram and its heat-absorbing and heat-rejecting profiles on a T-H diagram.

Step 1. Profile Matching

In addition to the four streams that constitute Case Study 4S1t, the service duty (in terms of raising intermediate pressure, IP, and low pressure, LP, steam for section X) is incorporated as two more cold streams as given below:

Stream C5 (IP steam @ 205°C): Load = 2248.4/3600 x 1921.4 = 1200 kW
Stream C6 (LP steam @ 145°C): Load = 4733/3600   x 2129.7 = 2800 kW

The GCC for this six-stream data set at ΔT_{min} = 20°C is shown in Figure 8.5. It indicates a threshold problem. Although the total amount of hot utility remains constant (5580 kW) for ΔT_{min} less than 25°C, the shape of the GCC varies with ΔT_{min}. Thus, the choice of ΔT_{min} affects the profile matching and may be critical to the overall CHP system performance. For preliminary design purposes, experience-based values or coarse optimization procedures may be used. For the present example, a ΔT_{min} of 20°C is employed.

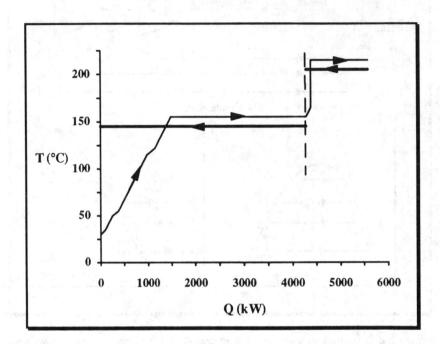

Figure 8.5 The targeting for the steam turbine may be done by matching the heat profiles for the back-pressure steam and the process.

Figure 8.5 shows a heat sink that extends from moderate to ambient temperature. By inspection and based on the LP and IP steam needs of section X, the turbine exhaust steam is placed at two levels on the GCC to meet the hot utility demand. Letting $\Delta T_{min,e} = -10$°C ensures production of back-pressure steam for direct use in section X with a "match ΔT_{min}" of 0°C. The heavy lines in Figure 8.5, which represent the engine profiles in terms of unshifted temperatures as before, show that totally 1300 kW of IP steam and 4280 kW of LP steam may be provided to the process at temperatures of 205°C and 145°C respectively. Thus, 100 kW of IP steam and 1480 kW of LP steam are available to the "4S1t process" at a "match ΔT_{min}" of 10°C.

Step 2. Calculation of Inlet-Outlet Conditions across Each Turbine Stage

Now, steam tables are used to obtain the relevant thermodynamic properties. At the turbine inlet (475°C and 9750 kPa), $H_{in,1} = 3320$ kJ/kg. As the outlet of stage 1 is at 1718 kPa (IP), the isentropic enthalpy change is found to be $\Delta H_{isen,1}$ = 440 kJ/kg. Since the turbine efficiency is 0.85, ΔH_1 = 440 x 0.85 = 374 kJ/kg and $H_{out,1}$ = 3320 − 374 = 2946 kJ/kg.

The outlet of stage 1 is the inlet of stage 2. With the outlet of stage 2 at 414.1 kPa (LP), $\Delta H_{isen,2}$ = 282.6 kJ/kg, ΔH_2 = 282.6 x 0.85 = 240.2 kJ/kg, and $H_{out,2}$ = 2946 − 240.2 = 2705.8 kJ/kg.

Finally, with the outlet of stage 3 at 10 kPa, $\Delta H_{isen,3}$ = 551.8 kJ/kg, ΔH_3 = 551.8 x 0.85 = 469 kJ/kg, and $H_{out,3}$ = 2705.8 − 469 = 2236.8 kJ/kg. The results are summarized in Table 8.1.

Table 8.1 Stagewise Inlet-Outlet Conditions of the Steam Turbine

Stage	T_{in}	T_{out}	P_{in}	P_{out}	H_{in}	H_{out}	S_{in}	S_{out}
1	475	205	9750	1718	3320	2946	6.573	6.698
2	205	145	1718	414.1	2946	2706	6.698	6.796
3	145	46	414.1	10	2706	2237	6.796	7.055

Units: temperature T in °C, pressure P in kPa, enthalpy H in kJ/kg, and entropy S in kJ/(kg °C).

Step 3. Calculation of Desuperheating Duty and Condensate Knockout

At the outlet pressure of each stage, the saturated vapor enthalpy and latent heat of vaporization are obtained from steam tables:

Stage	P_{out} (kPa)	H_{sat} (kJ/kg)	ΔH_{lat} (kJ/kg)
1	1718	2795.4	1921.4
2	414.1	2739.7	2129.7
3	10	2584.3	2392.3

As $H_{out,1} > H_{sat,1}$, the IP steam lies in the superheated region. Desuperheating is carried out to bring it to saturated conditions required for the process. Let $(1 - x)$ kg of superheated IP steam (H = 2946 kJ/kg) be mixed with x kg of feed water at 100°C (H = 418.9 kJ/kg) to give 1 kg of saturated steam at 1718 kPa (H = 2795.4 kJ/kg). By a simple enthalpy balance, the amount of desuperheating duty (x) is obtained as 0.0596 kg of feed water/kg of saturated steam.

As $H_{out,2} < H_{sat,2}$, the LP steam lies in the wet region. The wet LP steam goes to a knockout drum. Let $(1 + c)$ kg of wet LP steam (H = 2705.8 kJ/kg) be separated into c kg of condensate at 414.1 kPa (H = 610 kJ/kg) and 1 kg of saturated steam at 414.1 kPa (H = 2739.7 kJ/kg). As before, by a simple enthalpy balance, the amount of condensate knock out (c) is obtained as 0.0162 kg/kg of saturated steam.

For stage 3, $\Delta H_{subcool}$ = 2584.3 − 2236.8 = 347.5 kJ/kg. The wetness fraction is 18%. The enthalpy change in the condenser is given by ΔH_{cond} = 2392.3 − 347.5 = 2044.8 kJ/kg.

Step 4. Estimation of Power Availability Along with Back-pressure Flows

The flow rates of saturated steam can be determined by dividing the target loads (1300 kW for IP and 4280 kW for LP) by the latent heats. Thus,

Flow of saturated IP steam = 1300/1921.4 = 0.6766 kg/s; and
Flow of saturated LP steam = 4280/2129.7 = 2.0097 kg/s.

These flows must be converted to passout flows corresponding to conditions existing at the turbine exhaust. Therefore,

Flow of superheated IP steam = 0.6766 $(1 - x)$ = 0.6363 kg/s; and
Flow of wet LP steam = 2.0097 $(1 + c)$ = 2.0422 kg/s.

Finally, power availability may be estimated as follows:

Power generated in stage 1 = (2.0422 + 0.6363) 374 = 1001.8 kW;
Power generated in stage 2 = (2.0422) 240.2 = 490.5 kW; and
Total power generated = 1492.3 kW.

Step 5. Calculation of Flow through Economizers to Achieve Target Power

Because the objective is optimal energy performance, the use of economizers may be explored wherever thermodynamically feasible. The energy efficiency of the power-generating cycle can be improved by using exchangers (called economizers) to preheat the boiler feed water (BFW) before entering the steam drum (Figure 8.6). A part of the steam generated in the various headers (corresponding to the different turbine back-pressure stages) with passout flow rates, M_i, could be used in the economizers. A condensate handling system supplies BFW to the furnace after collecting condensate from the process exchangers, the economizers, and the condensing stage of the turbine. It also includes the blowdown from the steam drum (at 300°C) and the make-up water (at 25°C). There is a purge flow at 100°C. The value of ΔT_{min} in both economizers is chosen to be 20°C.

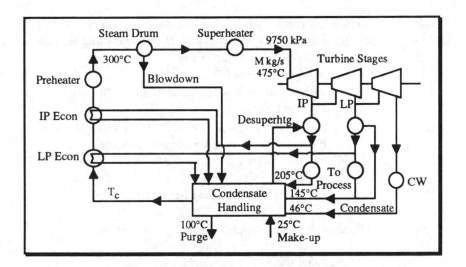

Figure 8.6 The above flowsheet for a practical steam Rankine cycle shows two back-pressure stages and a condensing stage. For better energy performance, economizer-interchangers are incorporated to exchange heat between IP/LP steam and BFW. (Temperature in °C)

A general model may be formulated for such a steam Rankine cycle flowsheet consisting of n turbine stages and m economizers in terms of a set of $(m + 2)$ equations. The unknowns will then be the m passout flow rates to the economizers, M_i; the turbine inlet flow rate, M; and the enthalpy of the BFW returning from the condensate handling system, H_c. The following quantities are required in the formulation:

b = blowdown (as fraction of feedwater) = 0.05

ΔH_j = enthalpy change across j-th turbine stage
         ($\Delta H_1 = 374$ kJ/kg, $\Delta H_2 = 240.2$ kJ/kg, and $\Delta H_3 = 469$ kJ/kg)

$\Delta H_{lat,i}$ = latent heat of steam at i-th turbine stage
         ($\Delta H_{lat,1} = 1921.4$ kJ/kg and $\Delta H_{lat,2} = 2129.7$ kJ/kg)

$\Delta H_{lat,p}$ = latent heat of steam at purge condition
         ($\Delta H_{lat,p} = 2256.5$ kJ/kg at 100°C)

H_b = liquid enthalpy of blowdown stream
         ($H_b = 1343.5$ kJ/kg at 300°C)

$H_{e,i}$ = enthalpy of BFW at outlet of i-th economizer
         ($H_{e,1} = 784.4$ kJ/kg at 185°C and $H_{e,2} = 524.3$ kJ/kg at 125°C)

H_i = liquid enthalpy at the level of steam exiting i-th turbine stage
($H_1 = 874.4$ kJ/kg at 205°C and $H_2 = 609.6$ kJ/kg at 145°C)

H_{mu} = liquid enthalpy of make-up stream
($H_{mu} = 103.3$ kJ/kg at 25°C)

H_n = liquid enthalpy at the level of steam exiting n-th turbine stage
($H_n = 192$ kJ/kg at 46°C)

H_p = liquid enthalpy of purge stream ($H_p = 418.9$ kJ/kg at 100°C)

M_{bi} = steam back pressure flow from i-th turbine stage
($M_{b1} = 0.6363$ kg/s and $M_{b2} = 2.0422$ kg/s)

x_i = desuperheating duty (x taken +ve) or condensate knockout
(x taken −ve) of steam exiting i-th turbine stage
($x_1 = 0.0596$ and $x_2 = -0.0162$ from step 3)

W = shaftwork to be generated by the cycle = 4500 kW

An enthalpy balance around the m-th economizer gives

$$[H_c - H_{e,m}] M/(1 - b) + \Delta H_{lat,m} M_m/(1 - x_m) = 0 \qquad (8.5a)$$

Similar enthalpy balances around the remaining $(m - 1)$ economizers give

$$[H_{e,i+1} - H_{e,i}] M/(1 - b) + \Delta H_{lat,i} M_i/(1 - x_i) = 0 \qquad \text{for } i = 1, \ldots, m - 1 \tag{8.5b}$$

A material balance around the total turbine system gives

$$M - \sum_{i=1}^{m} R_i M_i - \sum_{i=1}^{n-1} R_i M_{bi} - W \bigg/ \sum_{j=1}^{n} \Delta H_j = 0 \quad \text{where } R_i = 1 - \sum_{j=1}^{i} \Delta H_j \bigg/ \sum_{j=1}^{n} \Delta H_j \tag{8.5c}$$

Finally, an enthalpy balance around the condensate handling system gives

$$\frac{H_c M}{1 - b} - \frac{bM}{1 - b} \left[H_b + \left(1 - \frac{H_b - H_p}{\Delta H_{lat,p}} \right)(H_{mu} - H_p) \right] - \sum_{i=1}^{m} R_{Hi} M_i - \sum_{i=1}^{n-1} R_{Hi} M_{bi} - H_n W \bigg/ \sum_{j=1}^{n} \Delta H_j = 0$$

$$\text{where } R_{Hi} = \frac{H_i - x_i H_p}{1 - x_i} - H_n \sum_{j=1}^{i} \Delta H_j \bigg/ \sum_{j=1}^{n} \Delta H_j \tag{8.5d}$$

Substituting appropriate values in Equation 8.5 yields

$$[H_c - 524.3] \quad M/0.95 + 2129.7\, M_2/(1 + 0.0162) = 0$$
(enthalpy balance for LP economizer)

$$[524.3 - 784.4]\, M/0.95 + 1921.4\, M_1/(1 - 0.0596) = 0$$
(enthalpy balance for IP economizer)

$$M - [1 - 374/1083.2]\,(M_1 + 0.6363) - [1 - 614.2/1083.2]\,(M_2 + 2.0422)$$
$$- 4500/1083.2 = 0$$
(material balance for turbine system)

$$H_cM/0.95 - 0.05M/0.95\{1343.5 + [1 - (1343.5 - 418.9)/2256.5](103.3 - 418.9)\}$$
$$- [(874.4 - 0.0596 \times 418.9)/0.9404 - 192 \times 374/1083.2]\,(M_1 + 0.6363)$$
$$- [(609.6 + 0.0162 \times 418.9)/1.0162 - 192 \times 614.2/1083.2]\,(M_2 + 2.0422)$$
$$- 192 \times 4500/1083.2 = 0$$
(enthalpy balance for condensate handling)

The solution to the above set of four equations gives $M = 5.9638$ kg/s, $M_1 = 0.7992$ kg/s, $M_2 = -0.0337$ kg/s, and $H_c = 535.54$ kJ/kg. A negative value for the passout flow is invalid. It suggests that $T_c > 125°C$ and, consequently, the LP economizer is infeasible.

Equations 8.5 are now set up for the case of a single IP economizer to give

$$H_c M - 784.4\, M + 1941.01\, M_1 \quad = 0$$
(enthalpy balance for IP economizer)

$$M - 0.6547\, M_1 - 5.4552 = 0$$
(material balance for turbine system)

$$H_cM/0.95 - 60.906\, M - 836.976\, M_1 - 2346.59 = 0$$
(enthalpy balance for condensate handling)

Solving the above set of three equations gives $M = 5.9575$ kg/s, $M_1 = 0.7672$ kg/s, and $H_c = 534.45$ kJ/kg. The heat load on the IP economizer is 1567.5 kW (i.e., 1921.4 x 0.7672/0.9404). The temperature, T_c, of the BFW returning from the condensate handling system is now obtained as 127.2°C.

Finally, the net heat input to the system is obtained from the amount of external heating required for the BFW from the economizer outlet to the steam turbine inlet. Thus,

$$Q = (H_b - H_{e,m})\, M/(1 - b) + (H_{in,1} - H_b)\, M$$
$$= (1343.5 - 784.4)5.9575/0.95 + (3320 - 1343.5)5.9575 = 15281.1 \text{ kW}.$$

The net heat input must obviously equal the total output (within limits of accuracy) as shown below:

Output from cycle = 4500 (power) + 5580 (to process)
   + 5136.134 (condenser load) + 58.41 (purge)
   = 15274.5 kW.

This gives a cycle efficiency of 29.4%.

Example 8.2 Gas Turbine Cycle Design

Design an open gas turbine cycle to satisfy the shaft power and utility demands specified in Example 8.1. The schematic diagram of the gas turbine cycle with internal cycle heat exchange is shown in Figure 8.7 along with the T-H and T-S diagrams.

Given: ΔT_{min} contribution of exhaust stream $\quad = 20°C$
   Ambient temperature $\quad T_1 = 25°C$
   Ambient pressure $\quad P_1 = 100$ kPa
   Turbine isentropic efficiency $\quad \eta_t = 0.86$
   Compressor isentropic efficiency $\quad \eta_{co} = 0.83$
   Mechanical transmission efficiency $\quad \eta_m = 0.98$
   Combustion chamber temperature $\quad T_4 = 1100°C$
   Combustion chamber pressure loss $\quad \Delta P_{cc} = 5\%$
   Exhaust-side pressure loss in heat exchanger $\Delta P_{he} = 5$ kPa
   Air-side pressure loss in heat exchanger $\quad \Delta P_{ha} = 4\%$
   Mass flow ratio of exhaust to air $\quad M_e/M_a = 1.014$

Solution. Figure 8.7 shows the open cycle gas turbine (which is open to the ambient at the lowest heat rejection temperature). The ideal thermodynamic cycle consists of two isentropic processes and two isobaric processes and may be described as follows:
   $1 \to 2$: isentropic compression of a gas in a compressor;
   $2 \to 4$: heating of the gas at constant pressure, requiring heat input;
   $4 \to 5$: isentropic expansion of gas in a turbine producing work, W; and
   $5 \to 6$: cooling of the turbine exhaust at constant pressure, rejecting heat.
Point 5 is shown at a higher temperature than point 2 (as per the cycle pressure ratio, P_2/P_1) allowing the compressed air to be preheated by the hot turbine exhaust. In other words, the portion $6 \to 1$ (excluding segment $5 \to 6$ corresponding to internal cycle heat transfer with $2 \to 3$) is the gas turbine cycle exhaust profile that is usually matched against process sink profiles.

As in the previous example problem, the objective is first to match profiles and then to calculate the gas turbine cycle heat input/efficiency. To avoid iterative calculations, it is reasonable to assume a constant value of M_e/M_a.

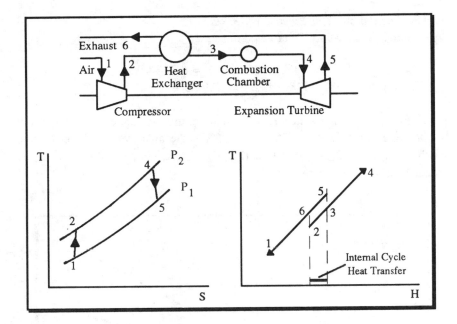

Figure 8.7 The thermodynamics of an open gas turbine (Joule-Brayton) cycle may be shown on a T-S diagram and its heat-absorbing and heat-rejecting profiles on a T-H diagram.

Step 1. Profile Matching

The GCC for the six-stream data set at ΔT_{min} = 20°C from Figure 8.5 is reproduced in Figure 8.8. Given the process GCC profile, the attempt to satisfy the entire heat sink with a single cycle would require a higher temperature level (and consequently a lower engine efficiency). Also, being an open cycle, some exhaust heat would have to be rejected to waste. Typically, there is a tradeoff between the amount of heat rejected to waste and the cycle efficiency, so the determination of the optimum profile is not straightforward and requires a marginal benefits analysis.

Based on a ΔT_{min} contribution of 20°C for the exhaust stream, Figure 8.8 shows the exhaust profile (in heavy line) that provides the closest approach to the process GCC. The two points of closest approach on the exhaust profile are X_1 and X_2 with coordinates (4380, 235) and (1480, 175) respectively. This gives a match ΔT_{min} of 30°C for process-to-exhaust matches. With the exhaust profile determined, the cycle calculations may be performed next.

Step 2. Calculation of the Exhaust Flow Rate
    From the exhaust profile targeted in step 1,

$$M_e C_{pe,E} = (4380 - 1480)/(235 - 175) = 48.33 \text{ kW/°C},$$
$$T_6 = 175 + (5580 - 1480)/48.33 = 259.83°C.$$

Note that T_6 is the highest heat rejection temperature for the exhaust. Over the temperature range T_6 to T_1 (259.83° − 25°C), the mean value of the exhaust gas heat capacity $C_{pe,E}$ is 1.026 kJ/(kg °C). Thus, the mass flow rate of the exhaust gas M_e is 47.1 kg/s.

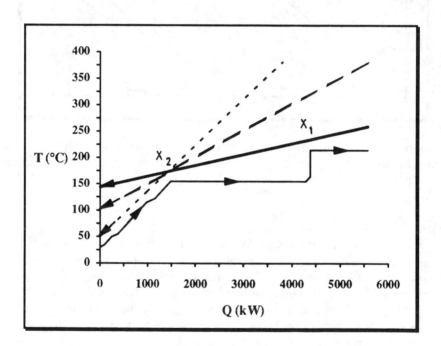

Figure 8.8 The targeting for the gas turbine may be done by matching the heat profiles for the exhaust and the process.

Step 3. Calculation of the Cycle Temperature and Pressure Conditions
    In the internal cycle heat exchanger, the driving force is at a minimum at the hot terminal and is in surplus (by T_s, say) at the cold terminal. Thus,

$$T_3 = T_5 - \Delta T_{min} \tag{8.6a}$$
$$T_2 = T_6 - \Delta T_{min} - T_s \tag{8.6b}$$

As T_s is not known a priori, it may be set to zero as a first approximation. Then, $T_2 = 259.83 - 30 = 229.83°C$.

The temperatures (absolute) and pressures across the compressor and expansion turbine are related by

$$(P_2/P_1)^{1 - 1/\gamma_a} - 1 = \eta_{co} (T_2/T_1 - 1) \tag{8.7a}$$
$$(T_5/T_4 - 1)/\eta_t = (P_5/P_4)^{1 - 1/\gamma_e} - 1 \tag{8.7b}$$

Noting that the mean value of the air heat capacity ratio γ_a is 1.39, the compression stagnation pressure ratio, P_2/P_1, is 5 as per Equation 8.7a. Consideration of the pressure drops gives

$$P_5 = P_1 + \Delta P_{he} \tag{8.8a}$$
$$P_4 = P_2 (1 - \Delta P_{ha} - \Delta P_{cc}). \tag{8.8b}$$

The above equations give $P_5 = 105$ kPa and $P_4 = 500 (1 - 0.04 - 0.05) = 455$ kPa. Since the mean value of the exhaust gas heat capacity ratio, γ_e, is 1.32, the temperature at the turbine outlet T_5 is 746.8°C as per Equation 8.7b. From Equation 8.6a, $T_3 = 716.8°C$.

The enthalpy balance across the exchanger, considering both the exhaust-side and the air-side, yields

$$M_e/M_a \, C_{pe,H} (T_5 - T_6) = C_{pa,H} (T_3 - T_2). \tag{8.9}$$

Using the appropriate mean specific heat capacity ($C_{pe,H} = 1.093$ kJ/kg°C and $C_{pa,H} = 1.084$ kJ/kg°C), the temperature T_2 is found to be 218.9 °C.

Substituting this value in Equation 8.6b gives the update for T_s as 10.92°C. The iterations are summarized in Table 8.2.

Table 8.2 Temperature Calculations for Gas Turbine Cycle

Iteration	T_2	P_2	P_4	T_5	T_3	$T_{2,update}$	T_s
1	229.83	500	455	746.8	716.8	218.91	10.92
2	218.91	466	424.1	761.0	731.0	218.59	11.24
3	218.59	465	423.2	761.4	731.4	218.59	11.24

Units: temperature T in °C and pressure P in kPa.

Calculations for each iteration proceed as follows: T_2 from Equation 8.6b, P_2 from Equation 8.7a, P_4 from Equation 8.8b, T_5 from Equation 8.7b, T_3 from Equation 8.6a, and update for T_2 from Equation 8.9.

Step 4. Calculation of Quantities of Interest
    The total power output is the difference between the turbine work and the compressor work. Thus,

$$W = M_e \left[C_{pe,T}(T_4 - T_5) - C_{pa,C}(T_2 - T_1)/(\eta_m M_e/M_a) \right] \qquad (8.10a)$$
$$= 47.1 \left[1.183(1100 - 761.4) - 1.021(218.59 - 25)/(0.98 \times 1.014) \right]$$
$$= 9498.2 \text{ kW.}$$

The net heat output of the cycle is given by

$$Q = W + M_e \left[C_{pe,E}(T_6 - T_1) + C_{pa,C}(T_2 - T_1)(1/\eta_m - 1) M_a/M_e \right] (8.10b)$$
$$= 9498.2 + 47.1 \left[1.026(259.83 - 25) + 1.021(218.59 - 25)(1/0.98 - 1)/1.014 \right]$$
$$= 21033.6 \text{ kW.}$$

The cycle efficiency is 45.2%.

Step 5. Establishing a New Exhaust Profile to Satisfy Power Demands Exactly
    More power is produced than the requirement of 4500 kW, so there is scope for decreasing the exhaust flow rate and simultaneously reducing the heat rejected to ambient. To achieve this without wastage of driving forces, the exhaust profile should approach the process sink profile as closely as possible, subject to the limitation set by the ΔT_{min} contribution. The concept of an exhaust pivot point may be used for this purpose, i.e., the exhaust profile may be rotated counter-clockwise using X_2 as pivot (see the heavy dashed line in Figure 8.8). It may be noted that the exhaust profile may be rotated clockwise using X_1 as pivot if there is a deficit in the power produced.
    The task reduces to finding a value of T_6 (by trial and error) so that the total power output of the cycle is strictly 4500 kW. This simply requires repeating the calculations in steps 3 and 4 with assumed values of T_6 (> 259.83°C). Table 8.3 summarizes the temperature calculations for T_6 = 381.4°C based on the following physical properties: γ_a = 1.38, γ_e = 1.322, $C_{pe,H}$ = 1.093 kJ/kg°C, $C_{pa,H}$ = 1.08 kJ/kg°C, $C_{pe,T}$ = 1.171 kJ/kg°C, $C_{pa,C}$ = 1.033 kJ/kg°C, and $C_{pe,E}$ = 1.039 kJ/kg°C. Then, $M_e C_{pe,E}$ = (5580 − 1480)/(381.4 − 175) = 19.86 kW/°C and M_e = 19.12 kg/s. Finally, Equations 8.10 yield the total power output and the net heat output of the cycle as

W = 19.12 [1.171 (1100 − 614.8) − 1.033 (345.29 − 25)/(0.98 x 1.014)]
  = 4497 kW.
Q = 4497 + 19.12[1.039(381.4−25) + 1.033(345.29−25)(1/0.98−1)/1.014]
  = 11704 kW.

The cycle efficiency is 38.4% with a compression ratio of 10.13 and the load on the internal cycle heat exchanger being 4877.6 kW (given by $M_e\, C_{pe,H}$ [$T_5 - T_6$]). The waste heat from the gas turbine to ambient is 1499.7 kW (given by 19.86 [100.5 − 25]).

Table 8.3 Temperature Calculations for New Exhaust Profile

Iteration	T_2	P_2	P_4	T_5	T_3	$T_{2,update}$	T_s
1	351.40	1046.8	952.6	609.3	579.3	345.43	5.97
2	345.43	1014.0	922.8	614.7	584.7	345.29	6.11
3	345.29	1013.3	922.1	614.8	584.8	345.28	6.12

Units: temperature T in °C and pressure P in kPa.

Step 6. Assessment of Alternative Design Based on Combined Cycles
    From the analysis so far, it may be concluded that the preferred solution to the CHP problem is the gas turbine cycle rather than the steam turbine cycle. At this stage, it may be worth assessing the performance of a combined gas turbine-steam turbine configuration relative to the single gas turbine cycle.
    An improvement may be attempted in the gas turbine cycle designed in step 5 by decreasing the flow rate of the exhaust from 19.12 kg/s to 11.04 kg/s (see exhaust profile given by dotted line in Figure 8.8). Consequently, the waste heat to the ambient from the gas turbine reduces from 1499.7 kW to 286.76 kW. However, there is a simultaneous reduction in the power output, which may be compensated by operating a back-pressure steam turbine. If the lost power cannot be totally compensated due to a low cycle efficiency for the steam turbine, then the deficit may be made up by a condensing stage. The combined gas turbine-steam turbine configuration is advantageous only if the heat rejection to ambient from this condensing stage is less than the reduction in waste heat gained from the gas turbine The quantitative implications of the modifications for the case study under consideration are given below:

Reduction in waste heat to ambient = 1499.7 − 286.8 = 1212.9 kW

Heat requirement of process sink not satisfied
$$= 5580 - 3801.4 = 1778.6 \text{ kW}$$
Power deficit due to size reduction in gas turbine cycle
(efficiency = 38.4%)
$$= (1212.9 + 1778.6)/(1/0.384 - 1) = 1864.8 \text{ kW}$$
Cycle efficiency of steam turbine
(estimate upper bound as Carnot efficiency)
$$= 1 - (145 + 273)/(300 + 273) = 0.27$$
Power produced by back pressure steam turbine
$$= 1778.6/(1/0.27 - 1) = 657.8 \text{ kW}$$
Cycle efficiency of steam turbine rejecting heat to ambient
$$= 0.35 \text{ (assumed)}$$
Heat rejected to ambient by condensing steam turbine stage
$$= (1864.8 - 657.8)(1/0.35 - 1) = 2241.6 \text{ kW}$$

As 2241.6 > 1212.9, the combined cycle does not yield any benefit. This effect is primarily due to the magnitude of the unsatisfied process sink (1778.6 kW) being larger than the waste heat reduction in the gas turbine (1212.9 kW).

Step 7. Synthesis of Heat Exchanger Network for Best Gas Turbine Cycle
The final step is to generate a HEN for the best CHP solution based on the single gas turbine cycle. In addition to the two hot streams and two cold streams in Case Study 4S1t, the LP and IP streams (as part of steam-raising) as well as the gas turbine exhaust stream are included (see Figure 8.9).

A utility pinch occurs at 155°C. Note that ΔT_{min} is 20°C except for matches with the exhaust stream (where a higher ΔT_{min} of 30°C is used). Figure 8.9 shows a possible MER design using the PDM (see Section 3.1). A ΔT_{min} violation of 0.5°C is tolerated at the cold end of match 5 so as to eliminate the necessity of splitting stream C3 below the utility pinch. The cooler on the turbine exhaust in Figure 8.9 (from 100.5° to 25°C) is a notional one as it corresponds to the waste heat rejected directly to the ambient.

By breaking two first-level loops (see Section 4.1), matches 2 and 3 may be eliminated from the MER design in Figure 8.9 to obtain a simplified six-unit network (not shown). Of course, additional matches corresponding to the gas turbine and the air preheater must be considered. It may be concluded that this preliminary design for the CHP system based on the gas turbine cycle appears attractive.

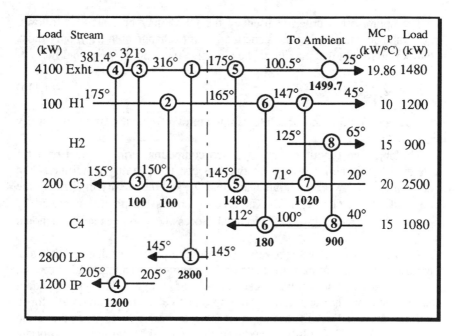

Figure 8.9 The exhaust stream is appended to the process stream set, and a HEN is generated by conventional methods for the CHP system based on the gas turbine cycle. (Temperature in °C and load in kW)

8.3 OPERATING LINE METHOD FOR CHP

A more recent approach to the synthesis of optimal heat and power cogeneration systems is the operating line method (OLM). The OLM (Irazoqui, 1986; Aguirre et al., 1989; Pavani et al., 1990) treats heat and power integration as a single inseparable problem that requires a simultaneous strategy for its solution, whereas the pinch method (PM) discussed in the previous section adopts a sequential strategy to solve the problem. The design parameter to be specified in the OLM is the slope, P, of the operating line whereas it is ΔT_{min} in the PM. Both methods consider power generation auxiliary to heat recovery. The methods yield different solutions in terms of the placement of power generating cycles, especially for the case of a GCC that is reentrant (i.e., has a pocket). The OLM in principle exploits the local heat sources and sinks that remain after heat integration to generate power. It predicts better energy performance on placing heat engines across the pinch, which would be considered inappropriate placement in the PM.

The optimization criterion used by the OLM may be stated as follows. Among all networks having the same total heat exchange area (A_e), only those networks are sought that maximize the objective function, G, given below:

$$G = Q_e (1 - T_o/T_v) + W_{hs} \qquad (8.11a)$$

Here, T_v and T_o are the absolute temperatures of the hot and cold utility respectively.

The objective function comprises two competing terms. The first term assesses thermal energy savings in two aspects: the quantity in terms of the heat exchanged between integrated streams, Q_e, and the quality in terms of the Carnot efficiency, $(1 - T_o/T_v)$. The second term is the power generated by coupling power-generating cycles to local process heat sources and is denoted by W_{hs}.

An increase in the heat recovery by process-to-process exchange leads to a decrease in the amount of work that can be generated. This tradeoff between heat integration and power generation may make the independent maximization of the two terms an unproductive exercise and call for a simultaneous optimization strategy. If several HENs exist with the same values of A_e and G (multiple maxima), then the OLM through appropriate constraints seeks those networks that maximize Q_e (i.e., preference is given to heat recovery over power generation).

The objective function, G, may be related to the rate of internal generation of entropy σ by

$$\sigma = K - G/T_o \qquad (8.11b)$$

where K is a constant for given inlet and outlet temperatures of process streams. Thus, an optimal solution corresponding to maximization of G will be equivalent to minimization of σ. The appropriate objective function for minimizing σ subject to a constant A_e constraint is given by

$$\sigma^* = \sigma + \lambda A_e \qquad (8.11c)$$

where λ is a Lagrange multiplier (a constant weighting factor to include situations with different values of the ratio A_e/G).

The optimization problem has been formulated so that it may be solved by principles of variational calculus. More details of the thermodynamic model and the optimization problem are provided by Irazoqui (1986); here, only the relevant final equations are presented in order to solve an illustrative problem.

Example 8.3 Operating Line Method for Heat and Power Cogeneration

Design an optimal HEN for a combined heat and power (CHP) system based on the stream data in Case Study 4S1. Use the operating line method (OLM). It is desired to generate 1000 kW of shaft power.

Given: $U = 0.1$ kW/(m^2 °C), hot utility temperature $T_v = 523$ K, and cold utility temperature $T_o = 298$ K.

Solution. The OLM may be used to generate a sequence of optimal solutions with different values of A_e and G that can be arranged in increasing order as per the ratio $\gamma = G/A_e$. Each solution in the sequence maximizes first the value of G (corresponding to A_e), next the value of Q_e, and finally the value of W_{hs}. Note that the ratio γ signifies the maximum energy recovery per unit heat exchanger area and is an indicator of the economic benefit gained from the CHP system with respect to the capital cost. When energy is expensive compared to exchange area, then relatively small values of γ should be used.

Step 1. Defining the Exchange Region and the Optimal Operating Line

The exchange region comprises all points on a T_h vs. T_c plot with feasibility of heat transfer from a hot composite stream of temperature T_h to a cold composite stream of temperature T_c. Figure 8.10 shows the exchange region for Case Study 4S1 bounded by heavy lines (318 K $\leq T_h \leq$ 448 K, 293 K $\leq T_c \leq$ 428 K, and $T_h \geq T_c$).

An operating line for the heat transfer process is defined as a function $T_h(T_c)$ that is confined to the exchange region and which specifies a matching policy between a hot composite stream of absolute temperature T_h and a cold composite stream of absolute temperature T_c. The structure of the HEN can be directly derived from the operating line. Both σ and A_e in the objective function (Equation 8.11c) are functions of the operating line, $T_h(T_c)$. Thus, the optimization problem involves determining the operating line, $T_h(T_c)$, that minimizes the functional $\sigma^*[T_h(T_c); \lambda] = \sigma[T_h(T_c)] + \lambda A_e[T_h(T_c)]$. The optimal operating lines have been obtained by Irazoqui (1986) and constitute a set of straight line segments corresponding to different values of λ as specified by the equation below:

$$T_h = P\,T_c$$
where $P = \{(2 + \lambda/U) + [(2 + \lambda/U)^2 - 4(1 - \lambda/U)]^{1/2}\}/[2(1 - \lambda/U)]$ (8.12a)

Here, U is the overall heat transfer coefficient. Furthermore, the operating line segment must lie in the optimality region defined by

$$(1 - T_h/T_v)\,(1 - T_c/T_h) > \lambda/U. \tag{8.12b}$$

Figure 8.10 shows the optimality region for $\lambda/U = 0.0017$ based on Equation 8.12a. The optimal operating line with a slope P of 1.06097 (using $\lambda/U = 0.0017$ in Equation 8.12a) is also shown. Note that λ/U is a design parameter that needs to be specified. There is usually an upper bound on the value of λ/U beyond which the operating line does not intersect the exchange region.

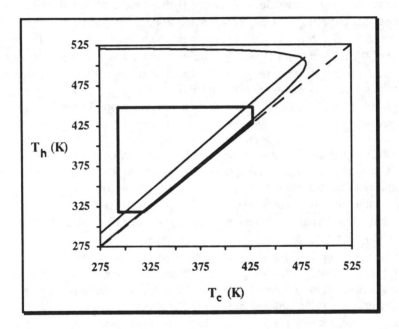

Figure 8.10 The first step in the application of the OLM is to identify the exchange region, the optimality region boundary, and the optimal operating line.

By plotting T_h/T_v vs. T_c/T_v for different values of λ/U, a dimensionless chart (Irazoqui, 1986) may be developed based on Equations 8.12. Here, the advantage is that the optimal operating lines and the corresponding optimality regions on such a chart are independent of the particular problem. The allowable range for λ/U for a given case study may be determined by superposing its exchange region on this fundamental chart. If energy costs dominate over exchange area costs, then small values of λ/U should be used.

Step 2. Obtaining the Integrated and Nonintegrated Pseudostreams

By drawing horizontal lines (corresponding to the hot stream temperatures) and vertical lines (corresponding to the cold stream temperatures) on Figure 8.10, the exchange region may be divided into subregions within which process streams may be matched in a countercurrent manner as per Equation 8.12a. For each subregion, the ratio of the heat capacity flow rates of the cold composite stream to the hot composite stream must equal the slope, P, of the operating line. This may be achieved by splitting the stream with the higher heat capacity flow rate. The process streams (or their fractions) that remain unmatched after integration are effectively the heat sources or sinks, whose T-Q profile provides the GCC for the OLM.

What is the shifted temperature axis to be used for the GCC in the case of the OLM? A cold stream temperature, T_c, will have a corresponding ordinate, PT_c, on the operating line whereas a hot stream temperature, T_h, will have a corresponding abscissa, T_h/P (see the first two columns in Table 8.4). Thus, on adding $\Delta T/2$ to T_c and subtracting $\Delta T/2$ from T_h as per the convention for GCC temperature shifting,

$$T_{gcc} = T_c + (PT_c - T_c)/2 = (1 + P)\,T_c/2 \tag{8.13a}$$
$$T_{gcc} = T_h - (T_h - T_h/P)/2 = (1 + 1/P)\,T_h/2 \tag{8.13b}$$

The third column in Table 8.4 may be filled simply by averaging the entries in the first two columns or by using Equations 8.13 with $P = 1.06097$.

Table 8.4 Determination of Pseudostreams for OLM

Temperature Intervals			$MC_{p,integrated\ streams}$				$MC_{p,nonintegrated\ streams}$			
T_h	T_c	T_{gcc}	H1	H2	C3	C4	H1	H2	C3	C4
454.10	428	441.05								
448	422.26	435.13							20	
408.47	385	396.74	10		10.6				9.4	
398	375.13	386.56	10		10.6				9.4	15
338	318.58	328.29	10	23	20	15	17			
332.08	313	322.54	10		10.6				9.4	15
318	299.73	308.86	10		10.6				9.4	
310.86	293	301.93							20	

Units: temperature in K and heat capacity flow rate in kW/°C.

Within each temperature interval in Table 8.4, streams are split such that the heat capacity flow rates for the integrated pseudostreams satisfy the OLM condition, $\Sigma MC_{p,c}/\Sigma MC_{p,h}$ = 1.06097. Only pseudostreams in the same temperature interval are thermally integrated. The integrated pseudostreams in each temperature interval are in perfect enthalpy balance, e.g., 35 (375.13 − 318.58) = 33 (398 − 338) in the fourth temperature interval. The MC_p-values of the nonintegrated branches of the split process streams are shown in the last four columns of Table 8.4 (against the lower temperature limit of the interval).

Step 3. Calculating the Work Generated from the GCC for the OLM

Figure 8.11 shows the GCC for the OLM based on the nonintegrated pseudostreams. The temperature axis is in accordance with Equation 8.13. Note that the slope of the GCC in an interval is given by the reciprocal of the total heat capacity flow rate of the unmatched pseudostreams.

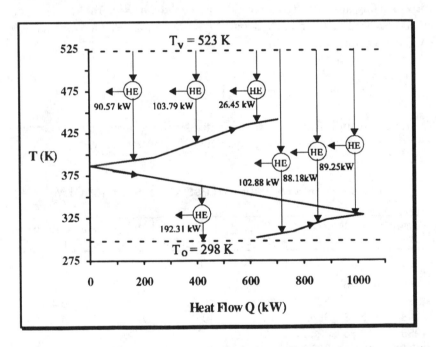

Figure 8.11   Power may be generated by coupling heat engines to the nonintegrated pseudostreams, according to the OLM.

After generating the GCC for the OLM, heat engines may be placed to avoid direct heat exchange between utilities and process streams so as to generate power at 100% marginal efficiency. Note that the splitting of streams (in step 2) favors the power generation by maximizing the temperature level of local heat sources (and minimizing the temperature level of local heat sinks) and simultaneously favors maximum process-to-process heat recovery consistent with a value of P or ΔT_{min} (Aguirre et al., 1989).

Figure 8.11 shows six heat engines absorbing heat from the hot utility (at $T_v = 523$ K) and discharging heat to the process sinks. If the process sink extends from $T_{c,i-1}$ K to $T_{c,i}$ K in the i-th temperature interval, then the power generated is given by

$$
\begin{aligned}
W_{h,i} &= MC_{pc,i} \int (T_v/T - 1)\, dT \\
&= MC_{pc,i} \left[T_v \ln(T_{c,i-1}/T_{c,i}) - (T_{c,i-1} - T_{c,i}) \right].
\end{aligned}
\tag{8.14a}
$$

Figure 8.11 also shows one heat engine absorbing heat from the process source (through a waste heat recovery boiler, say) and discharging heat to the cold utility (at $T_o = 298$ K). If the process source extends from $T_{h,i-1}$ K to $T_{h,i}$ K in the i-th temperature interval, then the power generated is given by

$$
\begin{aligned}
W_{c,i} &= MC_{ph,i} \int (1 - T_o/T)\, dT \\
&= MC_{ph,i} \left[(T_{h,i-1} - T_{h,i}) - T_o \ln(T_{h,i-1}/T_{h,i}) \right].
\end{aligned}
\tag{8.14b}
$$

Some sample calculations based on Equation 8.14 are given below:

$$W_{h,1} = 20\,[523\,\ln(428/422.26) - (428 - 422.26)] = 26.45\text{ kW.}$$
$$W_{c,4} = 17\,[(398 - 338) - 298\,\ln(398/338)] \qquad = 192.31\text{ kW.}$$

Summing the power from the various heat engines in Figure 8.11 yields the total power generated to be 693.43 kW. In order to satisfy the total power demand of 1000 kW, a stand-alone heat engine is required in addition to the above CHP system.

In other words, this is a dominant power load problem where the additional power may be generated by condensation turbines operating between high pressure steam and vacuum. Also, the fact that the cycle working fluid profile is not coincident with the GCC profile in practical heat engines (see Section 8.2) must be considered.

Step 4. Determining the Total Heat Input and Efficiency
    The total heat input from the hot utility to the CHP system may be obtained by adding the heat supplied to the nonintegrated process sink and the power generated. Thus,

$$Q_h = 693.43 + 20 (428-422.26) + 9.4 (422.26-385) + 24.4 (385-375.13)$$
$$+ 24.4 (318.58-313) + 9.4 (313-299.73) + 20 (299.73-293)$$
$$= 1794.1 \text{ kW}$$

This yields an efficiency of 38.7%.

Step 5. Calculating the Total Heat Exchange Area

The heat exchange area required by an optimal match between the integrated pseudostreams in the i-th temperature interval is given by

$$A_i = [U (P - 1)]^{-1} (\Sigma MC_{pc,i}) \ln(T_{c,i-1}/T_{c,i}) \tag{8.15}$$

Equation 8.15 can be derived by simplifying the uniform BATH formula (Equation 1.7) on using $\Sigma MC_{pc,i}/\Sigma MC_{ph,i} = (T_{h,i-1} - T_{h,i})/(T_{c,i-1} - T_{c,i}) = P$ and assuming uniform heat transfer coefficients. Substituting in Equation 8.15 from Table 8.4 and summing over all the temperature intervals gives

$$A_e = [0.1 \times 0.06097]^{-1} [10.6 \ln(422.26/385) + 10.6 \ln(385/375.13)$$
$$+ 35 \ln(375.13/318.58) + 10.6 \ln(318.58/313) + 10.6 \ln(313/299.73)]$$
$$= 1250.1 \text{ m}^2.$$

Step 6. Calculating Other Quantities of Interest

The total amount of heat exchanged between the integrated streams is obtained by calculating the enthalpy change in every temperature interval for either hot or cold streams. Thus,

$$Q_e = 10 (448 - 318) + 23 (398 - 338) = 2679.32 \text{ kW}.$$

Then, from Equation 8.11a, $G = 2679.32 (1-298/523) + 192.31 = 1344.98$ kW. Finally, $\gamma = G/A_e = 1344.98/1250.1 = 1.076 \text{ kW/m}^2$.

Step 7. Designing an Optimal HEN

The optimal HEN is derived from Table 8.4 and is shown in Figure 8.12. Three exchangers are required to match the integrated pseudostreams. Stream H2 is split further in the ratio 8.9:14.1 to ensure the OLM condition ($\Sigma MC_{p,c}/\Sigma MC_{p,h} = 1.06097$) is satisfied for matches 2 and 3. Heaters and coolers are shown to satisfy the cooling and heating requirements of the nonintegrated streams. Note that the OLM does not consider in its formulation the number of units required.

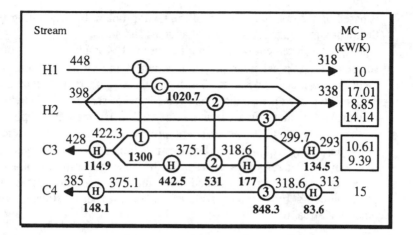

Figure 8.12 The heat exchanger network for the OLM may be derived from the pseudostreams represented in Table 8.4. (Temperature in K and load in kW)

Step 8. Comparing OLM and PM for CHP

To compare the OLM and the PM, the same total heat exchange area may be used as a basis. For an A_e of 1250 m^2, ΔT_{min} is found to be 22.4°C. Note that ΔT in the OLM solution varies from 18.3° to 25.7°C (given by $[T_h - T_c]$ of the integrated streams in Table 8.4).

Figure 8.13 shows the GCC for Case Study 4S1 at ΔT_{min} = 22.4°C. Process-to-process heat exchange occurs in the "pocket" of the GCC due to the local heat sink below the pinch. Four heat engines may be appropriately placed with none of them crossing the pinch. Using Equations 8.14, the total power generated is found to be 305.81 kW (lower than the 693.43 kW generated from the OLM). The total heat input from the hot utility to the CHP system is 1285.81 kW (lower than the 1794.1 kW supplied in the OLM). Thus, the efficiency from the PM is 23.8%, which is lower than that from the OLM (38.7%). Obviously, the heat exchanged between the integrated streams has increased from 2679.32 kW (in the OLM) to 2800 kW (in the PM). The PM requires less utility for process heating whereas the OLM saves on utility due to more efficient power production. The OLM provides the optimum solution considering this tradeoff.

To gain a better understanding of the OLM and the PM, the following differences must be appreciated (Irazoqui, 1986; Aguirre et al., 1989):
- The GCC for the PM (Figure 8.13) is based on the maximization of process heat recovery. However, the GCC for the OLM (Figure 8.11)

is based on the optimal solution of the CHP problem by accounting for the quality and quantity of the heat sources and sinks.

- Appropriate placement of heat engines in the PM is based on *net* heat sources and sinks whereas proper location in the OLM is based on *local* heat sources and sinks. Thus, the OLM may require that heat engines be placed across the pinch for a "pocket-shaped" GCC, which may prove advantageous from an overall CHP performance viewpoint.

- In the context of CHP, an efficient process does not imply merely minimum utility requirements, but also maximum potential to produce shaft power. Therefore, even when shaft power generation is not the objective, the analysis presented here may be used to assess the possibilities for further integration of the existing set of streams with other new streams.

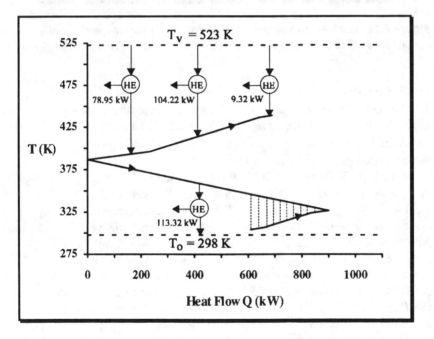

Figure 8.13 The power generated by coupling heat engines according to the PM may be compared with that from the OLM (see Figure 8.11).

8.4 OPTIMAL INTEGRATION OF HEAT PUMPS

The basic thermodynamic relations for a heat pump have been outlined in Section 8.1. Recall that energy is wasted if the heat pump is placed entirely on one side of the pinch, which is therefore considered inappropriate placement. For a heat pump to achieve its normal efficiency and thus be appropriately placed, it must be across the pinch. Even in this case, a load limit exists for the maximum utility savings imposed by the process source or sink during profile matching.

Figure 8.14 shows a schematic diagram of a mechanical heat pump and the relevant working fluid profiles on temperature-entropy (T-S) and temperature-enthalpy (T-H) diagrams. The ideal cycle may be described as follows:

2 → 3:   evaporation of the liquid at constant pressure absorbing heat at constant temperature;

3 → 4:   isentropic compression of the vapor leaving the evaporator to the required higher pressure;

4 → 1:   condensation at higher pressure rejecting heat at constant temperature (with cooling of the superheated vapor to its saturation point prior to condensation); and

1 → 2:   expansion (by engine or throttle valve) of the liquid from the condenser.

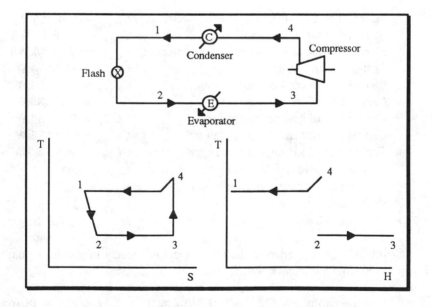

Figure 8.14 The thermodynamics of a simple heat pump cycle may be shown on a T-S diagram and its heat-absorbing and heat-rejecting profiles on a T-H diagram.

The thermodynamic understanding may be combined with the economics to assess the temperature lift and the optimum placement for a heat pump.

8.4.1 Economics of Heat Pumping

A general equation for the maximum economic lift, derived by Ranade (1988), provides insights into the characteristics for heat pumps on industrial sites.

Example 8.4 Maximum Economic Lift Target for Heat Pumping
Consider a heat pump operating on a site with two steam levels (IP steam at $T_{uh} = 478$ K and LP steam at $T_{ul} = 418$ K). Determine the minimum economic temperature, $T_{2,min}$, for the heat pump to accept heat under the following situations:
   (a) the LP steam has a positive marginal value since it can be exported from the process; and
   (b) the LP steam cannot be exported and therefore the waste heat from the process must be rejected to cooling water.

Given: Marginal cost of the driver utility (electricity) C_{md} = 0.05 $/kWh
        Marginal value of IP steam C_{mvh} = 0.016 $/kWh
        Marginal value of LP steam (export potential) C_{mvl} = 0.01 $/kWh
        Cost of cooling water C_{cw} = 0.002 $/kWh
        Capital cost of cooler (cooling water) CC_{cw} = 15 $/kW
        Capital cost of compressor CC_c = 500 $/kW
        Capital cost of heat pump condenser CC_{hpc} = 35 $/kW
        Capital cost of heat pump exchanger CC_{hpe} = 35 $/kW
        Capital cost of LP-raising exchanger CC_{lp} = 40 $/kW
        Thermodynamic efficiency of heat pump η_{hp} = 0.65
        Mechanical efficiency of heat pump η_m = 1.0
        Payback period P = 2 yrs
        Plant operation H = 8000 hrs/yr

Solution. On considering the mechanical efficiency of the heat pump, Equations 8.3 and 8.4 may be written as

$$W = (Q_1 - Q_2)/\eta_m = Q_1\,(1 - T_2/T_1)/(\eta_{hp}\eta_m) \qquad (8.16a)$$

The economics for heat pumping are given by the following expressions:

Annual Savings

= Value of IP steam saved − value of LP steam used − cost of driver utility

$$= [Q_1 C_{mvh} - Q_2 C_{mvl} - W C_{md}] H$$

$$= H Q_1 C_{mvh} - H Q_1 [1 - (1 - T_2/T_1)/\eta_{hp}] C_{mvl}$$

$$\qquad - H Q_1 [(1 - T_2/T_1)/(\eta_{hp}\eta_m)] C_{md} \qquad\qquad (8.16b)$$

Investment

$$I_{cc} = C_{hpc} + C_{hpe} + C_c + C_i - \Sigma\alpha_{hei} C_{hei} - \Sigma\alpha_{coi} C_{coi} \qquad (8.16c)$$

In Equation 8.16c, installation costs are included. Note that C_{hpc}, C_{hpe}, C_c, and C_i denote the capital costs for the heat pump condenser, the heat pump exchanger, the heat pump hardware (compressor), and the area increment in the rest of the plant. C_{hei} and C_{coi} are the capital costs associated with the i-th heater and cooler areas that are now in excess due to heat pumping whereas α denotes the corresponding fraction of the excess area that can be reused. For a conservative estimate during retrofitting, α may be set to zero.

Using Equations 8.16 along with the definition of the payback period, P, (as the ratio of investment to annual savings) yields the desired equation for the maximum economic lift target as

$$T_2/T_1 = 1 + \eta_{hp}\eta_m [I_{cc}/(HQ_1P) + C_{mvl} - C_{mvh}]/[C_{md} - \eta_m C_{mvl}]. \quad (8.16d)$$

For Case (a), substitution in the above equation (with $T_1 = T_{uh} = 478$ K and $C_{mvl} = 0.01$ \$/kWh) gives

$$T_2/478 = 1 + 0.65 [I_{cc}/(16000 Q_1) - 0.006]/0.04$$

The terms in the I_{cc} expression (Equation 8.16c) are given by

$$
\begin{aligned}
C_{hpc} &= CC_{hpc} Q_1 [1 - (1 - T_2/T_1)/\eta_{hp}] &= 35 Q_1 [1 - (1 - T_2/478)/0.65] \\
C_{hpe} &= CC_{hpe} Q_1 [1 - (1 - T_2/T_1)/\eta_{hp}] &= 35 Q_1 [1 - (1 - T_2/478)/0.65] \\
C_c &= CC_c Q_1 (1 - T_2/T_1)/\eta_{hp} &= 500 Q_1 (1 - T_2/478)/0.65 \\
C_{co} &= CC_{lp} Q_1 [1 - (1 - T_2/T_1)/\eta_{hp}] &= 40 Q_1 [1 - (1 - T_2/478)/0.65]
\end{aligned}
$$

Thus, Equation 8.16c (with $C_i = C_{hei} = 0$ and $\alpha_{coi} = 1$) simplifies to

$$I_{cc} = (35 + 35 - 40) Q_1 [1 - (1 - T_2/478)/0.65] + 500 Q_1 (1 - T_2/478)/0.65$$
$$\quad = 30 Q_1 + 470 Q_1/0.65 (1 - T_2/478)$$

Substituting for I_{cc} in the maximum economic lift equation finally gives

$$T_2/478 = 1 + (0.65 \times 30 + 470)/(0.04 \times 16000)$$
$$- 0.65 \times 0.006/0.04 - 470/(0.04 \times 16000)\ (T_2/478)$$

or $T_{2,min} = 459.5$ K.

For Case (b), the thermal energy used by the pump has a negative marginal value ($C_{mvl} = - C_{cw} = - 0.002$ \$/kWh) for it has no demand elsewhere in the plant and consequently incurs a cooling cost. This implies a change in the C_{co} expression for this scenario:

$$C_{co} = CC_{cw}\ Q_1\ [1 - (1 - T_2/T_1)/\eta_{hp}] = 15\ Q_1\ [1 - (1 - T_2/478)/0.65]$$

The rest of the formulation is identical to Case (a) and is given below:

$$T_2/478 = 1 + 0.65\ [I_{cc}/(16000\ Q_1) - 0.018]/0.052$$
$$I_{cc} = (35 + 35 - 15)\ Q_1\ [1 - (1 - T_2/478)/0.65] + 500\ Q_1\ (1 - T_2/478)/0.65$$
$$= 55\ Q_1 + 445\ Q_1/0.65\ (1 - T_2/478)$$
$$T_2/478 = 1 + (0.65 \times 55 + 445)/(0.052 \times 16000)$$
$$- 0.65 \times 0.018/0.052 - 445/(0.052 \times 16000)\ (T_2/478)$$

or $T_{2,min} = 421.3$ K.

The maximum economic lifts are 18.5 K (for Case a) and 56.7 K (for Case b); obviously, there is less scope for heat pumping in Case (a).

Note that the maximum theoretical economic lift or the minimum theoretical COP may be obtained from Equation 8.16d by evaluating the limit for very large payback period ($P \to \infty$). Thus,

$$(T_1 - T_2)_{max}/T_1 = \eta_{hp}\eta_m\ [C_{mvh} - C_{mvl}]/[C_{md} - \eta_m\ C_{mvl}] \qquad (8.17a)$$
$$COP_{min} = \eta_{hp}\eta_m T_1/(T_1 - T_2)_{max}$$
$$= [C_{md} - \eta_m\ C_{mvl}]/[C_{mvh} - C_{mvl}] \qquad (8.17b)$$

For Case (a), the maximum theoretical economic lift is 46.6 K and the minimum theoretical COP is 6.67. These define the bounds for the temperature range of the optimum lift and the COP of the optimum heat pump respectively.

Another interesting application of Equation 8.16d is to open systems (e.g., utility-to-utility heat pumps used to generate higher level steam from existing lower level steam). Here, the heat transfer area penalty term in Equation 8.16c is not significant (i.e., $I_{cc} \approx C_c$) and Equation 8.16d reduces to

$$T_2/T_1 = 1 + \eta_{hp}\eta_m[C_c/(HQ_1P) + C_{mvl} - C_{mvh}]/[C_{md} - \eta_m\ C_{mvl}] \qquad (8.17c)$$

For Case (a), Equation 8.17c yields 26.2°C as the maximum economic lift.

The above equations also form the basis of an algorithm (Ranade and Sullivan, 1988) that allows optimal integration of a closed type process-to-process heat pump for lifting heat from below the pinch to above the pinch, thereby saving on both hot and cold utility. After heat pump integration, there is an additional horizontal segment on the HCC (as the condenser acts as a hot stream above the pinch) as well as on the CCC (as the evaporator acts as a cold stream below the pinch). Consequently, the area requirements for the heater and cooler decrease and for the process-to-process heat transfer increase. The systematic search for the optimum is based on the tradeoff between the compressor costs (which increase with heat pump lift) and heat transfer area costs (which decrease with heat pump lift due to the increase in the driving forces in the condenser and evaporator). In this case, instead of specifying the payback period, the economics are assessed for various choices of the three independent variables (T_1, T_2, and Q_1). As this procedure for determining the optimal heat pump placement target is involved and demands many cycles of repeated calculations, only the key steps are outlined below.

1. Determine the energy and area targets as well as the area efficiency (Equation 7.1). The area efficiency should be used in Equation 8.16c during retrofitting.

2. Choose a value of T_1 from the feasible range. Note that $T_p + \Delta T_{min}/2 \leq T_1 \leq T_{1,max}$, where T_p is the interval pinch temperature on the process GCC and $T_{1,max}$ is decided by the process temperature range and the working fluid.

3. Choose a value of Q_1 from the feasible range. Note that $0 < Q_1 \leq Q_{1,max}$, where $Q_{1,max}$ is read from the process GCC corresponding to $T_{1,max} - \Delta T_{min}/2$.

4. Choose a value of T_2 from the feasible range. Note that $T_{2,min} \leq T_2 \leq T_{2,max}$, where $T_{2,min}$ is the minimum allowable working fluid temperature in the evaporator based on the process temperature range and the working fluid properties. $T_{2,max}$ and the corresponding load, $Q_{2,max}$, are obtained such that Q_2 calculated from Equation 8.16a (with $Q_1 = Q_{1,max}$) equals the value of Q_2 read from the process GCC corresponding to $T_2 + \Delta T_{min}/2$.

5. Calculate Q_2 from Equation 8.16a, annual savings from Equation 8.16b, investment from Equation 8.16c, and the payback period.

The above steps must be repeated by systematically varying T_2 (in the innermost loop), then Q_1, and finally T_1 (in the outermost loop). The optimum heat pump load and lift may be thus selected based on suitable economic criteria. In fact, the entire procedure may be repeated for different levels of heat integration (i.e., different process ΔT_{min} values) to establish the optimum combination of heat pumping and heat integration. Considering the iterative nature of the calculations, a software package (similar to the one described by Ranade and Sullivan [1988]) is ideally required for optimum heat pump placement targeting.

8.4.2 Refrigeration Systems

A heat pump where the rejected heat finally ends up in the ambient sink is commonly called a refrigerator. This subsection is devoted to refrigeration systems, which are especially important in low temperature processes.

The Carnot efficiency, η_c, plays an important role in refrigeration system design. Equation 8.3b may be rewritten as $W/Q_2 \geq T_1/T_2 - 1$. Then, T_1 may be chosen as the ambient temperature and T_2 as the refrigeration temperature. Note that η_c increases rapidly as T_2 tends to very low temperatures. In fact, η_c is infinitely large when T_2 becomes the absolute zero. This is the reason for the high cost of refrigeration duty in low temperature processes: the power requirement is very high for low refrigeration temperatures and is highly sensitive to system irreversibilities. This calls for a special effort in reducing the power requirement, which often results in complicated CHP designs for subambient plants.

Compression refrigeration and absorption refrigeration form two important types of refrigeration systems. Figure 8.14 represents a mechanical refrigeration system, provided the working fluid is condensed using cooling water.

In absorption refrigeration, the heat pumping is directly accomplished using heat (steam) without its actual conversion into shaftwork. The essential differences between an absorption system and a compression cycle are that the compressor in Figure 8.14 is substituted by a heat engine and a relatively non-volatile solvent is used in addition to the refrigerant. The absorption refrigeration machine is not discussed here as its use is limited on account of its high heat demand and cost. However, the absorption system may be employed when there is an above-ambient process source below the pinch that is hot enough and large enough to handle the refrigeration load (Linnhoff et al., 1982).

On the other hand, compression refrigeration is preferred when the above-ambient process source is limited and the refrigeration load is at a low temperature level. The case for compression refrigeration is further strengthened if significant power can be generated at high efficiency using the process sink above the pinch. Profile matching for such situations is discussed by Linnhoff et al. (1982).

Example 8.5   Matching Refrigeration Cycles against the Subambient Process Profile

Figure 8.15 shows the GCC for a subambient process. Discuss different refrigeration schemes based on a ΔT_{min} of 5°C.

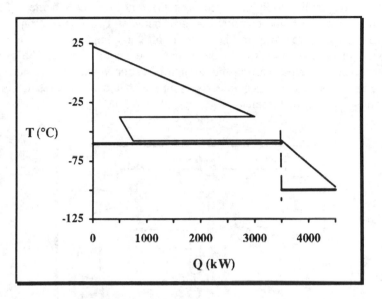

Figure 8.15 The subambient process GCC may be used effectively to select appropriate compression refrigeration schemes.

Solution. The process GCC data for Figure 8.15 are given below:

T_{int} (°C)	22.5	−37.5	−37.5	−57.5	−57.5	−97.5
Q (kW)	0	3000	500	750	3500	4500

The highest point in the GCC corresponds to the cooling water pinch temperature. This is because a utility pinch always occurs in any process at the temperature of the cooling water. Note that refrigeration is too expensive to be used for satisfying duties above the ambient temperature and is therefore used only below the cooling water pinch.

Figure 8.15 shows refrigeration supplied to meet the process duties at two levels (3500 kW at −60°C and 1000 kW at −100°C).   Note that the temperatures of the refrigeration levels are not shifted in Figure 8.15 for clarity. The selection of appropriate refrigeration schemes is broadly governed by the same principles discussed for profile matching in Section 8.2.1. In fact, all 4500 kW could have been supplied at a single level (–100°C); however, this would result in greater loss in driving forces between the refrigeration utility and the process GCC.

Figure 8.16 shows the schematic diagram for the compression refrigeration system corresponding to Figure 8.15. It is different from the scheme in Figure 8.14 in that the flash and compression now occur in two stages. Vapor from the first stage is returned to the suction of the compressor at the higher pressure, causing an increase in the vapor flow in this part of the system. However, there is a decrease in the amount of vapor flowing in the portion of the refrigeration cycle at the lower pressure with the overall result being a reduction in power requirement.

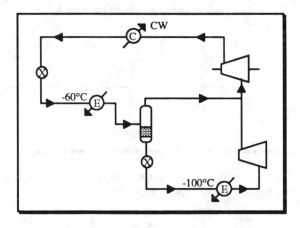

Figure 8.16 The above schematic diagram shows the compression refrigeration scheme corresponding to the profile matching in Figure 8.15.

Can the scheme in Figure 8.15 be further improved? Possibly, but only at the cost of additional complexity (Figure 8.17). Instead of supplying 3500 kW at –60°C (as in Figure 8.15), only 3000 kW may be supplied at –60°C and the remainder at 10°C. As discussed earlier, such a shifting of load to higher temperature levels results in a net power savings. Besides, the driving forces in the pocket of the GCC may be utilized better by adding a level at –35°C for recovery of refrigeration and another level at –20°C (say) of identical load to compensate for the lost process-to-process heat transfer. There is again a reduction in the power requirement. The increased load at –20°C is more than offset by the availability of a level at –35°C for rejecting heat. A compression refrigeration scheme corresponding to the profile matching in Figure 8.17 may be designed and a schematic diagram similar to Figure 8.16 developed. Although this scheme saves on power consumption compared to the system in Figure 8.16, it would be relatively complex. Therefore, the tradeoffs must be evaluated carefully before a final scheme is chosen.

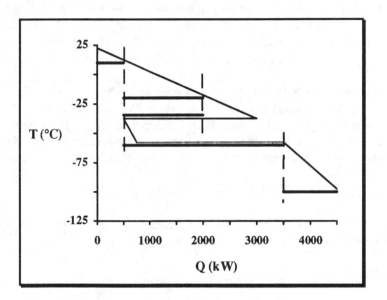

Figure 8.17 An alternative compression refrigeration scheme to that in Figure 8.15 may be devised that effectively uses the driving forces in the "GCC pocket."

Thus, pinch analysis allows targets to be established for the loads and levels of the refrigeration scheme using basic stream data and profile matching based on the subambient GCC. What would be advantageous in this connection is the ability to predict shaftwork requirements directly from stream data without designing either the refrigeration system or the HEN. Linnhoff and Dhole (1992) have proposed an elegant fundamental approach for this purpose based on a combination of pinch and exergy concepts. To appreciate their shaftwork targeting procedure, the concept of exergy must be understood.

8.5 THE CONCEPT OF EXERGY

The exergy of a system or a stream measures its work potential in relation to ambient conditions. Exergy is one of the measures (along with entropy and available energy) used in second law analysis to quantify losses or lost potential based on equilibrium thermodynamics. In fact, pinch analysis itself is a special case of second law analysis with a distinction. It provides a practical approach, incorporating the inevitable inefficiencies in available equipment, the complex effect of network interactions, and the economic tradeoffs (Linnhoff, 1989). Importantly, pinch analysis distinguishes between practically avoidable losses and inevitable losses. An MER network designed using pinch technology has only the inevitable exergy losses for the predetermined energy-capital tradeoff (i.e., ΔT_{min}). The example below demonstrates how these exergy losses are quantified (Linnhoff, 1989).

Example 8.6 Computation of Exergy Losses in HENs

Compute the exergy losses for Case Study 4S1 at a ΔT_{min} of 13°C from the composite curves. Show that the net exergy loss for the MER network in Figure 3.2 is identical to that obtained from the composite curves. Finally, determine the net exergy loss target for Case Study 4S1 as a continuous function of ΔT_{min}.

Given: Hot utility 180° - 179°C and cold utility 15° - 25° C.

Solution. If T_o is the ambient temperature, then the exergy change for a heat source/sink at constant temperature, T, is given by

$$\Delta Ex = \Delta Q \,(1 - T_o/T) \tag{8.18a}$$

If the temperature of the heat source/sink varies from T_i to T_f, then integration, assuming constant heat capacity flow rate, gives

$$\begin{aligned} \Delta Ex &= MC_p \int (1 - T_o/T)\, dT = MC_p\, [(T_f - T_i) - T_o \ln(T_f/T_i)] \\ &= \Delta Q\, (1 - T_o/T_{lmtd}) \end{aligned} \tag{8.18b}$$

When the above equations are used, temperatures must be absolute (in K).

Step 1. Exergy Calculations Based on Composite Curves

The composite curve segments represent heat sources and sinks, so Equations 8.18a and 8.18b may be applied to obtain the exergy change for each segment. Using the composite curve data for Case Study 4S1 ($\Delta T_{min} = 13$°C) in Table 2.8, the exergy changes based on composite curves may be obtained as in Table 8.5. Sample calculations are given below:

For the temperature interval 125°-175°C on the HCC,
$\Delta Ex = (3200 - 3700)\, [1 - 298 \ln(448/398)/(448 - 398)] = -147.34$ kW.

For the temperature interval 25°-40°C on the CCC,

ΔEx = (680 − 380) [1 − 298 ln(313/298)/(313 − 298)] = 7.31 kW.

Table 8.5 Computation of Exergy Changes

From Composite Curves				From MER design		From Stream Data	
HCC Itvl	ΔEx	CCC Itvl	ΔEx	Unit	Ex_{loss}	Stream	ΔEx
45 - 65	−18.24	15 - 20	−3.62	1	22.31	H1	−278.63
65 - 125	−565.25	20 - 25	−2.04	2	48.96	H2	−452.20
125 - 175	−147.34	25 - 40	7.31	3	49.50	C3	441.45
179 - 180	−122.92	40 -112	360.57	4	89.47	C4	154.53
		112 - 155	228.96	5	18.99	HU	$-0.341 Q_{hu}$
				6	33.33	CU	$-0.017 Q_{cu}$

Units: temperature in °C and exergy change in kW.

Now, the total system exergy loss, σT_o, may be obtained from

$$\sigma T_o = - \sum \Delta Ex(\text{sources \& sinks}) \qquad (8.18c)$$
$$= - (-18.24 - 565.25 - 147.34 - 122.92$$
$$- 3.62 - 2.04 + 7.31 + 360.57 + 228.96) = 262.57 \text{ kW}.$$

Step 2. Exergy Calculations Based on MER Network
The exergy loss for a heat exchanger may be calculated using

$$Ex_{loss} = T_o [MC_{p,c} \ln(T_{cf}/T_{ci}) - MC_{p,h} \ln(T_{hi}/T_{hf})]. \qquad (8.18d)$$

For example, application of Equation 8.18d to unit 1 in Figure 3.2 gives

$$Ex_{loss} = 298 [20 \ln(410/385) - 10 \ln(448/398)] = 22.31 \text{ kW}.$$

Calculating exergy losses for other units in a similar manner (Table 8.5) and summing up, the total exergy loss for the MER network is

$$\sigma T_o = 22.31 + 48.96 + 49.5 + 89.47 + 18.99 + 33.33 = 262.57 \text{ kW}.$$

This is precisely the total exergy loss obtained earlier in step 1, based on the composite curves. This may be contrasted against the total exergy loss for the network in Figure 1.2, which is given by

$\sigma T_o = 142.41 + 88.09 + 172.78 + 232.23 = 635.51$ kW.

Thus, pinch analysis enables targets to be established for the total inevitable exergy loss for an optimal network at a particular value of ΔT_{min}. Can the exergy loss target be established as a continuous function of ΔT_{min}? This is shown next.

Step 3. Exergy Calculations Based on Stream Data
Obviously, the streams themselves are heat sources and sinks, so Equations 8.18a and 8.18b are directly applicable to them. For example,

$$\Delta Ex = 10 (318 - 448) [1 - 298 \ln(448/318)/(448 - 318)]$$
$$= -278.63 \text{ kW (stream H1)}$$
$$\Delta Ex = 20 (428 - 293) [1 - 298 \ln(428/293)/(428 - 293)]$$
$$= 441.45 \text{ kW (stream C3)}$$
$$\Delta Ex = -Q_{hu,min} [1 - 298 \ln(453/452)/(453 - 452)]$$
$$= -0.3414 \, Q_{hu,min} \text{ (stream HU)}$$
$$\Delta Ex = Q_{cu,min} [1 - 298 \ln(298/288)/(298 - 288)]$$
$$= -0.0172 \, Q_{cu,min} \text{ (stream CU)}$$

The calculations are summarized in Table 8.5. Use of Equation 8.18c now gives

$$\sigma T_o = -[-278.63-452.2+441.45+154.53-0.3414 \, Q_{hu,min}-0.0172 \, Q_{cu,min}]$$
$$= 134.84 + 0.3414 \, Q_{hu,min} + 0.0172 \, (Q_{hu,min} - 80)$$
$$= 133.47 + 0.3586 \, Q_{hu,min} \text{ kW}$$

For $\Delta T_{min} = 13°C$, $\sigma T_o = 133.47 + 0.3586 (360) = 262.57$ kW (as before).

Having established σT_o as a function of $Q_{hu,min}$, the exergy loss target may be obtained as a continuous function of ΔT_{min}, provided the hot utility requirement is known for the complete ΔT_{min} range. For Case Study 4S1, this information is available in Table 2.6 and allows straightforward development of the exergy loss plot in Figure 8.18.

Finally, it must be remarked that the composite curves may be drawn with the ordinate given by the Carnot factor, η_c, (i.e., $1 - T_o/T$) rather than the temperature. Figure 8.19 shows such a plot for Case Study 4S1 at $\Delta T_{min} = 13°C$ starting with the data in Table 2.8. Umeda et al. (1979a) have referred to this plot as the energy availability diagram. On rewriting Equation 8.18b in the form $\Delta Ex = \int(1 - T_o/T) \, dQ = \int \eta_c \, dQ$, it is clear that the area between the balanced exergy composites in Figure 8.19 is the total exergy loss in the HEN.

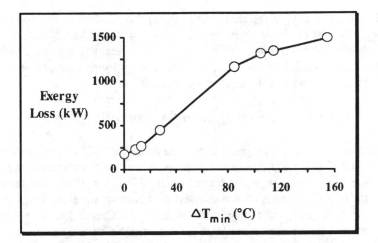

Figure 8.18 The total exergy loss (σT_o) target shows a piecewise linear variation with ΔT_{min}. Open circles denote significant shifts (see Table 2.6).

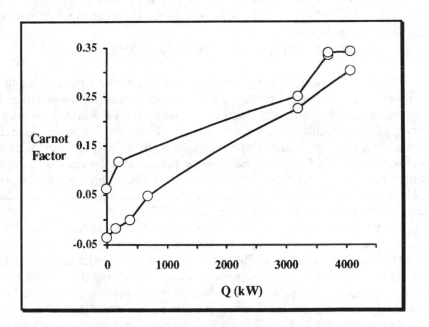

Figure 8.19 On an energy availability diagram, the area between the exergy composite curves (with the utilities included) is a true measure of the total exergy loss (σT_o) in the HEN.

The disadvantage of the energy availability diagram is that the composites no longer consist of linear segments, so, from a practical standpoint, the area between the balanced composites with the true temperature scale on the ordinate (see Figure 1.6) may still be considered a more convenient, though approximate, indicator of the total exergy losses.

8.5.1 The Exergy Grand Composite Curve

In principle, the exergy grand composite curve (EGCC) can be constructed from the GCC using a procedure analogous to that used for transforming the composite curves to exergy composite curves. However, if the area included within the EGCC (on considering utilities) is to correspond exactly to the total HEN exergy losses, then the ordinate must be the Carnot factor, calculated based on true temperatures (without the temperature shifting conventionally used in the GCC). The procedure to obtain the EGCC is given below.

Example 8.7 Plotting of the Exergy Grand Composite Curve

For the stream data in Case Study 4S1, plot the exergy grand composite curve (EGCC) for $\Delta T_{min} = 13°C$. Show how the total HEN exergy loss may be obtained from the EGCC.

Given: Hot utility 180° - 179°C and cold utility 15° - 25°C.

Solution.   Conventionally, the GCC is obtained by first shifting the composite curves vertically until they just touch each other (at the pinch) and then calculating the horizontal distance of a vertex from the other composite curve. This horizontal distance will be zero at the pinch points. Such a vertical shifting is not desirable when constructing the EGCC because the area enclosed by the EGCC should ideally be the same as that between the exergy composites for it to represent the HEN exergy losses. This may be simply achieved by using the true temperature scale and by maintaining the horizontal distances between the composite curves in the GCC as well.

The pinch tableau in Table 2.8 provides a convenient starting point. The hot and cold stream temperatures are merged and then sorted in ascending order to obtain the first row in Table 8.6. Temperatures underlined in Table 2.8 are ignored as these are interpolated values. For hot vertices, enthalpies are simply copied from Table 2.8 into the next row (labeled Q_h) of Table 8.6. In a similar fashion, cold stream enthalpies are copied from Table 2.8 into the third row (labeled Q_c) of Table 8.6. The "missing" enthalpies in the second and third rows of Table 8.6 are now entered by simple interpolation (and are shown underlined). The last row of Table 8.6 contains Q_{egcc}, namely (Q_c − Q_h). If T vs. Q_{egcc} is plotted, the area enclosed by it in the first quadrant will be identical to that between the composite curves. It may be noted that the "missing" enthalpies (after interpolation) at the ends of the Q_c and Q_h rows of Table 8.6 are filled by merely repeating the first and last known entries in the respective rows (e.g., the entries $Q_h = 0$ and $Q_c = 4060$ in Table 8.6).

On transforming the T values in Table 8.6 into η_c and plotting them against Q_{egcc}, the EGCC in Figure 8.20 is obtained. Note that linear interpolation is valid for T in Table 8.6 but not for η_c in Figure 8.20.

Table 8.6 Tableau for Generating EGCC

T	15	20	25	40	45	65	112	125	155	175	179	180
Q_h	0	0	0	0	0	200	2550	3200	3500	3700	3700	4060
Q_c	0	140	380	680	855	1555	3200	3460	4060	4060	4060	4060
Q_{egcc}	0	140	380	680	855	1355	650	260	560	360	360	0
ΔEx	--	-3.62	-2.04	7.31	9.70	45.6	-123	-93.1	83.4	-63.9	0	-123

$\sum \Delta Ex = -262.57$ kW.

Units: temperature in °C and enthalpy/exergy in kW.

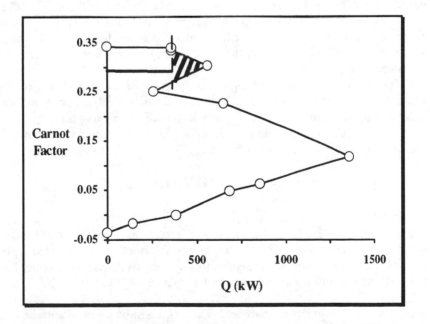

Figure 8.20 The exergy grand composite curve measures the total HEN exergy loss (σT_0) in terms of the area included within it and also allows the appropriate selection of utilities for minimization of the avoidable exergy losses.

As mentioned earlier, the total exergy loss in the HEN (including utility exchangers) is given by the area enclosed by the EGCC (including utilities) and the vertical axis. The exergy loss may be analytically computed by recognizing that each EGCC segment represents a heat source/sink and Equation 8.18b is therefore applicable. For example, the EGCC segment between 65° and 112°C yields

$$\Delta Ex = (650 - 1355) [1 - 298 \ln(385/338)/(385 - 338)] = -123.02 \text{ kW}.$$

Summing the entries in the last row of Table 8.6, the total HEN exergy loss is −262.57 kW (as before).

The EGCC provides a visual picture for distinguishing between avoidable and inevitable exergy losses based on an appropriate choice of utilities. Consider the region above the pinch (125°C). Exergy losses could be avoided by choosing a hot utility at a lower temperature (i.e., 149°-148°C rather than 180°-179°C) as shown by the heavy line in Figure 8.20. The hot utility line cannot be shifted down any further without a loss in the inevitable driving forces (i.e., the optimum ΔT_{min} based on economic tradeoffs), and so the exergy losses above the pinch are now inevitable. Note that the GCC pocket above the pinch represents inevitable exergy losses since the shaded area in Figure 8.20 cannot be easily reduced. Similar arguments hold below the pinch where the cold utilities could be chosen at the highest allowable temperatures to reduce exergy losses.

The philosophy behind the choice of utilities on the EGCC is virtually identical to that on the GCC (as described in Section 1.7.3). Hence, in most cases, it may be more convenient to do multiple utility targeting directly on the GCC. The EGCC, however, may be profitably used in systems with dominant shaftwork considerations, as discussed next.

8.6 SHAFTWORK TARGETING

In this section, a reliable procedure for shaftwork targeting (Linnhoff and Dhole, 1992) is described with special reference to subambient processes, where the power requirements are significant and need to be evaluated. The approach essentially involves understanding the exergy flows in the low temperature system by transforming the subambient GCC to the EGCC.

Example 8.8 Establishing Shaftwork Targets for Subambient Processes
Determine the exergy losses for the subambient process in Example 8.5 based on the two different refrigeration schemes represented by the GCCs in

Figures 8.15 and 8.17. Also, estimate the shaftwork savings if the scheme in Figure 8.15 (assumed to be the base case) is changed to the one in Figure 8.17.

Given: ΔT_{min} = 5°C and shaftwork consumption for base case (Figure 8.15) = 3500 kW.

Solution. Shaftwork targeting aids in the selection of a superior scheme for the HEN and refrigeration system of a subambient process. It allows for quick screening of various HEN/refrigeration system options. The selection of both the HEN and refrigeration system is done simultaneously without the tedious task of designing either of them. Importantly, the approach considers the complex interactions between the process, the HEN, and the refrigeration system during targeting.

Step 1. Generating the EGCC and Determining the Exergy Losses
The GCCs in Figures 8.15 and 8.17 are converted into EGCCs as shown in Figures 8.21 and 8.22 respectively. Given that the refrigeration levels form point utilities (represented by heavy lines), the EGCC for the utilities may be drawn separate from that for the process.

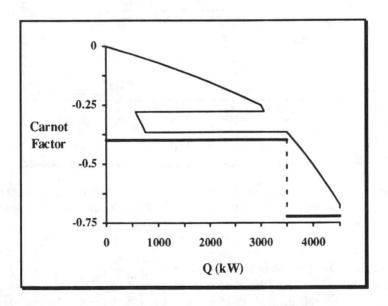

Figure 8.21 The area enclosed by the process EGCC and the utility EGCC equals the total exergy loss in the HEN ($\sigma T_{o,HEN}$).

The exergy balance yields

$$\Delta Ex(\text{HEN}) = \Delta Ex(\text{process}) + \Delta Ex(\text{refrigeration}) \qquad (8.19a)$$
$$= 1248.77 - 2119.25 = -870.48 \text{ kW (for Figure 8.21)}$$
$$= 1248.77 - 1834.87 = -586.10 \text{ kW (for Figure 8.22)}$$

Thus, $(\sigma T_o)_{HEN}$ is 870.48 kW and 586.10 kW in Figures 8.21 and 8.22 respectively. Note that the refrigeration system supplies exergy to the process via the HEN, where some exergy $(\sigma T_{o,HEN})$ is lost. In other words, $\sigma T_{o,HEN}$ is given by the area between the process EGCC and the utility (refrigeration) EGCC.

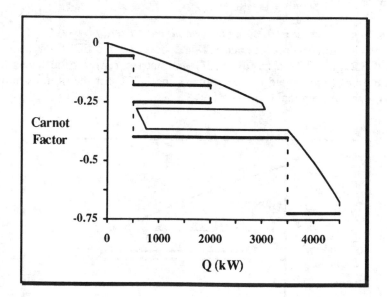

Figure 8.22 The EGCC plot provides easy visualization of the reduction in shaftwork requirement (compared to Figure 8.21) and the increased design complexity.

Step 2. Calculating the Shaftwork Reduction

Figures 8.21 and 8.22 have identical process EGCCs. The difference lies in the HEN and refrigeration system, but not in the process. On considering the exergetic efficiency of the refrigeration system, η_{ex}, the shaftwork input to the refrigeration system is given by

$$W = -\Delta Ex(\text{refrigeration})/\eta_{ex} = [\Delta Ex(\text{process}) + \sigma T_{o,HEN}]/\eta_{ex}. \quad (8.19b)$$

If η_{ex} is assumed to be unchanged between Figures 8.21 and 8.22, then the actual shaftwork reduction between the two cases is given by

$$\Delta W = \Delta[\sigma T_{o,HEN}]/\eta_{ex}. \quad (8.19c)$$

As $\sigma T_{o,HEN}$ is represented by the area between the process EGCC and the refrigeration EGCC, the shaftwork reduction is directly proportional to the decrease in the relevant areas between Figures 8.21 and 8.22.

Substitution in Equation 8.19c gives $\Delta W = (870.48 - 586.10)/(2119.25/3500) = 469.66$ kW. This reduction in shaftwork may now be assessed against the increase in design complexity. Also, note that it is possible to include compression discharge levels in the above analysis.

Linnhoff and Dhole (1992) have compared the shaftwork based on Equation 8.19c with that obtained through detailed system design/simulation and found the target to be accurate to within a few percentage points. The high accuracy is due to the approach being fundamental and rigorous. The constancy of the exergetic efficiency typically turns out to be an excellent assumption in practice, especially for similar refrigeration fluids and similar temperature ranges.

The process was not altered throughout the targeting in the above example. However, process modifications may be attempted (Linnhoff and Dhole, 1992), based on the EGCC, to reduce the shaftwork. For example, a change in the column pressure in the subambient plant will alter the process EGCC and may reduce ΔEx(process). If the refrigeration system is unchanged, then $\sigma T_{o,HEN}$ will increase as per Equation 8.19b. The new possibilities generated by the process change may now be exploited by suitably modifying the refrigeration system to reduce $\sigma T_{o,HEN}$ and consequently the shaftwork.

Air separation, ethylene, and gas processing provide some useful applications for this simple shaftwork targeting procedure.

8.7 TOTAL SITE TARGETING

The material presented thus far deals with a single process; in this section, a methodology (Dhole and Linnhoff, 1993a) is introduced that allows analysis of total factories including many processes serviced by a central utility system. The approach is based on total site profiles (TSPs), which allow setting of overall targets for fuel, power, steam, cooling, and emissions. It can be used effectively at both the grassroots and retrofit level to optimize the interactions between the various processes. The generation of TSPs is demonstrated first, followed by their use in targeting and process analysis.

Example 8.9 Plotting of Total Site Profiles

Obtain the total site profiles from the two sets of GCC data given below:

GCC1:	T_{int} (°C)	165	122	115	55	50	35	30
	Q (kW)	605	175	0	900	775	625	525
GCC2:	T_{int} (°C)	195	190	165	120	100		
	Q (kW)	700	400	0	200	500		

Note that GCC1 corresponds to the data for Case Study 4S1 at ΔT_{min} = 20°C. Let the global ΔT_{min} be 20°C also.

Solution.   The method for generating the total site profiles is demonstrated below for the case of only two processes, but the extension to a site with many processes is straightforward.

Step 1. Removal of the Nonmonotonic Parts from Individual GCCs

The nonmonotonic parts (pockets) on the GCCs, which represent process-to-process heat exchange, are eliminated first. It is observed that only GCC1 has a pocket (see shaded area in Figure 1.5), which may be removed to obtain the following modified GCC1 data:

GCC1':	T_{int} (°C)	165	122	115	80
	Q (kW)	605	175	0	525

The modification helps in minimizing inter-process interaction on the site, thus providing a higher degree of controllability and operability. GCC2 requires no modification as it does not have a pocket.

Step 2. Shifting of the Source and Sink Elements

The temperatures of the sink elements on the GCCs (entries above the pinch) are shifted upwards by $\Delta T_{min}/2$ whereas the temperatures of the source elements on the GCCs (entries below the pinch) are shifted downwards by $\Delta T_{min}/2$.

The shifted GCCs are shown (in heavy lines) in Figure 8.23, along with the original GCCs. Note that, on shifting, the entry corresponding to the original pinch temperature results in two entries (one above the pinch and another below it) as shown in Table 8.7.

Table 8.7 From Shifted GCCs to Total Site Profiles

Shifted GCC1		Shifted GCC2		Site Sink Profile		Site Source Profile	
T	Q	T	Q	T	Q	T	Q
175	605	205	700	125	0	155	0
132	175	200	400	132	175	110	−200
125	0	175	0	175	605	105	−275
105	0	155	0	200	1005	90	−725
70	525	110	200	205	1305	70	−1025
		90	500				

Units: temperature in °C and enthalpy in kW.

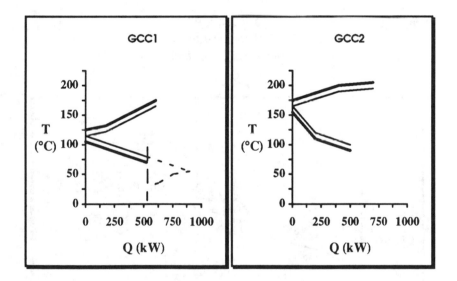

Figure 8.23 The individual process GCCs must be modified (by eliminating all pockets and shifting sink elements up and source elements down) before constructing the TSPs.

Step 3. Construction of the Total Site Profiles

The total site profiles (TSPs) are the composites of the shifted GCCs of the various processes. The site sink profile is constructed by combining the sink (above the pinch) portions of the two shifted GCCs. The procedure philosophically is similar to that adopted for the composite curves in Section 1.1.6. The site source profile is constructed in a similar manner to the site sink profile, except that it is placed on the negative enthalpy axis. The TSPs (also referred to as site source-sink profiles) based on Table 8.7 are shown in Figure 8.24.

Example 8.10 Establishing Targets Based on Total Site Profiles

Consider the utility system with two steam mains (Figure 8.25) for servicing the processes in Example 8.9. The central utility system generates very high pressure (VHP) steam at 475°C in a boiler. The VHP steam is supplied to steam turbines to generate power as well as intermediate pressure (IP) steam at 205°C and low pressure (LP) steam at 145°C. Assume cooling water is available at 25°C.

For this total site, obtain targets for VHP, IP, and LP steams, as well as for fuel, cooling and cogeneration (power).

Given: ΔT_{min} = 20°C, site power demand = 1000 kW
    ambient temp T_o = 25°C, theoretical flame temp T_{tf} = 1500°C
    cost (in \$/[kW.yr]): 438 (power), 130 (fuel), 10 (CW)

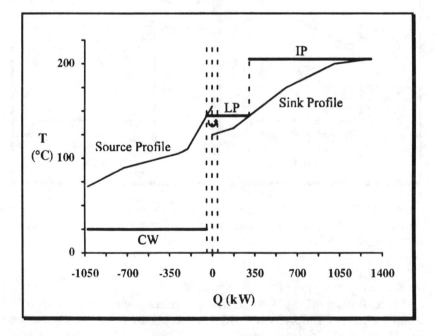

Figure 8.24 Total site profiles allow the establishment of overall targets for process heating and cooling demands.

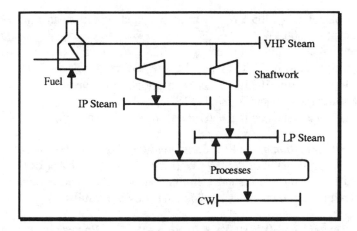

Figure 8.25 The utility system consists of two turbine stages that produce IP and LP steam from VHP steam, along with cogeneration of electrical power. The process heating duties are satisfied by IP and LP steam (supplemented by LP steam-raising) and the process cooling duties are met by cooling water.

Solution. The process heating and cooling demands are first established from the TSPs. Then, based on the utility system, the cogeneration potential and sitewide fuel requirement are determined.

Step 1. Requirement of LP Steam

The amount of LP steam required can simply be read off as the heat load at 145°C from the site sink profile in Figure 8.24. From Table 8.7, the LP steam requirement is obtained as 305 kW (i.e., 175 + [605 − 175] [145 − 132]/[175 − 132]) by interpolation for the heat load at 145°C on the second line segment of the site sink profile. Note that the temperature shifting in step 2 of Example 8.9 ensures a ΔT_{min} of 20°C.

However, from the site source profile in Figure 8.24, the potential for generating some LP steam can be observed. From Table 8.7, the amount of LP steam-raising possible is obtained as 44.44 kW (i.e., 200 [155 − 145]/[155 − 110]) by interpolation on the site source profile.

As 44.44 kW out of the total requirement of 305 kW can be generated in the process itself, the net requirement of LP steam is 260.56 kW.

Step 2. Requirement of IP Steam and Cooling Water

The total hot utility requirement is 1305 kW, which is the heat load at the highest temperature of 205°C as per the site sink profile in Table 8.7. As

305 kW out of this 1305 kW has been supplied by LP steam, the IP steam requirement is 1000 kW.

The total cold utility requirement is 1025 kW, which is the heat load at the lowest temperature of 70°C as per the site source profile in Table 8.7. As 44.44 kW out of this 1025 kW has been utilized in generating LP steam, the cooling water requirement is 980.56 kW.

The IP steam and cooling water demands are shown in Figure 8.24.

Step 3. Determination of the VHP Steam and the Cogeneration Targets

The steam demand of 260.56 kW (for LP) and 1000 kW (for IP) must be provided by the turbine system. Simulation of the turbine system using the procedure given in Example 8.1 yields the following results:

Stage	Target Load (kW)	Passout Flows (kg/s)	Power Generated (kW)
1	1000 (IP @ 205°C)	0.4894	229.55
2	260.56 (LP @ 145°C)	0.1243	29.86

Thus, the total cogeneration is 259.41 kW and the requirement for generating steam (VHP @ 475°C and 9750 kPa) is 1519.97 kW (see Figure 8.26).

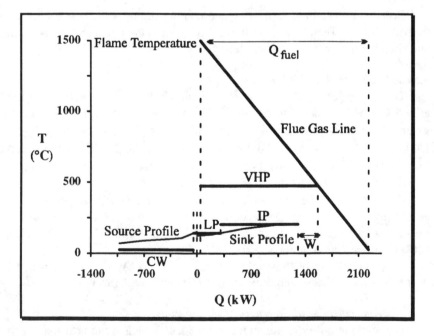

Figure 8.26 Total site profiles allow the establishment of sitewide targets for turbine loads, cogeneration, and fuel.

Step 4. Determination of the Fuel Target

The fuel target is now simply obtained by dividing the VHP steam requirement by the boiler efficiency. Thus, as shown in Figure 8.26,

$$Q_{fuel} = Q_{VHP} (T_{tf} - T_o)/(T_{tf} - T_{VHP})$$ (8.20)
$$= 1519.97 (1500 - 25)/(1500 - 475) = 2187.27 \text{ kW}$$

In fact, targets for fuel-related emissions (e.g., CO_2, SO_x, and particulates) can be determined at this stage, provided the fuel composition is known (see Section 8.8).

Step 5. Cost Estimation

As site power demand is 1000 kW and the cogeneration is only 259.41 kW, the rest of the power (740.59 kW) must be imported. Thus, the total utilities operating cost may be obtained as

Operating cost = 740.59 x 438 + 2187.27 x 130 + 980.56 x 10
$$= 618.53 \times 10^3 \text{ $/yr.}$$

Note that steam costs are strictly internal parameters (Linnhoff and Dhole, 1993) and therefore do not affect the cost estimates in total site targeting.

Step 6. Identifying Promising Modifications

It is observed that the LP steam raised by the site source is inadequate to meet the demands of the site sink. Thus, 260.56 kW of LP steam has to be generated by using a turbine stage. This may be avoided if the LP steam required is exactly balanced by the LP steam raised. This would call for either a process modification or a utility modification (say, an adjustment in the LP steam pressure level). For this case study, consider the temperature level of the LP steam to be adjusted to 129.53°C (obtained from $200 [155 - T]/45 = 175 [T - 125]/7$). As 113.2 kW of LP steam is raised and used, the IP steam and cooling water demands now become 1191.8 kW and 911.8 kW respectively.

Step 7. Cogeneration Target from Exergy TSPs

Instead of simulating the turbine system, the cogeneration target may be obtained from a Carnot factor vs. enthalpy plot (Dhole and Linnhoff, 1993a) as shown in Figure 8.27. Conceptually, the approach is no different from the shaftwork targeting in Example 8.8. The shaded area in Figure 8.27 represents the exergy difference between VHP steam and IP steam; therefore, it specifies the ideal shaftwork. The ideal shaftwork is multiplied by the exergetic efficiency, η_{ex}, of the turbine system to obtain the real shaftwork.

The exergetic efficiency may be estimated using the utility system in Figure 8.26 as a base case. Figure 8.26 may be replotted with a Carnot factor axis; then, the area corresponding to the ideal shaftwork target is found to be 463.05 kW. Thus, the exergetic efficiency is 0.56 (i.e., 259.41/463.05). Based on the shaded area in Figure 8.27,

$$W = 0.56 \,[(1191.8 + W)\,(1 - 298/748) - 1191.8\,(1 - 298/478)]$$

or $W = 226.59$ kW.

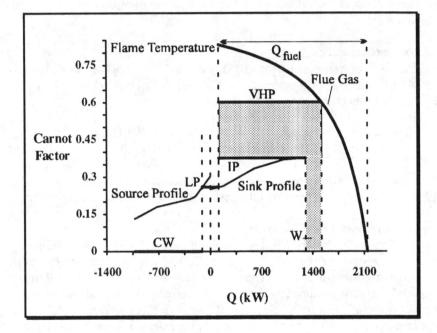

Figure 8.27   The exergy total site profiles allow the establishment of the sitewide cogeneration target directly in terms of the exergetic efficiency and the area shaded in the above η_c vs. Q plot.

Direct turbine simulation yields $W = 218.16$ kW. The difference in the shaftwork target (of about 4%) is due to the assumption of the constancy of the exergetic efficiency.

Step 8. Utility and Cost Evaluation of the Modifications
    The fuel target and the cost estimate for the modified utility system may be obtained as in steps 4 and 5. The results are summarized below for comparison purposes:

Figure	LP (kW)	IP (kW)	CW (kW)	VHP (kW)	W (kW)	Fuel (kW)	Cost ($/yr)
8.26	260.56	1000	980.56	1519.97	259.41	2187.27	618529
8.27	0	1191.8	911.8	1409.96	218.16	2028.97	615330

There is a decrease in steam, fuel, and cooling water demands due to the higher heat recovery at the LP level. Unfortunately, but expectedly, there is a corresponding reduction in the cogeneration. The net effect may not be beneficial due to the high cost of power and the tradeoffs involved. In this case study, it is seen that there is a small decrease in the total utilities operating cost. More importantly, capital investment should be reduced as the LP turbine stage is not necessary. The temperature driving forces between the LP steam and TSPs need to be evaluated to see the effect on heat exchanger area.

On the whole, the targets for the modified utility system appear favorable unless the LP mains pressure cannot be easily modified. In any case, TSPs form a powerful tool in the context of OPERA for screening the promising design alternatives at the targeting stage itself.

Some remarks regarding the total site targeting methodology are in order:

- The individual process GCCs used in constructing the TSPs could themselves be independently optimized prior to attempting total site targeting.
- TSPs may be profitably used to evaluate process and utility modifications as well as site expansion proposals (Dhole and Linnhoff, 1993a). For example, the appropriate choice of steam levels on a sitewide basis is suggested by the TSPs.
- Design modifications may be targeted for additional cogeneration (rather than fuel savings as in the above example) when the cost of power is high.

8.8 EMISSIONS TARGETING

Environmental emissions can be reduced through OPERA by efficient utilization of energy and raw materials in the process. Smith et al. (1990) have discussed this aspect with regard to waste minimization, flue gas emissions, and waste treatment options. The first step is to minimize wastes at source through process modifications and thus decrease raw material as well as waste treatment costs. The next step is to minimize flue gas emissions through improved heat integration and thereby define the process/utility system model. The third step is to evaluate waste treatment options within the overall process context and exploit possible integration opportunities. Economic tradeoffs must be assessed at each step, and iterations may be required before arriving at the final solution. In this section, the focus is on flue gas emissions from utility systems that are generated through combustion processes to supply the heat and power requirements of the plant.

8.8.1 Evaluating Flue Gas Emissions

Efficient use of energy leads to the effective use of fuel and, consequently, reduced fuel-related emissions. The approach may call for maximization of heat integration and process modifications. The energy targets may then be readily translated into flue gas emissions targets (Smith and Delaby, 1991) using simple thermodynamic models for combustion devices (like furnaces, steam boilers, steam turbines, gas turbines, and central power generation system). Supertargeting may be used to fix the appropriate economic level of heat recovery, and the GCC/TSPs may be used to assess different utility options at a local/sitewide level (Linnhoff and Dhole, 1993).

Example 8.11 Emissions Targets for Single Utilities without Cogeneration
Evaluate the minimum flue gas emissions (i.e., predict the pollutant rates at the local and global levels) for Case Study 4S1 at ΔT_{min} = 13°C. Consider the following two cases for the supply of the process hot utility requirement:
(a) a furnace using oil as a fuel; and
(b) a steam boiler using natural gas as a fuel and satisfying process duty through IP steam at 205°C (1718 kPa).

Given: theoretical flame temperature T_{tf} = 1500°C,
stack temperature T_{stack} = 160°C, and ambient temperature T_o is 25°C.

Assume the power requirement is 1000 kW. The central electricity generation system is a steam turbine power plant which uses coal as fuel and has an overall efficiency of 30%. It has an ash removal efficiency of 90%.

Given: ultimate analyses of various fuels (composition %)

	Carbon	Hydrogen	Oxygen	Nitrogen	Sulfur	Ash	NHV(kJ/kg)
Coal	75.8	5.1	8.2	1.5	1.6	7.8	31520
Fuel oil	87.26	10.94	0.64	0.28	0.84	0.04	39830
Nat. gas	76.0	22.8	0	1.1	0.1	0	51550

Solution. The pollutants targeted are CO_2 (that causes global warming), SO_x (that causes acid deposition), and particulates (like metal oxides and inorganic ash that are health hazards).
The targeting of fuel-related emissions is based on the procedure suggested by Smith and Delaby (1991). The first step is to estimate the fuel

requirement, Q_{fuel}, from the hot utility consumption, $Q_{hu,min}$. A simple thermodynamic model of the combustion device is used for this purpose. The next step is to use the following stoichiometric equation to model the emission rates of the various pollutants for a particular fuel:

$$M_{pol} = (Q_{fuel}/NHV)\, P_{pol}\, R_{pol} \tag{8.21}$$

where M_{pol} is the mass flow rate of the pollutant (e.g., CO_2 and SO_x), NHV is the net heating value of the fuel, P_{pol} is the mass percentage of the pollutant in nonoxidized form (e.g., carbon or sulfur) in fuel, and R_{pol} is the ratio of the molar masses of the oxidized form to the nonoxidized form of the pollutant. Note that R_{pol} is 3.67 (i.e., 44/12) for carbon dioxide and 2 (i.e., 64/32) for sulfur dioxide. It is assumed that all the carbon, sulfur, metals, and inorganic ash in the fuel are oxidized and emitted. Equation 8.21 may also be used for particulates with $R_{pol} = 1$. It is required that any pretreatment of fuel be accounted for in the mass analysis.

Equation 8.21 is not applicable for NO_x as the oxidation reaction does not reach equilibrium in the combustion zone and, therefore, the amount of nitrogen oxidized cannot be calculated accurately (as it depends on mass transfer and kinetic effects). However, experience-based conversion factors (Smith and Delaby, 1991) may be used to obtain rough estimates.

Step 1. Determining the Hot Utility Requirement
The process heat duty for Case Study 4S1 at $\Delta T_{min} = 13°C$ is 360 kW.

Step 2. Estimating Fuel Consumption Based on Simple Furnace Model
The furnace could be appropriately placed using the procedure discussed in Example 3.4 and the fuel target could be subsequently determined using Equation 3.4b. An estimate of the fuel consumption may be obtained by dividing the hot utility requirement by the furnace efficiency. Thus,

$$Q_{fuel} = Q_{hu,min}\,(T_{tf} - T_o)/(T_{tf} - T_{stack}) \tag{8.22}$$
$$= 360\,(1500 - 25)/(1500 - 160) = 396.27\ \text{kW}$$

Step 3. Predicting Pollutant Rates
Equation 8.21 may be used along with the mass analysis data and the net heating value for the fuel oil to yield the following results:

Local M_{CO_2} $\quad = (396.27/39830)\ 0.8726 \times 3.67 \times 3600 = 114.60$ kg/hr
Local M_{SO_x} $\quad = (396.27/39830)\ 0.0084 \times 2 \times 3600 \quad = 0.602$ kg/hr
Local $M_{particulates} = (396.27/39830)\ 0.0004 \times 1 \times 3600 \quad = 0.014$ kg/hr

Note that emissions are quantified in terms of the flow rate (i.e., kg/hr) rather than the (more conventionally used) concentration (i.e., ppm). This provides a true measure of the emissions and appropriately considers pollution as a global problem (Smith and Delaby, 1991).

Step 4. Assessing Global Emissions

The local emission rates generated by the on-site utilities have been calculated so far. However, the process has a power requirement of 1000 kW, which must be imported from the central power station. As the central power station emits pollutants in producing this load, it is appropriate to define global emissions as (Smith and Delaby, 1991)

Global emissions = [local emissions]
+ [emissions relating to power imported from central station]
− [emissions relating to power exported from site]   (8.23)

Global emissions provide a more authentic description of the total environmental burden.

The emissions corresponding to the 1000 kW of power imported from the central station (with $Q_{fuel} = 1000/0.3 = 3333.33$ kW) are given as

$$M_{CO_2,i} = (3333.33/31520) \, 0.758 \times 3.67 \times 3600 = 1058.12 \text{ kg/hr}$$
$$M_{SO_x,i} = (3333.33/31520) \, 0.016 \times 2 \times 3600 = 12.183 \text{ kg/hr}$$
$$M_{particulates,i} = (3333.33/31520) \, 0.078 \times 0.1 \times 3600 = 2.970 \text{ kg/hr}$$

Thus, the global emissions as per Equation 8.23 may be reported as

$$\text{Global } M_{CO_2} = 114.60 + 1058.12 - 0 = 1172.72 \text{ kg/hr}$$
$$\text{Global } M_{SO_x} = 0.602 + 12.183 - 0 = 12.785 \text{ kg/hr}$$
$$\text{Global } M_{particulates} = 0.014 + 2.970 - 0 = 2.984 \text{ kg/hr}$$

It is observed that the global emissions flow rates are significantly higher than the local ones, partly due to the low efficiency of 0.3 (which includes distribution) at the central power station. From an emissions viewpoint, this is motivation to explore on-site cogeneration (see Example 8.12).

Step 5. Comparing Furnace Using Oil with Steam Boiler Using Natural Gas

The fuel requirement is obtained through a simple thermodynamic model of the steam boiler as described below. The IP steam at 205°C is generated by let down from VHP steam, which is itself generated by combustion of natural gas. As the VHP steam is throttled, it is necessary to maintain the quality of the steam by desuperheating through the addition of boiler feed water (BFW).

Let $(1 - x)$ kg of superheated VHP steam at 475°C and 9750 kPa ($H =$ 3320 kJ/kg) be mixed with x kg of feed water at 100°C ($H = 418.9$ kJ/kg) to give 1 kg of saturated steam at 1718 kPa ($H = 2795.4$ kJ/kg). By a simple enthalpy balance, the amount of desuperheating duty (x) is obtained as 0.1808 kg of BFW/kg of saturated steam (i.e., $[3320 - 2795.4]/[3320 - 418.9]$).

Then, the fuel required in the boiler is given by

$$Q_{fuel} = (Q_{hu,min}/\Delta H_{lat,p}) (1 - x)$$
$$(\Delta H_{lat,b} + \Delta H_{sup,b}) (T_{tf} - T_o)/(T_{tf} - T_b - \Delta T_b) \qquad (8.24)$$

where $\Delta H_{lat,p}$ = latent heat of steam satisfying process duty = 1921.4 kJ/kg
$\Delta H_{lat,b}$ = latent heat of steam at boiler condition = 1405.1 kJ/kg
$\Delta H_{sup,b}$ = enthalpy of superheat at boiler condition = 565 kJ/kg
T_b = steam condensing temperature at boiler condition = 300°C
ΔT_b = approach temperature in boiler = 40°C

Substitution in Equation 8.24 gives

$$Q_{fuel} = (360/1921.4) (1 - 0.1808)$$
$$(1405.1 + 565) (1500 - 25)/(1500 - 300 - 40) = 384.5 \text{ kW}$$

The procedure for predicting local pollutant rates and assessing global emissions is as described previously in steps 3 and 4. The results for two different combustion devices (furnace and steam boiler) and two different fuels (oil and natural gas) are summarized below for comparison purposes:

Local emissions (kg/hr)	M_{CO_2}	M_{SO_x}	$M_{particulates}$
Furnace (fuel oil)	114.60	0.602	0.0143
Steam boiler (nat. gas)	74.83	0.054	0.0
Furnace (nat. gas)	77.12	0.055	0.0
Steam boiler (fuel oil)	111.19	0.584	0.0139

It is observed that there is a reduction of over 32% in CO_2 emissions on switching from fuel oil to natural gas. This is because of the relatively low carbon content and high *NHV* of gaseous fuels. Natural gas also results in substantially lower SO_x emissions and no particulates.

Note that the emissions ($M_{CO_2,i}$, $M_{SO_x,i}$, and $M_{particulates,i}$) corresponding to the 1000 kW of power imported from the central station remain unaltered in this case where there is no cogeneration.

Example 8.12   Emissions Targets for Single Utilities with Cogeneration

Estimate the pollutant rates at the local and global levels for Case Study 4S1 at ΔT_{min} = 13°C. Consider the following two cases for the supply of the process hot utility requirement:

(a) a gas turbine; and

(b) a steam turbine satisfying process duty through IP steam at 205°C (1718 kPa).

Oil is the fuel used in both cases. Assume the power requirement is 1000 kW, and use relevant data from Example 8.11.

Given: combustion chamber temperature (turbine inlet) $T_{g,in}$ = 1100°C
         turbine outlet temperature $T_{g,out}$ = 760°C.

Solution. Both the gas turbine and the steam turbine are devices that generate heat as well as power, so these cogeneration devices reduce the power imported from the central station. If the power generated by the turbine systems is in excess of the requirement (1000 kW for this example), then the power is exported.

The procedure to determine the emissions targets is identical to that in Example 8.11, except that appropriate thermodynamic models will have to be used to estimate the fuel requirements for the turbines.

Step 1. Estimating Fuel Consumption Based on Simple Gas Turbine Model

The Carnot factor as per Equation 8.1c is given as

$$\eta_c = 1 - (760 + 273)/(1100 + 273) = 0.2476.$$

The fuel requirement is then related to the process utility consumption by

$$Q_{fuel} = Q_{hu,min}/(1 - \eta_e \eta_c) \; (T_{g,out} - T_o)/(T_{g,out} - T_{stack}) \qquad (8.25)$$
$$= 360/(1 - 0.9 \times 0.2476) \; (760 - 25)/(760 - 160) = 567.47 \text{ kW}$$

As recommended by Smith and Delaby (1991), η_e is chosen to be 0.9. The power delivered by the gas turbine may be calculated from Equation 8.2 as

$$W = 0.9 \times 0.2476 \times 567.47 = 126.46 \text{ kW.}$$

Step 2. Predicting Pollutant Rates and Global Emissions for Gas Turbine

The local emissions calculated in step 3 of Example 8.11 will be increased by a factor of (567.47/396.27). Thus, Local M_{CO_2} = 164.10 kg/hr, Local M_{SO_x} = 0.862 kg/hr, and Local $M_{particulates}$ = 0.0205 kg/hr. From the central station, only 873.54 kW of power (rather than 1000 kW) need to be imported. The emissions corresponding to the 873.54 kW imported (using Q_{fuel} = 873.54/0.3 = 2911.82 kW) are given by $M_{CO_2,i}$ = 924.31 kg/hr, $M_{SO_x,i}$ = 10.642 kg/hr, and $M_{particulates,i}$ = 2.594 kg/hr. These emissions are obtained by decreasing the corresponding values in step 4 of Example 8.11 by a factor of (2911.82/3333.33). The global emissions as per Equation 8.23 now become

$$\text{Global } M_{CO_2} \quad = 164.10 + 924.31 - 0 = 1088.41 \text{ kg/hr}$$
$$\text{Global } M_{SO_x} \quad = 0.862 \ + 10.642 - 0 = 11.504 \ \text{kg/hr}$$
$$\text{Global } M_{particulates} = 0.0205 + 2.594 \ -0 = 2.615 \ \ \text{kg/hr}$$

It is observed that the global emissions decrease by about 8% on the average when a gas turbine is used instead of a furnace.

Step 3. Estimating Fuel Consumption Based on Simple Steam Turbine Model
    In principle, the simulation procedure employed in Example 6.1 may be used. Assuming an isentropic efficiency of 0.85 for the turbine, the power generated may be calculated from

$$W = 0.85 \{\Delta H_{lat,b}(1-T_{ps}/T_b) + \Delta H_{sup,b}[1-T_{ps}\ln(T_{sup}/T_b)/(T_{sup}-T_b)]\} \quad (8.26)$$
$$= 0.85\{1405.1(1-478/573) + 565\ [1-478\ln(748/573)/(748-573)]\}$$
$$= 328.65 \text{ kJ/ kg of process steam}$$

In Equation 8.26, temperatures are absolute (in K), and T_{ps} and T_{sup} denote the temperatures of the saturated process steam (at 205°C) and superheated boiler steam (at 475°C) respectively. Now, the enthalpy of the steam after the turbine expansion is 2933.35 kJ/kg (i.e., 3320 − 328.65/0.85). Next, the amount of desuperheating duty (x) is obtained as 0.0549 kg of BFW/kg of saturated steam [i.e., (2933.35 − 2795.4)/(2933.35 − 418.9)]. Then, Equation 8.24 gives

$$Q_{fuel} = 384.5 \ (1-0.0549)/(1-0.1808) = 443.6 \text{ kW.}$$

The mass flow rate of the steam to be produced in the boiler to satisfy the process duty is given by

$$M_{steam} = (360/1921.4) \ (1-0.0549) = \ 0.177 \text{ kg/s}$$

Thus, the total power produced by the steam turbine is

$$W = 0.177 \times 328.65 = 58.2 \text{ kW.}$$

Step 4. Predicting Pollutant Rates and Global Emissions for Steam Turbine

The local emissions calculated in step 3 of Example 8.11 will now be increased by a factor of (443.6/396.27). Thus, Local M_{CO_2} = 128.28 kg/hr, Local M_{SO_x} = 0.674 kg/hr, and Local $M_{particulates}$ = 0.016 kg/hr. The emissions corresponding to the 941.8 kW imported from the central station (using Q_{fuel} = 941.8/0.3 = 3139.33 kW) are given by $M_{CO_2,i}$ = 996.53 kg/hr, $M_{SO_x,i}$ = 11.474 kg/hr, and $M_{particulates,i}$ = 2.797 kg/hr. Thus, the global emissions as per Equation 8.23 are

$$
\begin{aligned}
\text{Global } M_{CO_2} &= 128.28 + 996.53 - 0 = 1124.81 \ \text{kg/hr} \\
\text{Global } M_{SO_x} &= 0.674 + 11.474 - 0 = 12.148 \ \text{kg/hr} \\
\text{Global } M_{particulates} &= 0.016 + 2.797 - 0 = 2.813 \ \text{kg/hr}
\end{aligned}
$$

It is observed that the emissions for the steam turbine are higher (by about 3 to 8%) than those for the gas turbine on a global basis although the reverse is true on a local basis. This emphasizes the fact that emissions may be reduced by increasing the on-site generation of power because of the large inefficiencies usually associated with the central power station. In practical situations, it may also be worthwhile to assess the reduction in global emissions possible by decreasing the outlet pressure of the steam turbine or the outlet temperature of the gas turbine. Note that decreasing the outlet temperature of the gas turbine increases its local emissions. Also, gas turbines appear to produce less emissions than steam turbines except when VLP steam is used to meet the process heating demands (Smith and Delaby, 1991).

In the above example, the comparison is on the basis of the same heat output. Note that the power outputs of the two devices are different. Comparisons for cogeneration devices may be based on the same heat output or the same power output. For choosing between steam and gas turbines from an emissions standpoint, it may be more appropriate to compare these cogeneration devices on the basis of heat to power ratio as suggested by Smith and Delaby (1991). The heat-to-power ratio in the above example is 0.36 (i.e., 360/1000) for the process, 2.85 (i.e., 360/126.46) for the gas turbine, and 6.19 (i.e., 360/58.2) for the steam turbine. Typical heat-to-power ratios range from 2.5 to 4 for chemical processes, from 0.85 to 3 for gas turbines, and from 3.5 to 10 for steam turbines. Thus, it is observed that the process heat-to-power ratio in the example is very low necessitating the import of power.

An alternative way of comparison is in terms of cost as demonstrated next. This is probably more meaningful as it also allows for optimization of ΔT_{min}.

Example 8.13 Costs and Emissions for Different Fuel/Utility Selections

Determine the optimum value of ΔT_{min} (to an accuracy of 1°C) for Case Study 4S1 that results in minimum total annual cost of heat, power, and HEN

capital. Assume the power requirement to be 1000 kW and use relevant data from Examples 8.11 and 8.12. Consider the following three cases:
(a) a gas turbine using oil as fuel (cost = 350 $/[kW.yr]);
(b) a gas turbine using natural gas as fuel (cost = 400 $/[kW.yr]); and
(c) a furnace using oil as fuel (cost = 175 $/[kW.yr]).
Given cost of electricity (in $/[kW.yr]): 675 (for import) and 550 (for export).

Solution. Targets for utility, area, and units are established first using the procedures described in Chapter 1. Next, the power is determined for cogeneration devices as in Example 8.12, using simple thermodynamic models. The total annual cost can then be simply calculated.

Consider Case Study 4S1 at ΔT_{min} = 13°C with the hot utility requirement of 360 kW met by a gas turbine. The units target is six. The area target is 1550.31 m^2, given that the gas turbine exhaust goes from 760° to 160°C. The power delivered by the gas turbine is 126.46 kW (see Example 8.12). Therefore, for the case where oil is the fuel, the total annual cost of heat, power, and HEN capital is given by

$$\text{Cost} = 360 \times 350 + 675 (1000 - 126.46)$$
$$+ 6 (30000 + 750 (1550.31/6)^{0.81}) (1 + 0.1)^5/5$$
$$= 903.97 \times 10^3 \text{ \$/yr.}$$

The cost calculation may be repeated for different values of ΔT_{min} to determine the optimum.

The above procedure may be applied to the other fuel/utility choices to obtain the following results:

	Optimum ΔT_{min} (in° C)	Cost (in 10^3 $/yr)	Power (in kW)	Global M_{CO_2} (in kg/hr)
Gas turbine (oil)	13	903.97	126.46	1088.41
Gas turbine (nat. gas)	12	921.82	119.45	1036.03
Furnace (oil)	11	925.26	0	1159.98

Cogeneration is observed to be a promising option from both a cost and emissions viewpoint. The gas turbine with oil shows the lowest cost whereas the one with natural gas shows the lowest global emissions. In other words, a reduction of 419040 kg/yr in CO_2 emissions costs 17850 $/yr. It would be necessary to associate penalty costs for emissions (and the individual pollutants, if possible) in order to optimize on a more objective basis. This currently does not appear to be straightforward and is the reason for minimizing costs and not emissions directly.

The procedure described above can be generalized to the case of multiple utilities (e.g., flue gas with steam or turbine) by allocating utilities based on Section 1.7.3.

QUESTIONS FOR DISCUSSION

1. Show that an alternative form of Equation 8.3 (in terms of the heat absorbed) is $W/Q_2 \geq \eta_c$, where $\eta_c = T_1/T_2 - 1$. Note that Equation 8.3 is in terms of the heat rejected.

2. Appropriate placement of a heat engine results in 100% efficient conversion of heat to work. Explain how this is compatible with the second law of thermodynamics. Would it be more appropriate to say that the "first law" efficiency of appropriately placed heat engines approaches 100%? Note that the conversion is not truly 100% due to some heat losses in practical systems, yet the 100% conversion is an ideal but useful conceptualization.

3. What would be the appropriate placement with respect to the pinch for (a) the vapor recompression scheme of a distillation column and (b) a condensing turbine?

4. A heat sink on a GCC appears as increasing heat flow with increasing temperature whereas a heat source on a GCC appears as increasing heat flow with decreasing temperature. Is this the normal convention for heat sink and source representation? Are there any advantages to this "reverse" convention?

5. Draw a sketch to illustrate the load vs. level conflict that may sometimes arise during profile matching. Specifically, show that a cycle profile may absorb more heat from the process source profile below the pinch but at a lower temperature than another profile that absorbs a lower heat but with a higher cycle efficiency.

6. Why is *superheated* steam usually sent to the inlet of a turbine when the objective is to raise HP or MP steam?

7. What is "bled steam"? How does it help improve the efficiency of steam power stations? Can pinch analysis help in optimizing the bleed steam flow rates?

8. Discuss the tradeoff between cycle efficiency and waste of heat to ambient in the case of a gas turbine exhaust profile based on Figure 8.8.

9. Comment on the following aspects that influence the selection between steam and gas turbine cycles:

- A gas turbine cycle may be preferred when the process sink profile above the pinch is steep and ranges from high to low temperature.
- A steam turbine cycle may be preferred when the process sink profile is at moderate temperatures and is roughly horizontal.
- A gas turbine cycle or a combined cycle system may be preferred when the heat-to-power ratio is low.
- A steam Rankine cycle may be preferred when the heat-to-power ratio is high.
- The upper cycle temperature possible in a gas turbine is much higher than that in a steam turbine.

10. How can the optimum air flow split ratio be obtained for a combined cycle power station? Specifically, discuss how the power output, capital cost of steam-raising, and heat loss to ambient vary for the limiting cases of all air being sent to the boiler and all air going through the gas turbine.

11. What is the optimality region and the optimal operating line for the OLM in the limit $\lambda/U \to 0$ and $\lambda/U \to 1$? Explain the results physically.

12. How can the operating line be defined for the pinch method in the case where the GCC is not reentrant (does not have a pocket)?

13. Show that heat exchange in the case of the OLM occurs with a ΔT proportional to the absolute temperatures of the integrated process streams. Why does this variation in ΔT exclude the possibility of process-to-process heat exchange in the "pocket" of the GCC?

14. Discuss the extension of the OLM to the more general case of multiple steam levels (see Aguirre et al., 1990).

15. Is exergy loss (which measures the reduction in a system's ability to deliver work) a more useful concept than entropy gain (which requires the consideration of the universe)? Does exergy decrease with temperature as well as pressure?

16. Is pinch analysis a true synthesis tool for the design of integrated systems? On the other hand, is second law analysis (exergy analysis) an evolutionary tool because it requires a base case design?

17. Equation 8.18b is an expression for simply calculating the exergy change. Show that the analogous expression for the entropy gain is given by $(\Delta Q/T_{lmtd})$. Discuss the accuracy of these expressions.

18. A turbine consumes 15500 kW of fuel and delivers 4500 kW of work. For the same fuel consumption and operating conditions, an ideal heat engine would deliver 6250 kW of work. In another case, a heating system consumes

15500 kW of fuel and produces 10000 kW of useful heat. For the same fuel consumption and operating conditions, an ideal heat pump would produce 6250 kW of useful heat. Compare the two systems using efficiencies based on the first and second law.

19.   Would it be preferable to save 500 kW of refrigerant exergy or 500 kW of steam exergy? Substantiate your answer in terms of fuel exergy savings.

20.   Consider 1000 kW of heat available at 175°C vs. 666.67 kW available at 250 kW.   Calculate the exergy for the two heat sources.   Although thermodynamics suggests that loads and levels are ideally interchangeable in this case, does this hold in chemical process design?

21.   Discuss the extension of the shaftwork targeting procedure in Section 8.6 to above-ambient CHP systems.   What would be the analog of Equation 8.19b for a power plant?

22.   Discuss the tradeoffs in satisfying heating demands through a single central power station (where the site main pressure levels may not be appropriate for individual processes) vs. local power-generating systems (where on-site steam-raising and direct-drive steam turbines may be appropriate for satisfying the large heat and power demand of an individual process).

23.   List the possibilities that exist when the ratio of heat to power demand of a site is high (e.g., power may be exported or steam-raising package boilers may be used) and when it is low (e.g., power generation efficiency may be improved or power may be imported or condensing sets may be used).

24.   A simplified tariff structure for imported electricity is given as follows: 0.06 $/kWh (daytime) and 0.04 $/kWh (nighttime). Is it profitable to import electricity or produce it in-house using oil as the fuel (NHV = 39830 kJ/kg and cost = 0.485 $/kg): (a) against the condensing portion of a power cycle (efficiency = 28%); and (b) against process sinks (efficiency = 85%)?

25.   A gas turbine (power output = 4500 kW, cycle efficiency = 38%, power conversion efficiency = 97%, and combustion chamber efficiency = 97%) is integrated with a steam power station (resulting in a marginal boiler efficiency of 85%). If oil (NHV = 39830 kJ/kg and cost = 0.485 $/kg) is the marginal fuel, then calculate the marginal savings against imported electricity (which costs 0.06 $/kWh during daytime and 0.04 $/kWh during nighttime).

26. Total site analysis establishes targets that could be achieved over a period of time, especially when investments are made for expansion, replacement, and maintenance. Explain (see Linnhoff and Dhole, 1993).

27. Can the emissions as defined in Section 8.8 become negative? Under what conditions will this happen?

28. Example 8.12 compares steam and gas turbines on a local as well as global basis for the same heat output. How would these comparisons appear on the basis of (a) same power output and (b) same fuel consumption?

PROBLEMS

8.A Design of Steam Rankine Cycle

Design a CHP system using a steam turbine to obtain 65000 kW of shaft power and meet a process hot utility demand of 21280 kW for Case Study 3T1 (the front end of an ICI tonnage chemical process) discussed by Townsend and Linnhoff (1983). In addition, 19600 kg/hr of 1900 kPa steam (IP at 210°C) and 80700 kg/hr of 362 kPa steam (LP at 140°C) are required to be produced for use in another section of the process. Thus, the total requirement is 18840 kW of IP steam and 60870 kW of LP steam. Note that the ΔT_{min} contribution is 25°C for process streams and 5°C for evaporative changes. Only the region below 210°C may be matched against the engine exhaust. A ΔT_{min} of 20°C may be used for the economizers. The enthalpy of the BFW at the outlet of the economizers is given by $H_{e,IP} = 807.4$ kJ/kg at 190°C and $H_{e,LP} = 504.1$ kJ/kg at 120°C. The liquid enthalpy of the blowdown stream is 1427 kJ/kg at 314°C.

Follow the procedure used in Example 8.1. Given: Turbine inlet at 500°C and 10450 kPa (H_{in} = 3367 kJ/kg) and each turbine stage has an isentropic efficiency of 0.85.

Stage	P_{out} (kPa)	H_{sat} (kJ/kg)	ΔH_{lat} (kJ/kg)	ΔH_{isen} (kJ/kg)
1	1900	2798	1901	459
2	362	2734	2145	339.8
3	8	2576.4	2402.6	554

Answers:
Turbine inlet flow rate = 81.766 kg/s
Flow rate to IP economizer = 11.007 kg/s
Total heat input to cycle = 211954 kW
Note the LP economizer is not feasible.

8.B Design of Gas Turbine Cycle

Design an open gas turbine cycle to satisfy the shaft power and utility demands specified in Problem 8.A. Assume there is a heat source available for raising 10350 kW of IP steam and a portion (35850 kW) of the LP steam (Townsend and Linnhoff, 1983). So, only 12230 kW of LP steam need to be raised.

Follow the procedure used in Example 8.2.

Given: ΔT_{min} for heat exchanger $= 40°C$
   Ambient temperature $T_1 = 25°C$
   Ambient pressure $P_1 = 100$ kPa
   Turbine isentropic efficiency $\eta_t = 0.87$
   Compressor isentropic efficiency $\eta_{co} = 0.85$
   Mechanical transmission efficiency $\eta_m = 0.99$
   Combustion chamber temperature $T_4 = 1100°C$
   Combustion chamber pressure loss $\Delta P_{cc} = 5$ %
   Exhaust-side pressure loss in heat exchanger $\Delta P_{he} = 4$ kPa
   Air-side pressure loss in heat exchanger $\Delta P_{ha} = 3$ %
   Mass flow ratio of exhaust to air $M_e/M_a = 1.0137$

Answers:

The two points of closest approach on the exhaust profile are X_1 and X_2 with coordinates (37780, 284) and (12790, 160) respectively. The point of highest heat rejection for the exhaust is (77760, 452.61). Then, the total power output is found to be 50850 kW and the net heat cycle output is 137780 kW (giving an efficiency of 36.91%). The load on the heat exchanger is 19510 kW, and the mass flow rate of the exhaust gas is 191.8 kg/s. The following physical properties may be assumed: $\gamma_a = 1.378$, $\gamma_e = 1.321$, $C_{pe,H} = 1.11$ kJ/kg°C, $C_{pa,H} = 1.08$ kJ/kg°C, $C_{pe,T} = 1.187$ kJ/kg°C, $C_{pa,C} = 1.03$ kJ/kg°C, and $C_{pe,E} = 1.051$ kJ/kg°C.

The exhaust profile may be rotated clockwise using X_1 as pivot to satisfy exactly the power demand of 65000 kW. For $T_6 = 414.4°C$, it is found that the total power output is 64998 kW and the net heat cycle output is 167340 kW (giving an efficiency of 38.84%). The load on the heat exchanger is 48090 kW, and the mass flow rate of the exhaust gas is 248.6 kg/s. The following physical properties may be assumed: $\gamma_a = 1.388$, $\gamma_e = 1.319$, $C_{pe,H} = 1.113$ kJ/kg°C, $C_{pa,H} = 1.082$ kJ/kg°C, $C_{pe,T} = 1.192$ kJ/kg°C, $C_{pa,C} = 1.023$ kJ/kg°C, and $C_{pe,E} = 1.048$ kJ/kg°C.

8.C Emissions Targets for Cogeneration Systems

Estimate the global pollutant rate of CO_2 and the power generated for the aromatics process discussed by Smith and Delaby (1991). The data are identical to those for Case Study 9A2 except that the outlet temperature of stream H1 is 45°C (and not 40°C). Consider a gas turbine using:

(a) oil as the fuel (NHV = 42000 kJ/kg and carbon = 86.2%) and ΔT_{min} = 24°C; and

(b) natural gas as the fuel (NHV = 51600 kJ/kg and carbon = 75.38%) and ΔT_{min} = 21°C.

Assume the power requirement is 8400 kW. The central electricity generation system is a steam turbine power plant, which uses coal as fuel (NHV = 30000 kJ/kg and carbon = 74.5%) and has an overall efficiency of 28%.

Given: combustion chamber temperature (turbine inlet) $T_{g,in}$ = 1027°C

turbine outlet temperature $T_{g,out}$ = 720°C

stack temperature T_{stack} = 160°C

ambient temperature T_o is 15°C

Answers:

For oil, Global M_{CO_2} = 10705 kg/hr and power generated = 8379 kW.

For natural gas, Global M_{CO_2} = 7782 kg/hr and power generated = 7790 kW.

Note that Smith and Delaby (1991) use η_e = 0.9 only for the work calculation (and not for fuel requirement computation in Equation 8.25).

CHAPTER 9

Energy and Resource Analysis
of Various Processes

Of time you would make a stream
upon whose bank you would sit and watch its flowing.

Kahlil Gibran

Though the focus of the book is on HENS, many of the concepts described so far have a broader application. With this expanded scope and as discussed at the start of the book, OPERA is the more appropriate term than pinch analysis. Targeting before design allows screening of alternative options during the preliminary stages of the conceptual design of the entire process (rather than merely the heat recovery system). This chapter deals with the application of the principles of heat integration developed in earlier chapters to various processes of practical interest. These include batch processes, flexible processes, distillation, evaporation, separation using mass separating agents, and reaction processes.

9.1 BATCH PROCESSES

While the HENS problem has been analyzed extensively in the case of continuous processes, it has received relatively little attention for batch processes.

Early methods in the case of batch processes used a pseudo-continuous approach, which involved averaging of heat loads over the entire batch period (Clayton, 1986) and may be referred to as the time average model (TAM). Obeng and Ashton (1988) discuss both a time-independent and a time-dependent approach for solving batch process problems. An appropriate time-dependent method of analysis was proposed by Kemp and Macdonald (1988) and was followed by a series of three papers by Kemp and Deakin (1989b). This work, along with the article of Kemp (1990a), provides methods for obtaining energy targets as well as designing networks and is presented in detail below. It includes a discussion for the rescheduling of streams.

9.1.1 Energy Targets by Time-Cascade Analysis

The PTA discussed in Chapter 1 can be extended to the energy targeting of batch processes, provided the time-dependence of the streams is accounted for. In other words, streams that do not coexist in a time interval cannot take part in direct heat exchange although a sufficient temperature difference may exist between the streams. Initial attempts at energy targeting of batch processes used a pseudo-continuous method or the time average model (TAM). TAM assumes that all streams exist at all times with the heat loads averaged over the batch period to allow for streams existing over longer periods to be weighted relative to those existing for shorter periods. The other approach is the time slice model (TSM), where the batch period is divided into time intervals within which the process behaves like a continuous one. The PTA then is simply applied to each interval to yield the energy targets. Importantly, the TSM allows exploration of the possibilities of heat storage between time intervals, when direct exchange is not possible. The TAM is described briefly below, while the TSM is discussed in detail.

Example 9.1 Energy Targeting by Time Average Model

Consider the stream data in Case Study 4S1 with the following modification: the streams do not exist over the entire batch period, but only over certain time intervals and with MC_p values as specified below:

Stream	H1	H2	C3	C4
Start time (h)	0.6	0.2	0.3	0.0
Finish time (h)	1.0	1.0	0.5	0.1
MC_p (kW/°C)	25	50	100	150

Let this modified data be referred to as Case Study 4S1b. The example process corresponds to two exothermic batch reactors, each with a single feed and a single product, as given below:

Reactor 1 (Holding time: 0.1 to 0.2 h)
Feed Stream C4 (MC_p = 150 kW/°C)   Product Stream H2 (MC_p = 50 kW/°C)
Heated from 40° to 112°C   Cooled from 125° to 65°C
Filling time from 0.0 to 0.1 h   Discharge time from 0.2 to 1.0 h

Reactor 2 (Holding time: 0.5 to 0.6 h)
Feed Stream C3 (MC_p = 100 kW/°C)   Product Stream H1 (MC_p = 25 kW/°C)
Heated from 20° to 155°C   Cooled from 175° to 45°C
Filling time from 0.3 to 0.5 h   Discharge time from 0.6 to 1.0 h

For the stream data and the time intervals in Case Study 4S1b, determine the minimum utility requirements predicted by the time average model (TAM). The batch period is 1 hour and $\Delta T_{min} = 20°C$.

Solution. The first step is to weight streams existing over a longer time period with respect to those existing for a shorter one. This is simply done by multiplying the MC_p value of the stream by the time for which the stream exists. For example for stream H1, $MC_p \, dt = 25 \, (1.0 - 0.6) = 10 \text{ kWh/°C}$. Similarly, the values of $MC_p \, dt$ for streams H2, C3, and C4 are 40, 20, and 15 kWh/°C respectively. Then, on the assumption that all streams exist throughout the batch period, the PTA is applied as in Example 1.1. On the construction of the infeasible cascade (Q_{cas} as in column D of Table 1.1) and the feasible cascade (R_{cas} as in column E of Table 1.1), the hot and cold utility requirements are obtained as 605 kWh and 525 kWh respectively. The reader may note that Case Study 4S1b is so constructed that, when streams are weighted in terms of their time intervals of existence, the data is identical to that in Case Study 4S1.

The TAM does not consider the change in stream population in different time intervals as is done in the different temperature intervals. Thus, the TAM predicts energy targets based on direct heat exchange, which is physically impossible between streams existing in two different time intervals. For example in the case of a single batch, no direct heat exchange is possible between the feed and product streams in the above reactors as they do not have any overlap in time. The TAM targets are often over-optimistic and not practically achievable as they predict higher heat recovery levels by neglecting time constraints. However, the TAM targets specify the ideal limit (upper bound) to energy recovery and provide motivation for process changes (rescheduling) to obtain better performance. Also, the heat released in one time interval can be stored and released to a later interval, which amounts to indirect heat exchange using storage units. The TAM is inadequate for prediction of such indirect heat exchange possibilities. These difficulties associated with the TAM are overcome by the TSM described below.

Example 9.2 Energy Targeting by Time Slice Model

Rework Example 9.1 using the time slice model (TSM), i.e., determine the minimum utility requirements for Case Study 4S1b for a batch period of 1 hour and $\Delta T_{min} = 20°C$.

Solution. The TSM divides the batch period into time intervals so that the stream population does not change within each interval. This is simply done by sorting the start and finish times of all the streams in ascending order, omitting repeated entries. Then, two consecutive values of time constitute a time interval (or time slice). The streams existing in each time interval are determined as in Table 9.1.

The PTA is applied to each of the time intervals (i.e., a total of five times, as the interval 0.1-0.2 h contains no streams), considering only those streams that exist within a particular interval. Thus, for the fourth time interval, the data for streams H2 and C3 are considered. Five tables similar to the one in Table 1.1 are generated with the following minor modification. The Q_{cas} and R_{cas} columns (i.e., columns D and E in Table 1.1) are multiplied by the appropriate time interval since it is more meaningful to obtain the utility requirements in terms of heat loads (kWh) rather than heat flows (kW). Table 9.1 shows the individual infeasible cascades ($Q_{cas,k}$, where k denotes the time interval and runs from 1 through 6). These individual cascades have infeasibilities (negative entries) which can be removed by adding a hot utility (of magnitude equal to the most negative entry in the $Q_{cas,k}$ column) to the top of the cascade. The feasible cascades thus obtained are shown in Table 9.2.

Table 9.1 Infeasible Cascades for the Various Time Intervals

Time	0	0.1	0.2	0.3	0.5	0.6	1.0
Streams	C4			H2	H2, C3	H2	H1, H2
T_{int}	$Q_{cas,1}$	$Q_{cas,2}$	$Q_{cas,3}$	$Q_{cas,4}$	$Q_{cas,5}$	$Q_{cas,6}$	
165	0		0	0	0	0	
122	0		0	−860	0	430	
115	−105		0	−1000	0	500	
55	−1005		300	−1600	300	2300	
50	−1080		300	−1700	300	2350	
35	−1080		300	−2000	300	2500	
30	−1080		300	−2100	300	2500	

Units: time in h, temperature in °C, and heat load in kWh.

As in Table 1.1, the minimum hot utility requirement and the minimum cold utility requirement for each time interval appear as the first and last entries in the $R_{cas,k}$ column. Summing over all the time intervals gives the total hot utility as 3180 kWh and the total cold utility as 3100 kWh. It is observed that these targets are 2575 kWh greater than those in Example 9.1, indicating that the TAM predictions are over-optimistic.

Each time interval is independent of the others in the TSM. This allows use of different ΔT_{min} for the various time intervals. Also, the TSM gives results which may be termed maximum heat exchange (MHX) targets since it considers only direct heat exchange and external utilities but no storage between intervals. Before introducing the idea of storage, a new algorithm (Golwelker, 1994) for implementation of the TSM is described below because it appears to provide certain advantages.

Table 9.2 Feasible Cascades for the Various Time Intervals

Time	0	0.1	0.2	0.3	0.5	0.6	1.0
Streams	C4		H2	H2, C3	H2	H1,H2	
T_{int}	$R_{cas,1}$	$R_{cas,2}$	$R_{cas,3}$	$R_{cas,4}$	$R_{cas,5}$	$R_{cas,6}$	
165	1080		0	2100	0	0	
122	1080		0	1240	0	430	
115	975		0	1100	0	500	
55	75		300	500	300	2300	
50	0		300	400	300	2350	
35	0		300	100	300	2500	
30	0		300	0	300	2500	

Total hot utility requirement = 1080 + 2100 = 3180 kWh
Total cold utility requirement = 300 + 300 + 2500 = 3100 kWh

Units: time in h, temperature in °C, and heat load in kWh.

Example 9.3 Implementation of Time Slice Model Using Stream Cascade Superposition (SCS) Principle

Determine the infeasible cascades for the various time intervals in the TSM for Case Study 4S1b for a batch period of 1 hour and $\Delta T_{min} = 20°C$. Attempt not to use the PTA repeatedly as in Example 9.2.

Solution. Example 9.2 requires repeated application of the PTA. If the number of time intervals is large, then implementation of TSM tends to be tedious, especially if calculations are performed manually. The algorithm below appears computationally advantageous, especially when the number of streams is small relative to the number of time intervals. It is based on the heat cascades for the individual streams which may then be superposed to obtain the infeasible cascade for each time interval.

Step 1. Determination of Temperature Intervals (T_{int})

The entire problem is divided into temperature intervals such that every inlet and outlet temperature defines a boundary. Note that all streams in a batch process are considered at this stage. For a given ΔT_{min}, the interval boundary temperatures are specified $\Delta T_{min}/2$ below hot stream temperatures and $\Delta T_{min}/2$ above cold stream temperatures as conventionally done in the PTA. This gives the first column in Table 9.3.

Table 9.3 Heat Cascades for the Individual Streams

T_{int}					$S_{cas,H1}$	$S_{cas,H2}$	$S_{cas,C3}$	$S_{cas,C4}$
165					0	0	0	0
122					1075	0	−4300	0
115					1250	0	−5000	−1050
55					2750	3000	−11000	−10050
50					2875	3000	−11500	−10800
35					3250	3000	−13000	−10800
30					3250	3000	−13500	−10800
Stream	H1	H2	C3	C4				
MC_p	25	50	100	150				

Units: temperature in °C, heat capacity flow rate in kW/°C, and enthalpy in kW.

Step 2. Determination of Heat Cascade for Each Stream (S_{cas} in Table 9.3)
     The net enthalpy in a temperature interval for a stream is obtained by multiplying the MC_p of the stream (with the appropriate sign for hot and cold) by the temperature difference for that interval. This is then cascaded down from a higher interval to a lower one to obtain the stream heat cascade. Thus, the column is generated by use of the following formula:

If the stream exists in temperature interval between Row $r - 1$ and Row r, then

$$S_{cas,Row\ r} = S_{cas,Row\ r-1} \pm MC_{p,stream}\ (T_{int,Row\ r-1} - T_{int,Row\ r}) \quad (9.1a)$$

If the stream does not exist in temperature interval between Row $r - 1$ and Row r, then

$$S_{cas,Row\ r} = S_{cas,Row\ r-1} \quad (9.1b)$$

Note that the positive sign is used for hot streams while the negative sign is used for cold streams in the above equation. Also, $S_{cas,Row\ 1} = 0$. Thus, use of Equation 9.1 for stream H1 gives

$$S_{cas,Row\ 2} = 0 + 25\ (165 - 122) = 1075$$
$$S_{cas,Row\ 3} = 1075 + 25\ (122 - 115) = 1250, \text{ and so on.}$$

The S_{cas} columns for all four streams are generated in this manner and shown in Table 9.3. The streams are represented by vertical bars in the table to ascertain their existence in a temperature interval.

Step 3. Determination of Infeasible Cascades for Various Time Intervals
    (Q_{cas} in Table 9.1)

The determination of infeasible cascades is now straightforward especially in comparison with the method adopted in Example 9.2. It simply involves use of the following formula for each time interval:

$$Q_{cas,k,Row\ r} = \Sigma S_{cas,Row\ r}\ dt \tag{9.2}$$

where the summation runs over all the streams that are present in the time interval k. For example, summing the S_{cas} columns for streams H2 and C3 in Table 9.3 and multiplying by the fourth time interval (namely, $0.5 - 0.3 = 0.2$) gives the $Q_{cas,4}$ column in Table 9.1. Thus, Table 9.3 allows generation of Table 9.1 with less computational effort. The use of stream-dependent ΔT_{min} based on a minimum flux specification as in Section 2.2.1 is possible through the stream cascade superposition principle by developing diverse heat cascades for the individual streams. This is more meaningful than the use of different ΔT_{min} for the various time intervals, which is possible in the methodology of Example 9.2. As will be shown later, the SCS principle helps in stream rescheduling as it precisely gives the heat flow contribution of a particular stream at a specified temperature in a given time interval.

9.1.2 Reducing Utility Requirements through Heat Storage

The MHX targets in Example 9.2 do not incorporate heat storage between time intervals. The heat available at a particular temperature in a time interval may be released to a later time interval at the same or lower temperature by storing the heat content of a stream in an insulated unit for the required time period. To obtain the heat storage and utility targets, the individual infeasible cascades (Table 9.1) have to be operated upon. For heat recovery via storage, there can be various considerations. Two important objectives could be minimization of storage time and maximization of temperature driving force. Also, the heat stored may be within a single batch or between repeated batches.

Example 9.4 Targets with Heat Storage for a Single Batch

Determine the storage and utility targets for Case Study 4S1b for a single batch with a period of 1 hour. Given: ΔT_{min} = 20°C and minimization of storage time is an important consideration.

Solution. The heat available in the closest time interval is considered first for storage and then, if required, previous intervals are searched for heat availability. In other words, priority is given to heat available in a closer time interval for minimization of storage time.

The basic idea is to remove all the negative entries (infeasibilities) in Table 9.1 by shifting heat from the positive entries (i.e., by giving priority to heat storage over external utility). In Table 9.1, the Q_{cas} columns for all the time intervals together form an m x n matrix. The infeasible heat cascade matrix is scanned row by row till the first negative entry (say, Q_{ij} where i is the row number and j is the column number) is encountered. On reaching an infeasibility, the row is rescanned in a reverse direction, i.e., the ith row is scanned from column $(j - 1)$ to column (1). The first positive number encountered (say, Q_{ik} where $1 \le k < j$) provides potential for heat storage. Let S_{ijk} = min(Q_{ik}, |Q_{ij}|). Then, it is possible to shift heat equal to S_{ijk} in magnitude from (i, k) to (i, j). Note that S_{ijk} indicates the amount of heat that needs to be stored at the ith temperature level between the kth and jth time intervals. Recalling that the columns in the matrix correspond to heat cascades, shifting S_{ijk} units of heat implies adding S_{ijk} to elements in column j with row number $\ge i$ and subtracting S_{ijk} from elements in column k with row number $\ge i$. When S_{ijk} = |Q_{ij}|, the negative entry will become zero at the end of the heat-shifting; however, when S_{ijk} = Q_{ik}, the infeasibility is not removed completely and the row needs to be scanned further for available heat (i.e., another positive entry). Scanning of the row in reverse continues till the negative entry gets replaced by a zero. If the first column is reached, the search for a positive number is terminated. A hot utility needs to be supplied at the top of the cascade to remove the infeasibility as there is no heat available in the previous time intervals.

Two precautions need to be observed (Golwelker, 1994) to ensure minimum storage for a given utility usage. When adding a hot utility at the top (of, say, the jth column) to remove an infeasibility Q_{ij}, a check must be performed on the elements in the jth column above Q_{ij} to determine if they have been made positive by adding heat via storage. If this is the case, then the amount of the hot utility must be subtracted from the storage. If the hot utility is greater than the amount of heat stored, then the storage may be set to zero. In other words, care must be exercised to ensure that external hot utility is not added to remove an infeasibility at a temperature level lower than that of the storage because it leads to an excess of heat storage for the same utility target.

A check is also required to avoid the undesirable situation where the storage causes a negative entry further down the column that requires an external hot utility for its satisfaction. In this case, the storage must be reduced so as not to call for hot utility usage below the storage temperature level. Consider shifting S_{ijk} units of heat. Then, S_{ijk} is subtracted from elements in column k with row number $\geq i$. If this causes negative entries, then the rows corresponding to these should be scanned. If Q_{pk} is one such negative entry, then the pth row is scanned. If the sum S_p of all positive entries in the pth row for earlier time intervals is greater than or equal to Q_{pk}, the storage of S_{ijk} is valid; otherwise, an amount equal to $(S_{ijk} - (|Q_{pk}| - S_p))$ should be stored instead of S_{ijk}.

The above procedure is repeated for each negative entry in every row until all the negative entries are removed in the infeasible heat cascade matrix. Although the whole procedure appears complicated at first glance, its actual implementation is relatively straightforward as shown below on Case Study 4S1b. Table 9.4 shows the calculation at every step.

The infeasible heat cascades (see Table 9.1) for Case Study 4S1b form a 7 x 6 matrix (seven temperature intervals and six time intervals). Column 2 is blank as no stream exists from 0.1-0.2 h. The first row does not have a negative entry. The first infeasibility is $Q_{24} = -860$. As no heat is available in the previous columns, it is not possible to remove the infeasibility except by adding a hot utility of 860 units at the top of column 4. The next infeasibility is $Q_{31} = -105$. Since it is in the first column, the only option is to add 105 units of hot utility to column 1. $Q_{34} = -140$ (i.e., $-1000 + 860$) is the next infeasibility, which again is removed by adding 140 as hot utility since no heat for storage is available. $Q_{41} = -900$ (i.e., $-1005 + 105$) is also satisfied by 900 units of hot utility. The next infeasibility is $Q_{44} = -600$ (i.e., $-1600 + 860 + 140$), which has $Q_{43} = 300$ as the closest positive entry. Thus, $S_{443} = 300$ which implies that 300 units are added to Q_{44}, Q_{54}, Q_{64}, and Q_{74}, while 300 units are subtracted from Q_{43}, Q_{53}, Q_{63}, and Q_{73}. The infeasibility is not completely removed, and Q_{44} now has a value of -300 (i.e., $-600 + 300$). The only option to make Q_{44} zero is to add 300 units of hot utility to column 4. The next infeasibility of $Q_{51} = -75$ (i.e., $-1080 + 105 + 900$) is removed by adding hot utility. The remaining infeasibilities in column 4 have to be satisfied by hot utility of 500 (i.e., $2100 - 1600$) since no heat is available in columns 1 and 2.

Table 9.4 Operations Performed on Infeasible Heat Cascade Matrix (Single Batch)

Time	0		0.1	0.2	0.3		0.5
		C4		H2	H2, C3		
T_{int}	Q			Q	Q		
165	0+105+900+75			0	0+860+140	+300+500	
122	0+105+900+75			0	−860+860+140	+300+500	
115	−105+105+900+75			0	−1000+860+140	+300+500	
55	−1005+105+900+75			300−300	−1600+860+140+300+300+500		
50	−1080+105+900+75			300−300	−1700+860+140+300+300+500		
35	−1080+105+900+75			300−300	−2000+860+140+300+300+500		
30	−1080+105+900+75			300−300	−2100+860+140+300+300+500		

Steps:
1. $Q_{24} = -860$ No heat available in row. Add 860 as hot utility to column 4.
2. $Q_{31} = -105$ No heat available in row. Add 105 as hot utility to column 1.
3. $Q_{34} = -140$ No heat available in row. Add 140 as hot utility to column 4.
4. $Q_{41} = -900$ No heat available in row. Add 900 as hot utility to column 1.
5. $Q_{44} = -600$ $Q_{43} = 300$; $S_{443} = 300$ Shift 300 from (4,3) to (4,4).
6. $Q_{44} = -300$ No heat available in row. Add 300 as hot utility to column 4.
7. $Q_{51} = -75$ No heat available in row. Add 75 as hot utility to column 1.
8. $Q_{54} = -100$ No heat available in row. Add 100 as hot utility to column 4.
9. $Q_{64} = -300$ No heat available in row. Add 300 as hot utility to column 4.
10. $Q_{74} = -100$ No heat available in row. Add 100 as hot utility to column 4.

Units: time in h, temperature in °C, and heat load in kWh.

The final overall cascade (which features no infeasibilities) for a single batch with heat storage is shown in Table 9.5. It indicates that 300 kWh may be recovered by storage, leading to a reduction in hot and cold utility consumptions by the same amount.

Can the utility requirements be further reduced by increasing the storage beyond 300 kWh? Table 9.1 shows that heat is available at high temperatures at the end of the batch and required at low temperatures at the start of the batch. Although this exchange is not possible within a single batch, it may be exploited when several batches are processed in succession without overlap (repeated batches). Then, the heat available in an interval of an earlier batch may be stored and released to some interval of a later batch.

Table 9.5 Overall Feasible Cascade with Storage for a Single Batch

Time	0		0.1	0.2	0.3	0.5	0.6	1.0
Streams		C4		H2	H2, C3	H2	H1,H2	

T_{int}	$Q_{final,1}$	$Q_{final,2}$	$Q_{final,3}$	$Q_{final,4}$	$Q_{final,5}$	$Q_{final,6}$
165	1080		0	1800	0	0
122	1080		0	940	0	430
115	975		0	800	0	500
55	75		0	500	300	2300
50	0		0	400	300	2350
35	0		0	100	300	2500
30	0		0	0	300	2500

Total hot utility requirement = 1080 + 1800 = 2880 kWh
Total cold utility requirement = 300 + 2500 = 2800 kWh
Storage requirements: S_{443} = 300 kWh

Units: time in h, temperature in °C, and heat load in kWh.

Example 9.5 Targets with Heat Storage for Repeated Batches

Rework Example 9.4 for the case of repeated batches, i.e., determine the storage and utility targets for Case Study 4S1b, given that ΔT_{min} = 20°C and minimization of storage time is an important consideration.

Solution. The approach is similar to that adopted in Example 9.4, except for a minor modification required to account for the possibility of heat storage between repeated batches. The process of scanning rows from left to right for a negative entry is as before. However, while scanning in reverse from right to left for a positive entry, the row is scanned from column $(j - 1)$ to (1) and then further scanned from column (n) to $(j + 1)$. Thus, if the closest positive entry encountered is Q_{ik}, then k can assume any value between 1 and n (of course, $k \neq j$). But for this change, the rest of the procedure is as in Example 9.4.

For Case Study 4S1b, the first infeasibility in Table 9.1 is $Q_{24} = -860$ for which the closest positive number is $Q_{26} = 430$. Therefore, $S_{246} = 430$, which implies that 430 is added to column 4 and subtracted from column 6 (at and below row 2). This leaves an infeasibility of -430 at Q_{24}, which has to be satisfied by hot utility. The next infeasibility is $Q_{31} = -105$ and the closest positive entry is $Q_{36} = 500 - 430 = 70$. Using $S_{316} = 70$, an infeasibility of -35 remains to be satisfied by hot utility. The next negative number is $Q_{34} = -1000 + 430 + 430 = -140$, which again needs a hot utility. The next infeasibility is $Q_{41} = -900$. The closest positive number is $Q_{46} = 1800$, which is greater than the infeasibility; hence, 900 units are transferred via storage using $S_{416} = 900$. The infeasibility of -600 at Q_{44} is met by transferring 300 units via storage from Q_{43} and Q_{46}. All infeasibilities are similarly removed. The calculations and steps are summarized in Table 9.6.

Table 9.6 Operations Performed on Infeasible Heat Cascade Matrix (Repeated Batches)

Time 0	0.1 C4	0.2 H2	0.3 H2, C3	0.5 H2	0.6 H1, H2 1.0
T_{int}	Q	Q	Q	Q	Q
165	0 +35	0	0 +430+140	0	0
122	0 +35	0	−860+430+430+140	0	430−430
115	−105+70+35	0	−1000+430+430+140	0	500−430−70
55	−1005+70+35 +900	300−300	−1600+430+430+140 +300+300	300	2300−430−70−900 −300
50	−1080+70+35 +900+75	300−300	−1700+430+430+140 +300+300+100	300	2350−430−70−900 −300−75−100
35	−1080+70+35 +900+75	300−300	−2000+430+430+140 +300+300+100+300	300	2500−430−70−900 −300−75−100−300
30	−1080+70+35 +900+75	300−300	−2100+430+430+140 +300+300+100+300 +100	300	2500−430−70−900 −300−75−100−300 −100

Steps:
1. $Q_{24} = -860$ $Q_{26} = 430$; $S_{246} = 430$ Shift 430 from (2,6) to (2,4).
2. $Q_{24} = -430$ No heat available in row. Add 430 as hot utility to column 4.
3. $Q_{31} = -105$ $Q_{36} = 70$; $S_{316} = 70$ Shift 70 from (3,6) to (3,1).
4. $Q_{31} = -35$ No heat available in row. Add 35 as hot utility to column 1.
5. $Q_{34} = -140$ No heat available in row. Add 140 as hot utility to column 4.
6. $Q_{41} = -900$ $Q_{46} = 1800$; $S_{416} = 900$ Shift 900 from (4,6) to (4,1).
7. $Q_{44} = -600$ $Q_{43} = 300$; $S_{443} = 300$ Shift 300 from (4,3) to (4,4).
8. $Q_{44} = -300$ $Q_{46} = 900$; $S_{446} = 300$ Shift 300 from (4,6) to (4,4).
9. $Q_{51} = -75$ $Q_{56} = 650$; $S_{516} = 75$ Shift 75 from (5,6) to (5,1).
10. $Q_{54} = -100$ $Q_{56} = 575$; $S_{546} = 100$ Shift 100 from (5,6) to (5,4).
11. $Q_{64} = -300$ $Q_{66} = 625$; $S_{646} = 300$ Shift 300 from (6,6) to (6,4).
12. $Q_{74} = -100$ $Q_{76} = 325$; $S_{746} = 100$ Shift 100 from (7,6) to (7,4).

Units: time in h, temperature in °C, and heat load in kWh.

The final overall feasible cascades for repeated batches with heat storage are shown in Table 9.7. The hot and cold utility targets are 605 kWh and 525 kWh respectively, while storage is 2575 kWh. It is seen on comparison with the TAM targets (Example 9.1) that the repeated batches achieve the same energy targets. This is possible because there is no time constraint on heat transfer: heat storage is permitted between any two time intervals, regardless of their order of occurrence in time.

Table 9.7 Overall Feasible Cascade with Storage for Repeated Batches

Time	0		0.1		0.2		0.3		0.5		0.6		1.0
Streams		C4				H2		H2, C3		H2		H1,H2	

T_{int}	$Q_{final,1}$	$Q_{final,2}$	$Q_{final,3}$	$Q_{final,4}$	$Q_{final,5}$	$Q_{final,6}$
165	35		0	570	0	0
122	35		0	140	0	0
115	0		0	0	0	0
55	0		0	0	300	600
50	0		0	0	300	475
35	0		0	0	300	325
30	0		0	0	300	225

Total hot utility requirement = 35 + 570 = 605 kWh
Total cold utility requirement = 300 + 225 = 525 kWh
Storage requirements: $S_{246} = 430$ kWh; $S_{316} = 70$ kWh; $S_{416} = 900$ kWh;
$S_{443} = 300$ kWh; $S_{446} = 300$ kWh; $S_{516} = 75$ kWh;
$S_{546} = 100$ kWh; $S_{646} = 300$ kWh; $S_{746} = 100$ kWh.

Units: time in h, temperature in °C, and heat load in kWh.

In the heat storage model for repeated batches, all heat available within the system is utilized and minimum use of external utility is made. The TSM has an advantage here in that it clearly specifies the temperature and time for storage (through the S_{ijk} values in Table 9.7), which is not possible in the TAM.

In Table 9.7, there exists a constant pinch temperature of 115°C, which is the same as the TAM pinch. Recall that the pinch is the temperature below which heat is released and above which heat is absorbed. It may be noted then that, in a batch process, each time interval has a pinch of its own. From Table 9.2, it is seen that the pinch locus for the TSM does not have any fixed behavior since the time intervals are independent of each other. When heat storage between intervals is considered, the time intervals get linked with one another. When storage occurs within a single batch, the pinch of an interval depends on the previous intervals only. Heat transfer through storage leads to a lowering of the pinch of the donor interval. Thus, the pinch in a single batch can only move higher up in temperature level with time. For repeated batches, all time constraints on heat transfer are removed and so all the heat available is recovered either by direct heat exchange or through storage. The process behaves in a pseudo-continuous manner and a constant overall pinch is observed, which is identical to the TAM pinch.

The above algorithms may be used with small modifications for other considerations (Golwelker, 1994). When considering the maximum temperature driving force criterion, all intervals with excess heat that can be stored need to be compared. The search for a positive entry is done as before. On encountering a positive entry, the column is scanned upwards to locate the highest temperature at which heat is available. This is done for all positive entries in the row being scanned, and the time interval releasing heat at the highest temperature is given priority. For example, consider the infeasibility Q_{44} in Table 9.1. Heat is available in time intervals 3, 5, and 6. For repeated batches, heat may be transferred from any of these intervals. On scanning each of these columns upwards, it is seen that the highest temperature at which heat is available in columns 3 and 5 is 55°C whereas it is 122°C in column 6. Therefore, column 6 provides the maximum temperature driving force, but there may be a tradeoff between storage time and temperature. If heat is stored at a high temperature for a long period of time, heat losses may be significant. Thus, the temperature drop due to heat losses is a function of both time and temperature of storage and should be used to decide the interval that would actually provide the maximum driving force.

To minimize storage for a certain utility requirement, the following two stage approach may be used. In the first stage, all columns (from the first to the last) are scanned downwards. On encountering a negative entry, the scanning of a row for a positive entry is carried out as described earlier. All possible heat transfer via storage is explored. No utility is added at this stage; thus, infeasibilities would still exist in the cascades. Now, the row-by-row scanning is repeated to remove infeasibilities using hot utility. This would fix the hot utility requirement for each interval. The second stage requires starting all over again with the infeasible heat cascade matrix. The predetermined hot utility is added to each interval first. Then, the infeasibilities in the resulting matrix are removed by adding heat via storage according to a storage criterion decided a priori. If this two-stage approach is used, then the infeasibility of -105 at Q_{31} is fully taken care of by storage from Q_{36}. Then, it may be shown (left as an exercise to the reader) that storage occurs between the following time intervals: 3 and 4 (300 kWh), 6 and 1 (1080 kWh), and 6 and 4 (1195 kWh). Compared to Table 9.7, the storage time is reduced as an additional 35 kWh is shifted from interval 6 to 1 (rather than to 4).

The heat storage model (HSM) in this section aims at minimizing energy consumption by maximum utilization of the heat available within the system. However, storage means additional capital cost and complexity; hence, it may not necessarily be viable. Moreover, the HSM assumes ideal conditions that need not be practically achievable, but it provides the designer an objective to work towards. More realistic targets are discussed in the next section.

9.1.3 Practical Heat Storage

Some means of heat storage (Kemp, 1990b) are hot wells, stratified storage tanks, thermal regenerators (cowper stoves), and systems using latent heat of fusion/evaporation. Thus, heat transfer between time intervals via storage may be classified into two cases:
  (a) the hot process stream may itself be stored in an insulated unit so as to exchange heat in a later time interval; or
  (b) the hot process stream may exchange heat with an external stream that could be stored and could later exchange heat with a cold process stream.

Ideal storage in case (a) assumes no heat losses during the storage period and heat storage at the highest temperature at which the stream is available. In reality, there is a temperature drop due to heat losses. Also, the final temperature of storage is not the highest temperature of the hot stream but the mean of the temperatures at which the stream exists. This causes a decrease in the actual temperature driving force; hence, an additional minimum temperature of approach (ΔT_s) is required in case of practical heat storage when compared to direct heat exchange. Such an additional ΔT_s is also required for the two-stage heat transfer in case (b).

To obtain the storage temperature and the amount of heat available for storage, the GCCs of the time intervals involved in the storage must be used. A method to obtain practical storage targets based on these concepts is given below.

Example 9.6 Establishing Practical Storage Targets

Establish practical storage targets for Case Study 4S1b allowing for an additional minimum approach temperature, ΔT_s, of 10°C for storage. Specifically, determine the temperature and the amount of heat available for storage, starting with the HSM targets established for a single batch (Table 9.5) and for repeated batches (Table 9.7).

Solution. Practical storage model (PSM) targets are first determined for a single batch and then for repeated batches.

The GCCs of the time interval releasing heat and of the time interval accepting heat are drawn on the same plot (Figure 9.1). The heat load and temperature corresponding to the point of intersection represent ideal heat storage. The region to the left of the intersection point is the one where heat transfer via storage can take place.

Note that the GCCs are plotted using shifted temperatures, so a temperature difference of ΔT between two GCCs implies an actual temperature difference of $\Delta T + \Delta T_{min}$. A storage match will actually have a minimum approach temperature of $\Delta T_s + \Delta T_{min}$ as compared to ΔT_{min} for a direct match. The heat available for storage varies with the value of ΔT_s. If T_s is the storage temperature, then the storage receives heat at $T_s + \Delta T_s/2$ and releases heat at $T_s - \Delta T_s/2$.

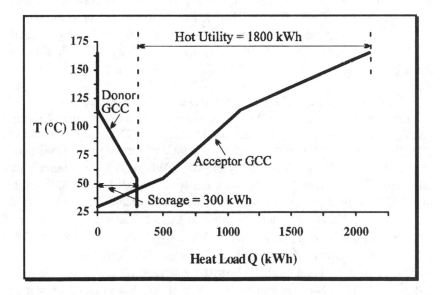

Figure 9.1 The GCC of the time interval donating heat is superimposed on the GCC of the time interval accepting heat to determine the optimal heat storage.

The objective is to find the value of Q_s for which the vertical temperature difference between the two GCCs is ΔT_s. The storage temperature is then the mean of the two temperatures obtained on the GCCs corresponding to Q_s. For this purpose, it is convenient to divide the GCCs into enthalpy intervals using vertical lines at every vertex.

Step 1. Determination of Intersection Point for GCCs Involved in Storage

From Table 9.5, 300 kWh of heat needs to be stored. The interval (0.2 - 0.3 h) is the time interval donating heat while the interval (0.3 - 0.5 h) is the time interval accepting heat. The GCC data for these intervals are available in Table 9.2 under the column headings T_{int}, $R_{cas,3}$, and $R_{cas,4}$. The data are plotted in Figure 9.1. The point of intersection of the two GCCs is (300, 45).

Step 2. Determination of Storage Enthalpy Interval

The closest vertex to the left of the intersection point is (300, 50) on the donor GCC. At this vertex, the vertical temperature difference between the two GCCs is 5°C (namely, 50 − 45). Since this is less than the specified ΔT_s of 10°C, heat storage is not possible in this interval. Note that the donor GCC is vertical below 55°C because no stream exists below this temperature. However, the heat available at 55°C can be used at any lower temperature.

The vertical temperature difference at the next vertex (300, 55) is exactly 10°C (the specified ΔT_s) and so storage is possible. Thus, 300 kWh of heat is available at a mean temperature of 50°C (average of 45° and 55°C).

Step 3. Practical Storage Model (PSM) Targets for Single Batch

On comparing with the HSM targets, it is seen that the amount of storage possible is the same, but the temperature is different. The heat is available at 55°C; however, it is not released at the same temperature as in the HSM but at 45°C. In Table 9.4, the infeasibility of −600 kWh at 55°C in column 4 was satisfied by a storage of 300 kWh and a utility of 300 kWh, but PSM predicts that 300 kWh of storage can release heat to a cold stream at or below 45°C. So, the −600 kWh at 55°C must be satisfied by utility alone (i.e., a utility of 600 kWh would be required instead of 300 kWh). The −100 kWh at 50°C would need a utility of 100 kWh as before. The infeasibility (Q_{64}) of −300 kWh at 35°C is now satisfied partly by storage (200 kW) and partly by utility (100 kW). The remaining −100 kWh at 30°C is satisfied by storage. The total hot utility requirement of 1800 kWh in time interval 4 remains unchanged, but, importantly, the hot utility of 300 kWh is now required at a higher temperature.

Step 4. Practical Storage Model (PSM) Targets for Repeated Batches

The same procedure is essentially followed to obtain PSM targets for repeated batches. From Table 9.7, it is seen that storage occurs between the following time intervals: 3 and 4 (300 kWh), 6 and 1 (1045 kWh), and 6 and 4

(1230 kWh). The storage follows this definite sequence in accordance with the fixed order of operations in the batch process. When a single interval releases heat to more than one interval, then the occurrence in time is important (i.e., storage between 6 and 1 would occur first and the remaining heat would then be given to 4).

The storage between time intervals 3 and 4 has already been discussed and is identical to the single batch case. On considering the GCC data for intervals 6 and 1 (see columns T_{int}, $R_{cas,6}$, and $R_{cas,1}$ in Table 9.2), the point of intersection is found to be (816.67, 104.44) on Figure 9.2. The closest vertex to the left of the intersection point is (500, 115) on the donor GCC. At this vertex, the vertical temperature difference between the two GCCs is 31.67°C, so heat storage is possible in this enthalpy interval. The vertical temperature difference between the two GCCs is exactly 10°C at $Q_s = 716.67$ kWh. Thus, PSM predicts a heat storage of 716.67 kWh at a mean temperature T_s of 102.78°C (average of 97.78° and 107.78°C). Note that the total heat storage predicted by HSM from interval 6 to 1 is different (1045 kWh at various temperatures, both above and below T_s). Thus, the PSM calls for an increase in the hot utility target compared to the HSM.

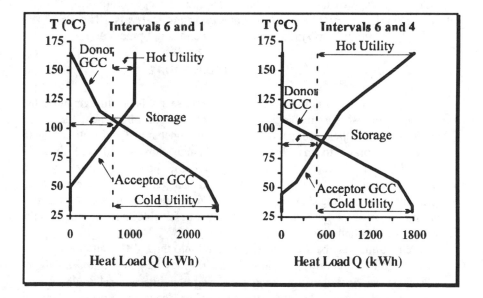

Figure 9.2 The intersection point of the donor and acceptor GCCs shows heat recovery via storage at a single temperature level. However, an additional temperature penalty must be imposed during practical heat storage to account for the heat transfer into and out of the store.

Finally, consider the storage between time intervals 6 and 4. At this stage, interval 4 has already received heat via storage from interval 3, and interval 6 has released 716.67 kWh to interval 1. Therefore, the GCC data of both intervals 4 and 6 will get modified as follows. 300 kWh will be added to the infeasible cascade of interval 4 below 45°C in Table 9.1. The resulting cascade for interval 4 is still infeasible and needs 1800 kWh of hot utility at the top. For interval 6, 716.67 kWh needs to be subtracted for temperatures below 107.78°C. Heat flows at and above this temperature are zero. The modified GCC data are given below:

T_{int} (°C)	165	122	115	107.78	55	50	45	35	30
GCC$_{donor}$ (kWh)	0	0	0	0	1583.3	1633.3		1783.3	1783.3
GCC$_{accept}$ (kWh)	1800	940	800	--	200	100	0	0	0

The rest of the procedure is the same. The point of intersection is obtained as (545.83, 89.58). The storage predicted by PSM is 470.83 kWh at a mean temperature of 87.08°C. Again, the storage is less than that predicted by HSM (1230 kWh) and, consequently, there will be an increase in the utility requirement.

The advantage of the PSM is that it requires a single storage unit to be used at a single temperature level between two time intervals. The HSM, on the other hand, predicts too many storage units and is not practical.

9.1.4 Process Rescheduling

It is possible to minimize the capital expense on heat storage units by rescheduling the streams in batch processes to enhance direct heat recovery. Though the overall heat recovery target remains unchanged, the ratio of heat recovery by storage to that by direct exchange may be altered through rescheduling. Thus, for the rescheduled process, the TAM targets would remain the same, but the TSM targets would change. The choice between storage and rescheduling depends on the individual process flowsheet and the economics. Rescheduling may not always be feasible due to constraints imposed by the actual process requirements.

Rescheduling may be conveniently divided into three different types:
- Overall batch periods of parallel batches may be rescheduled for more effective heat transfer without altering the internal schedule of the streams within a batch. This type of rescheduling poses minimum practical difficulties as it does not affect the process itself.
- A stream may be shifted in time so as to exist in a different time interval for greater possibility of direct heat exchange without altering

its flow rate and hence its duration. Here, major plant changes may be needed which may not be feasible.

- The duration of a stream may be changed by altering its flow rate for matching it with another stream. This would require changing pump speeds and recalculating the performance of heat exchangers.

A combination of the above types of rescheduling may be attempted and the advantages of rescheduling analyzed through the cascade analysis (Kemp and Deakin, 1989b). An alternative approach (Golwelker, 1994) using the SCS principle (Example 9.3) is described below. The advantage of using individual stream cascades is that it allows easy identification of streams capable of releasing heat and those requiring heat.

Example 9.7 Rescheduling Using Stream Cascade Superposition Principle

Attempt rescheduling the process in the case of a single batch for Case Study 4S1b so that storage is replaced by direct heat exchange.

Solution. Rescheduling for a single batch is done using the feasible cascades for HSM (Table 9.5) as a starting point. Since storage occurs between intervals 3 and 4, the streams in these intervals (H2 and C3) need to be suitably altered. As shown in Table 9.8, the individual stream cascades (Table 9.3) for H2 and C3 are appropriately multiplied by 0.1 hr (for interval 3) and 0.2 hr (for interval 4).

Table 9.8 Stream Heat Cascades Before and After Rescheduling

T_{int}	Before Rescheduling			After Rescheduling			
	Intvl. 3 $S_{cas,H2}$	Intvl. 4 $S_{cas,H2}$	Intvl. 4 $S_{cas,C3}$	Intvl. 3 $S_{cas,H2}$	Intvl. 3 $S_{cas,C3}$	Intvl. 4 $S_{cas,H2}$	Intvl. 4 $S_{cas,C3}$
165	0	0	0	0	0	0	0
122	0	0	−860	0	−286.7	0	−573.3
115	0	0	−1000	0	−333.3	0	−666.7
55	300	600	−2200	300	−733.3	600	−1466.7
50	300	600	−2300	300	−766.7	600	−1533.3
35	300	600	−2600	300	−866.7	600	−1733.3
30	300	600	−2700	300	−900	600	−1800

Units: temperature in °C and heat loads in kWh.

There is a heat deficiency in interval 4 because of the existence of stream C3. Instead of storing and using the excess heat of stream H2 from interval 3, an alternative is to promote direct heat exchange by making stream C3 exist in interval 3 as well. Accordingly, the flow rate of stream C3 is decreased so that it exists for a longer period (0.2-0.5 h) instead of (0.3-0.5 h). Its MC_p now becomes 66.67 kW/°C (i.e., 100 x 0.2/0.3) in place of 100 kW/°C. With this new value of MC_p, the heat cascades for intervals 3 and 4 after rescheduling stream C3 are recalculated (Table 9.8).

It is observed that, with this changed stream population, a hot utility of 600 kWh is required in interval 3 and 1200 kWh in interval 4. The other intervals remain unaffected. The total hot utility target is 2880 kWh (same as in Table 9.5), but storage has been replaced by direct heat exchange through rescheduling (i.e., changing the duration of a stream).

Though rescheduling has been demonstrated above for the case of a single batch, it may be carried out for repeated batches also. However, total replacement of storage may not always be possible due to practical constraints of the process. Obeng and Ashton (1988) claim that TAM targets may be achieved by rescheduling alone without storage; however, Kemp (1988) argues that this may be impossible as the rescheduled process may no longer be operable. For Case Study 4S1, it is not possible to achieve TAM targets with rescheduling alone because all hot streams cannot exchange heat with all cold streams at all times (i.e., the product is obtained in practice with some time lag after the feed).

9.1.5 Network Design for Batch Processes

Kemp and Deakin (1989b) have discussed network design for the TAM, TSM, and HSM. The methods are similar to those used for continuous processes; however, for a single network to operate throughout the batch period, the exchangers need to operate in certain time periods only and be of variable capacity.

Example 9.8 Designing Networks for Heat Recovery in Batch Processes

For Case Study 4S1b, design networks at ΔT_{min} = 20°C for the following cases:

  a) TAM
  b) TSM
  c) PSM -- single batch
  d) PSM -- repeated batches
  e) the rescheduled process.

Solution. Networks may be designed for all the cases discussed in the previous sections based on the targeting results obtained thus far.

a) Network for TAM

The network based on the TAM achieves maximum energy recovery and hence is called the MER network. The TAM (Example 9.1) assumes that all streams exist over the entire batch period and that there is an overall pinch corresponding to the continuous process at which the problem may be divided. However, in order to achieve the MER targets, direct heat exchange alone is not sufficient and storage is necessary. Hence, TAM matches have two components, one due to direct heat exchange (Q_d) and another due to heat storage (Q_s).

Noting that the pinch is at 115°C, the above- and below-pinch designs are carried out separately using the conventional PDM. The structures of possible MER networks remain unchanged from those given in Figures 4.9 and 4.20. Of course, the loads are now in kWh and not kW. The next task is to determine the Q_d and Q_s components of each match. Consider the match between streams H2 and C3 in Figure 4.20. The two streams coexist for a period of 0.2 h (from 0.3 to 0.5 h). With an isothermal mixing junction, the heat loads of the two streams being matched are 600 kWh (for H2 from [125 − 65] x 50 x 0.2) and 1425 kWh (for C3 from [105 − 33.75] x 100 x 0.2). Q_d corresponds to the lower of these two values, namely, 600 kWh. Therefore, Q_s is 825 kWh. Similar calculations may be performed for all matches. In this case study, hot and cold streams do not coexist in any of the other time intervals; hence, the Q_d component of all other matches is zero.

b) Network for TSM

The TSM approach requires design of independent networks for the individual time intervals, followed by merging of these networks to obtain the final network (Kemp and Deakin, 1989b). However, merging may not be easy and may lead to a higher number of units. Giving preference to a match that already exists in an earlier time interval is a heuristic that may help alleviate this problem. Kotjabasakis and Linnhoff (1987b) and Marechal and Kalitventzeff (1989) propose a similar criterion for the design of flexible HENs. The other option could be to design all possible networks for a given time interval and then choose the one with minimum cost or the one which has matches in common with networks for the other time intervals.

For Case Study 4S1b, the stream population is sparse in the different time intervals. There is only one interval (0.3-0.5 h) with both a hot and cold stream (H2 and C3) and where a process-process exchanger (of load 600 kWh) is possible. All the other intervals have either hot or cold streams (but not both) and hence only utilities are required. Thus, the individual networks can be designed very easily and simply merged. The final network (Figure 9.3) shows the total load of a match over the entire batch period and the total utility required by each stream. The hot utility requirement is 3180 kWh and the cold utility requirement is 3100 kWh, as predicted by the TSM targets (Table 9.2). This network should have all the required matches and would have to work for the variable conditions in the different time intervals through the incorporation of bypasses and control valves.

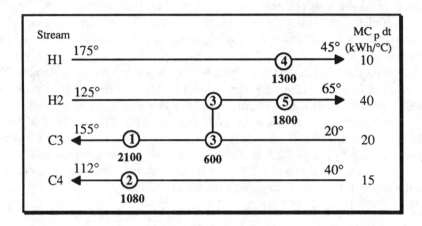

Figure 9.3 The TSM network is obtained by merging the independent designs for the individual time intervals. (Temperature in °C and load in kWh)

Operational details for the network in Figure 9.3 are provided below:

Interval	Unit	Load (kWh)	$T_{in,h}$	$T_{out,h}$	$T_{in,c}$	$T_{out,c}$
1) 0.0-0.1 h	2	1080	180	179	40	112
3) 0.2-0.3 h	5	300	125	65	15	25
4) 0.3-0.5 h	3	600	125	65	20	50
	1	2100	180	179	50	155
5) 0.5-0.6 h	5	300	125	65	15	25
6) 0.6-1.0 h	4	1300	175	45	15	25
	5	1200	125	65	15	25

c) Network for PSM -- Single Batch

To incorporate storage in a network, the interval releasing heat must have an additional cold stream and the interval accepting heat must have an additional hot stream. The rest of the procedure is similar to that for the TSM where networks for different time intervals are designed and then merged. As the idealized HSM is often difficult to implement in a network because of its several temperature levels, the single temperature level PSM target is used below.

The PSM target obtained for a single batch is 300 kWh at a mean storage temperature of 50°C between intervals 3 and 4. The actual temperature of the hot stream (releasing heat) is $(\Delta T_{min} + \Delta T_s)/2$ (i.e., 15°C) above the mean storage temperature (50°C) and the cold stream (accepting heat) is 15°C below 50°C. The networks for these two intervals are modified as in Figure 9.4 by adding a fictitious cold stream to interval 3 and a fictitious hot stream to interval 4.

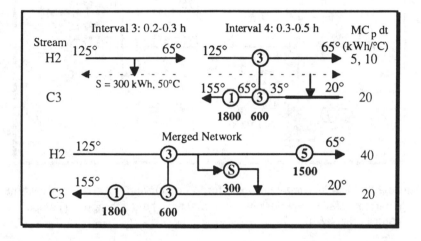

Figure 9.4 The PSM network is obtained by introducing fictitious streams in the donor and acceptor intervals, and then merging the designs for these intervals. (Temperature in °C and load in kWh)

Only streams H2 and C3 are affected. Specifically, the H2-C3 match shifts to a temperature range of 35°-65°C from the earlier range of 20°-50°C (see unit 3 in Figure 9.3). The merged network in Figure 9.4 shows that the total hot utility will be 2880 kWh and the total cold utility will be 2800 kWh (a decrease of 300 kWh as compared to the TSM network in Figure 9.3).

d) Network for PSM -- Repeated Batches

Network design for storage between batches is complex because the storage units usually increase in number and the order in which they store/release heat becomes important. From the PSM for repeated batches, storage occurs from interval 3 to 4, from interval 6 to 1, and from interval 6 to 4. The networks for the individual intervals are designed first (Figure 9.5a).

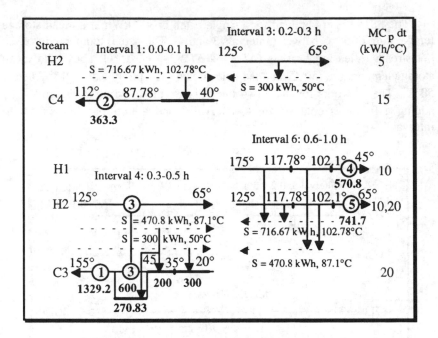

Figure 9.5a The individual networks for the donor and acceptor intervals must be generated first when designing for practical heat storage. (Temperature in °C and load in kWh)

300 kWh is transferred via storage from interval 3 to 4 at a mean temperature of 50°C, raising the temperature of C3 to 35°C. Across intervals 6 and 1, 716.67 kWh are stored at a mean temperature of 102.78°C. This means that the heat content of the hot streams above 117.78°C is stored and utilized to heat stream C4 to 87.78°C. Across intervals 6 to 4, 470.83 kWh may be stored at a mean temperature of 87.08°C. However, there exists a direct match for which the inlet temperature of C3 must not be greater than 45°C because the hot stream outlet temperature is 65°C and a ΔT_{min} of 20°C must be observed. As the storage from interval 3 has already raised the temperature of stream C3 to 35°C, storage can be used from 35° to 45°C and only 200 kWh can be transferred before splitting stream C3. Thus, the heat content of 470.83 kWh from the hot streams (H1 and H2) in the temperature range 117.78°-102.08°C is stored and transferred to interval 4.

The final network (Figure 9.5b) shows hot and cold utility demands of 1692.5 kWh and 1612.5 kWh respectively.

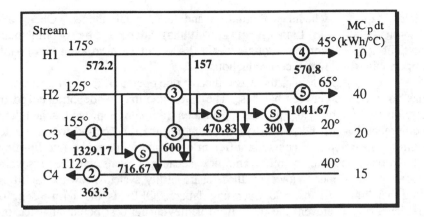

Figure 9.5b The PSM network for repeated batches is obtained by merging the independent designs for the individual time intervals in Figure 9.5a. (Temperature in °C and load in kWh)

e) Network for Rescheduled Process

The concepts for network design described above obviously apply for the rescheduled process. The appropriate modification is made in the stream data to eliminate the 300 kWh of storage: stream C3 has a new MC_p value of 66.67 kW/°C and goes from 0.2 to 0.5 h. Using the procedure adopted earlier for the TAM, it is seen that the network for a single batch has the same structure as in Figure 9.3. The heat load on unit 3 increases to 900 kWh and the loads on units 1 and 5 show a corresponding decrease of 300 kWh each. The total hot utility is 2880 kWh and the total cold utility is 2800 kWh (identical to the PSM for a single batch). Thus, rescheduling has eliminated the storage unit at the cost of additional area on the direct match between H2 and C3.

Conventional evolution (loop breaking and path relaxation) may be performed in general on the networks as described by Kemp and Deakin (1989b).

9.1.6 Overall Plant Bottlenecks in Batch Processes

Obeng and Ashton (1988) and Linnhoff et al. (1988a) argue that it is inappropriate to treat energy in isolation during batch process integration. Energy costs often form only a fraction of the total batch operating costs while costs associated with raw materials, yields, labor, handling, wastes, and reworks tend to be significant. Thus, the problem needs to be addressed in a broader context considering interactions and tradeoffs between energy, capital, yields, capacity, schedules, flexibility, and labor requirements. Obeng and Ashton (1988) and Linnhoff et al. (1988a) advocate the overall plant bottleneck (OPB) concept, along with the TAM and TSM, to identify valuable opportunities for overall cost reductions.

In a single batch line, there is typically one operation that is tight, i.e., it has no free time between batches. The batch line may be debottlenecked by appropriately modifying process conditions (through improved scheduling, equipment design, and chemistry) and reducing the time period for this rate-limiting operation. Of course, a different operation will now be rate limiting. The problem is usually more complex because a plant will have several different batch lines. However, these share utility systems, rework facilities, feed handling systems, and operating labor. OPBs is the term used for bottlenecks that prevent the plant from achieving its best potential when the total plant interdependencies are considered. A useful representation for tackling OPBs is the time-event chart (Linnhoff, 1993a) that captures the essential time factors controlling the batch process. These are the waiting times for material to flow in and out, for temperatures to reach prescribed values, for equipments to be accessible, and for shift labor to be available. In addition to time-event charts (Figure 9.6), equipment occupancy charts/Gantt charts may also prove useful. As the heat flow is a key factor, energy analysis could yield a multifold cost benefit in terms of capacity, yield, product quality, and utility requirements (probably in this order).

Figure 9.6 shows the existing scenario for Case Study 4S1b, with both reactors (R1 and R2) having a minimum residence time of 0.1 h. However, assuming this schedule to be sacrosanct and not challenging the constraints imposed by it can lead to missed opportunities for overall cost savings as discussed below. Scenario 1 corresponds to the rescheduling done in Example 9.7 to eliminate the storage OPB, but any plant has several OPBs and consequently various scenarios are possible. Instead of slowing down the charging of reactor R2 by decreasing the flow rate of C3 (Scenario 1), reactor R2 can be simply shifted in time (i.e., started 0.1 h earlier). Now, the discharge of reactor R1 must be speeded up because it is limiting. Scenario 2 shows that starting reactor R2 early and speeding up stream H2 reduces the total cycle time to 0.9 h and eliminates the capacity OPB.

Scenario 1 has a hot utility consumption of 2880 kWh, but this is much above the TAM target (605 kWh is the best achievable when there are no constraints whatsoever) and so there is an energy OPB. This may be debottlenecked by making the hot and cold streams exist over the same time periods. For example, if stream H2 is discharged at four times its initial rate, then it coincides with stream C3 (Scenario 3 in Figure 9.6). Using stream cascade superposition, the TSM targets for the rescheduled process show that the new hot utility requirement is 2080 kWh (a savings of 800 kWh from

Scenario 1).   Interestingly, the discharge operation of reactor R2 is now limiting, and reduction in cycle time may be achieved by speeding up stream H1. Of course, all rescheduling is subject to process constraints.

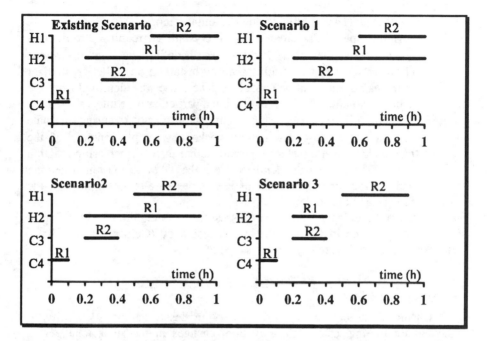

Figure 9.6 Various possible schedules are shown using time-event charts, which provide a convenient representation for batch process debottlenecking.

In this manner, the OPB concept may be used profitably to approach the ultimate target set by the TAM, using the TSM as the energy analysis tool to evaluate each scenario. In this case study, it is not possible to achieve the TAM target exactly for a single batch due to obvious constraints on the reactor feed and product.

In addition to rescheduling, improved designs may help overcome OPBs. Consider the common example of charging a reactor completely and then heating it using a steam jacket. Alternatives could be explored: heating using a steam jacket while charging the reactor and modeling the varying-heat-transfer-area batch process (Natarajan and Shenoy, 1992), or heating the feed using an external exchanger while filling occurs (as in Case Study 4S1b). Linnhoff (1993a) provides another example where the reactor feed is heated by the reactor product (which is cooled) using an external exchanger along with a simple storage tank for decoupling in time. Linnhoff et al. (1988a) report that such an approach increased the overall capacity of a grain whisky distillery where the feed water was preheated by the cooker discharge resulting in increased time availability of the cooker and eliminating the necessity for vacuum cooling.

Linnhoff et al. (1988a) provide other examples, too:

- A throughput OPB created by a crystallizer requiring refrigerant cooling was debottlenecked by rescheduling cooling operations elsewhere on the plant (rather than by installing a parallel crystallizer or increased refrigeration capacity). The cause was identified from the TSM as too much cooling required in a particular time interval.
- A yield OPB created by a high temperature reactor requiring alternate heating and cooling was debottlenecked by simple repiping of the recirculating heat transfer oil circuit (rather than by increasing reaction time). The cause was identified from the TAM as too much heating and cooling for the same oil as it was passed alternately through the furnace and the cooler.

Note that time-dependency of processes may also require consideration of start-up and shut-down (Kemp, 1991) and resiliency (Colberg et al., 1989). However, these topics are beyond the scope of the book.

9.2 FLEXIBLE PROCESSES

Chemical, petrochemical, and refinery processes require flexibility to account for fouling, catalyst deactivation, changes in feedstock, changes in product demand, changes in product specifications, and varied seasonal operation. Changes in process operating conditions will typically require changes in the design of various units (including heat exchangers, reactors and separators). Although the discussion that follows is restricted to HENS (as has been the case with much of the research in the design of flexible processes), it is still valuable since the basic methodology may be extended to overall flexible process design. Also, temperature control is an important determining factor for operability. Flexibility requirements make significant cost demands and are relevant to both grassroots designs as well as retrofits.

9.2.1 Better Designs in Flexible HENS

The operating conditions in a HEN could change in three possible ways:

- The supply temperatures of some streams may change.
- The flow rates of various streams may change (e.g., in debottlenecking, plant flow rates increase by a certain percentage).

- The *UA* values of some exchangers may change (e.g., heat transfer coefficients may change due to fouling).

The flexibility problem involves determining the most effective change in some of the exchangers to compensate for the above three possibilities. This must be done under various constraints such as maintaining the target temperatures at certain specified values or the loads of heaters/coolers within certain limits.

The approach discussed below is broadly applicable although for specificity the practical problem of fouling in industries is considered. Fouling adversely affects plant operation and translates into significant additional costs. From the standpoint of flexible process design, the decrease in *UA* with time of operation must be accounted for. An extreme case of fouling of practical importance is the removal of certain exchangers for cleaning, in which case the value of *UA* goes to zero. The traditional solution to fouling is to overdesign the heat exchangers which get fouled. Kotjabasakis and Linnhoff (1987a) proposed an alternative strategy for improved HEN design based on overall network sensitivities to compensate for fouling in a cost-effective manner. Their philosophy is diametrically opposite to that in conventional design practice. They suggested overdesigning the nonfouling exchangers and leaving the fouling ones intact. An illustration of their suggestion is given in the example below, though the methodology adopted is original.

Example 9.9 Effective Design for Fouling in HENS

Consider the network in Figure 3.1 (an MER design based on Case Study 4S1 for ΔT_{min} = 13°C). Assume the target temperature for stream H2 is not critical. However, the remaining target and supply temperatures must meet the specifications in Figure 3.1. The temperatures (important in the flexibility analysis to follow) are labeled T_1 through T_{13} in Figure 9.7 (following the convention of Kotjabasakis and Linnhoff (1986), wherein temperatures are marked starting from the inlet of the streams and proceeding to the outlet). These temperatures for the initial network are given by the vector:

$$\mathbf{T}^t = [175, 125, 73, 125, 51.4, 125, 87.67, 20, 112, 137, 40, 74.67, 112]$$

All streams have a heat transfer coefficient of 0.2 kW/(m^2 °C). Stream C4 contains a foulant, which is active when the temperature exceeds 90°C. The fouling is in accordance with the classic asymptotic model with the heat transfer coefficient of stream C4 reaching 0.12 kW/(m^2 °C) after six months of operation. Assume annual maintenance shutdowns.

The task is to determine the most cost-effective location for the additional heat transfer area necessary due to the fouling of unit 3. Note that stream H2 is split in the MC_p-ratio of 25:15. Assume countercurrent heat exchangers.

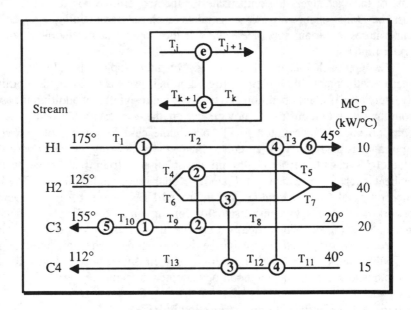

Figure 9.7 The above network shows the base-case design. The optimum for fouled conditions is different from the optimum for clean conditions since a three-way tradeoff exists between energy, capital, and flexibility. (Temperature in °C and load in kW)

Solution. The method involves the rigorous simulation of the network by the setting up of a matrix (system of linear equations) based on the *N-R-P* (dimensionless) quantities defined for heat exchangers in Chapter 6.

Step 1. Evaluation of R and P for All Exchangers
The following notation is convenient to follow. Let an exchanger be denoted by *e*, the inlet temperature of the hot stream through it by subscript *j*, and the inlet temperature of the cold stream by subscript *k* (see inset of Figure 9.7). Each exchanger in the network is thus characterized by the two subscripts, *j* and *k*, as observed in Figure 9.7. For unit 1, $e = 1$, $j = 1$, and $k = 9$.

Two useful dimensionless quantities, *P* and *R*, may be defined for each exchanger *e* as follows (see Equation 6.5):

$$R_e = (T_j - T_{j+1})/(T_{k+1} - T_k) = M_c C_{pc}/M_h C_{ph} \qquad (9.3a)$$
$$P_e = (T_{k+1} - T_k)/(T_j - T_k) \qquad (9.3b)$$

From Equations 9.3, P and R for all the exchangers may be determined (Table 9.9), provided the terminal temperatures (given by the vector, $\mathbf{T}^t$) are known.

A relation of the form $N(R, P)$ exists for a given heat exchanger type, as discussed in Chapter 6. For a countercurrent exchanger, N may be calculated from the expressions given in Problem 6.D(a). From $N \equiv UA/M_c C_{pc}$, the area for each exchanger in the base-case network may be calculated using $U = 0.1$ kW/(m^2 °C). The calculations are summarized in Table 9.9 (the last row will be generated in the next step). The areas obtained obviously agree with those calculated earlier in Table 4.20.

Table 9.9 Analysis of Exchangers in Base-Case Design

Unit	j	k	R	P	N	A
1	1	9	2	0.3968	1.0726	214.53
2	4	8	0.8	0.8762	4.4093	881.86
3	6	12	1	0.7417	2.8718	430.77
4	2	11	1.5	0.4078	0.8443	126.65
3f	6	12	1	0.6829	2.1539	430.77

Units: area in m^2.

Step 2. Analysis of Fouling Exchanger

From Equation 1.16a, the overall heat transfer coefficient for unit 3 in the fully fouled state is $U_f = [1/0.2 + 1/0.12]^{-1} = 0.075$ kW/(m^2 °C). To maintain the outlet temperatures of unit 3 in the presence of fouling, UA must be invariant in the clean and fouled states, i.e., $A_f = 0.1 \times 430.77/0.075 = 574.36$ m^2. Thus, unit 3 needs 143.59 m^2 of additional heat transfer area in accordance with the conventional practice of oversizing the fouling exchanger.

In the absence of the additional area, unit 3 will not deliver the expected outlet temperatures. In terms of the N-R-P notation, this implies a decrease in P for unit 3, which may be calculated as follows. For unit 3 in the fouled state and without overdesigning, $N = 0.075 \times 430.77/15 = 2.1539$ and $R = 1$ (as before). As discussed in Chapter 6, a relation of the form $P(N, R)$ exists for a given heat exchanger type. For a countercurrent exchanger, P may be calculated from the expressions given in Problem 6.D(a). Thus, $P = 0.6829$ for unit 3 (fouled, but not oversized). This information is added to Table 9.9 (under the label 3f) for ready reference.

The goal now is to assess the possibility of changing the areas of some of the other exchangers (units 1, 2, and 4) to meet the specified target temperatures.

Thus, the problem is tackled from the systems engineering viewpoint rather than the individual unit operation viewpoint.

Step 3. Calculation of the Minimum Number of Changes Required

The question that needs to be addressed first is "what is the minimum number of changes required to meet the required target temperatures?" In other words, is it possible to have an operable network that meets the design specifications by oversizing only one of the exchangers? In Figure 9.9, one of the cold streams (C3) has a heater, whose heat duty could be manipulated to meet the target temperature of 155°C for stream C3. Similarly, one of the hot streams (H1) has a cooler, where the conditions of the cooling medium may be adjusted to obtain the desired target temperature of 45°C. As stream H2 has no critical constraint on its outlet temperature, the only target temperature that must be met is that of stream C4. Thus, it may be argued that one degree of freedom is necessary, i.e., one change minimally needs to be made. In general, the number of active constraints (N_{ac}) is given by

$$N_{ac} = N_{con} - N_{hc} \qquad (9.4)$$

where N_{con} is the number of constraints in terms of streams with specified target temperatures and N_{hc} is the number (subset) of the constrained streams with heater or coolers on them. For this example, Equation 9.4 gives $N_{ac} = 3 - 2 = 1$. Note that N_{ac} corresponds to the minimal number of exchangers which require a change in area.

Step 4. Generation of the Flexibility Simulation Matrix

Equations 9.3 may be rewritten in the following convenient form:

$$T_j - T_{j+1} + R_e T_k - R_e T_{k+1} = 0 \qquad (9.5a)$$
$$P_e T_j + (1 - P_e) T_k - T_{k+1} = 0 \qquad (9.5b)$$

For example, Equations 9.5 for unit 1 on substituting for R, P, j, and k give

$$T_1 - T_2 + 2 T_9 - 2 T_{10} = 0$$
$$0.3968 T_1 + 0.6032 T_9 - T_{10} = 0$$

Similar equations may be written for the other three process exchangers (units 2, 3f, and 4), giving a total of eight equations. The coefficients of these equations form the first eight rows of the flexibility simulation matrix, F, in Table 9.10. In the matrix, the R-equations (as per Equation 9.5a) for all the four process exchangers are written first and the P-equations (as per Equation 9.5b) are written next.

Table 9.10 Flexibility Simulation Matrix for the HEN

$$
\begin{bmatrix}
1 & -1 & 0 & 0 & 0 & 0 & 0 & 0 & 2 & -2 & 0 & 0 & 0 \\
0 & 0 & 0 & 1 & -1 & 0 & 0 & 0.8 & -0.8 & 0 & 0 & 0 & 0 \\
0 & 0 & 0 & 0 & 0 & 1 & -1 & 0 & 0 & 0 & 0 & 1 & -1 \\
0 & 1 & -1 & 0 & 0 & 0 & 0 & 0 & 0 & 0 & 1.5 & -1.5 & 0 \\
0.397 & 0 & 0 & 0 & 0 & 0 & 0 & 0 & 0.603 & -1 & 0 & 0 & 0 \\
0 & 0 & 0 & 0.876 & 0 & 0 & 0 & 0.124 & -1 & 0 & 0 & 0 & 0 \\
0 & 0 & 0 & 0 & 0 & 0.683 & 0 & 0 & 0 & 0 & 0 & 0.317 & -1 \\
0 & 0.408 & 0 & 0 & 0 & 0 & 0 & 0 & 0 & 0 & 0.592 & -1 & 0 \\
1 & 0 & 0 & 0 & 0 & 0 & 0 & 0 & 0 & 0 & 0 & 0 & 0 \\
0 & 0 & 0 & 1 & 0 & 0 & 0 & 0 & 0 & 0 & 0 & 0 & 0 \\
0 & 0 & 0 & 0 & 0 & 0 & 0 & 1 & 0 & 0 & 0 & 0 & 0 \\
0 & 0 & 0 & 0 & 0 & 0 & 0 & 0 & 0 & 0 & 1 & 0 & 0 \\
0 & 0 & 0 & 0 & 0 & 0 & 0 & 0 & 0 & 0 & 0 & 0 & 1 \\
0 & 0 & 0 & 1 & 0 & -1 & 0 & 0 & 0 & 0 & 0 & 0 & 0
\end{bmatrix}
$$

The next four rows of the flexibility simulation matrix (i.e., rows 9 through 12) correspond to the specified supply temperatures. Appropriately, ones are placed in the columns corresponding to T_1, T_4, T_8, and T_{11}. Finally, the constraint equations are written down. Here, the last rows in matrix F indicate the target temperature for stream C4 is to be maintained (i.e., $T_{13} = 112$) and the splitting junction on stream H2 is to be accounted for (i.e., $T_4 - T_6 = 0$).

Therefore, any network can be simulated in the form:

$$F\,T = S \tag{9.6}$$

where F is the flexibility simulation matrix (Table 9.10), T is the column vector of labeled temperatures to be calculated on the network (in this problem, T_1 through T_{13}), and S is the column vector consisting of zeroes and the specified temperatures. In this example,

S^t = [0, 0, 0, 0, 0, 0, 0, 0, 175, 125, 20, 40, 112, 0]

In general, for a network consisting of e exchangers, N_{ps} streams, and s splitting junctions, T will be a column vector with $(2e + N_{ps} + s)$ elements. The flexibility simulation matrix will consist of R-equations (e in number), P-equations (e in number), supply temperature specification equations (N_{ps} in number), and constraint equations ($N_{ac} + s$ in number). Thus, there are N_{ac} more equations than unknowns in the system given by Equation 9.6. To provide for the necessary degrees of freedom, some equations (N_{ac} in number) must be dropped from the system. On eliminating the appropriate number of P-equations (in this case, $N_{ac} = 1$) from matrix F, the system of linear equations given by Equation 9.6 may be solved.

Step 5. Solution of the Linear System of Equations
    Before solving the system, the P-equation to be dropped must be decided. Consider that the P-equation corresponding to unit 4 (row 8 in matrix F) is dropped. Then, the set of linear equations may be simply solved for the 13 unknown temperatures (say, by Gaussian elimination) to yield the following:

T^t = [175, 125, 59, 125, 51.4, 125, 97, 20, 112, 137, 40, 84, 112]

Step 6. Calculation of the Additional Area
    From Equation 9.3b, the value of P for unit 4 (the dropped equation in step 5) can be calculated as

P_4 = (84 − 40)/(125 − 40) = 0.5176

The value of N for unit 4 (in counterflow) is

N_4 = ln[(1 − 0.5176)/(1 − 1.5 × 0.5176)]/(1.5 − 1) = 1.5383

Thus, for the fouled system to operate as per specifications, unit 4 must have an area A_4 = 1.5383 × 15/0.1 = 230.74 m^2. Hence, the additional area requirement on unit 4 is 104.09 m^2 (27.5% less than for compensation in unit 3). Note that T_{10} has not changed its value from 137°C; hence, the hot utility requirement of 360 kW remains unchanged. However, the new value of T_3 is 59°C, leading to a reduced cold utility requirement of 140 kW. This effect on unit 6 (cooler) is a direct consequence of oversizing unit 4. Figure 9.7 shows that such a favorable effect is not possible by overdesigning unit 3. If the network corresponding to the new conditions is drawn, it is observed that ΔT_{min} = 13°C. However, the outlet temperature of stream H2 is given by

(15 x 97 + 25 x 51.4)/40, i.e., **68.5°C**.   Therefore, the savings of 140 kW in cold utility is really at the sacrifice of not achieving the original outlet temperature for stream H2, as is clear from the following calculation: $40(68.5 - 65) = 140$ kW.

Step 7. Determining the Most Cost-effective Location for Additional Area

Unit 4 was manipulated (by dropping its P-equation and increasing its area) in the above steps. Steps 5 and 6 may be repeated for the other units and the cost-effectiveness of the manipulations evaluated.   The results are summarized in Table 9.11. Note that it is not possible to manipulate unit 2 and obtain an operable network because ΔT_{min} becomes negative and P_2 becomes greater than unity, indicating infeasibility.   An interesting situation occurs while manipulating unit 1. Its area requirement decreases whereas the overall utility requirements increase. This capital-energy tradeoff may be evaluated using the following equation:

$$
\begin{aligned}
\text{Total additional cost} = \ &\text{Annualized cost of additional capital} \\
&+ \text{Annual cost of energy after manipulation} \\
&- \text{Annual cost of energy before manipulation} \quad (9.7)
\end{aligned}
$$

Table 9.11 Cost Analysis for Best Placement of Additional Area

Unit Manipulated	Additional Area	Hot Utility Savings	Cold Utility Savings	Total Additional Cost
1	−150.43	−228.85	−88.85	32.4
2				Infeasible
3	143.59	0	0	13.5
4	104.09	0	140	9.2

Units: area in m^2, utility in kW, and cost in 10^3 $/yr.

Using Equation 9.7, the total additional cost is estimated in Table 9.11 based on the cost data in Example 1.9. Note that it is possible to manipulate multiple exchangers using the above methodology (by dropping more than one P-equation at a time).   In the above example, the modification of multiple exchangers does not prove cost effective. Thus, it is observed that it is best to manipulate unit 4 (based on Table 9.11) and not overdesign the fouling exchanger (unit 3).

Oversizing the nonfouling exchanger circumvents the following disadvantages associated with the overdesign of the fouling exchanger (Kotjabasakis and Linnhoff, 1987a):

- The extra capital investment is not efficiently utilized when the fouled exchanger is removed for cleaning if the plant is operated in these cleaning periods.
- The extra capital investment (in excess heat transfer area) is not thoroughly utilized until complete fouling occurs.
- The additional area usually leads to decreased velocities, lower heat transfer coefficients, and accelerated fouling.

Note that the preferred locations for additional investment (area) for fouling compensation are typically different from those for the clean base-case design.

Importantly, it may be recognized that integrated processes do not necessarily suffer from flexibility/operability problems. In fact, they can provide better system design opportunities (e.g., to reduce fouling costs or to handle different operating cases efficiently).

9.2.2 Sensitivity Tables and Downstream Paths

Sensitivity tables (Kotjabasakis and Linnhoff, 1986) are an alternative to rigorous simulation using the flexibility matrix (see Table 9.10). They help in determining the amount of flexibility that is cost effective, and the network design that provides the cheapest flexibility. It may not be always meaningful to achieve flexibility at all costs or rigidly to specify a certain requirement of flexibility (in terms of deviation ranges). Instead, it must be recognized that there is a three-way tradeoff between energy, capital, and flexibility.

Downstream paths (Linnhoff and Kotjabasakis, 1986) show how disturbances propagate in a network and help in the understanding of flexibility in terms of the process structure.

The example below illustrates how sensitivity tables and downstream paths aid in designing for good operability of integrated processes.

Example 9.10 Flexible HENS Using Sensitivity Tables and Downstream Paths

The network in Figure 3.1 (an MER design based on Case Study 4S1 for $\Delta T_{min} = 13°C$ forming the base-case operation) is to be retrofitted to permit an alternative operating case for four months/year. Flexibility is necessary to allow an alternative mode of operation, wherein the inlet temperature of stream H1 is 155°C (not 175°C) and the MC_p of stream C4 is 18 kW/°C (not 15 kW/°C). Assume the target temperature for stream H2 is not critical. The remaining temperatures and MC_ps remain unchanged from their values in the base-case operation.

Design a flexible HEN for both operation modes that minimizes the total additional cost of retrofitting. Also, discuss how possible savings may be achieved if small deviations in target temperatures could be tolerated.

Solution. Ideally, flexibility must be treated as a part of the optimization process. If contingency (whose cost is frequently high) for alternative operating cases is added as an afterthought to an already optimized base-case design, then there is a strong possibility that the original optimization no longer holds.

Step 1. Conceptualization of Flexibility as a Control Problem

Certain concepts used in control studies (usually associated with short-term variations) may be used for flexibility analysis (typically associated with long-term variations). Thus, the alternative operation mode may be considered as a disturbance from the base case (i.e., a long-term deviation of the process operation in terms of changes in its supply temperatures and heat capacity flow rates). Furthermore, some parameters may be considered controlled (in this case, target temperatures should not change). Denoting disturbances by Δ_D and controlled parameters by Δ_C, the flexibility problem for this example may be described as follows:

$$\Delta_D (T_{in,H1}) = -20°C \quad \text{and} \quad \Delta_D (MC_{p,C4}) = +20\%$$
$$\Delta_C (T_{out,H1}) = \Delta_C (T_{out,H2}) = \Delta_C (T_{out,C3}) = \Delta_C (T_{out,C4}) = 0$$

Figure 9.8 shows the four controlled parameters and the two disturbances. How will the disturbances propagate through the system? Downstream paths allow determination of the controlled parameters that are affected by the disturbances. On the other hand, sensitivity tables allow determination of the magnitude of the effects.

Both these aspects are discussed next since they help in devising strategies for the elimination of undesired effects.

Step 2. Using Downstream Paths to Study the Passive Response of the Network

A disturbance will affect a controlled parameter only if the two points on the network are connected by a path that is totally downstream. Figure 9.8 shows two paths from the disturbance on stream C4 to the controlled parameter on stream H1. The path through unit 4 is completely downstream. On the other path, the disturbance cannot travel from unit 3 to unit 2 as it would mean going upstream (against the natural flow of stream H2).

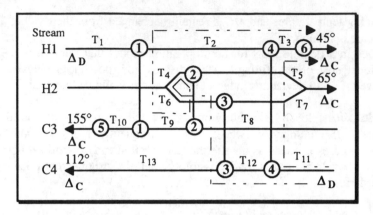

Figure 9.8 A disturbance (Δ_D) can affect a controlled parameter (Δ_C) only if a downstream path exists between the two points. Of the two paths from $\Delta_D(MC_{p,C4})$ to Δ_C ($T_{out,H1}$), only the path through unit 4 is fully downstream. (Temperature in °C)

Using this logic for downstream paths, it is easy to tabulate (as shown below) whether the two disturbances in Figure 9.8 will affect the four controlled parameters.

	$\Delta_C (T_{out,H1})$	$\Delta_C (T_{out,H2})$	$\Delta_C (T_{out,C3})$	$\Delta_C (T_{out,C4})$
$\Delta_D (T_{in,H1})$	Yes	Yes	Yes	Yes
$\Delta_D (MC_{p,C4})$	Yes	Yes	No	Yes

To ensure operability for the alternative case, possible design changes are (Linnhoff and Kotjabasakis, 1986):

- A downstream path may be broken (e.g., removal of unit 4 in Figure 9.8, if possible, would prevent the disturbance on stream C4 from reaching the controlled parameter on stream H1).
- An upstream element may be introduced (e.g., swapping the locations of units 3 and 4 on stream C4, if possible, would prevent the disturbance on stream H1 from reaching the controlled parameter on stream H2).
- Simple manipulations may be attempted (e.g., a bypass on any of the four exchangers would keep the controlled parameter on stream C4 at its required value since there are downstream paths from all four exchangers to it).

Note that it is not necessary for an exchanger to lie on a direct path between the disturbance and the controlled parameter to allow for design changes. Thus, units 1 and 2 could be used for manipulations to offset the effect of Δ_D $(MC_{p,C4})$ on Δ_C $(T_{out,C4})$, as they have downstream paths to Δ_C $(T_{out,C4})$ even though they do not lie on stream C4. In other words, potential exchangers for contingency are those with downstream paths to the controlled parameter because a manipulation in an exchanger (i.e., a modification in its effective UA or MC_p) may also be considered a disturbance (although a desirable one).

Step 3. Generation of Sensitivity Tables

The changes in network temperatures caused by changes in operating conditions (i.e., supply temperatures, flow rates, and UA values) may be tabulated conveniently in sensitivity tables. Equations 9.5 are linear in temperature (on rewriting the equations in terms of $N \equiv UA/M_cC_{pc}$ and $R \equiv M_cC_{pc}/M_hC_{ph}$); hence, a single calculation suffices to correlate the change in each supply temperature to the responses in network temperatures starting with the base case. On the other hand, the equations are nonlinear in terms of MC_p and UA changes; therefore, several calculations (that include small ranges) are required as demonstrated below.

First, the sensitivity tables are generated based on Figure 9.8 with respect to all the supply temperatures (considered individually). The calculations for R, P, N, and UA for each exchanger are performed (see Table 9.9) from the base-case data. The flexibility simulation matrix is also determined as in Table 9.10.

Consider that the supply temperature of stream H1 is changed by 1°C (i.e., from 175° to 176°C). Neither the values of R, N, P, and UA nor the flexibility simulation matrix will change (except that P for unit 3 will be 0.7417 rather than the fouled value of 0.6829). Only the column vector S in Equation 9.6 will change as given below:

$$S^t = [0, 0, 0, 0, 0, 0, 0, 0, 176, 125, 20, 40, 112, 0]$$

On dropping the equation for the maintenance of the target temperature for stream C4 (i.e., $T_{13} = 112$, or row 13 of the flexibility simulation matrix), the set of linear equations is solved for the thirteen unknown temperatures to yield:

$$T^t = [176, 125.206, 73.08, 125, 51.4, 125, 87.727, 20, 112, 137.397, 40,$$
$$74.75, 112.023]$$

The sensitivity factor is defined as the ratio of the change in the network response to the change in the operating condition (in this case, supply temperature). Thus, the sensitivity vector is given by

$$\delta T^t(T_1) = [1, 0.206, 0.08, 0, 0, 0, 0.061, 0, 0, 0.397, 0, 0.084, 0.023]$$

The sensitivity vectors for the four supply temperatures (T_1, T_4, T_8, and T_{11}) are collected in Table 9.12 to form the sensitivity table for changes in supply temperatures. The table is accurate for combinations of supply temperature changes and over any range. The nonzero entries in the sensitivity table indicate the existence of a downstream path from the supply temperature to the network temperature whereas the zero entries suggest the absence of a downstream path.

Table 9.12 Sensitivity Table for Changes in Supply Temperatures

	T_1	T_2	T_3	T_4	T_5	T_6	T_7	T_8	T_9	T_{10}	T_{11}	T_{12}	T_{13}
T_1	1	.206	.080	0	0	0	.061	0	0	.397	0	.084	.023
T_4	0	.695	.270	1	.299	1	.467	0	.876	.529	0	.283	.817
T_8	0	.098	.038	0	.700	0	.028	1	.124	.075	0	.040	.012
T_{11}	0	0	.612	0	0	0	.438	0	0	0	1	.592	.154

Units: temperature change in °C.

Next, the sensitivity tables may be generated with respect to each MC_p change. For example, consider increasing the MC_p value of stream C4 from 15 kW/°C by 20% to 18 kW/°C. Then, $R_3 = 1.2$, $R_4 = 1.8$, $P_3 = 0.6554$, and $P_4 = 0.3498$. After making the necessary changes for R and P in the flexibility simulation matrix, the set of linear equations is solved to yield:

$$T^t = [175, 125, 71.48, 125, 51.4, 125, 81.53, 20, 112, 137, 40, 69.73, 105.95]$$

The sensitivity factor is simply defined as the temperature change in the network response. Hence, the sensitivity vector is given by

$$\delta T^t(MC_{p,C4}) = [0, 0, -1.52, 0, 0, 0, -6.13, 0, 0, 0, 0, -4.93, -6.05]$$

The sensitivity vectors for $MC_{p,C4}$ changes from +20% to −20% (in steps of 10%) are collected in Table 9.13 to form the sensitivity table for heat capacity flow rate changes in stream C4. Similar tables for the MC_p changes of the other streams may be generated, if necessary.

Sensitivity tables for effective UA changes in each exchanger may be generated in an analogous manner. Table 9.14 shows an abbreviated sensitivity table for UA changes in exchanger 3. Based on the concept of downstream paths, it is clear from Figure 9.8 that only T_7 and T_{13} change in terms of the network response.

Table 9.13 Sensitivity Table for Changes in Heat Capacity Flow Rate of Stream C4

	T_1	T_2	T_3	T_4	T_5	T_6	T_7	T_8	T_9	T_{10}	T_{11}	T_{12}	T_{13}
+20%	0	0	-1.52	0	0	0	-6.13	0	0	0	0	-4.93	-6.05
+10%	0	0	-0.83	0	0	0	-3.27	0	0	0	0	-2.65	-3.06
-10%	0	0	1.02	0	0	0	3.75	0	0	0	0	3.10	3.08
-20%	0	0	2.26	0	0	0	8.07	0	0	0	0	6.78	6.03

Units: temperature in °C.

Table 9.14 Sensitivity Table for Changes in *UA* of Exchanger 3

	-80%	-40%	-20%	20%	40%	80%	100%	120%	140%	180%	200%
T_7	18..97	5.48	2.26	-1.68	-2.97	-4.84	-5.54	-6.12	-6.62	-7.43	-7.77
T_{13}	-18.97	-5.48	-2.26	1.68	2.97	4.84	5.54	6.12	6.62	7.43	7.77

Units: temperature in °C.

Note that sensitivity tables may be prepared with a simple computer program, using the base-case network without any knowledge of the actual changes. Then, they may be used for quick screening of all available options through simple hand calculations, as shown below, to decide on the most promising change(s). The tables enable the process designer to gain insight into the sensitivities and available flexibility in the network.

Step 4. Determination of Network Responses Using Sensitivity Tables
The target temperatures of streams H1 and C3 may be maintained easily by regulating the cooling water and steam flow rates in the cooler and heater. As the target temperature of stream H2 is not critical, the task reduces to maintaining the outlet temperature of stream C4 at 112°C.

Assuming that the *UA* values of all exchangers remain unaltered, the change in T_{13} (due to the disturbances created by the alternative operation) may be estimated using Tables 9.12 and 9.13 as follows:

$$\delta T_{13} = \delta T_{13}(\Delta_D(T_{in,H1})) + \delta T_{13}(\Delta_D(MC_{p,C4}))$$
$$= 0.023\,(-20°C) + (-6.05°C) = -6.51°C$$

The above computation assumes that the total network response is the sum of the individual responses from each disturbance. This is clearly an approximation since it ignores interaction effects and the nonlinear response from MC_p changes. However, the estimated network response is accurate enough (Kotjabasakis and Linnhoff, 1986) for preliminary screening of flexible design options.

Step 5. Design Changes for Compensation Using Sensitivity Tables

The decrease in T_{13} must be compensated through design changes if it is to be kept constant at 112°C. The UA value of an exchanger (say, unit 3) may be modified to nullify the temperature decrease as given below:

$$\delta T_{13} = -6.51 + \delta T_{13}(\Delta_D(UA_3)) = 0$$

For $\delta T_{13}(\Delta_D(UA_3)) = +6.51°C$, $\Delta_D(UA_3) = 135.6\%$ from Table 9.14 by interpolation. Similar calculations based on the sensitivity tables for the effective UA changes of the other exchangers show that it is not possible to obtain the required compensation from exchangers 1, 2, and 4. With exchanger 1 completely bypassed (i.e., $\Delta_D(UA_1) = -100\%$), δT_{13} is only 5.27°C. With the size of exchanger 2 quadrupled (i.e., $\Delta_D(UA_2) = +300\%$), δT_{13} is only 1.03°C. Also, the maximum δT_{13} achievable with very large increases in the UA of exchanger 4 is 5.68°C.

Therefore, the design change identified involves installing additional area on exchanger 3. The additional area is 584.12 m^2 from the sensitivity table calculations, which compares well with 573.79 m^2 according to rigorous simulation (see Figure 9.9). For the network in Figure 9.9 to operate under normal base conditions (Figure 3.1), the steam flow through the heater should be decreased, the cooling water should be increased, and exchanger 3 should be bypassed. The total additional cost for the network in Figure 9.9 is obtained as 67.42 x 10^3 \$/yr from Equation 9.7. Note that the change in utility loads affects both energy and capital costs.

In this example, there is only one possible scheme for design modification if the specified flexibility requirements are rigorously enforced. In general, the choice of scheme (when options exist) depends on the period of operation of the alternative case and on the flexibility desired. If the target temperature of stream C4 need not be strictly 112°C, then a scheme with complete bypassing of exchanger 1 may be explored. The total additional cost for this network (see Figure 9.10) is 41.65 x 10^3 \$/yr. Note that adding area and bypassing exchanger 3 (Figure 9.9) seems the more apparent design change compared to the bypassing of exchanger 1 (Figure 9.10); however, it turns out to be more costly by about 62%. The target temperature of stream H2 in Figure 9.10 is

65.8°C (very close to 65°C), but that of stream C4 is 109.57°C (rather than 112°C). Thus, cost savings is possible by sacrificing some flexibility.

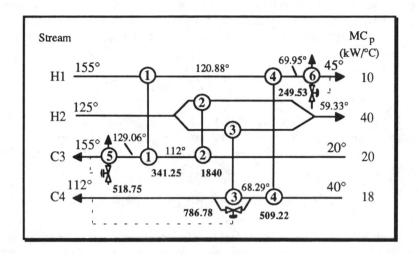

Figure 9.9 A flexible network may be designed to perform under two sets of operating conditions where target temperatures are controlled by using a bypass on process exchangers or by varying utility flow rates. (Temperature in °C and load in kW)

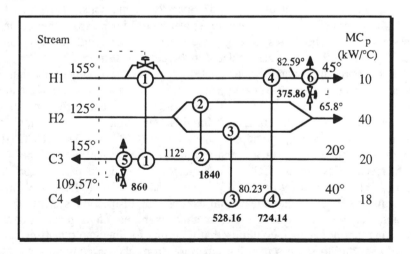

Figure 9.10 An alternative network to the one in Figure 9.9 provides a cheaper solution by considering the three-way tradeoff between flexibility, energy, and capital. (Temperature in °C and load in kW)

Finally, it must be noted that the above formulation of the flexibility simulation matrix in terms of R and P (based on Equations 9.3) has an advantage.   It can be used for any heat exchanger type as long as the appropriate N-R-P expressions are employed.   Also, the system of equations represented by the flexibility simulation matrix may be solved in a sequential manner for simple networks, provided two of the four terminal temperatures for every exchanger are known.

9.3 DISTILLATION PROCESSES

Distillation is the most popular separation process in the chemical industry, especially in the petroleum refining and organic chemicals industries. Being a highly energy-intensive process, it usually provides ample scope for energy optimization.

9.3.1 The Column Grand Composite Curve

A distillation column may be simply represented on a temperature-enthalpy (T-H) diagram as a trapezium (see Figure 9.11), whose horizontal lengths correspond to the reboiler and condenser loads.   If the loads on the reboiler and condenser are equal (approximately), then the simpler representation is a rectangle rather than a trapezium.   As the horizontal lengths represent the column heat loads, the trapezium can be moved horizontally on the T-H diagram without changing the operating conditions.   However, for vertical movement of the trapezium upwards or downwards, the column pressure needs to be correspondingly increased or decreased.   Such column modifications will be discussed in detail in the next subsection.

For the present, it may be noted that the area enclosed by the trapezium is a measure of the energy consumed by the process. In fact, a portion of the area within the trapezium corresponds to the minimum work needed for separation. The curve that defines the boundary for this "minimum work" area may be referred to as the column grand composite curve (CGCC).   The CGCC (see Figure 9.11) corresponds to a column with infinite stages and infinite side exchangers.   How is the CGCC constructed?   A reliable method for this purpose has been recently proposed by Dhole and Linnhoff (1992, 1993b) that holds for complicated, nonideal, multicomponent distillation systems (with variable relative volatilities).   The method is based on a practical near-minimum thermodynamic condition that accounts for inevitable inefficiencies (i.e., losses due to sharp separation, pressure drop, chosen configuration, and feed) through an actual column simulation. Previous methods (Fonyo, 1974; Ho and Keller, 1987) could truly compute the CGCC only for ideal binary mixtures.   The method given by Dhole and Linnhoff (1992, 1993b) is illustrated below through an example.

Example 9.11 Construction of the Column Grand Composite Curve

Generate the column grand composite curve (CGCC) and the column composite curves for a column with the following specifications for feed (F), distillate (D) and bottoms (B):

Properties	Feed	Distillate	Bottoms
Pressure	200	100	102.5
Temperature	100	40	125
Molar flow	1000	599.9	400.1
Mole fraction liquid	0.5795	1.0	1.0
Mole fraction component 1 (c1)	0.2000	0.3328	0.0008
Mole fraction component 2 (c2)	0.2000	0.3279	0.0081
Mole fraction component 3 (c3)	0.2000	0.3225	0.0163
Mole fraction component 4 (c4)	0.2000	0.0157	0.4764
Mole fraction component 5 (c5)	0.2000	0.0011	0.4984

Units: pressure in kPa, temperature in °C, and molar flow in kmol/h.

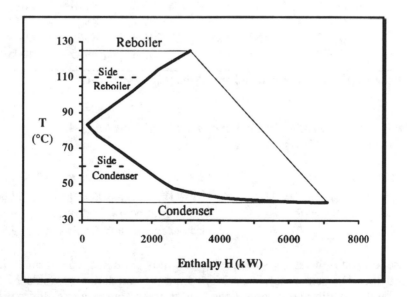

Figure 9.11 The CGCC represents the practical near-minimum thermodynamic condition and provides more information compared to the trapezium representation of a distillation column.

Solution. The CGCC construction requires data from a converged simulation of the distillation column. Table 9.15 shows the relevant results of the simulation for ten stages (feed stage = 5) based on a total condenser (with the condenser temperature specified at 40°C). Consider a light and heavy key model with components c4 and c5 grouped together as heavy keys. Such grouping is effective in refinery columns. Heavy keys are denoted by subscript H and light keys by subscript L. Also, X and Y denote the mole fractions in the liquid (L) and vapor (G) respectively.

Table 9.15 Simulation Results for Column with Ten Stages

Stage No.	X_H*	Y_H*	$L*$	$G*$	H_L*	H_G*
1	0.0167	0.0167	246.3	--	384	--
2	0.2212	0.0168	196.7	846.3	510	8226
3	0.4965	0.0673	162.4	796.6	697	8406
4	0.6342	0.1189	149.3	762.4	789	8593
5	0.6858	0.1457	624.2	222.8	3553	2601
6	0.7066	0.1700	625.4	224.2	3728	2686
7	0.7575	0.2305	627.2	225.3	4118	2861
8	0.8412	0.3747	636.4	227.1	4873	3254
9	0.9238	0.6150	655.8	236.3	5866	4009
10	0.9748	0.8439	400.1	255.8	3995	5005

Units: flows ($L*$, $G*$) in kmol/h and enthalpies (H_L*, H_G*) in kW.

Step 1. Calculation of Minimum Vapor and Liquid Flow Rates
   The operating line is given by

$$G Y_L - L X_L = D_L \qquad \text{Light key above feed stage} \qquad (9.8a)$$
$$G Y_H - L X_H = D_H \qquad \text{Heavy key above feed stage} \qquad (9.8b)$$
$$L X_L - G Y_L = B_L \qquad \text{Light key at/below feed stage} \qquad (9.8c)$$
$$L X_H - G Y_H = B_H \qquad \text{Heavy key at/below feed stage} \qquad (9.8d)$$

These equations must be solved simultaneously with the equilibrium line equations. The stagewise compositions from the simulation (denoted by *) provide the equilibrium compositions. Using these compositions, and noting that the equilibrium and operating curves are coincident at the minimum thermodynamic condition, yields

$$G_{min} Y_L{}^* - L_{min} X_L{}^* = D_L \quad \text{Light key above feed stage} \qquad (9.9a)$$
$$G_{min} Y_H{}^* - L_{min} X_H{}^* = D_H \quad \text{Heavy key above feed stage} \qquad (9.9b)$$
$$L_{min} X_L{}^* - G_{min} Y_L{}^* = B_L \quad \text{Light key at/below feed stage} \qquad (9.9c)$$
$$L_{min} X_H{}^* - G_{min} Y_H{}^* = B_H \quad \text{Heavy key at/below feed stage} \qquad (9.9d)$$

Equations 9.9 define the thermodynamic minimum vapor and liquid flows at each stage temperature. The sample calculation for stage 3 (which is above the feed) in the ten-stage column is given below:

$$
\begin{aligned}
L_{min} &= (D_L Y_H{}^* - D_H Y_L{}^*)/(Y_L{}^* X_H{}^* - Y_H{}^* X_L{}^*) \\
&= (589.822 \times 0.0673 - 10.078 \times 0.9327)/(0.9327 \times 0.4965 - 0.0673 \times 0.5035) \\
&= 70.8 \text{ kmol/h.} \\
G_{min} &= (D_L X_H{}^* - D_H X_L{}^*)/(Y_L{}^* X_H{}^* - Y_H{}^* X_L{}^*) \\
&= (589.822 \times 0.4965 - 10.078 \times 0.5035)/(0.9327 \times 0.4965 - 0.0673 \times 0.5035) \\
&= 670.8 \text{ kmol/h.}
\end{aligned}
$$

Calculations for other stages are done similarly (see Table 9.16).

Table 9.16 CGCC Data for Column with Ten Stages

Stage No.	L_{min}	G_{min}	$H_{L,min}$	$H_{G,min}$	H_{def}	H_{cas}
1	--	--	--	--	--	7091
2	0.0	599.90	0	5834	−5029	2062
3	70.84	670.77	305	7079	−5969	1122
4	119.00	718.93	630	8104	−6669	422
5	614.15	214.08	3494	2498	−6959	132
6	600.03	199.95	3576	2396	−6775	316
7	565.08	167.04	3711	2120	−6364	727
8	514.65	114.58	3942	1640	−5653	1438
9	466.22	66.15	4171	1122	−4906	2185
10	400.10	0.0	3995	0	−3960	3131

Units: flows in kmol/h and enthalpies in kW.

Step 2. Calculation of Minimum Vapor and Liquid Enthalpies
Assuming molar proportionality for enthalpies, the minimum enthalpies corresponding to the minimum flows may be calculated from

$$H_{L,min} = H_L^* (L_{min}/L^*) \qquad\qquad (9.10a)$$
$$H_{G,min} = H_G^* (G_{min}/G^*) \qquad\qquad (9.10b)$$

The sample calculation for stage 3 is shown below:

$$H_{L,min} = 697 \, (70.84/162.4) = 305 \, kW$$
$$H_{G,min} = 8406 \, (670.77/796.6) = 7079 \, kW$$

Step 3. Calculation of Net Heat Deficit at Each Stage Temperature
The enthalpy deficit on each stage is given by

$$H_{def} = H_{L,min} - H_{G,min} + H_D \qquad \text{above feed stage} \qquad (9.11a)$$
$$H_{def} = H_{L,min} - H_{G,min} + H_D - H_F \quad \text{at/below feed stage} \qquad (9.11b)$$

The calculation for stage 3 gives $H_{def} = 305 - 7079 + 805 = -5969$ kW. Note that $H_D = 805$ kW and $H_F = 8760$ kW.

Step 4. Cascading the Heat Deficits
This is done by simply adding the condenser load (7091 kW) to the H_{def} on each stage. The resulting cascade (H_{cas} in Table 9.16) may be plotted against the stage temperature (or stage number, if the proper feed condition is to be ascertained) to arrive at the CGCC in Figure 9.11. The CGCC typically exhibits a pinch near the feed stage since its shape is distinctly affected by the feed enthalpy. Like the GCC, the CGCC provides a picture of the temperature levels at which heat is required to be supplied and rejected in the column. As in the case of multiple utilities, a portion of the heat could be supplied at a temperature lower than that of the reboiler and a portion could be rejected at a temperature higher than that of the condenser.

The CGCC construction procedure may be extended (Dhole and Linnhoff, 1993b) to complicated column arrangements, multiple feeds, multiple products, and different key components in different column sections.

Step 5. Construction of the Column Composite Curves
Figure 9.12 shows the CGCC for the same column, but with five stages (feed stage = 3) rather than ten stages. The column composite curves (Figure 9.12) may be obtained from the CGCC on realizing that the horizontal distance between the composites must equal that between the CGCC and the temperature axis. The area enclosed by the composite curves is a measure of the heat and mass transfer losses (Dhole and Linnhoff, 1993b). The composite curves provide a stagewise picture of the upward and downward flows of the vapor and liquid in the column. The capital costs (in terms of the number of stages) and the internal processes (in terms of the available driving forces) may

be judged on the column composite curves whereas the energy costs (in terms of external heat sources and sinks) may be assessed on the CGCC.

9.3.2 Appropriate Placement and Integration of Columns

Distillation columns absorb heat at the reboiler temperature and then reject heat at a lower (condenser) temperature. From a process integration viewpoint, they really run on temperature and not heat (Linnhoff, 1983). Columns may be integrated with the background process and/or with one another to save on utility consumption.

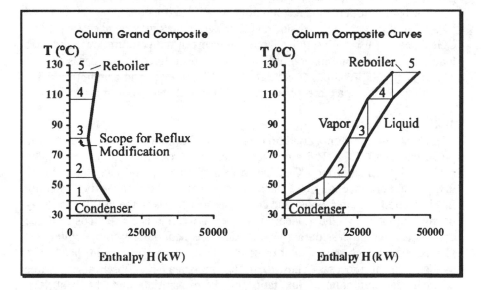

Figure 9.12 The column composites are related to the CGCC in the same way as the composite curves are related to the GCC in Chapter 1.

The thermodynamic arguments for appropriate placement relative to the pinch were put forth in Section 8.1.1 for heat engines. Similar arguments are possible for distillation columns. Appropriate placement (Linnhoff et al., 1983) requires placing a column either entirely above or below the pinch of the background process. If a column is placed above the pinch, then the condenser can reject heat into the process sink. If a column is placed below the pinch, then the reboiler can accept heat from the process source. In both cases, there is a decrease in the hot and cold utility consumptions. If the reboiler and condenser loads are equal, then the column can be run for free (i.e., the total energy requirements for the column and process together are identical to that for the process alone). The column utilizes the temperature driving forces available in the process and, in a sense, consumes temperature (and not heat). If the column is placed across the pinch, then the reboiler consumes heat from the process sink above the pinch and the condenser rejects heat into the process source below the pinch. So, the total energy requirements for the column and process together are no different from that for the separate systems; thus, there is no gain in terms of energy savings from the integration. Finally, there is a heat load limit (like in Section 8.1.1) for distillation columns which requires that no infeasibility (negative heat flow) must occur in the cascade for the process over the temperature range spanned by the column.

For optimum performance, modifications related to pressure, reflux ratio, feed preheating/cooling, and side condensers/reboilers may be necessary. The CGCC may be used to establish targets in terms of temperature level and heat loads for the best combination of these options prior to column/process design. The column modifications (Linnhoff et al., 1983; Dhole and Linnhoff, 1993b) that might be attempted are given next (in their probable order of priority).

Modifying Reflux Ratio

The minimum reflux condition for the distillation column is given by the pinch on the CGCC. The scope for reflux ratio reduction is defined by the horizontal distance of the CGCC pinch from the temperature axis (see Figure 9.12). This distance may be decreased by lowering the reflux ratio, resulting in a reduction in both reboiler and condenser loads. Then, the utility costs may be insignificant and the tradeoff may be between the capital costs of the column and the process during integration. It is beneficial from a capital cost viewpoint to match the reboiler or condenser duties perfectly with heat loads available in the process. Thus, the reflux ratio may be chosen at any value between the minimum reflux ratio for the separation and the heat load available in the cascade of the process.

Changing Pressure

The pressure may be changed so that the distillation column does not cross the pinch of the background process (Linnhoff et al., 1983). If the pressure is increased, then the separation becomes more difficult due to the decrease in relative volatility. The increase in the reflux ratio or the number of plates needed is normally counterbalanced by the decrease in the latent heat of vaporization or the column diameter. However, an upper limit exists for increasing pressures governed by various factors including tendency of the products to decompose at high temperatures and the temperature of the hottest utility. An analogous lower limit exists for decreasing pressures dictated by the unwillingness to use refrigeration or operate under vacuum.

Preheating or Cooling the Feed

A distinct enthalpy change in the CGCC near the feed stage suggests an improper feed condition. This is more conspicuous on a CGCC plot of stage number vs. enthalpy. If a feed is excessively subcooled (and leads to abrupt quenching), then it causes a steep enthalpy change at the CGCC nose on the reboiler side (whose magnitude indicates the scope for feed preheating and for reduction in the reboiler duty on nearly a one-to-one basis). Similarly, the condenser load may be decreased through feed cooling.

Note that steep enthalpy changes in the CGCC may also be caused by improper location of the feed stage, and this must be corrected first.

Using Side Exchangers

The CGCC provides the upper bound on the heat load that can be shifted to a side exchanger at a particular temperature. The dotted lines in Figure 9.11 show that 1662 kW may be shifted onto a side reboiler at 110°C and 1304 kW may be shifted to a side condenser at 60°C. In practice, side condensing/reboiling for stand-alone columns is economically viable only when the feed is a wide-boiling mixture and a large temperature difference occurs across the column.

If a column is forced to be placed across the pinch during integration with the background process, then side exchangers may be used to minimize the total utility required for the overall process. Further, side reboilers may be used below the pinch and side condensers above the pinch (even when the column is not placed across the pinch) to guarantee positive heat flows in the cascade (recall the load limit). This is necessary if the heat flow in the cascade at certain temperature levels is inadequate to accommodate the total load.

As side reboilers/condensers increase the capital costs (due to extra equipment and increase in the number of stages), the tradeoff against the energy savings must be examined.

Splitting Column Loads

The column feed may be split and more than one column may be used if integration of the total column load is not possible due to the cascade heat flow not being sufficiently large (recall the load limit). The operating pressure may be selected to prevent any column from being placed across the pinch. The split increases the capital cost, and the tradeoff must be assessed.

The above splitting is a generalization of the multiple effect principle (Ho and Keller, 1987), wherein the condenser of the high-pressure column is integrated with the reboiler of the low-pressure column. However, the multiple effect principle may be inappropriate during column integration if it causes heat to be transferred across the pinch.

Exploring Thermal Coupling

In addition to splitting column loads, thermal coupling (side-stream strippers/rectifiers) may also be explored for large distillation loads in the case of multiproduct, multicolumn configurations. The coupling through liquid and vapor side-streams decreases the total load and eliminates reboiler(s) and/or condenser(s). However, the constraint that the columns operate at the same pressure may force heat to be transferred across the pinch and make thermal coupling detrimental.

The above modifications on the CGCC may be used for integrated as well as stand-alone columns and are aimed at energy savings. They imply a decrease in driving forces and an increase in capital costs (number of stages) as observed through the column composites (see Figure 9.12). Thus, the energy-capital tradeoff may be evaluated through the simultaneous use of the CGCC and the column composites. To avoid high capital cost penalties, modifications may be made where driving forces on column composites are not tight.

The GCC (and the pinch) of the background process before and after column integration may not be the same because the temperatures of the feed, top product, and bottom product form part of the stream data for the background process. If the effect of this change is critical (typically it is not), then an iterative method will have to be employed.

For process integration purposes, the CGCC should be superposed on the GCC of the background process with minimum temperature overlap. It is convenient to plot the CGCC in the reverse enthalpy direction (which is valid considering the relative nature of the enthalpy axis) as in Figure 9.13.

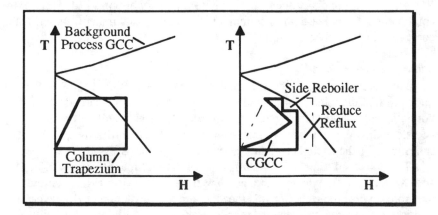

Figure 9.13 The CGCC provides a more accurate picture for column targeting and design modifications compared to the trapezium representation.

In order to eliminate overlap of the trapezium, the reduction of the column pressure may be considered. However, based on the CGCC, the reduction of reflux ratio and the use of a side reboiler appear suitable. Thus, targets for the column coupled with those for the background process enable identification of favorable modifications for column integration.

9.4 EVAPORATION PROCESSES

Evaporation, like distillation, is an energy-intensive process in which the latent heat of vaporization must be provided for. Multistage systems, where vapor from a stage is used for evaporation in the next stage, are often employed to reutilize the latent heat of the vaporized material. The greater the number of stages, the lower is the operating cost due to the larger evaporation per unit of steam admitted to the first stage. This reduction in operating cost is offset by the increased capital cost. The optimum number of stages is obtained by assessing the tradeoff between the reduced energy consumption and the increased capital cost on increasing the stages. However, a different design will usually result if the evaporation process is heat integrated with the background process. In this section, evaporation processes are considered in isolation as well as in combination with background processes accounting for the capital-energy tradeoff. The discussion is essentially based on the work of Smith and Jones (1990).

9.4.1 Stand-alone Evaporators

The design below assumes constant latent heat, negligible boiling point rise, and saturated conditions for both inlet and outlet solutions. The starting point is the result of Nishimura (1980) for the minimum heat transfer area in HENs. Consider a single hot stream at temperature t in enthalpy balance with n cold streams at temperature T_j, or a single cold stream at temperature t in enthalpy balance with n hot streams at temperature T_j. Then, the system has minimum area if the temperature difference $(T_j - t)$ of every stream is everywhere inversely proportional to the square root of the overall heat transfer coefficient U_j as given by

$$\sqrt{U_1}\,(T_1 - t) = \sqrt{U_2}\,(T_2 - t) = \dots\dots\dots = \sqrt{U_n}\,(T_n - t). \tag{9.12a}$$

Application of the above equation to a multistage evaporation process yields

$$\Delta T_{i,jc+1}/\Delta T_{i+1,jc+2} = \sqrt{U_{i+1,jc+2}}/\sqrt{U_{i,jc+1}} \tag{9.12b}$$

where $\Delta T_{i,jc+1}$ and $U_{i,jc+1}$ respectively denote the temperature difference and the overall heat transfer coefficient between hot stream i and cold stream (jc + 1). Note that subscript i stands for the hot stream (vapor) exiting the i-th stage while jc stands for the cold stream (vaporization) in the i-th stage.

The goal is to minimize the capital cost in a stand-alone evaporation process. If a, b, and c are the typical cost coefficients (see Section 1.5), then the installed costs for the evaporator and the heat exchanger are given by

$$CC_{ev} = a_{ev} + b_{ev} A^{c_{ev}} \qquad (9.13a)$$
$$CC_{hx} = a_{hx} + b_{hx} A^{c_{hx}} \qquad (9.13b)$$

From cost considerations, the heat exchange between the final stage vapor and the cold utility will typically be in an external shell-and-tube exchanger rather than in a calandria. A smaller ΔT may be used for the cheaper exchanger area to allow a larger ΔT for the expensive calandria area. A modified U (denoted by U^* and given by Equation 1.37b) is required for the exchanger to obtain its capital cost using the cost law for the evaporator as a reference.

Now, Equation 9.12b may be rewritten to obtain the ΔT between the various stages that will result in minimum capital cost rather than minimum area. Thus,

$$\Delta T_i / \Delta T_{N/cw} = \sqrt{U^*}_{N/cw} / \sqrt{U_i} \qquad (9.14a)$$

where ΔT_i denotes the ΔT between the condensation and vaporization duties in the i-th stage while $\Delta T_{N/cw}$ denotes the ΔT between the final stage condensation and cooling water.

If the arithmetic mean rather than the logarithmic mean is used for the temperature difference of the cooling water, then the following equation is obtained:

$$\Sigma \, \Delta T_i + \Delta T_{N/cw} = T_{steam} - T_{cw} = \text{overall driving force.} \qquad (9.14b)$$

Here, N is the number of stages, T_{steam} is the steam temperature, T_{cw} is the mean temperature of the cooling water, and the summation for the temperature differences runs over all the N stages. Combining Equations 9.14a and 9.14b gives the final equation as

$$\Delta T_i = (T_{steam} - T_{cw}) / [\Sigma_{j=1}^{N} (\sqrt{U_i} / \sqrt{U_j}) + \sqrt{U_i} / \sqrt{U^*}_{N/cw}]. \qquad (9.14c)$$

The use of these equations is demonstrated in the example below.

Example 9.12 Optimum Number of Stages for Stand-alone Evaporators

A sugar industry requires an evaporator installation that has the capacity for concentrating 1380 kg/hr from 13° Brix (degrees Brix is the % by weight of sugar in the solution) to 60° Brix (the concentration at which it will be decolorized). Find the total cost and the optimum number of stages for a stand-alone evaporator system with the following specifications:

For the evaporator duty, the film heat transfer coefficients for condensation and vaporization may both be assumed to be 2.5 kW/m^2 °C.

Steam at 180°C is available at a cost (C_{steam}) of 120 $/(kW.yr) while cooling water (15° - 25°C) is available at a cost of 10 $/(kW.yr). The film heat transfer coefficient for condensing steam is 2.5 kW/m^2 °C and for cooling water is 0.2 kW/m^2 °C. The cost of electricity (C_{elec}) is 438 $/(kW.yr).

The capital cost laws for the evaporator and exchanger are given by CC_{ev} (in $) = $30000 + 6000A^{0.81}$ and CC_{hx} (in $) = $30000 + 750A^{0.81}$, where A is the heat exchange area in m^2. The plant lifetime may be assumed to be five years and the rate of interest as 10%. The vapors exiting from the final stage are cooled by cooling water in an external heat exchanger.

Solution. If $U_1 = U_2 = \ldots = U_N \equiv U_{ev}$ in Equation 9.14c, then it is best to distribute the temperature difference equally (i.e., $\Delta T_1 = \Delta T_2 = \ldots = \Delta T_N \equiv \Delta T_{ev}$) and the following simple equation holds.

$$\Delta T_{ev} = (T_{steam} - T_{cw})/(N + \sqrt{U_{ev}}\sqrt{U^*_{N/cw}}) \tag{9.14d}$$

Step 1. Calculation of the Temperature Differences in Various Stages

Equations 1.16a, 1.37b, 9.14d, and 9.14a are used here in that order.

$$
\begin{aligned}
U_{ev} &= (1/2.5 + 1/2.5)^{-1} = 1.25 \text{ kW/m}^2 \text{ °C}\\
U_{N/cw} &= (1/2.5 + 1/0.2)^{-1} = 0.185 \text{ kW/m}^2 \text{ °C}\\
U^*_{N/cw} &= U_{N/cw} (b_r/b_s)^{1/cr} A^{1-cs/cr}\\
&= 0.185 (6000/750)^{1/0.81} A^{1-1} = 2.41 \text{ kW/m}^2 \text{ °C}\\
\Delta T_{ev} \text{ (°C)} &= (180 - 20)/(N + \sqrt{1.25}/\sqrt{2.41}) = 160/(N + 0.72)\\
\Delta T_{N/cw} \text{ (°C)} &= \sqrt{1.25}/\sqrt{2.41} \times 160/(N + 0.72) = 115.16/(N + 0.72)
\end{aligned}
$$

Step 2. Calculation of Evaporator and Exchanger Areas

The heat duty is calculated by multiplying the mass to be evaporated with the latent heat of vaporization (2257 kJ/kg). Dividing this heat duty by ($U \Delta T$) gives the area. Now, 0.3 kg/s of water needs to be evaporated in the evaporation system. Thus,

Total evaporator area for an N-stage evaporator (m^2)
$A_{ev} = 0.3 \times 2257 \, (N + 0.72)/(1.25 \times 160) = 3.39 \, (N + 0.72)$
Area of the exchanger (m^2)
$A_{hx} = 0.3/N \times 2257 \, (N + 0.72)/(0.185 \times 115.16) = 31.75 \, (1 + 0.72/N)$

Step 3. Calculation of Capital Costs
Equations 9.13 are used here.

$CC_{ev} \, (\$) = N \, [a_{ev} + b_{ev} \, (A_{ev}/N)^{cev}]$
$\qquad\qquad = N \, \{30000 + 6000 \, [3.39 \, (1 + 0.72/N)]^{0.81}\}$
$CC_{hx} \, (\$) = a_{hx} + b_{hx} \, A_{hx}{}^{chx} = 30000 + 750 \, [31.75 \, (1 + 0.72/N)]^{0.81}$
Total annual capital cost (\$/yr)
$\qquad = [30000 \, (N + 1) + (16111.9 \, N + 12344.5) \, (1 + 0.72/N)^{0.81}] \, (1.1)^5/5$

Step 4. Calculation of Utility Costs
The heat duty of each stage is $677.1/N$ kW. This corresponds to the heat to be supplied by the steam in the first stage as well as the heat to be removed by the exchanger using cooling water.

Total utility cost (\$/yr) $= 130 \, (0.3 \times 2257)/N = 88023/N$.

Step 5. Calculation of Total Annual Costs for Different Number of Stages
The formula for the total annualized cost is

$TAC \, (\$/yr) = [9663(N + 1) + (5189.7 \, N + 3976.2) \, (1 + 0.72/N)^{0.81}]$
$\qquad\qquad\qquad\qquad\qquad\qquad\qquad\qquad\qquad + 88023/N.$

The annual costs (in 10^3 \$/yr) for $N = 1$ through 5 are given in the table below:

Number of stages	Capital cost	Utility cost	Total cost
1	33.55	88.02	121.57
2	47.40	44.01	91.41
3	61.92	29.34	91.26
4	76.60	22.01	98.61
5	91.35	17.60	108.95

From the above results, it is observed that the optimum number of stages may be chosen to be three (or two).

Step 6. Economics of Electrically-driven Vapor Recompression Heat Pumping
If Q is the heat to be supplied, then the costs involved will be $C_{steam} \, Q$ (for steam usage) and $C_{elec} \, Q \, (1 - T_2/T_1)/\eta_e$ (for heat pumping as per

Equation 8.4).   Therefore, heat pumping is economical if $T_1 < T_2/(1 - \eta_e \, C_{steam}/C_{elec})$. If $\eta_e = 0.65$ and $T_2 = 298$ K, then $T_1 < 89.6°C$ for heat pumping to be more economical than steam.   If the annualized capital cost of the compressor is included, then the upper limit on T_1 will further decrease. Thus, in this example, heat pumping is not economical when compared with steam (which provides a greater temperature difference across the evaporator).

9.4.2 Integrated Evaporators

As in distillation, appropriate placement of evaporators involves placing them either completely above or completely below the pinch.   In this case, they can run for free (i.e., no additional energy is needed over the requirements for the background process).   Thus, evaporators utilize the temperature driving forces available in the process and consume temperature (not heat).   If the inlet and outlet duties of the evaporator are integrated across the pinch, then the integrated system shows no benefit in terms of energy savings over the stand-alone system.   Thus, the important rule is not to place an evaporator across the pinch of the background process.   Note that this is contrary to the rule for the proper placement of heat pumps (see Section 8.1.1) and, therefore, heat pumping and integrated evaporators do not go together.

The inlet vaporization and outlet condensation duties will appear at different temperatures on the GCC (in terms of shifted temperatures) although they actually occur at the same temperature.   Thus, the predominant latent heat loads for an evaporator may be represented conveniently by a rectangle on the GCC.   Many of the concepts for column integration (see Section 9.3) apply to evaporator integration.   For example, if the evaporator operating pressure is increased or decreased, then the rectangle will move upwards or downwards on the GCC plot.   The temperatures of the feed, condensate, and concentrated product will be affected by this pressure change, leading to a change in the background process GCC shape.   This effect may be neglected in the preliminary phase because its assessment requires iterative calculations.

Example 9.13 Integration of Evaporators with Background Process

Integrate the evaporator system whose specifications appear in Example 9.12 with the background process whose stream data appear in Case Study 4S1.

Solution.   As long as evaporators are not integrated across the pinch, they should have zero energy costs.   However, the choice of ΔT_{min} is critical to account correctly for the capital-energy tradeoff.   It is important to recognize that the temperature differences in the background process reduce over the range from which heat is removed through the evaporator (between the first stage vaporization and the last stage condensation). This reduction changes the optimum ΔT_{min} for the process on integration of the evaporator, which also affects the shape of the GCC. Thus, the initial hot stream process temperature (T_{he}) and the initial cold stream process temperature (T_{ce}) chosen for integration will not continue to be correct. This implies that, in principle, simultaneous optimization for ΔT_{min}, T_{he}, and T_{ce} is necessary.

So long as the evaporator can be fully integrated with the process and the change in the GCC shape is not drastic, a simplified procedure (Smith and Jones, 1990) may be employed as given below:

a) Start with a ΔT_{min} (in the range in which the optimum lies).
b) Find the GCC for this ΔT_{min}.
c) Decide on the number of evaporator stages and determine whether the evaporator can fit in the GCC or not (i.e., find T_{he} and T_{ce} from the GCC).
d) Include the first stage vaporization and the final stage condensation among the other streams.
e) Find the cost of the new background process, the evaporator, and the total cost for the integrated system.
f) Repeat steps b) through e) for various ΔT_{min} values to find the optimum ΔT_{min} at which the cost is minimum.

It may be advisable to start with the optimum ΔT_{min} for the background process (13°C for Case Study 4S1 as per Example 1.10) and to initialize the temperature differences in the evaporator stages as per Equation 9.14.

For illustration, the cost of a three-stage evaporator integrated in the below-pinch region of the GCC (for Case Study 4S1) is calculated below:

a) Let ΔT_{min} be 13°C to start with.
b) The GCC at $\Delta T_{min} = 13$°C for the background process (including the utility streams as dummy process streams) is shown in Figure 9.14.
c) The three-stage evaporator with a load of 225.7 kW (for each stage) is now fitted into the below-pinch region of the GCC. In this case,

$$T_{he} = 103.45 + 13/2 = 109.95\text{°C and } T_{ce} = 28.29 - 13/2 = 21.79\text{°C.}$$

Next, the temperature differences between the process and the first stage vaporization $(\Delta T_{pr/ev1})$ as well as between the final stage condensation and the process $(\Delta T_{evN/pr})$ must be set. Based on Equation 9.12b,

$$\Delta T_{pr/ev1}/\Delta T_{ev} = \sqrt{U_{ev}}/\sqrt{U_{pr/ev1}}$$

Now, U_{ev} = 1.25 kW/m^2 °C whereas $U_{pr/ev1}$ = 2.5 kW/m^2 °C if the resistance to heat transfer on the process side is neglected (Smith and Jones, 1990). Note that the process-side contribution is included in the background process. Also, though the h-value for the process in this example is simply 0.2 kW/m^2 °C (being the same for all streams), it is generally not easy to determine as the process stream for the evaporator integration is not known a priori. Thus,

$$\Delta T_{pr/ev1}/\Delta T_{ev} = 0.7071.$$

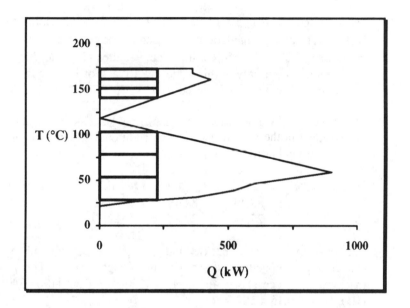

Figure 9.14 A three stage evaporator may be integrated either above the pinch or below the pinch of the GCC for the background process.

As the match between the final stage condensation and the process follows different cost laws, the approach in the previous example of using a modified U (denoted by U^*) must be adopted. Thus,

$$\Delta T_{evN/pr}/\Delta T_{ev} = \sqrt{U_{ev}}/\sqrt{U^*_{evN/pr}}.$$

Now, $U^*_{evN/pr} = U_{evN/pr}\, (b_r/b_s)^{1/cr}\, A^{1-cs/cr}$
$$= 2.5\,(6000/750)^{1/0.81}\, A^{1-1} = 32.57 \text{ kW/m}^2 \text{ °C}.$$

Again, the process-side heat transfer resistance is neglected since, in general, it is not easy to determine and is accounted for in the background process to some extent. Thus,

$$\Delta T_{evN/pr}/\Delta T_{ev} = 0.1959.$$

The overall driving force is given by summing the various temperature differences (see Equation 9.14b) as

$$0.7071 \, \Delta T_{ev} + (N - 1) \, \Delta T_{ev} + 0.1959 \, \Delta T_{ev} = T_{he} - T_{ce}.$$

Substituting $N = 3$, $T_{he} = 109.95°C$ and $T_{ce} = 21.79°C$ yields $\Delta T_{ev} = 30.4°C$. The above calculations are based on the process streams (used in the integration) being at constant temperatures (T_{he} and T_{ce}). This assumption is used only at this stage due to lack of knowledge on the actual matches and can be adjusted later.

d) The first-stage vaporization and the final-stage condensation are incorporated in the stream data as two additional streams to form the new background process:

Stream Label	T_{in} (°C)	T_{out} (°C)	MC_p (kW/°C)	h (kW/m² °C)
HE	88.48	88.48	225.7	2.5
CE	27.74	27.74	225.7	0.2

e) The cost of the new background process is determined by the conventional procedure described in Chapter 1. Thus, TAC (background process) = $46000 + 8 \, [30000 + 750 \, (1965.3/8)^{0.81}]$ $(1.1)^5/5 = 290.15 \times 10^3$ $/yr.
The cost of the evaporator is 35.71×10^3 $/yr, so the total cost for the integrated process is 325.86×10^3 $/yr. This is not very different from the cost (328.26×10^3 $/yr) for the nonintegrated system comprising the stand-alone three-stage evaporator (from Example 9.12) and the process at $\Delta T_{min} = 13°C$. The capital-energy tradeoff could now be reoptimized.

f) To determine the optimum that results in the minimum total cost for the integrated system, the calculations may be repeated for various values of ΔT_{min}, for placement in the above-pinch region (see Figure 9.14), and for different number of stages (to manipulate the system heat flow). A split solution with integration in both the above- and below-pinch regions may also be attempted.

9.5 PROCESSES USING MASS SEPARATING AGENTS

Separation systems in the chemical process industries use either energy or mass as the separating agent, and this may be used as a basis for their classification (King, 1980). Distillation and evaporation are energy-based separation processes. The synthesis of networks for separations using mass separating agents, which includes processes like absorption, desorption, ion-exchange, leaching, adsorption and liquid-liquid extraction, forms the subject matter of this section.

Representations and techniques originally developed for HENS may be applied to mass exchanger network synthesis (MENS) based on two approaches: thermodynamic (El-Halwagi and Manousiouthakis, 1989) and mathematical programming (El-Halwagi and Manousiouthakis, 1990a; Gupta and Manousiouthakis, 1993). For this purpose, a mass exchanger may be defined as any countercurrent, direct-contact, mass-transfer operation that uses a mass separating agent (MSA) like a solvent, an adsorbent, or an ion-exchange resin. In MENS, the mass transfer of a species from a set of rich process streams to a set of lean process streams is maximized for a specified minimum allowable composition difference (ε). This is analogous to HENS, where heat recovery between a set of hot process streams and a set of cold process streams is maximized for a specified minimum allowable temperature difference (ΔT_{min}). Whereas external utilities (such as cooling water) are used in HENS, auxiliary lean streams (external MSAs) act as utilities in MENS.

Since concepts and techniques in HENS are well developed, their extension to MENS would certainly prove worthwhile. That is the endeavor in this section. A first effort in this direction was attempted by El-Halwagi and Manousiouthakis (1989); however, their analogy was not complete as it resulted in (N_{ls} + 1) composition scales (for N_{ls} lean process streams) in contrast to the two temperature scales (one hot and one cold) that always result in HENS. Further, representations such as the composite curves and the grand composite curve developed in pinch analysis then fail to have true analogs in MENS (note that Figure 4 in El-Halwagi and Manousiouthakis [1989] has several composition axes).

9.5.1 Mass Exchanger Network Synthesis (MENS)

This subsection develops a simple, top-level transformation that maps a single (key) component MENS problem to an equivalent HENS problem, for which the technology is well established. Separation targets are often defined in terms of one key component in industrial practice even though several components are actually transferred.

Example 9.14 MENS for Zinc Recovery from Metal Pickling Plant

Synthesize a minimum utility cost MEN for recovering zinc chloride from spent pickle liquor (rich stream R1) and rinse wastewater (rich stream R2) for ε = 0.0001, using the stream data given in Table 9.17. Two mass exchange operations, solvent extraction and ion exchange, are to be utilized to effect the recovery. Five MSAs are available: the three solvents are tributyl phosphate (S1 costing 0.02 $/kg), triisooctyl amine (S2 costing 0.11 $/kg), and di-2-ethyl hexyl phosphoric acid (S3 costing 0.04 $/kg) whereas the two ion exchange resins are a strong base resin (S4 costing 0.05 $/kg) and a weak base resin (S5 costing 0.13 $/kg). The equilibrium relationships for zinc chloride in each of the lean streams are as follows: $y = 0.845\, x_1$, $y = 1.134\, x_2 + 0.01$, $y = 0.632\, x_3 + 0.02$, $y = 0.376\, x_4 + 0.0001$, and $y = 0.362\, x_5 + 0.002$. Also, the target compositions for the lean streams may truly be regarded as upper bounds.

Table 9.17 Stream Data for Zinc Recovery - MENS Problem

Rich Streams				Lean Streams			
Stream	y_{in}	y_{out}	G	Stream	x_{in}	x_{out}	L
R1	0.08	0.020	0.2	S1	0.0060	0.060	∞
R2	0.03	0.001	0.1	S2	0.0100	0.020	∞
				S3	0.0090	0.050	∞
				S4	0.0001	0.010	∞
				S5	0.0040	0.015	∞

Units: flow rate (L and G) in kg/s and composition (x and y) in w/w %.
Data taken from El-Halwagi and Manousiouthakis (1990a)

Solution. El-Halwagi and Manousiouthakis (1990a) have provided the flow diagram for the process and solved the problem using mathematical programming. Here, the problem is solved by simple pinch analysis concepts.

Step 1. Transformation to Equivalent HENS Problem

The first step is to transform the lean stream data and obtain the equivalent HENS problem. The operating line equation for a mass exchanger transferring a component from a rich stream to a lean stream is given by

$$G\,(y_{in} - y_{out}) = L\,(x_{out} - x_{in}) = \text{mass flow of transferred component.} \quad (9.15a)$$

It is desired to reduce the rich stream composition from y_{in} to y_{out} and maximize the lean stream composition (x_{out} if x_{in} is specified, and vice versa) by varying L/G. Theoretically speaking, the maximum is achieved when the operating line touches the equilibrium line; however, this requires an infinitely large exchanger. Thus, practically speaking, a minimum difference (ε) is required between the operating and equilibrium compositions of the lean stream. If a linear equilibrium relationship (say, $y = m x + b$) for the distribution of the key component among the various streams holds over the operating range or its subintervals, then the minimum allowable composition difference (ε) may be mathematically expressed as

$$x = (y - b)/m - \varepsilon. \tag{9.15b}$$

The above equation may be rearranged to give

$$y = [m x + b + (m - 1)\varepsilon] + \varepsilon. \tag{9.15c}$$

Letting $T_y = y$ and $T_x = m x + b + (m - 1)\varepsilon$ yields $T_y = T_x + \varepsilon$. This expression is in the form $T_h = T_c + \Delta T_{min}$ as required by a HENS problem (where the equilibrium relation is $T_h = T_c$). Equation 9.15a may be now rewritten as

$$G (T_{in,y} - T_{out,y}) = (L/m) (T_{out,x} - T_{in,x}). \tag{9.15d}$$

Letting $MC_{p,G} = G$ and $MC_{p,L} = L/m$ yields an equation that is identical in form to the energy balance in HENS. Essentially, if the inlet/outlet compositions and the flow rates of all lean process streams/MSAs are transformed using $T_x = m x + b + (m - 1)\varepsilon$ and $MC_{p,L} = L/m$ respectively, then the given MENS problem results in an equivalent HENS problem.

The transformed data for this example are given in Table 9.18, and a few sample calculations appear below:

$$T_{in,x} = 0.845 \times 0.006 + 0 + (0.845 - 1) 0.0001 = 50.545 \times 10^{-4} \text{ (for S1)}$$
$$T_{out,x} = 1.134 \times 0.02 + 0.01 + (1.134 - 1)0.0001 = 326.934 \times 10^{-4} \text{ (for S2)}$$

Note that $MC_{p,L} = L/m \to \infty$ for all the lean streams because they are to be treated as external MSAs (utilities).

Step 2. Development of Problem Table for Transformed Data

The PTA is applied to the transformed data in the conventional manner (see Section 1.1.2). The problem is divided into a series of intervals whose boundaries (T'_{int}) are obtained by subtracting $\varepsilon/2$ from the rich stream compositions and adding $\varepsilon/2$ to the lean stream compositions. A material balance on the solute is performed for each interval, and the mass surplus is then cascaded from the highest interval to the lowest to obtain the cumulative flow of mass of solute available for transfer (Q'_{cas}). Just sufficient mass flow must be added to the first interval to ensure that none of the mass flows between intervals are negative.

The problem table for the transformed zinc recovery example (Table 9.19) shows a threshold problem.

Table 9.18 Data for Zinc Recovery - Equivalent HENS Problem

	Rich Streams				Lean Streams		
Stream	$T_{in,y}$	$T_{out,y}$	$MC_{p,G}$	Stream	$T_{in,x}$	$T_{out,x}$	$MC_{p,L}$
R1	800	200	0.2	S1	50.545	506.845	∞
R2	300	10	0.1	S2	213.534	326.934	∞
				S3	256.512	515.632	∞
				S4	0.752	37.976	∞
				S5	33.842	73.662	∞

Units: flow rate ($MC_{p,L}$ and $MC_{p,G}$) in kg/s and composition (T_x and T_y) in 10^{-4} w/w %.

Table 9.19 Problem Table for Transformed Zinc Recovery Example

Interval i			T'_{int}	$MC'_{p,int}$	Q'_{int}	Q'_{cas}	R'_{cas}
0			799.5	0	0	0	0
1			299.5	−0.2	−0.01	0.01	0.01
2			199.5	−0.3	−0.003	0.013	0.013
3			9.5	−0.1	−0.0019	0.0149	0.0149
Stream	R1	R2					
MC'_p	0.2	0.1					

Units: composition (T'_{int}) in 10^{-4} w/w % and flow rates in kg/s.

In general, the first entry in the last column of Table 9.19 (0 kg/s) corresponds to the excess capacity of the process lean streams to remove the solute while the last entry (0.0149 kg/s) corresponds to the minimum amount of the solute to be removed by the external MSAs. Which of the five MSAs should be used for this purpose? The problem now is identical to the allocation of multiple utilities in HENS (see Section 1.7.3).

Step 3. Optimum Usage of MSAs

The GCC for the process is shown in Figure 9.15 based on the T'_{int} vs. R'_{cas} data in Table 9.19.

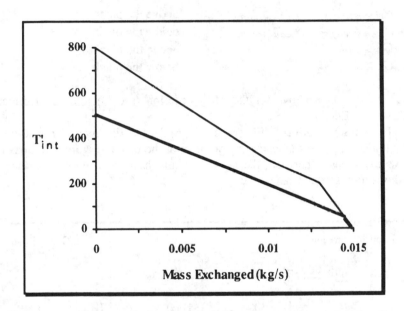

Figure 9.15 The allocation of MSAs (represented by heavy lines above) during MENS may be done on the GCC in the same manner as the allocation of multiple utilities in HENS. Note the units of T'_{int} are 10^{-4} w/w %.

As S1 is the cheapest MSA available, its use is maximized. On the GCC, 51.045×10^{-4} (shifted $T_{in,x}$ for S1) corresponds to 0.014485 kg/s, which is the maximum mass that can be exchanged with S1. Therefore, the flow rate of S1 is 0.26823 kg/s (i.e., 0.014485/[0.06 – 0.006]).

The level at which the remaining mass of zinc chloride $(4.1545 \times 10^{-4}$ kg/s) has to be exchanged precludes the use of S3. The next cheapest MSA (S4) is used with a flow rate of 0.04196 kg/s (i.e., $4.1545 \times 10^{-4}/[0.01 – 0.0001]$).

Step 4. Designing the Minimum Utility MEN

A utility pinch occurs on Figure 9.15 at the point where the utility GCC touches the process GCC (i.e., $T'_{int} = 51.045 \times 10^{-4}$). In other words, this pinch is at 51.545×10^{-4} for the rich streams, at 60×10^{-4} for S1, and at 134.26×10^{-4} for S4. Note that the inverse transformation is required to obtain the pinch compositions for the lean streams.

The pinch design method may now be used to synthesize the minimum utility MEN. The feasibility criteria at the pinch (see Section 3.1.1) become

$N_{rich\ streams} \leq N_{lean\ streams}$	above the pinch
$N_{rich\ streams} \geq N_{lean\ streams}$	below the pinch
$G \leq L/m$	above the pinch
$G \geq L/m$	below the pinch

The analogs given above for the MC_p criterion (see Equation 3.2) directly follow from Equation 9.15d.

The minimum utility MEN (Figure 9.16) for the zinc recovery example is easy to design by decomposing the problem at the pinch. Figure 9.16 does not show any loops; otherwise, the network could have been evolved using the methods discussed in Chapter 4.

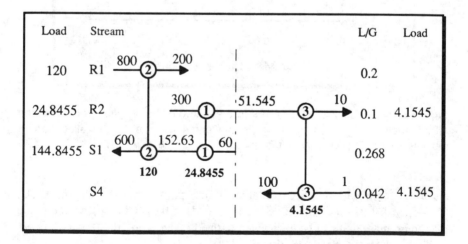

Figure 9.16 A minimum utility MEN that meets the targets can be synthesized using the pinch design method (with minor modifications in the feasibility criteria). Note the units of composition and mass-exchange loads are 10^{-4} w/w % and 10^{-4} kg/s respectively.

The focus in this section has been on developing the relationship between HENS and MENS (with single component targets). There appears to be a growing interest in MENS with applications towards waste minimization, material recovery, feed preparation, and product separation/finishing. This is evident from the various articles that have recently appeared in the literature. These articles are based on mathematical programming and are beyond the scope of this book; however, they are listed below for the convenience of the interested reader.

- The problem of simultaneous synthesis of mass exchange and regeneration networks for waste minimization involving both once-through and regenerable MSAs has been solved by an MINLP/MILP approach by El-Halwagi and Manousiouthakis (1990b). They have considered the example of removal and recovery of phenols from the waste streams of a coal conversion plant.

- The problem of synthesizing reactive mass exchange networks (REAMENs) involving both physical and chemical MSAs has been addressed by El-Halwagi and Srinivas (1992). A LP formulation has been devised for REAMEN problems with linear and/or convex equilibrium relations. They have considered the example of desulfurization of the gaseous wastes from a rayon plant.

- The problem of synthesizing cost-effective REAMENs featuring MSAs with general nonlinear (convex and/or nonconvex) equilibrium functions has been discussed by Srinivas and El-Halwagi (1994a). Problems in this category show two peculiar features: the mass exchange pinch location is not restricted to the inlet compositions of the streams and the mass exchangers may be placed across the pinch in a minimum utility cost network. They have considered the example of desulfurization of the gaseous wastes of a coal-to-methanol plant.

- The problem of synthesizing combined heat and reactive mass exchange networks (CHARMENs) has been addressed by Srinivas and El-Halwagi (1994b). They have considered the desulfurization of gaseous emissions with simultaneous heat recovery for a pulp and paper plant.

- The general problem of synthesizing MENs for waste minimization based on a hyperstructure representation and an MINLP formulation has been solved by Papalexandri et al. (1994). Both network operating and investment cost have been simultaneously optimized without decomposition of the problem at the pinch. Several of the examples presented earlier by El-Halwagi and coworkers have been reexamined. Reactive and regeneration networks have also been considered.

It may be preferable to handle MENS with multicomponent targets through mathematical programming formulations. However, pinch analysis may be exploited to tackle MENS problems for multiple components when there is a single MSA. Wastewater minimization, which is discussed in the next subsection, is an important problem in this category where water is the only MSA.

9.5.2 Wastewater Targeting and System Design

Wastewater treatment is an important industrial activity, especially in the petroleum refineries. Wang and Smith (1994) have recently discussed how wastewater may be minimized from processes and utilities so as to reduce the costs of freshwater, effluent treatment, pumping, and piping. Three possible ways to achieve this are given below:

- *Reuse*: Wastewater from a process may be directly utilized in other processes provided the contamination level is below certain acceptable limits.
- *Regeneration reuse*: Wastewater from a process may be treated to partly remove contaminants and facilitate its use in other processes.
- *Regeneration recycle*: Here, the regenerated wastewater may be recycled to processes where it was earlier used. At times, recycling may not be permitted to prevent the accumulation of contaminants not removed during regeneration.

Reuse (with or without regeneration) may call for blending with freshwater and/or wastewater from other processes.

Targeting and design methodologies for minimum wastewater for single and multiple contaminants are presented below for the case of reuse only.

Example 9.15 Water Target and Network for Single Contaminant

Consider four processes involving a single contaminant with the following limiting water profile data:

Process Label	c_{in} (ppm)	c_{out} (ppm)	M_c (kg/h)	M_w (kg/h)
P1	0	200	4	20000
P2	150	300	7.5	50000
P3	200	600	16	40000
P4	200	300	2	20000

Note that M_c and M_w denote the mass load of contaminant and mass flow rate of water respectively, whereas c_{in} and c_{out} denote the maximum inlet and outlet concentrations of contaminants.

The goal is to determine the minimum freshwater requirement and design appropriate water networks to meet the minimum wastewater target. Assume dilute systems where the mass transferred varies linearly with concentration.

Solution. The transfer of a contaminant from a process stream to water is somewhat analogous to the transfer of heat from a hot stream to a cold stream. The streams may be conveniently represented on a concentration-mass diagram (similar to the temperature-enthalpy diagram in HENS).

The highest allowable inlet concentration of contaminant in the water required by a process depends on certain process-specific factors and must be first ascertained if the capacity for water reuse from other processes is to be maximized. Then, the outlet concentration may be maximized, corresponding to this inlet concentration, to obtain the minimum flow rate of water required by the process. The maximum allowable inlet and outlet concentrations depend on constraints related to the minimum mass transfer driving force, maximum solubility, corrosion, fouling, and prevention of precipitation/settling of material.

In other words, each process has a minimum water requirement, which is dictated by the maximum possible inlet and outlet concentrations of contaminant. This defines a limiting water profile (Wang and Smith, 1994) that is unique to each process and that accounts for the minimum mass transfer driving force as well as other constraints. The limiting water profile provides an upper bound to the water supply line to be employed in the actual design. Use of the limiting water profile instead of the process stream data itself makes the approach independent of the type of mass transfer operation as well as various other constraints (mentioned earlier). For example, the limiting water profile for process P1 is given by the line segment whose end points are (0, 0) and (4, 200).

Step 1. Target without Water Reuse

Consider the case where freshwater is used for each individual process and there is no water reuse. The water supply line (whose slope corresponds to the minimum freshwater flow rate) is then given by a segment whose end points are (0, 0) and (M_c, c_{out}). For this operation without water reuse, the total freshwater flow rate is calculated by summing individual process contributions.

Thus, freshwater flow rate $= (4/200 + 7.5/300 + 16/600 + 2/300) \times 10^6$
$= 78333.33$ kg/h.

Step 2. Target with Water Reuse

Now, consider the case where there is water reuse and the attempt is to minimize the water flow rate in the overall context. The limiting water profiles for the different processes may be combined (within concentration intervals defined by the c_{in} and c_{out} values) to obtain the limiting composite curve (LCC) for the total system as given below. Table 9.20 is easily generated by following the procedure for the construction of the composite curves from Example 1.2. Based on Table 9.20, the LCC may be obtained by plotting c vs. $CumM_c$ as in Figure 9.17.

Table 9.20 Data for Limiting Composite Curve

Interval					c	$SumM_w$	M_c	$CumM_c$
0					0	0	0	0
1					150	20000	3	3
2					200	70000	3.5	6.5
3					300	110000	11	17.5
4					600	40000	12	29.5
Process	P1	P2	P3	P4				
M_w	20000	50000	40000	20000				

Units: concentration in ppm and M_w and M_c in kg/h.

The x-axis is now rotated until it just touches the LCC to obtain the matching water supply line (represented by the heavy line in Figure 9.17). The water supply line starts at the origin (assuming freshwater is initially used) and the reciprocal of its slope gives the minimum water requirement. For this example, this minimum water target with reuse is 58333.33 kg/h (i.e., $17.5/[300 \times 10^{-6}]$), which is 25.5% less than the water requirement for the case without reuse. Note that the water supply line can actually touch the composite curve (to define the water pinch at 300 ppm) since the necessary process constraints are already incorporated into the LCC.

Step 3. Design for Maximum Driving Forces and Minimum Wastewater

Here, the concentration driving forces are maximized by dividing the plot in Figure 9.17 into mass load intervals through drawing vertical lines at every vertex (where there is a change in the slope of the LCC). The four vertical intervals that result are already available in Table 9.20. The concentrations on the water supply line corresponding to the boundaries of these intervals are

51.43 ppm, 111.43 ppm, 300 ppm, and 505.71 ppm. These values are simply obtained by multiplying the $CumM_c$ entries in Table 9.20 by (300/17.5). The water is split between the various processes that exist within each interval in the ratio of their limiting water flow rates. The network is shown in Figure 9.18 in the conventional grid representation with the flow rates for the water splits (defined by the LCC stream population) indicated below the matches.

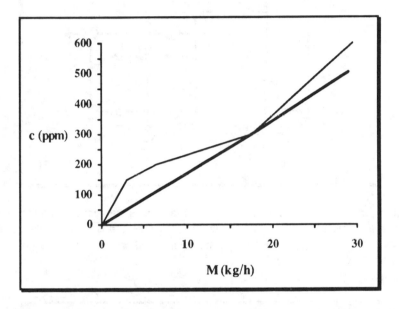

Figure 9.17 The minimum water target is simply given by the reciprocal of the slope of the water supply line. The point where the water supply line touches the limiting composite curve is the water pinch.

The water network in Figure 9.18 is similar to the spaghetti design in HENS and may be simplified by loop breaking (see Section 4.1). The Euler minimum unit target (Equation 1.26) for this example is 4, so there are three loops in Figure 9.18. To break the three first-level loops, the mass loads of the two units involved in the loop may be simply added together. Figure 9.19 shows a simplified design after combining match 2 with match 1 and match 4 with match 3 (in Figure 9.18). It is seen that the new design does not violate the concentration specifications in the limiting water profile data. However, it is not possible to break the loop crossing the pinch by combining matches 5 and 7 without causing a violation (i.e., without exceeding the maximum concentrations specified). The violation may be removed and the concentration restored by incurring a penalty, i.e., by using water in excess of the minimum. The tradeoff between the increase in water requirement and the decrease in design complexity may be assessed to determine a final design.

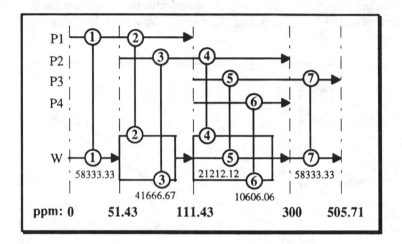

Figure 9.18 The above water network maximizes the concentration driving forces and minimizes the water flow rate. Within the vertical intervals, the concentration is defined by the water supply line. (Contaminant concentration in ppm and flow rate in kg/h)

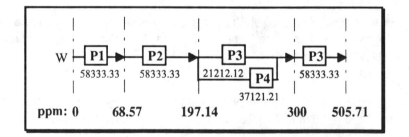

Figure 9.19 The above water network (shown as a conventional flowsheet) is obtained by simplifying the design in Figure 9.18 through loop breaking. (Contaminant concentration in ppm and flow rate in kg/h)

Step 4. Design for Minimum Number of Water Sources and Wastewater

    A design with the minimum number of matches is usually advantageous in water networks. The strategy of stream splitting, mixing and bypassing proposed by Wood et al. (1985) and discussed in Section 4.3 may be profitably used here. Wang and Smith (1994) have generalized the strategy to provide a robust method for generating an MNU design for pinched problems without incurring a penalty. Rather than using mass load intervals and maximizing driving forces (as in step 3), an alternative strategy to designing water networks is to use concentration intervals and minimize the number of water sources for every process (Figure 9.20). The concentration intervals are obtained by drawing horizontal lines at every vertex on the LCC and are directly specified by the c column in Table 9.20. In each match, driving forces are minimized and just sufficient water is supplied for feasibility. Excess water is bypassed.

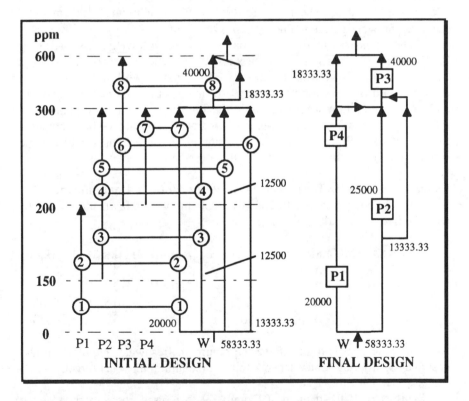

Figure 9.20 The above water network minimizes the water flow rate as well as the number of water sources. Horizontal intervals (with the concentrations defined by the LCC) form the basis. The final design is evolved from the initial design by loop breaking. (Contaminant concentration in ppm and flow rate in kg/h)

In each concentration interval, the processes that exist and the associated mass loads of contaminant may be tabulated first.

Interval	1	2	2	3	3	3	4
Process	P1	P1	P2	P2	P3	P4	P3
M_c (kg/h)	3	1	2.5	5	4	2	12

To remove 3 kg/h of contaminant from process P1 and achieve the outlet concentration of 150 ppm in interval 1, the freshwater requirement is 20000 kg/h (i.e., $3/[150 \times 10^{-6}]$). The remaining freshwater is bypassed. Now, to achieve the concentration of 200 ppm in interval 2, the water requirements are 20000 kg/h (i.e., $1/[50 \times 10^{-6}]$) for process P1 and 12500 kg/h (i.e., $2.5/[200 \times 10^{-6}]$) for process P2. Continuing in this manner, the initial network design in Figure 9.20 is obtained.

To break the four loops in the initial design, the following mass loads may be combined: matches 1 and 2 for process P1; matches 3, 4 and 5 for process P2; and matches 6 and 8 for process P3. The final network design shown in conventional flowsheet form in Figure 9.20 satisfies the minimum water target and the concentration specifications in the limiting water profile data by an interesting scheme involving splitting, mixing and bypassing. When compared to the network in Figure 9.19, the final design in Figure 9.20 uses only a single water source for each process and may therefore be more practical.

Example 9.16 Water Target and Network for Multiple Contaminants
Consider two processes (P1 and P2) involving two contaminants (A and B) with the following limiting water profile data:

Process p	Contaminant	$q c_{in}$ (ppm)	c_{out} (ppm)	M_c (kg/h)	M_w (kg/h)
P1	A	0	200	4	20000
P1	B	50	150	2	20000
P2	A	175	525	10.5	30000
P2	B	50	150	3	30000

As before, the goal is to determine the minimum freshwater requirement and design an appropriate water network to meet the minimum water target.

Solution. If the method of Example 9.15 is used considering A as the key contaminant, the minimum water target is 23750 kg/h. On the other hand, considering B as the key contaminant yields the target of 33333.33 kg/h. Targeting based on the worst possibility is not adequate. A method that considers both contaminants simultaneously is required and is given below. The minimum wastewater targeting method for multiple contaminants with reuse only (i.e., no regeneration) presented below is a modified version of the algorithm presented by Wang and Smith (1994).

Step 1. Establishing Concentration Intervals Based on Reference Contaminant

Any contaminant may be chosen as the reference (say, A in this example). The limiting water profiles may be then constructed as in Figure 9.21 based on contaminant A. Process P1 is represented by the line segment from (0, 0) to (4, 200) whereas process P2 by the line segment from (4, 175) to (14.5, 525).

The initial concentration intervals are obtained by arranging all the c_{in} and c_{out} values of the reference contaminant in ascending order. Here, the concentration intervals are given by 0, 175, 200, 525.

Step 2. Shifting Concentration Intervals

To account for the simultaneous transfer of the other contaminant B, it is here assumed (Wang and Smith, 1994) that the transfer of the various contaminants occurs in a proportionate manner, i.e., the proportion of the reference contaminant transferred equals the proportion (of the total load) of the other contaminant transferred. Any other appropriate relation (e.g., Kremser equation) may be used to correlate the contaminant transfers.

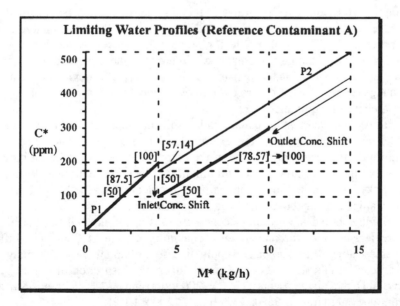

Figure 9.21 Targeting for minimum water for multiple contaminants usually requires shifts in the inlet and outlet concentrations of the limiting water profiles.

Based on proportional transfer of contaminants, the corresponding concentrations of B may be easily calculated (see numbers in brackets in Figure 9.21). If freshwater is utilized for process P1, then $c_{B,P1}$ is 87.5 ppm when $c_{A,P1}$ is 175 ppm. This implies that water from process P1 cannot be reused in process P2 as the latter requires $c_{B,P2}$ to be 50 ppm at this point.

The water from process P1 can be reused at the inlet to process P2 if $c_{B,P1}$ is 50 ppm, which may be achieved through an inlet concentration shift of process P2 with respect to process P1 as shown in Figure 9.21. This does not completely solve the water reuse problem: at the outlet of process P1 ($c_A = 200$ ppm), $c_{B,P1}$ is 100 ppm whereas $c_{B,P2}$ is 78.57 ppm. If reuse is to be possible, then $c_{B,P2}$ should be 100 ppm.

A possible solution is to attempt another inlet concentration shift; however, this would result in surplus driving force at the inlet to process P2 and a water flow rate in excess of the minimum. To maximize the potential for water reuse, one of the contaminants (say, B) must be limiting all through (i.e., both at the inlet to process P2 and at the outlet of process P1). This may be achieved through an outlet concentration shift given by

$$c_{A,P2} = 100 + (200 - 100)(150 - 50)/(100 - 50) = 300 \text{ ppm.}$$

With this shift, the axes (M^* and c^*) in Figure 9.21 must be regarded as reference axes, providing an absolute scale for process P1 but a relative scale for process P2. The final profile for process P2 (shown by the heavy line) appears compressed on the reference axis in terms of the mass load transferred; however, the actual mass transferred in process P2 obviously remains unchanged except that it occurs over a lower concentration range causing an increased water requirement.

It may be noted that the concentration shift in some cases may be upwards (rather than downwards as in this example). This is required if no component is limiting. It is necessary for one of the components always to be limiting in order that the water flow rate be minimum. In fact, concentration shifting is the tricky step in multiple contaminant problems, and both upward and downward shifts may be required in a given problem.

Having shifted the necessary concentrations at the interval boundaries to ensure feasibility of the nonreference contaminants, the LCC for the shifted processes along with the water supply line for minimum flow rate may be constructed at this stage. For this simple example, inspection of Figure 9.21 shows that the pinch occurs at $c^* = 200$ ppm and $M^* = 7$ kg/h. So, the minimum water target is 35000 kg/h (i.e., $7/[200 \times 10^{-6}]$).

An alternative approach useful for complicated multiple contaminant problems and suitable for computer implementation follows.

Step 3. Calculating the Lower Bound for the Water Target

If p is the index for the set of processes and q is the index for the set of contaminants, then the lower bound for the minimum water flow rate is

$$M_L = \min_q \{ [\Sigma_p M_{w,p} (c_{out,pq} - c_{in,pq})]/\max_p (c_{out,pq})\} \qquad (9.16)$$
$$= \min [14.5/(525 \times 10^{-6}), 5/(150 \times 10^{-6})]$$
$$= \min [27619.05, 33333.33] = 27619.05 \text{ kg/h}.$$

Step 4. Calculating the Cumulative Mass of the Reference Contaminant

The intervals after the concentration shifts in step 2 are given by 0, 100, 200, 300. Process P1 exists from 0 to 200 ppm whereas process P2 from 100 to 300 ppm. So, the cumulative mass of the reference contaminant is:

$$20000 \,(100 \times 10^{-6}) \quad = \;2 \text{ kg/s (for the} \quad 0 \text{ - } 100 \text{ ppm interval)}$$
$$50000 \,(100 \times 10^{-6}) + 2 = \;7 \text{ kg/s (for the } 100 \text{ - } 200 \text{ ppm interval)}$$
$$30000 \,(100 \times 10^{-6}) + 7 = 10 \text{ kg/s (for the } 200 \text{ - } 300 \text{ ppm interval).}$$

A plot of this cumulative mass for the reference contaminant in the different intervals yields the multicomponent LCC.

Step 5. Locating the Pinch

The flow rate deficit in each interval is given by

$$27619.05 - 2/(100 \times 10^{-6}) \;= \;7619 \text{ kg/h (for the 0 - 100 ppm interval)}$$
$$27619.05 - 7/(200 \times 10^{-6}) \;= -7381 \text{ kg/h (for the 100-200 ppm interval)}$$
$$27619.05 - 10/(300 \times 10^{-6}) = -5714 \text{ kg/h (for the 200-300 ppm interval).}$$

The largest deficit indicates the pinch. In this case, it corresponds to 200 ppm.

Step 6. Determining the Minimum Water Target

The minimum water flow rate required is obtained on dividing the cumulative mass by the concentration at the pinch. So, the water target is 35000 kg/h (i.e., $7/[200 \times 10^{-6}]$).

Note that the calculations are simplified if the limiting contaminant is chosen to be the reference, but this is not usually easy to determine a priori. In the above example, choosing B as the reference component yields the same final water target but reduces computations. The limiting contaminant may not be unique in complex problems.

Step 7. Design for Minimum Number of Water Sources and Wastewater

Two different water networks (as in Example 9.15) may be designed for multiple contaminant problems based either on maximum driving forces or on minimum number of water sources. Figure 9.22 shows a network for the latter case designed by the procedure given in step 4 of Example 9.15.

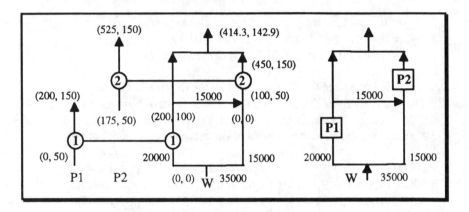

Figure 9.22 The above water network for the case of multiple contaminants (whose concentrations are shown in parentheses) minimizes both the water flow rate and the number of water sources. (Contaminant concentration in ppm and flow rate in kg/h)

Having established the basic water network structure, the detailed design may be performed through more complex models.

Though not discussed here, the methodology may be extended (Wang and Smith, 1994) to provide a systematic assessment of the overall opportunities for regeneration in terms of location and type.

9.6 REACTION PROCESSES

Heat integration of process subsystems (that exchange energy/work) into the overall process can often help further reduce the energy consumption. Reactors, which fall into this category, are often considered as part of the background process and excluded from heat integration. This is acceptable for fixed reactor performance, but reactor parameters may be meaningfully changed (within certain limits) for process improvement as discussed below.

Heat integration of reactors (Glavic et al., 1988, 1990) involves the following three steps: decomposition, modeling, and thermodynamic analysis. Of course, this would finally lead to total flowsheet integration.

In the decomposition of reactors, its energy functions (enthalpy change of the reaction, transfer of this enthalpy from/to the fluid, interchange of energy between fluids entering the reactor at different temperatures, and exchange with the surroundings) are considered separately. Then, the reaction enthalpy is represented by a reactor profile and the other energy functions by a process profile; then, the reactor may be investigated apart from the rest of the process and HEN.

Modeling of reactors is necessary when there are two or more inlet streams and/or several contacting patterns. Careful modeling of a reactor provides a better representation for structural and parametric optimization, which is the focus during the thermodynamic analysis.

Thermodynamic analysis of reactors includes proper placement of reactors relative to the pinch and matching of reactor and process profiles to reduce exergy losses (load and level analysis).

Proper placement depends on the reactor type. For an exothermic reactor to be appropriately placed, it must be above the pinch. This provides maximum economy because the heat evolved from the reaction is then supplied to the process sink, resulting in a corresponding reduction in the hot utility consumption (with no change in the cold utility requirement). An exothermic reactor placed below the pinch causes an increase in the cold utility consumption (with no change in the hot utility requirement). Placement of an exothermic reactor across the pinch is better than placement below the pinch but worse than placement above the pinch. Analogous arguments hold for an endothermic reactor. Essentially, for an endothermic reactor to be appropriately placed, it must be below the pinch. Then, the cold utility requirement is decreased. An endothermic reactor placed above the pinch causes an increase in the hot utility consumption. If placed across the pinch, there is a partial reduction in cold utility and a partial increase in hot utility. For appropriate placement, the conventional process GCC may be used.

For structural and parametric optimization, an alternative representation is convenient that allows assessment of driving force losses. Here, it is important to be able to view the reactor in a manner similar to a heat exchanger on a temperature-enthalpy diagram (T-Q plot). Reactors being complex, by way of operation, need to be "mapped" onto a heat exchanger. For example, consider an exothermic reaction in an adiabatic plug flow reactor with a single inlet stream. The reactant has chemical energy which is released as the enthalpy of the reaction. This released energy is transferred to the process fluid (at nearly 100% efficiency and practically instantly in the case of homogeneous reactions) thereby raising its temperature. This enthalpy interchange is conveniently visualized on the reactor-process profile (RPP), which forms the basis for energy integration of reactors. Here, the reactor profile is represented by a linear segment with a positive slope and is treated as a chemical utility (a fictitious hot stream drawn separately to show the potential available for heat integration). The real cold stream through the exothermic reactor also has the same profile since the end temperatures of the two streams are the same (with a $\Delta T = 0$). It is included in the process profile, which is obtained by the conventional addition of all the cold and hot process streams. Construction of the RPP by essentially superposing reactor and process profiles is given below.

Example 9.17 Construction of the Reactor-Process Profile

Consider the reactor subsystem in Figure 1.1. The subsystem consists of only two streams (H1 with an MC_p of 10 kW/°C going from 175° to 45°C and C3 with an MC_p of 20 kW/°C going from 20° to 155°C). Along with the reactor stream (CR with an MC_p of 20 kW/°C going from 155° to 175°C), let the data for these three streams form Case Study 4S1r. Construct the reactor-process profile (RPP) for Case Study 4S1r with $\Delta T_{min} = 20$°C.

Solution. The procedure for the construction of the process profile is similar to that for the GCC. The difference is in the handling of the reactor stream. Rather than adding a temperature contribution to the cold stream end temperatures (as in the PTA), the temperature contribution needs to be subtracted from the end temperatures of the cold reactor stream. The rationale lies in the fact that the cold stream, through the exothermic reactor, is assumed to exchange heat (equal in magnitude to the reaction enthalpy) with the fictitious chemical utility (hot stream). From a heat integration viewpoint, this heat is to be used elsewhere, so the cold stream must be looked upon as a "hot stream" for later use, thereby justifying the deduction of the temperature contribution term from the end temperatures.

For the construction of the process profile, the PTA (see Section 1.1.2) is employed with the temperature contribution deducted from the cold reactor stream. The results of these calculations are given in Table 9.21.

Table 9.21 Problem Table for Reactor-Process Profile

Interval i				T_{int}	$MC_{p,int}$	Q_{int}	Q_{cas}	R_{cas}
0	|	|	|	165	0	0	0	1800
1	|	|	|	145	30	600	−600	1200
2	|	|		35	10	1100	−1700	100
3			|	30	20	100	−1800	0
Stream	H1	C3	CR					
MC_p	10	20	20					

Units: temperature in °C, heat capacity flow rate in kW/°C, and enthalpy in kW.

The process profile is obtained by plotting T_{int} vs. R_{cas} from Table 9.21. The reactor profile goes from (400, 165) to (0, 145) and is superposed on the process profile to yield the RPP (Figure 9.23).

In addition, hot utility (say, steam at 180°C) is required to satisfy the remaining 1400 kW of heat duty as shown by the heavy horizontal line in Figure 9.23.

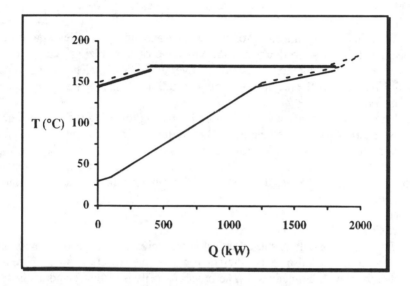

Figure 9.23 The reactor-process profile consists of the reactor/utility profile (shown by the heavy line) superposed on the process profile. It may be used to examine the effect of changes in the reactor operating parameters.

What is the significance of the RPP? The RPP provides a measure of the exergy losses from the system. On a transformed RPP plot with the ordinate as $(1 - T_o/T)$, the area enclosed between the reactor (utility) and process profiles is the exergy loss from the system, as discussed in Section 8.5. For this example, the exergy loss is 328.21 kW. The use of the RPP is explored further in the next example.

Example 9.18 Assessment of Reactor Modifications Using the RPP

For Case Study 4S1r, assume a first order, reversible, exothermic reaction occurring in an adiabatic plug flow reactor. The various reaction model parameters are given below:

Forward reaction: $k_{o1} = 1.1 \times 10^6$ s^{-1} and $E_1 = 83.7$ kJ/mol.
Backward reaction: $k_{o2} = 6.6 \times 10^9$ s^{-1} and $E_2 = 167.4$ kJ/mol.
Heat of reaction: 30 kJ/mol.   Specific heat of reactant: 0.04 kJ/mol.
Residence time in reactor: 230 s.

Analyze the effect on the reactor profile of increasing the feed temperature, increasing the inerts in the feed, and decreasing the reactor length. How is the exergy loss affected in each case?

Solution. The matching of the reactor and utility profiles against the process profile must be done with the objective of minimizing exergy losses (for efficient heat integration with the overall process). The reduction in exergy losses may be achieved through small modifications in the reactor parameters.

The reactor profile may be modified by raising/lowering the reactor curve or by altering its slope/length. This requires changes in the reactant temperature, composition, or flow rate. The process profile may also be changed by modifying the HEN. It may be noted that a change in the reactor profile necessarily changes the process profile whereas the reverse is obviously not true.

Increasing the Feed Temperature
The reactor profile is shifted upwards on increasing the feed temperature. The profile slope also gets altered because the reaction rate increases exponentially with temperature. The process profile is also affected as there is an increase in the enthalpy of the cold reactor stream.

If the feed temperature is increased from 155°C (to 160°C), then the reactor model (Fogler, 1992) predicts the outlet temperature to be 192.7°C. Figure 9.23 shows the changed RPP (in dotted lines). It is noted that there is an increase in both the conversion (43.6%) and exergy loss (339.88 kW) with feed temperature. The hot utility requirement decreases to 1323 kW. These values may be compared with the corresponding values of 175°C (outlet temperature), 26.79% (conversion), 328.21 kW (exergy loss), and 1400 kW (hot utility) for the base case.

Increasing the Inerts in the Feed
If the fraction of inerts in the feed is increased to 30% (from no inerts initially), then the outlet temperature decreases to 167°C and the reactor profile gets shortened. Thus, the effect of increased inert content in the feed is to decrease both exergy loss (324.15 kW) and conversion (21.99%).

Decreasing the Reactor Length

The effect of decreasing reactor length is qualitatively similar to increasing the inerts fraction. If the reactor length is decreased by 10%, then the reactor profile gets shortened (outlet temperature becomes 171.97°C). Also, there is a decrease in both exergy loss (326.59 kW) and conversion (22.62%).

The base case and the effect of the three modifications are summarized in the first four rows of Table 9.22. It may be more meaningful at times to employ a combination of changes rather than to consider them individually.

Table 9.22 Effect of Reactor Modifications

Inlet Temp.	Frac. Inc. Inerts	Frac. Dec. Length	Outlet Temp.	Residence Time	Conversion	Hot Utility	Exergy Loss
155	0	0	175	230	0.2679	1400.0	328.21
160	0	0	192.7	230	0.4360	1323.0	339.88
155	0.3	0	167	230	0.2199	1480.1	324.15
155	0	0.1	171.97	207	0.2262	1430.3	326.59
160	0.3	0	177.44	230	0.3197	1475.6	329.60
160	0.3	0.1	174.95	207	0.2741	1500.5	328.18

Units: temperature in °C, residence time in s, utility in kW, and exergy loss in kW.

Attempting Multiple Modifications

Increasing the feed temperature by 5°C and the fraction of inerts to 30% results in an outlet temperature of 177.44°C, a conversion of 31.97%, and an exergy loss of 329.6 kW. If the reactor length is also decreased by 10% (i.e., all the three modifications are simultaneously performed), then the outlet temperature becomes 174.95°C, the conversion becomes 27.41%, and the exergy loss becomes 328.18 kW (close to the base case).

Clearly, several modifications are possible and their net effect is often complex. The tradeoff between increased conversion and increased exergy loss must be assessed along with operating and capital costs. It may be concluded that the RPP is a useful tool for reactor integration: it recognizes that reactors form a part of the process subsystem and the utility subsystem simultaneously. Thus, reactor modifications affect both the subsystems, and the RPP is an analysis tool that appropriately accounts for these interactions.

9.7 THE OVERALL PROCESS

It would be appropriate to close our discussion on OPERA by looking at the overall process. Overall chemical process design is complex since there are a large number of possible options for the reactor, separation system, HEN, and utility system. Furthermore, there is a tremendous interplay of the design factors governing these different subsystems. With regard to the final flowsheet design, it is important to consider structural as well as parametric optimization.

Structural optimization calls for understanding and creativity on the part of the process engineer. A systematic approach toward structural optimization may be devised based on the OPERA tools discussed thus far in the book. Parametric optimization may then be attempted through mathematical programming (see Chapter 10) by specifying the objective function and constraints.

The onion model (Linnhoff et al., 1982) provides an explicit picture of the hierarchy of process design. A slightly modified (more detailed) onion model is presented below based on the elements identified by Westerberg et al. (1979) for an energy integrated approach to process design. Starting with the reaction path (chemistry), the core of the onion will involve reactor design considering selectivity and capacity. The reactor design and product specifications prescribe the design of the separation (and recycle) subsystem, which forms the next layer of the onion. The reaction and separation tasks define the energy requirements and, therefore, the third layer involves HENS. As heat recovery between process streams is typically insufficient to meet the total energy needs, external utilities are required. Therefore, the fourth layer of the onion is utility system design. To account for flexibility, controllability, and resiliency, the final layer involves the design of the control system. Structural and parametric optimization is necessary at various stages to arrive at a promising final flowsheet, which may be assessed through simulation and costing. It must be emphasized that no layer of the onion can be designed prior to specifying the inner layers; at the same time, the inner layers cannot be assessed without designing the complete onion and considering interactions. The targeting procedures discussed so far serve the express purpose of reducing/eliminating the effort and time required for generating complete designs. They help evaluate cost tradeoffs by decomposition (Smith and Linnhoff, 1988) in accordance with the onion model.

In the previous chapters, the core of the onion was considered sacrosanct (i.e., the reactor and separation subsystems were assumed fixed). The next subsection presents a powerful principle for modifying the onion core in the context of the total process in an attempt to reduce overall utility requirements.

9.7.1 Process Modifications Based on the Plus/Minus Principle

The appropriate placement principle and its importance have already been stressed for various units including heat engines, heat pumps, distillation columns, evaporators, and reactors.   For example, any separator (e.g., distillation column or evaporator) that absorbs heat at a higher temperature and rejects it a lower temperature should be appropriately placed either entirely above or entirely below the pinch. Is there a general fundamental principle behind the appropriate placement concept? Indeed, there is such a principle (Linnhoff and Parker, 1984) and it may be stated as follows.

Energy optimization is possible either by shifting hot streams from below the pinch to above the pinch or by shifting cold streams from above the pinch to below the pinch. The explanation for this principle is straightforward. Consider the shifting of cold streams from above the pinch to below the pinch. This causes the total cold stream load above the pinch to decrease, thereby reducing the hot utility requirement. At the same time, the total cold stream load below the pinch increases, thereby reducing the cold utility requirement. Simply speaking from a heat integration standpoint, it is invariably advantageous to keep hot streams hot and cold streams cold (Linnhoff et al., 1982). Thus, energy savings is possible through any process modification that:

- increases the total hot stream heat load above the pinch; or
- decreases the total hot stream heat load below the pinch; or
- increases the total cold stream heat load below the pinch; or
- decreases the total cold stream heat load above the pinch.

The above rule is referred to as the plus/minus principle (Linnhoff and Vredeveld, 1984; Linnhoff and Parker, 1984). The principle is very general in that it applies to all layers of the onion model and helps in proper heat integration of reactors, evaporators, distillation columns, dryers, etc. The use of the principle is demonstrated below through an example.

Example 9.19 Process Modifications for Energy Savings

It is desired to integrate the following three streams (associated with a distillation column) with the four streams in Case Study 4S1.

Stream Label	T_{in} (°C)	T_{out} (°C)	MC_p (kW/°C)
Hcon	110.1	110	54000
Creb	195	195.1	54000
Cfv	115	115.1	81000

What will be the utility requirements for such an integration? Discuss possible process modifications for saving energy. Assume $\Delta T_{min} = 20$°C.

Solution. The first step is to plot the composite curves for the seven-stream problem. As shown by the dotted lines in Figure 9.24, the hot and cold utility requirements are 14105 kW and 5925 kW respectively. Since the hot utility for Case Study 4S1 was 605 kW, it is observed that there is an increase of 13500 kW (i.e., 5400 + 8100). The cold utility for Case Study 4S1 was 525 kW, so there is an increase of 5400 kW. Thus, the increases simply correspond to the loads of the three new streams.

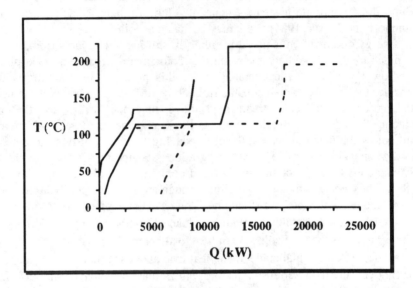

Figure 9.24   The utility consumption may be reduced through a suitable process modification (e.g., appropriately place the distillation column above the pinch by increasing its pressure).

Figure 9.24 shows the pinch occurs at 125°C for hot streams and at 105°C for cold streams; therefore, the distillation column is inappropriately placed across the pinch initially (as shown by the dotted lines). The column may be shifted above the pinch by increasing its pressure such that the Hcon stream now goes from 135.1° to 135°C and the Creb stream goes from 220° to 220.1°C (see solid lines in Figure 9.24). This results in the condensing stream (a hot stream) being shifted from below the pinch to above the pinch. The pinch remains at 125°C/105°C; however, the hot utility requirement reduces to 8705 kW (given by 605 + 5400 + 8100 − 5400) and the cold utility requirement reduces to 525 kW. This is in accordance with the plus/minus principle since the total hot stream heat load has increased above the pinch and decreased below the pinch.

Instead of shifting the column (Figure 9.24), an alternative strategy (Smith and Linnhoff, 1988) to place the column appropriately is to shift the pinch by decreasing the pressure of the vaporizer (such that the Cfv stream now goes from 90° to 90.1°C) as in Figure 9.25. The new pinch is at 110°C/90°C, and the hot and cold utility requirements now are 8480 kW and 300 kW respectively. On the HCC, heat load has been subtracted below the pinch and added above the pinch as per the plus/minus principle.

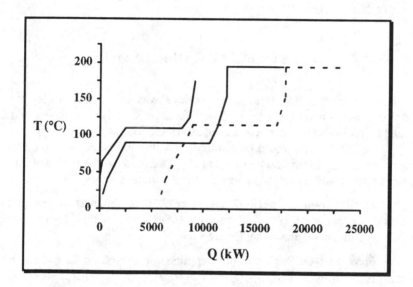

Figure 9.25 Another process modification (an alternative to increasing the column pressure as in Figure 9.24) involves shifting the pinch by decreasing the pressure of the vaporizer.

Process modifications were attempted in the previous example based on the plus/minus principle in order to reduce utility costs. Figures 9.24 and 9.25 show that the modifications lead to a decrease in temperature driving forces and a resulting increase in capital costs. In fact, the optimum ΔT_{min} and the proper choice of utilities must be reestablished after process modifications, using the targeting procedures in Chapter 1. Linnhoff et al. (1990) have discussed in detail the optimization for such heat integration considering a three-way tradeoff between process modifications, utility selection, and capital/energy.

It may be concluded that the heat integration principles originally developed for HENS have the potential of being extended to other systems and processes. Optimization of individual unit operations is usually not the right approach since better design solutions are possible by a proper appreciation of the various tradeoffs. Targeting procedures hold the key to optimizing the different layers of the process onion within the context of total processes and with reasonable computational effort. As of now, a reasonably well-developed approach for the optimization of processes by energy and resource analysis (OPERA) exists, as described in this chapter. Hopefully, improved tools for OPERA will emerge in the near future, especially with regard to reactor and control system design.

QUESTIONS FOR DISCUSSION

1. In Example 9.9, one of the options was to install 143.59 m^2 of additional area on unit 3 (the fouling exchanger). Discuss how a bypass around unit 3 would be operated to account for the fouling.

An alternative would be to place an additional heater at the outlet of unit 3 (the fouling exchanger) on Stream C4. Discuss how the hot utility usage in this heater would be varied to account for the fouling.

2. Flexibility in the network in Figure 9.10 is obtained by completely bypassing exchanger 1 (i.e., by reducing UA). Explain this design change physically and discuss the tradeoffs involved.

3. Show that the net work consumption for separation in a distillation column is given by $W = (Q_r - Q_c) + T_o (Q_c T_r - Q_r T_c)/(T_r T_c)$ where Q_r = reboiler load, Q_c = condenser load, T_r = reboiler temperature in K, T_c = condenser temperature in K, and T_o = ambient temperature in K. If the reboiler and condenser loads are equal ($Q_r = Q_c = Q$), then show that the work consumption is proportional to the area (i.e., $Q [T_r - T_c]$) of the rectangle that represents the column on a T-H plot.

4. Does the reboiler have to be integrated for appropriate placement of a distillation column above the pinch? Does the condenser have to be integrated for appropriate placement of a distillation column below the pinch? Does integrating either the condenser or the reboiler (but not both) prove advantageous from an operability viewpoint? Are there cases where it is necessary to integrate both the condenser and reboiler?

5. Why is feed conditioning a preferred design modification for distillation columns compared to side condensing/reboiling?

6. Comment briefly on the following rules of thumb (Ho and Keller, 1987) for distillation column modifications:

- Feed subcooling and side condensing may be considered for feeds that contain mainly high-boiling components.
- Vapor feed and side reboilers may be considered for feeds that contain mainly low-boiling components.

7. Equations 9.14 are based on the arithmetic mean temperature for the cooling water. How will the equations change if the more accurate logarithmic mean is used? Will the calculations then become iterative?

8. The analysis for the economics of electrically driven heat pumping in step 6 of Example 9.12 does not include the capital cost of the compressor. If $CC_{comp} = a_{comp} + b_{comp} \, W^{c_{comp}}$, then derive an expression for the upper bound on T_1.

9. The design of integrated evaporation systems in Section 9.4 does not consider (a) condensate heat recovery within the evaporator itself, (b) allowance for the boiling point rise, and (c) latent heat variations with pressure. Discuss how these effects may be accounted for.

10. Discuss the limitations of the transformation (in Section 9.5.1) for mapping a MENS problem to a HENS problem. Can only linear equilibrium relationships and single species be handled? Name some industrial processes where single component targets suffice.

11. External MSAs should not be used above the pinch during MENS. Why?

12. During the synthesis of a minimum utility MEN, how does one ensure that sufficient driving forces exist at all points in the network (say, in Figure 9.16)? Is it advisable actually to design the MEN using the transformed variables for the lean streams?

13. In Figure 9.23, the reactor profile is drawn linear. Is it truly linear? How can one account for the nonlinear evolution of enthalpy along a reactor? Would the reactor profile be horizontal in the case of an isothermal reactor?

14. Is it true that heat integration will invariably gain from hotter condenser streams and colder reboiler streams for distillation columns? Is this a special case of a more general principle?

15. Discuss heat integration principles for dryers. Why can dryers not be treated in a manner similar to evaporators or distillations columns?

PROBLEMS

9.A Targeting for Batch Processes

Consider the stream data in Case Study 4L2 with the following modification: the streams do not exist for the entire batch period, but only over certain time intervals and with MC_p values as specified below:

Stream	H1	H2	C3	C4
Start time (h)	0.25	0.3	0.5	0.0
Finish time (h)	1.0	0.8	0.7	0.5
MC_p (kW/°C)	4	3	10	8

For this batch process stream data (taken from Kemp and Deakin, 1989b), determine the minimum utility requirements predicted by the TAM and the TSM. Also, predict the minimum utility targets and the amount of heat storage based on the HSM for a single batch and for repeated batches.

Assume the batch period is 1 hour and $\Delta T_{min} = 10°C$.

Answers:

TAM targets: 20 kWh of hot utility and 60 kWh of cold utility

TSM targets: 198 kWh of hot utility and 238 kWh of cold utility

HSM targets (single batch): 134 kWh of hot utility,

174 kWh of cold utility, and 64 kWh recovered via heat storage

HSM targets (repeated batch): 20 kWh of hot utility,

60 kWh of cold utility, and 178 kWh recovered via heat storage

9.B MENS for COG Sweetening Example

Synthesize a minimum utility MEN for the case study discussed by El-Halwagi and Manousiouthakis (1989) on the sweetening of coke oven gas (COG). The stream data are given below.

Rich Streams				Lean Streams			
Stream	y_{in}	y_{out}	G	Stream	x_{in}	x_{out}	L
R1	0.0700	0.0003	0.9	S1	0.0006	0.0310	2.3
R2	0.0510	0.0001	0.1	S2	0.0002	0.0035	∞

Units: flow rate (L and G) in kg/s and composition (x and y) in w/w %.

The problem involves removing hydrogen sulfide from two rich process streams (the sour COG stream R1 and the tail gases stream R2) for $\varepsilon = 0.0001$ using a lean process stream (the aqueous ammonia stream S1) and an external MSA (the chilled methanol stream S2). The equilibrium solubility relationships for hydrogen sulfide are $y = 1.45\ x_1$ (in aqueous ammonia) and $y = 0.26\ x_2$ (in methanol). Specifically, determine the minimum flow rate of S2 required to remove hydrogen sulfide from R1 and R2. Also, determine the reduced flow rate of S1 compatible with the minimum flow rate of S2. Sketch the rich composite curve and the lean composite curve.

Answers:

Flow rate of S2 = 0.2227 kg/s; new flow rate of S1 = 2.2068 kg/s.

Note that the entry that specifies the hot utility in HENS signifies the excess capacity of the process lean streams (in this case, aqueous ammonia) to remove the solute (in this case, hydrogen sulfide).

The mass exchange pinch is at 0.001015 on the rich stream scale. A possible minimum utility network appears in Figure 8 (El-Halwagi and Manousiouthakis, 1989).

Above Pinch	*Below Pinch*
R1 - S1a (0.062087 kg/s)	R1 - S2 (0.000644 kg/s)
R2 - S1b (0.004999 kg/s)	R2 - S2 (0.000092 kg/s)

9.C MENS for Copper Recovery Example

Synthesize a minimum utility MEN for the case study discussed by El-Halwagi and Manousiouthakis (1990a) on copper recovery in an etching plant. The stream data are given below.

	Rich Streams				Lean Streams		
Stream	y_{in}	y_{out}	G	Stream	x_{in}	x_{out}	L
R1	0.13	0.10	0.25	S1	0.030	0.07	∞
R2	0.06	0.02	0.1	S2	0.001	0.02	∞

Units: flow rate (L and G) in kg/s and composition (x and y) in w/w %.

The problem involves recovering copper from the spent etchant stream (R1) and the rinse water stream (R2) by extraction using an aliphatic α-hydroxyoxime (S1 costing 0.01 \$/kg) and an aromatic β-hydroxyoxime (S2 costing 0.12 \$/kg). The equilibrium solubility relationships and minimum allowable composition differences for each rich-lean stream pair are

$$y_1 = 0.734\ x_1 + 0.001 \quad \text{and} \quad \varepsilon_{11} = 0.0844 \quad \text{(for R1-S1)};$$
$$y_1 = 0.111\ x_2 + 0.008 \quad \text{and} \quad \varepsilon_{12} = 0.0061 \quad \text{(for R1-S2)};$$
$$y_2 = 0.734\ x_1 + 0.001 \quad \text{and} \quad \varepsilon_{21} = 0.00006 \quad \text{(for R2-S1)};$$
$$y_2 = 0.148\ x_2 + 0.013 \quad \text{and} \quad \varepsilon_{22} = 0.0052 \quad \text{(for R2-S2)}.$$

Specifically, determine the minimum flow rates for S1 and S2.

Answers:

Flow rate of S1 = 0.1875 + 0.09234 = 0.27984 kg/s.

Flow rate of S2 = 0.01613 kg/s.

Given that the ε values are rich-lean-stream-pair dependent, two separate GCCs (one for R1 and another for R2) may be constructed. Now, $\varepsilon_{ij} = \varepsilon_i + \varepsilon_j$ where i denotes the rich stream and j the lean stream. For convenience, ε_i may be chosen to be zero.

A utility pinch occurs at 0.023064 on the rich stream scale. A possible minimum utility network appears in Figure 10 (El-Halwagi and Manousiouthakis, 1990a).

Above Pinch	Below Pinch
R1 - S1a (0.0075 kg/s)	R2 - S2 (0.000306 kg/s)
R2 - S1b (0.003694 kg/s)	

9.D Water Targeting for Single Contaminant

Consider four processes involving a single contaminant with the following limiting water profile data (Wang and Smith, 1994):

Process Label	c_{in} (ppm)	c_{out} (ppm)	M_c (kg/h)	M_w (te/h)
P1	0	100	2	20
P2	50	100	5	100
P3	50	800	30	40
P4	400	800	4	10

Determine the minimum freshwater requirement with and without reuse.

Answers:

Water flow rate is 90 te/h (with reuse) and 112.5 te/h (without reuse).

9.E Water Targeting for Multiple Contaminants

Consider two processes (P1 and P2) involving two contaminants (A and B) with the following limiting water profile data (Wang and Smith, 1994):

Process	Contaminant	c_{in} (ppm)	c_{out} (ppm)	M_c (kg/h)	M_w (te/h)
P1	A	0	100	4	40
P1	B	25	75	2	40
P2	A	80	240	5.6	35
P2	B	30	90	2.1	35

Determine the minimum flow rate of water required considering reuse only.

Answers:

Minimum water flow rate = 54 te/h.

CHAPTER 10

Mathematical Programming Formulations for HENS

The mere formulation of a problem is far more often essential than its solution, which may be merely a matter of mathematical or experimental skill.

Albert Einstein

The systematic evolutionary approach based on pinch technology (PT), presented in the book so far, employs sequential design strategies. The other approach uses simultaneous strategies and is based on mathematical programming (MP) techniques. It has also received a lot of attention from HENS researchers over the last two decades. However, its use has been essentially restricted to academia because the MP models for HENS are relatively complex (to formulate and implement) and the MP solutions do not provide sufficient physical insight. The PT approach has been the industrial favorite on account of its apparent simplicity and its numerous successes. In order to provide a complete treatment of HENS, this chapter presents MP formulations and highlights some of those tasks where the MP solution is superior to that provided by PT.

In terms of mathematical programming, the first significant contribution was a linear programming (LP) transshipment model for energy targeting by Papoulias and Grossmann (1983). The minimum units problem was formulated by them as a mixed integer linear programming (MILP) model. Floudas et al. (1986) have presented a nonlinear programming (NLP) model for generating an optimum network. In more recent contributions, Colberg and Morari (1990) have presented an NLP model for finding the minimum area requirement for specified energy targets with forbidden matches and unequal heat transfer coefficients. An NLP model for simultaneous optimization of energy and area targets, as well as a mixed integer NLP (MINLP) model for synthesis of optimum HENs, have been presented in a series of three papers (Yee and Grossmann, 1990; Yee et al., 1990a, 1990b). The other important issue addressed by MP has been the retrofitting of HENs (Ciric and Floudas, 1990b, 1990c; Yee and Grossmann, 1991).

While pinch technology remains the industrial standard, mathematical programming does offer advantages in certain cases. Apart from being rigorous (more algorithmic and less heuristic), it can easily handle constraints like forbidden and preferred matches. While MP can determine the true minimum area target, it may be recalled that the area target obtained by PT in Section 1.2 is the true minimum only when the heat transfer coefficients of all streams are equal. As pointed out by Sagli et al. (1990), another drawback in network synthesis using PT is the inappropriate assumption that the best solution for a given number of units can also be evolved to give the best solution with one unit less. There are also limitations with the MP approach. It requires very large computer times and storage to solve typical large-sized industrial problems. Also, nonconvexities in the objective function and constraints may lead to local optima, especially in MINLPs whose solution is very difficult with current solvers. However, the probable reason that MP methods are not popular is the lack of easy-to-use software.

The first section of this chapter presents a utility targeting LP and a unit targeting MILP based on the formulations of Papoulias and Grossmann (1983). Using the Colberg and Morari (1990) formulation as a basis, an area targeting NLP model is described in Section 10.2. An implementation of the superstructure model of Yee and Grossmann (1990) is presented in Section 10.3. Based on an original representation of a network, two NLP models for load optimization and data reconciliation of HENs are developed in Sections 10.4 and 10.5. Section 10.6 discusses an MINLP model for evolution of HENs with minimum energy penalty.

All the above models are implemented in an advanced programming language called GAMS (general algebraic modeling system) that has built-in solvers for LP, NLP, MILP, and MINLP formulations. The reader not familiar with this language should refer to the GAMS User's Guide (Brooke et al., 1992). Every section of this chapter follows the same structure: a brief description of the model is provided first, and this is followed by the GAMS listing of its implementation. (The GAMS listing without comments is included in the diskette accompanying this book.) Examples illustrating the use of the GAMS programs are given in each section. A student version of GAMS accompanies the book under a special arrangement with GAMS Development Corporation (Washington, D.C.); thus, the reader can gain hands-on experience with GAMS using the listings provided.

10.1 UTILITY TARGETING AND MER NETWORKS

The utility targeting model (Papoulias and Grossmann, 1983) is essentially an LP implementation of the PTA. It can find energy targets for a specified value of ΔT_{min} with multiple utilities and forbidden/preferred match constraints. The minimum number of units is determined by extending this LP to an MILP, as shown later.

10.1.1 The Transshipment Model

As in the PTA, the problem is divided into a series of temperature intervals (TIs). Two temperature axes, one for hot streams/utilities and another for cold streams/utilities, are maintained. Each TI boundary temperature on the hot axis equals either a hot stream supply temperature or a cold stream supply temperature plus ΔT_{min}. Similarly, the TI boundary temperatures on the cold axis correspond either to a cold stream inlet temperature or a hot stream inlet temperature minus ΔT_{min}. This ensures a minimum driving force of ΔT_{min} between the hot and cold streams.

In each TI obtained by the above procedure, energy balance constraints are imposed. On regarding the energy recovery problem as a transshipment model as in Figure 10.1, five kinds of heat flows may be identified for each interval. The hot streams can transfer heat to cold streams and cold utilities. There also exists the possibility of heat transfer from hot utilities to cold streams. Further, the hot streams can cascade their residual heat to lower TIs.

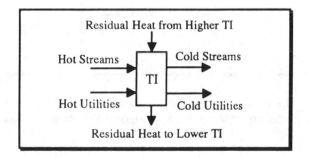

Figure 10.1 An analogy between the HENS problem and the transshipment model exists: heat acts as the commodity to be shipped, hot streams/utilities act as sources, cold streams/utilities act as destinations, and temperature intervals act as intermediate warehouses.

In the formulation below, the utility targets are first determined for the specified ΔT_{min} and then used to find the minimum number of units in a HEN. The unit targeting MILP allows only one exchanger between any pair of hot and cold streams.

10.1.2 GAMS Listing for Energy Targets and MER Networks

The formulation for utility and unit targeting using the transshipment model is presented below. The LP for utility targeting essentially minimizes the total utility cost subject to energy balance constraints on every stream in every TI. The variables are the utility requirements, UTILHOT(JHU,I) and UTILCOLD(JCU,I); the residual heat flows, RES(JH,I) and RESHU(JHU,I); and the heat transfer rates, Q(JH,JC,I), QHU(JHU,JC,I), and QCU(JH,JCU,I). These variables for the various heat flows in a typical TI appear in Figure 10.2.

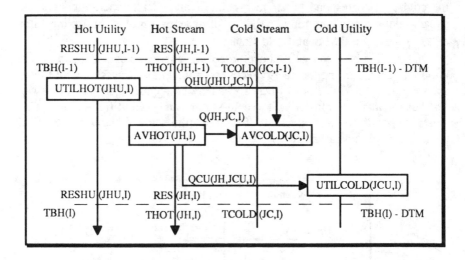

Figure 10.2 The heat flows (and the variables used) in a typical TI are shown using a representative hot utility, hot stream, cold stream, and cold utility.

Although the GAMS listing that follows is specific to Case Study 4S1t for a ΔT_{min} of 25°C, it may be easily modified for other problems. The program is commented for ease of understanding.

$TITLE UTILITY TARGETING LP AND UNIT TARGETING MILP
* This program determines targets for the minimum utility and units required
* in a HEN for a specified value of the minimum approach temperature
* (DTM). It can be used for problems with multiple utilities and allows
* constraints on forbidden matches. Although based on the work of Papoulias
* and Grossmann (1983), the listing below is a different implementation of the
* formulation.

```
* Program Name: HXUTIL.GMS
* Program developed by Yogesh Makwana and Uday V. Shenoy
$onnestcom inlinecom // // eolcom ##
```

// The first step is to define sets of hot and cold streams as well as utilities. These sets form the domain for the variables. A series of temperature intervals (TIs) is formed with each boundary temperature (TBH) corresponding to a hot inlet temperature or a cold inlet temperature plus DTM. In Case Study 4S1t, there are two hot streams, two cold streams, one hot utility, and one cold utility. //

```
SETS
JH  Hot streams /1*2/
JC  Cold streams /1*2/
JHU  Hot utilities /1*1/
JCU  Cold utilities /1*1/
I Temperature boundaries + 1 /1*7/
DATA /TIN,TOUT,MCP,H/;

    TABLE HOTS(JH,DATA)   Hot streams data
        TIN    TOUT   MCP    H
1     175.00  45.00  10.00  0.2000
2     125.00  65.00  15.00  2.0000
;
    TABLE HUTIL(JHU,DATA)   Hot utilities data
        TIN    TOUT   MCP    H
1     180.00 179.00   0.00  4.0000
;
    TABLE COLDS(JC,DATA)  Cold streams data
        TIN    TOUT   MCP    H
1     20.00  155.00  20.00  0.2000
2     40.00  112.00  15.00  2.0000
;
    TABLE CUTIL(JCU,DATA)  Cold utilities data
        TIN    TOUT   MCP    H
1     15.00   25.00   0.00  2.0000
;
PARAMETERS
DTM           Minimum approach temperature
NL            Left temperature boundary of the subnetwork
NR            Right temperature boundary of the subnetwork
CSTS(JHU)     Costs of hot utilities
```

CSTW(JCU)   Costs of cold utilities
TBH(I)   Temperature at the Ith boundary;
DTM =   25.00;
NL = 1;   NR = 6;
CSTS('1') =   120.00;   CSTW('1') =   10.00;

TBH('1') =   180.00;   TBH('4') =   65.00;
TBH('2') =   175.00;   TBH('5') =   45.00;
TBH('3') =   125.00;   TBH('6') =   40.00;

// If the match between two streams or between a stream and a utility is
forbidden, the index MHC, MSC, or MHW is appropriately set to zero. No
match is forbidden in the problem considered. Hence, all indices are 1. //

* Forbidden match data for hot and cold streams
   TABLE MHC(JH,JC)
       1   2
1   1   1
2   1   1;
* Forbidden match data for hot utilities and cold streams
   TABLE MSC(JHU,JC)
       1   2
1   1   1;
* Forbidden match data for hot streams and cold utilities
   TABLE MHW(JH,JCU)
       1
1   1
2   1;

* Alias declares an alternate name for a set
ALIAS   (I,NOI1);

PARAMETERS
TBC(I)   Boundary temperatures at the cold ends
STAGE(I)   Existence of a stage
HUEX(JHU,I)   Coefft for real existence of hot utility, not just cascading
H(JH,I)   Existence coefficient for hot stream JH in interval I
HU(JHU,I)   Existence coefficient for hot utility JHU in interval I
C(JC,I)   Existence coefficient for cold stream JC in interval I
CU(JCU,I)   Existence coefficient for cold utility JCU in interval I
AVHOT(JH,I)   Heat available in hot stream JH in interval I
AVCOLD(JC,I)   Heat required by cold stream JC in interval I

QHCMAX(JH,JC)   Maximum possible heat exchange between JH and JC
QHWMAX(JH,JCU) Maximum possible heat exchange between JH and JCU
QSCMAX(JHU,JC)   Maximum possible heat exchange between JHU and JC
THOT(JH,I)   Temp of hot stream JH crossing the boundary I
TCOLD(JC,I)   Temp of hot stream JC crossing the boundary I
LOAD(JH,JC)   Load on match between hot stream JH & cold stream JC
LOADH(JHU,JC)   Load on match between hot utility & cold stream JC
LOADC(JH,JCU)   Load on match between hot stream JH & cold utility
RESTOT(I)   Total residual heat from interval I
QHUMIN(JHU)   Requirement of JHU as determined by energy targeting
QCUMIN(JCU)   Requirement of JCU as determined by energy targeting;

TBC(I)$(ORD(I) LE NR) = TBH(I) - DTM;
STAGE(I)$(ORD(I) GT NL AND ORD(I) LE NR) = 1;
HUEX(JHU,I)$(TBH(I-1) LE HUTIL(JHU,'TIN') AND
        TBH(I-1) GT HUTIL(JHU,'TOUT')) = 1;

// For hot streams (and utilities), the stream existence coefficient in an interval
is 1 if a stream exists or can cascade heat to that interval. For a cold stream (or
utility), the existence coefficient is 1 in an interval only if it exists in that
interval. All existence coefficients can be calculated by the following four
equations at this stage, as the boundary temperatures are known. //

H(JH,I)$(TBH(I) LT HOTS(JH,'TIN') AND STAGE(I)) = 1;
HU(JHU,I)$(TBH(I) LT HUTIL(JHU,'TIN') AND STAGE(I)) = 1;
C(JC,I)$(COLDS(JC,'TIN') LE TBC(I) AND
        COLDS(JC,'TOUT') GE TBC(I) AND STAGE(I)) = 1;
CU(JCU,I)$(CUTIL(JCU,'TIN') LE TBC(I) AND
        CUTIL(JCU,'TOUT') GE TBC(I) AND STAGE(I)) = 1;

// The heat available in hot streams (and the heat deficits in cold streams) in an
interval is given by the product of the difference between the temperatures at
which the stream crosses the boundaries of the interval and its heat capacity
flow rate (MCP). //

AVHOT(JH,I)$((ORD(I) GT NL) AND (ORD(I) LE NR) AND H(JH,I)) =
        (MAX(TBH(I-1),HOTS(JH,'TOUT'))
        - MAX (TBH(I), HOTS(JH,'TOUT')))*HOTS(JH,'MCP');

AVCOLD(JC,I)$((ORD(I) GT NL) AND (ORD(I) LE NR) AND C(JC,I)) =
        (MIN(TBC(I-1),COLDS(JC,'TOUT'))
        - MIN ((TBH(I)-DTM),COLDS(JC,'TOUT')))*COLDS(JC,'MCP');

* Declaring variables positive automatically ensures nonnegativity constraint.

POSITIVE VARIABLES

UTILHOT(JHU,I)	Hot utility JHU required in interval I
UTILCOLD(JCU,I)	Cold utility JCU required in interval I
Q(JH,JC,I)	Heat transferred by hot stream JH to cold stream JC in interval I
RES(JH,I)	Heat residual of hot stream JH in interval I
RESHU(JHU,I)	Heat residual of hot utility JHU in interval I
QHU(JHU,JC,I) *	Heat transferred by hot utility JHU to cold stream JC in interval I
QCU(JH,JCU,I) *	Heat transferred by hot stream JH to cold utility JCU in interval I
AVHU(JHU,I)	Hot utility JHU required in interval I;

// The binary variables for counting the units are declared below. These variables are 1 if the match exists and 0 if it does not exist. This formulation assumes that any pair of streams can be matched only once in a network. //

BINARY VARIABLES

NHC(JH,JC)	For a match between JH and JC
NHW(JH,JCU)	For a match between JH and JCU
NSC(JHU,JC)	For a match between JHU and JC;

* The free variables to be minimized are declared below.
VARIABLES

SUTIL	Sum of hot and cold utilities required
SUNITS	Number of units;

* The equations (constraints and objective function) are declared below.
EQUATIONS
ENBALH(JH,I)
ENBALHU(JHU,I)
ENBALC(JC,I)
ENBALCU(JCU,I)
HUINT(JHU,I)
HUAV(JHU,I)
RESCON1(JH)
RESCON2(JH)
RESCON3(JH,I)
RESHUCON1(JHU)
RESHUCON2(JHU)
RESHUCON3(JHU,I)

UCON1(JH,JC)
UCON2(JH,JCU)
UCON3(JHU,JC)
HUREQ(JHU)
CUREQ(JCU)
OBJEN   Objective function for energy
OBJUNITS   Objective function for units;

* The constraints are given next.

// Each stream and utility must satisfy the energy balance constraints given below. For a hot stream, five components of heat transferred can be written in each interval, namely the residue from previous interval, heat available in the interval, heat transferred to the cold streams in the interval, heat transferred to the cold utilities in the interval, and the residue going to the next interval. This is shown in equation ENBALH. Similar balances on hot utilities, cold streams, and cold utilities give the other balance equations. //

ENBALH(JH,I)$(STAGE(I) AND H(JH,I))..
  AVHOT(JH,I) + RES(JH,I-1) =E=
SUM(JC$(MHC(JH,JC) AND C(JC,I)),Q(JH,JC,I)) +
SUM(JCU$(MHW(JH,JCU) AND CU(JCU,I)),QCU(JH,JCU,I)) + RES(JH,I);

ENBALHU(JHU,I)$(STAGE(I) AND HU(JHU,I))..
  AVHU(JHU,I) + RESHU(JHU,I-1) =E=
  SUM(JC$(MSC(JHU,JC) AND C(JC,I)),QHU(JHU,JC,I)) + RESHU(JHU,I);

ENBALC(JC,I)$(STAGE(I) AND C(JC,I))..
    AVCOLD(JC,I) =E=
    SUM(JHU$(MSC(JHU,JC) AND HU(JHU,I)),QHU(JHU,JC,I)) +
    SUM(JH$(MHC(JH,JC) AND H(JH,I)),Q(JH,JC,I));

ENBALCU(JCU,I)$(STAGE(I) AND CU(JCU,I))..
    UTILCOLD(JCU,I) =E=
    SUM (JH$(MHW(JH,JCU) AND H(JH,I)),QCU(JH,JCU,I));

HUINT(JHU,I)$(STAGE(I) AND HU(JHU,I))..
    UTILHOT(JHU,I) =E=
    SUM (JC$(MSC(JHU,JC) AND C(JC,I)),QHU(JHU,JC,I));

HUAV(JHU,I)$(STAGE(I) AND HUEX(JHU,I))..
    AVHU(JHU,I) =E=

```
        SUM((JC,NOI1)$(C(JC,NOI1) AND MSC(JHU,JC) AND
                STAGE(NOI1)), QHU(JHU,JC,NOI1))*
        (TBH(I-1) - MAX(TBH(I),HUTIL(JHU,'TOUT')))/
        (HUTIL(JHU,'TIN') - HUTIL(JHU,'TOUT'));
```

// The RESCON and RESHUCON equations ensure that no residuals (for hot streams/utilities) are cascaded to or from the ends of the overall cascade. //

```
RESCON1(JH)..
    RES(JH,'1') =E= 0;
RESCON2(JH)..
    SUM(NOI1$(ORD(NOI1) EQ NR),RES(JH,NOI1)) =E= 0;
RESCON3(JH,I)$(H(JH,I) EQ 0)..
    RES(JH,I) =E= 0;

RESHUCON1(JHU)..
    RESHU(JHU,'1') =E= 0;
RESHUCON2(JHU)..
    SUM(NOI1$(ORD(NOI1) EQ NR),RESHU(JHU,NOI1)) =E= 0;
RESHUCON3(JHU,I)$(HU(JHU,I) EQ 0)..
    RESHU(JHU,I) =E= 0;
```

// Additional constraints for unit targeting are given. QHCMAX, QHWMAX, and QSCMAX are computed after solving the utility targeting LP. //

```
UCON1(JH,JC)$MHC(JH,JC)..
    SUM(I$(STAGE(I) AND H(JH,I) AND C(JC,I)), Q(JH,JC,I))
        =L= QHCMAX(JH,JC) * NHC(JH,JC);

UCON2(JH,JCU)$MHW(JH,JCU)..
    SUM(I$(STAGE(I) AND H(JH,I) AND CU(JCU,I)), QCU(JH,JCU,I))
        =L= QHWMAX(JH,JCU) * NHW(JH,JCU);

UCON3(JHU,JC)$MSC(JHU,JC)..
    SUM(I$(STAGE(I) AND HU(JHU,I) AND C(JC,I)), QHU(JHU,JC,I))
        =L= QSCMAX(JHU,JC) * NSC(JHU,JC);

HUREQ(JHU)..
    SUM(I$STAGE(I), UTILHOT(JHU,I)) =E= QHUMIN(JHU);

CUREQ(JCU)..
    SUM(I$STAGE(I), UTILCOLD(JCU,I)) =E= QCUMIN(JCU);
```

```
* OBJECTIVE FUNCTION FOR MINIMUM UTILITY REQUIREMENT
OBJEN..
    SUTIL =E=
        SUM(JHU,CSTS(JHU)*SUM(I$STAGE(I),UTILHOT(JHU,I))) +
        SUM(JCU,CSTW(JCU)*SUM(I$STAGE(I),UTILCOLD(JCU,I)));

* OBJECTIVE FUNCTION FOR MINIMUM UNITS
OBJUNITS..
    SUNITS =E= SUM((JH,JC)$MHC(JH,JC), NHC(JH,JC)) +
                SUM((JH,JCU)$MHW(JH,JCU), NHW(JH,JCU)) +
                SUM((JHU,JC)$MSC(JHU,JC), NSC(JHU,JC));

* First the minimum utility requirements are found.

MODEL UTIL /ENBALH, ENBALHU, ENBALC, ENBALCU,
HUINT, HUAV, RESCON1, RESCON2, RESCON3,
RESHUCON1, RESHUCON2, RESHUCON3, OBJEN /;
SOLVE UTIL USING DNLP MINIMIZING SUTIL;
```

// The following equations determine the temperature at which the hot and cold streams cross the temperature boundaries. As the hot streams can cascade heat, the temperature of a hot stream crossing the boundary can be greater than the boundary temperature. However the cold streams cannot cascade heat and hence the temperature at which the cold stream crosses the boundary must equal the boundary temperature. //

```
THOT(JH,I) = MAX(TBH(I),HOTS(JH,'TOUT'))
                + RES.L(JH,I)/HOTS(JH,'MCP');
TCOLD(JC,I) = MIN(TBC(I), COLDS(JC,'TOUT'));

RESTOT(I) = SUM(JH, RES.L(JH,I)) + SUM(JHU,RESHU.L(JHU,I));
DISPLAY RESTOT;
```

* The maximum possible heat exchange between any pair of streams/utilities is
* then calculated.

```
QHCMAX(JH,JC)$MHC(JH,JC)
    = MIN(HOTS(JH,'MCP')*(HOTS(JH,'TIN') - HOTS(JH,'TOUT')),
        COLDS(JC,'MCP')*(COLDS(JC,'TOUT') - COLDS(JC,'TIN')));
QHWMAX(JH,JCU)$MHW(JH,JCU)
    = MIN(HOTS(JH,'MCP')*(HOTS(JH,'TIN') - HOTS(JH,'TOUT')),
        SUM(I$STAGE(I),UTILCOLD.L(JCU,I)));
```

```
QSCMAX(JHU,JC)$MSC(JHU,JC)
  = MIN(SUM(I$STAGE(I),UTILHOT.L(JHU,I)),
      COLDS(JC,'MCP')*(COLDS(JC,'TOUT') - COLDS(JC,'TIN')));

QHUMIN(JHU) = SUM(I$STAGE(I),UTILHOT.L(JHU,I));
QCUMIN(JCU) = SUM(I$STAGE(I),UTILCOLD.L(JCU,I));
```

* Now the MILP for unit targeting is solved. The solution may be profitably
* used to obtain the MER networks (see examples below).

```
MODEL UNIT / ENBALH, ENBALHU, ENBALC, ENBALCU,
HUINT, HUAV, RESCON1, RESCON2, RESCON3, RESHUCON1,
RESHUCON2, RESHUCON3, UCON1, UCON2, UCON3, HUREQ, CUREQ,
OBJUNITS/;

SOLVE UNIT USING MIP MINIMIZING SUNITS;

THOT(JH,I) =
        MAX(TBH(I), HOTS(JH,'TOUT')) + RES.L(JH,I)/HOTS(JH,'MCP');
TCOLD(JC,I) = MIN(TBH(I), COLDS(JC,'TOUT'));

LOAD(JH,JC) = SUM(I$STAGE(I),Q.L(JH,JC,I));
LOADH(JHU,JC) = SUM(I$STAGE(I),QHU.L(JHU,JC,I));
LOADC(JH,JCU) = SUM(I$STAGE(I),QCU.L(JH,JCU,I));
DISPLAY THOT, TCOLD, LOAD, LOADH, LOADC;
```

* END OF GAMS LISTING FOR UTILITY/UNIT TARGETING

Example 10.1 Utility Targets by LP and MER Networks by MILP

a) Determine the utility as well as unit targets for Case Study 4S1t with $\Delta T_{min} = 25°C$.

b) Determine the utility as well as unit targets for Case Study 4S1t with $\Delta T_{min} = 25°C$ and the match between streams H2 and C4 forbidden.

c) Determine the utility targets for Case Study 4S1 with $\Delta T_{min} = 20°C$.

d) Determine the targets for Case Study 4S1 with $\Delta T_{min} = 20°C$ when the match between streams H2 and C3 is forbidden.

e) Find the optimum usage levels when four utilities as in Example 1.13 are used (for Case Study 4S1 with $\Delta T_{min} = 20°C$).

f) Determine the targets when four utilities as in Example 1.13 are used and the match between streams H2 and C3 is forbidden (for Case Study 4S1 with $\Delta T_{min} = 20°C$).

Determine the MER networks for all cases from the results of unit targeting.

Solution. The program HXUTIL.GMS may be used to solve the problem for the minimum utilities, the minimum number of exchanger units, the optimum mix for the multiple utilities, and the targets for restricted matches. To execute the program, type the following at the DOS prompt: gams hxutil.gms tabin 8. The results of the unit targeting MILP in the GAMS output file (HXUTIL.LST) may then be interpreted to generate the relevant MER networks.

In fact, the general command to run any of the GAMS programs on the accompanying disk is: gams filename.gms tabin 8.   The complete GAMS output then appears in the file with the .LST extension.

Note that the cost coefficients may be set to one (when cost data is not available) to obtain the minimum utility consumption directly, provided only one hot utility and one cold utility are considered.

a) Threshold Problem

The listing given above for HXUTIL.GMS may be directly used to solve this part of the problem to give the total cost of utilities as 1.896×10^5 $/yr, with the hot utility requirement at 1580 kW.   No cold utility is required, indicating a threshold problem. The minimum number of units as found by the MILP is 4. The utility and unit targets agree with those obtained using PT in Table 2.11. In fact, the match loads are displayed in the HXUTIL.LST file as LOAD(1,2) = 180, LOAD(1,1) = 1120, LOAD(2,2) = 900, and LOADH(1,1) = 1580. These may be used to synthesize the MER network in Figure 10.3.

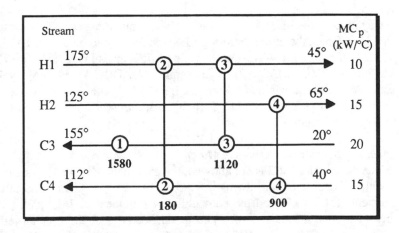

Figure 10.3 The above network for a threshold problem may be synthesized from the results of the unit targeting MILP since it directly provides the matches with their heat loads. (Temperature in °C and load in kW)

b) Threshold Problem with Forbidden Match

In the listing given above for HXUTIL.GMS, MHC(2,2) is set to 0 to forbid the match between streams H2 and C4 (Stream C4 is cold stream 2 in the GAMS listing). The utility and unit targets remain unchanged at 1580 kW and 4 respectively. The match loads from the unit targeting MILP correspond to those in Figure 3.9 (obtained earlier by the fast matching algorithm).

c) Pinched Problem

The following changes are required in the listing given above to run HXUTIL.GMS for Case Study 4S1 at ΔT_{min} = 20°C.

   (i)   Change the stream data to that for Case Study 4S1 (i.e., MCP for stream H2 is 40). It is not necessary to change the heat transfer coefficients because they are irrelevant for utility and unit targeting.

   (ii) Set DTM to 20.

   (iii) Also, TBH(4) = 60, TBH(5) = 40, and TBH(6) = 35.

Then, HXUTIL.GMS yields the total cost of utilities as 7.785 x 10^4 $/yr, with the hot utility requirement at 605 kW and the cold utility requirement at 525 kW. The utility and operating cost targets agree with those obtained using PT in Examples 1.1 and 1.9. As the parameter RESTOT(3) is zero, the pinch is at the third temperature boundary, i.e., TBH(3) = 125°C.

To synthesize the MER network for the above-pinch region, NR is set to 3. The utility targeting LP gives the cost to be 7.26 x 10^4 $/yr for 605 kW of hot utility. The unit targeting MILP gives LOAD(1,1) = 500, LOADH(1,1) = 500, and LOADH(1,2) = 105, which corresponds to the above-pinch design in Figure 4.9.

To synthesize the MER network for the below-pinch region, NL is set to 3 and NR to 6. Then, the utility targeting LP gives the cost to be 5.25 x 10^3 $/yr for 525 kW of cold utility. The unit targeting MILP gives LOAD(2,1) = 1700, LOAD(2,2) = 700, LOAD(1,2) = 275 and LOADC(1,1) = 525, which corresponds to the below-pinch design in Figure 4.9. Note that splits need to be ascertained through the variable Q(JH,JC,I) and unnecessary splits may be eliminated by inspection.

d) Pinched Problem with Forbidden Match

Since the match H2-C3 is not allowed, the parameter MHC(2,1) is set to 0. This yields the cost of utilities as 1.9485 x 10^5 $/yr, with the hot utility requirement at 1505 kW and the cold utility requirement at 1425 kW. The number of units obtained is five. The match loads from the unit targeting MILP are: LOAD(1,1) = 1195, LOAD(1,2) = 105, LOAD(2,2) = 975, LOADH(1,1) = 1505, and LOADC(2,1) = 1425. This corresponds to the network in Figure 4.31.

e) Multiple Utilities Problem

Besides adding the utility data, the following changes are needed:

(i) Change the set definitions for JHU and JCU to have two elements (i.e., make /1*1/ to /1*2/).

(ii) Give the cost of the two new utilities through the parameters CSTS and CSTW. It may be noted that priority indices can be given, if costs are not known [i.e., for this example, CSTS('1') = 2, CSTS('2') = 1, CSTW('1') = 2, CSTW('2') = 1].

(iii) Give the forbidden match data for new utilities (i.e., add a row of ones to Table MSC and a column of ones to Table MHW).

(iv) Add the boundary temperatures for the new utilities (i.e., add 135°C and 120°C to TBH) and modify the definition of the boundary temperature set I (i.e., make /1*7/ to /1*9/) .

(v) Change the value of parameter NR (i.e., change 6 to 8).

Now, as per HXUTIL.LST, the utility loads are 400 kW for HU1, 205 kW for HU2, 450 kW for CU1, and 75 kW for CU2. These values agree with the results of Example 1.13. Based on the zeroes in the parameter RESTOT, the pinches are at the third, fourth, and fifth temperature boundaries, i.e., at TBH of 135°, 125°, and 120°C. Now, each of the four subnetworks may be designed by appropriately setting the NL and NR parameters. The subnetworks are shown in Figures 10.4a and 10.4b.

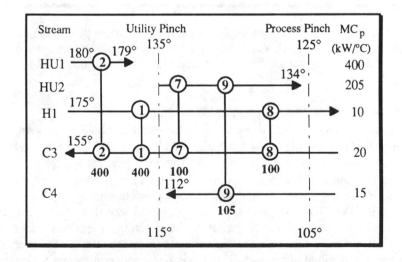

Figure 10.4a An MER network for a multiple utility - multiple pinch case may be designed using the HXUTIL.GMS program. The two subnetworks for the region above the process pinch are shown. (Temperature in °C and load in kW)

The subnetwork above the 135°C (hot) pinch is identical to that in Figure 3.4a, and the subnetwork below the 120°C (hot) pinch is identical to that in Figure 3.4b. However, the remaining two subnetworks (between two pinches) are different from those shown earlier in Figure 3.4.

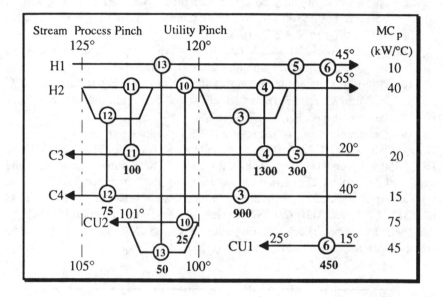

Figure 10.4b An MER network for a multiple utility - multiple pinch case may be designed using the HXUTIL.GMS program. The two subnetworks for the region below the process pinch are shown. (Temperature in °C and load in kW)

f) Multiple Utilities Problem with Forbidden Match

Since the match H2-C3 is not allowed, the parameter MHC(2,1) is set to 0. From HXUTIL.LST, the utility loads are 400 kW for HU1, 1105 kW for HU2, 1300 kW for CU1, and 125 kW for CU2. There is a pinch at a TBH of 135°C. Figure 10.5 shows the design of the two subnetworks. It must be appreciated that such a design is not easy to obtain through pinch technology.

Thus, HXUTIL.GMS provides a very powerful tool that allows handling of a variety of cases. Not only does it do the targeting, but it simultaneously generates networks for achieving maximum energy recovery. However, the process of generation of networks is not interactive and, consequently, does not provide insights to the designer. Also, the entire range of possible designs for a particular case (e.g., see CHENs in Section 4.5) may not be obvious.

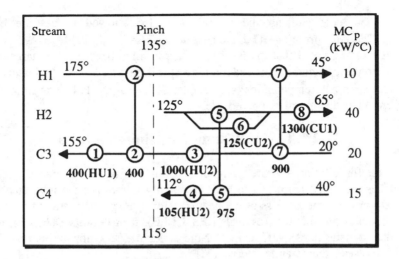

Figure 10.5 An MER network may be designed with relative ease using the HXUTIL.GMS program for complex cases involving multiple utilities as well as forbidden matches. (Temperature in °C and load in kW)

10.2 AREA TARGETING

The energy targeting LP from Section 10.1 may be extended to obtain a nonlinear programming model for area targeting (Colberg and Morari, 1990). The NLP model can find the true minimum area requirement for a fixed utility level and for unequal heat transfer coefficients as well as constrained matches. In their formulation, Colberg and Morari (1990) allow point utilities by introducing new sets. To keep the formulation in this section simple, a 1° approximation is used for such streams and utilities. This introduces a negligible error (less than 0.1%) in the area target for most problems.

10.2.1 The Transshipment NLP Model

The area targeting NLP differs from the energy targeting LP in the following two ways:

- The TIs, from the utility targeting LP, are further divided into enthalpy intervals (EIs) based on the target temperatures of the streams and utilities. This ensures that the composite curves will be reproduced by appropriately distributing the loads within the EIs; consequently, the area target based on composite curves will be recovered when the heat transfer coefficients are equal.

- The cold streams and utilities are also allowed to cascade heat "deficits" from one TI to the next hotter TI.

It must be noted that the NLP in principle minimizes the area target for specified utility requirements and not for a specified ΔT_{min}. The NLP is based on dual approach temperatures (see Chapter 5) in that the HRAT is fixed and the EMAT is the optimization variable.

10.2.2 GAMS Listing for Minimum Area

The formulation for area targeting using the nonlinear transshipment model is presented below. The NLP essentially minimizes the total area for process-process and process-utility heat transfer in a spaghetti structure where the variables are the heat residuals RES(JH,I,E), heat deficits D(JC,I,E), and the heat transfer rates Q(JH,JC,I,E). Note E denotes an enthalpy interval (EI) and I denotes a temperature interval (TI).

The GAMS listing that follows is specific to Case Study 4S1 for utility requirements of 605 kW (hot utility) and 525 kW (cold utility), which correspond to a ΔT_{min} of 20°C; however, it is easily modified to solve problems with different stream/utility data.

```
$TITLE AREA TARGETING NLP
* This program determines the target for the minimum network area when
* utilities are specified. It can be used for problems with multiple utilities,
* unequal heat transfer coefficients, and forbidden matches.
* Program Name: HXAREA.GMS
* Program developed by Yogesh Makwana and Uday V. Shenoy
$onnestcom inlinecom // // eolcom ##
```

// As in energy targeting, sets are defined which form domains for the variables. The hot streams and utilities are combined into one set. Likewise, the cold streams and utilities form one set. The term "stream" now represents both the streams and utilities in the original problem. The problem is again divided into TIs with temperature boundaries given by the supply temperatures of streams. Each TI is further divided into several enthalpy intervals (EIs) depending on the target temperatures of streams.

The area targeting NLP optimizes the approach temperatures at the various TI and EI boundaries to get a minimum area. To avoid numerical difficulties, a lower bound of 0.1 is specified on the temperature difference and used to compute the initial TIs. //

```
SETS
JH  Hot streams and utilities  /1*3/
```

JC Cold streams and utilities /1*3/
I   Temperature boundaries + 1 /1*7/
E   Enthalpy intervals /1*4/
DATA /TIN,TOUT,MCP,H/;

TABLE HOTS(JH,DATA) Hot streams data

	TIN	TOUT	MCP	H
1	175.00	45.00	10.00	0.2000
2	125.00	65.00	40.00	0.2000
3	180.00	179.00	605.00	0.2000
;				

TABLE COLDS(JC,DATA) Cold streams data

	TIN	TOUT	MCP	H
1	20.00	155.00	20.00	0.2000
2	40.00	112.00	15.00	0.2000
3	15.00	25.00	52.50	0.2000
;				

SCALAR NOI No of temperature boundaries;
NOI = 6;

PARAMETERS
TBH(I)   Temperature at the Ith boundary
IMAX(I)   Number of EIs in TI I ;

TBH('1') = 180.00;   TBH('4') =   40.10;
TBH('2') = 175.00;   TBH('5') =   20.10;
TBH('3') = 125.00;   TBH('6') =   15.10;

IMAX('1') = 0;   IMAX('5') = 2;
IMAX('2') = 2;   IMAX('6') = 1;
IMAX('3') = 2;   IMAX('7') = 1;
IMAX('4') = 4;

* For forbidden matches, the same convention as in energy targeting is used.

TABLE MHC(JH,JC) Forbidden match data

	1	2	3
1	1	1	1
2	1	1	1
3	1	1	0
;			

* Alias declares an alternate name for a set
ALIAS (I,NOI1)
       (E,EOI1);

PARAMETERS
MINDTM	Lower bound on temperature difference
TBC(I)	Boundary temperatures at the cold ends
H(JH,I)	Existence coefficient for hot stream JH in interval I
C(JC,I)	Existence coefficient for cold stream JC in interval I
AVHOT(JH,I)	Heat available from hot stream JH in interval I
AVCOLD(JC,I)	Heat required by cold stream JC in interval I;

MINDTM = 0.1;
TBC(I)$(ORD(I) LE NOI) = TBH(I) - MINDTM;

* The existence coefficients are calculated as in energy targeting.
* The only difference lies in the fact that cold streams can also cascade heat.

H(JH,I)$(TBH(I) LT HOTS(JH,'TIN')) = 1;
C(JC,I)$(TBC(I-1) GT COLDS(JC,'TIN')) = 1;

* The heat available in each interval from the hot streams (AVHOT) and the
* heat required by the cold streams (AVCOLD) are calculated as given below.

AVHOT(JH,I)$(H(JH,I) AND TBH(I-1) GT HOTS(JH,'TOUT')) =
(TBH(I-1) - MAX(TBH(I),HOTS(JH,'TOUT')))*HOTS(JH,'MCP');

AVCOLD(JC,I)$(C(JC,I) AND TBC(I) LT COLDS(JC,'TOUT')) =
(MIN(TBC(I-1),COLDS(JC,'TOUT'))- TBC(I))*COLDS(JC,'MCP');

POSITIVE VARIABLES
Q(JH,JC,I,E)	Heat transfer by hot stream JH to cold stream JC in EI E of TI I
RES(JH,I,E)	Heat residual of hot stream JH in EI E of TI I
D(JC,I,E)	Heat deficit of cold stream JC in EI E of TI I
THOT(JH,I,E)	Temp of hot stream JH at colder boundary of EI E in TI I
TCOLD(JC,I,E)	Temp of cold stream JC at colder boundary of EI E in TI I
THOTH(JH,I,E)	Temp of hot stream JH at hotter boundary of EI E in TI I
TCOLDH(JC,I,E)	Temp of cold stream JC at hotter boundary of EI E in TI I;

* The free variable to be minimized is declared below.
VARIABLES
SUMAREA   Sum of area requirements to be minimized;

* The equations (constraints and objective function) are declared below.

```
EQUATIONS
THEQ(JH,I,E)
TCEQ1(JC,I,E)
TCEQ2(JC,E)
TCEQ3(JC,I)
THHEQ1(JH,I,E)
THHEQ2(JH,I)
THHEQ3(JH)
TCHEQ1(JC,I,E)
TCHEQ2(JC,I)
TCHEQ3(JC)
ENBALH1(JH,I)
ENBALH2(JH,I,E)
ENBALC1(JC,I)
ENBALC2(JC,I,E)
RESCON1(JH,E)
RESCON2(JH)
RESCON3(JH,I,E)
DCON1(JC)
DCON2(JC,E)
DCON3(JC,I,E)
DTCON1(JH,JC,I,E)
DTCON2(JH,JC,E)
OBJ        Objective function;
```

* The constraints and the objective function are given below.

// The temperatures at which hot streams and cold streams cross the EI boundaries are computed first. This is done for both hot and cold boundaries of each EI. Since the hot as well as the cold streams can cascade heat, the temperature of a hot stream at a TI boundary must be greater than the boundary temperature (TBH) itself and that of a cold stream must be less than the boundary temperature at the lower end (TBC). These computations are done in the TH, TC, THH, and TCH equations given below. //

```
THEQ(JH,I,E)$((ORD(I) GT 1) AND (ORD(I) LE NOI)
              AND (ORD(E) LE IMAX(I)))..
THOT(JH,I,E) =E=
MAX(TBH(I),HOTS(JH,'TOUT')) + (RES(JH,I,E))/(HOTS(JH,'MCP'));
```

TCEQ1(JC,I,E)$((ORD(I) GT 2) AND (ORD(I) LE NOI)
                    AND (ORD(E) LT IMAX(I)))..
TCOLD(JC,I,E) =E=
MIN(TBC(I-1),COLDS(JC,'TOUT')) - D(JC,I,E+1)/COLDS(JC,'MCP');

TCEQ2(JC,E)$((ORD(E) LT IMAX('2')))..
TCOLD(JC,'2',E) =E=
COLDS(JC,'TOUT') - D(JC,'2',E+1)/COLDS(JC,'MCP');

TCEQ3(JC,I)$((ORD(I) GT 1) AND (ORD(I) LE NOI))..
SUM(EOI1$(ORD(EOI1) EQ IMAX(I)),TCOLD(JC,I,EOI1)) =E=
MIN((TBH(I)-MINDTM),COLDS(JC,'TOUT')) -
                    D(JC,I+1,'1')/COLDS(JC,'MCP');

THHEQ1(JH,I,E)$((ORD(E) GT 1) AND (ORD(E) LE IMAX(I))
                    AND (ORD(I) GT 1) AND (ORD(I) LE NOI))..
THOTH(JH,I,E) =E= THOT(JH,I,E-1);

THHEQ2(JH,I)$((ORD(I) GT 2) AND (ORD(I) LE NOI))..
THOTH(JH,I,'1') =E=
SUM(EOI1$(ORD(EOI1) EQ IMAX(I-1)),THOT(JH,I-1,EOI1));

THHEQ3(JH)$(H(JH,'2') EQ 1)..
THOTH(JH,'2','1') =E= HOTS(JH,'TIN');

TCHEQ1(JC,I,E)$((ORD(E) GT 1) AND (ORD(E) LE IMAX(I))
                    AND (ORD(I) GT 1) AND (ORD(I) LE NOI))..
TCOLDH(JC,I,E) =E= TCOLD(JC,I,E-1);

TCHEQ2(JC,I)$((ORD(I) GT 2) AND (ORD(I) LE NOI))..
TCOLDH(JC,I,'1') =E=
SUM(EOI1$(ORD(EOI1) EQ IMAX(I-1)),TCOLD(JC,I-1,EOI1));

TCHEQ3(JC)$C(JC,'2')..
TCOLDH(JC,'2','1') =E= COLDS(JC,'TOUT');

// The heat balances in each EI on each stream are now written. All the heat available from a hot stream is placed in the hottest EI in a TI (i.e., E = 1). Similarly, all the heat deficit of a cold stream in a TI is placed in the coldest EI of the TI [i.e. E = IMAX(I)]. Therefore, the heat balance for a hot stream in the hottest EI of TI will have four components, namely the heat available, the residue from the previous EI, the heat transfer to the cold streams, and the residue to the next EI. All other EIs of the TI will not have the heat available term. The heat balances on cold streams can also be written similarly. //

ENBALH1(JH,I)$(ORD(I) GT 1 AND H(JH,I) AND (ORD(I) LE NOI))..
    AVHOT(JH,I)
    + SUM(EOI1$(ORD(EOI1) EQ IMAX(I-1)),RES(JH,I-1,EOI1)) =E=
    SUM(JC$(MHC(JH,JC) AND C(JC,I)),Q(JH,JC,I,'1')) + RES(JH,I,'1') ;

ENBALH2(JH,I,E)$((ORD(E) GT 1) AND (ORD(E) LE IMAX(I))
        AND (ORD(I) GT 1) AND H(JH,I) AND (ORD(I) LE NOI))..
    RES(JH,I,E-1) =E=
    SUM(JC$(MHC(JH,JC) AND C(JC,I)),Q(JH,JC,I,E)) + RES(JH,I,E) ;

ENBALC1(JC,I)$((ORD(I) GT 1) AND C(JC,I) AND (ORD(I) LE NOI))..
    AVCOLD(JC,I) + D(JC,I+1,'1') =E=
    SUM(JH$(MHC(JH,JC) AND H(JH,I)),
    (SUM(EOI1$(ORD(EOI1) EQ IMAX(I)),Q(JH,JC,I,EOI1))))
    + SUM(EOI1$(ORD(EOI1) EQ IMAX(I)),D(JC,I,EOI1));

ENBALC2(JC,I,E)$((ORD(E) LT IMAX(I)) AND (ORD(I) GT 1)
        AND C(JC,I) AND (ORD(I) LE NOI))..
    D(JC,I,E+1) =E=
    SUM(JH$(MHC(JH,JC) AND H(JH,I)),Q(JH,JC,I,E)) + D(JC,I,E);

// The residues from the ends of the cascades are set to zero next. Also, the residues from intervals where a stream does not exist are set to zero. //

RESCON1(JH,E)$(ORD(E) LE IMAX('1'))..
      RES(JH,'1',E) =E= 0;
RESCON2(JH)..
      SUM(NOI1$(ORD(NOI1) EQ NOI),
      SUM(EOI1$(ORD(EOI1) EQ IMAX(NOI1)),
      RES(JH,NOI1,EOI1))) =E= 0;
RESCON3(JH,I,E)$(H(JH,I) EQ 0)..
      RES(JH,I,E) =E= 0;

DCON1(JC)..
      D(JC,'2','1') =E= 0;
DCON2(JC,E)..
      SUM(NOI1$(ORD(NOI1) EQ (NOI+1)),D(JC,NOI1,E)) =E= 0;
DCON3(JC,I,E)$(C(JC,I) EQ 0)..
      D(JC,I,E) =E= 0;

// Constraints are placed on the temperatures at which the streams cross the EI boundaries so as to ensure feasible heat transfer. DTCON1 is for TIs other than the first TI, and DTCON2 is for the first TI. //

DTCON1(JH,JC,I,E)$((ORD(E) LT IMAX(I)) AND (ORD(I) GT 2) AND H(JH,I) AND C(JC,I) AND MHC(JH,JC) AND (ORD(I) LE NOI))..
  MAX(TBH(I),HOTS(JH,'TOUT')) + RES(JH,I,E)/HOTS(JH,'MCP') =G=
  MIN(TBC(I-1),COLDS(JC,'TOUT')) - D(JC,I,E+1)/COLDS(JC,'MCP')
  + MINDTM;

DTCON2(JH,JC,E)$((ORD(E) LT IMAX('2')) AND H(JH,'2') AND C(JC,'2') AND MHC(JH,JC))..
  MAX(TBH('2'),HOTS(JH,'TOUT')) + RES(JH,'2',E)/HOTS(JH,'MCP') =G=
  COLDS(JC,'TOUT') - D(JC,'2',E+1)/COLDS(JC,'MCP') + MINDTM;

// The objective function is the sum of heat transfer areas required for exchange of Q(JH,JC,I,E) heat between hot stream JH and cold stream JC. The conventionally used logarithmic mean temperature difference is approximated by the sum of two-thirds of the geometric mean and one-third of the arithmetic mean (Paterson, 1984). //

OBJ..
  SUMAREA =E= SUM(JH,SUM(JC$MHC(JH,JC),
  (1/HOTS(JH,'H') + 1/COLDS(JC,'H')) *
  (SUM(I$((ORD(I) GT 1) AND (ORD(I) LE NOI)
  AND H(JH,I) AND C(JC,I)),
  SUM(E$(ORD(E) LE IMAX(I)), (Q(JH,JC,I,E)/
  ((2./3.)*SQRT((THOTH(JH,I,E) - TCOLDH(JC,I,E))
  *(THOT(JH,I,E) - TCOLD(JC,I,E))+1E-4)
  + (1./3.)*0.5*(THOTH(JH,I,E) - TCOLDH(JC,I,E)
  + THOT(JH,I,E) - TCOLD(JC,I,E)+1E-8)))))))));

MODEL AREA /ALL/;

// Upper bounds on RES, D, and Q can be placed as the total heat content of the respective streams. //

RES.UP(JH,I,E) = HOTS(JH,'MCP')*(HOTS(JH,'TIN') - HOTS(JH,'TOUT'));
D.UP(JC,I,E) = COLDS(JC,'MCP')*(COLDS(JC,'TOUT')-COLDS(JC,'TIN'));
Q.UP(JH,JC,I,E) =
          MIN((HOTS(JH,'MCP')*(HOTS(JH,'TIN') - HOTS(JH,'TOUT'))),
          (COLDS(JC,'MCP')*(COLDS(JC,'TOUT') - COLDS(JC,'TIN'))));

SOLVE AREA USING DNLP MINIMIZING SUMAREA;

* END OF GAMS LISTING FOR AREA TARGETING

Example 10.2 Area Targets by NLP

Determine the area targets for the following case studies:

a) Case Study 4S1 with $\Delta T_{min} = 20°C$
b) Case Study 4S1 with $\Delta T_{min} = 13°C$
c) Case Study 4S1t with $\Delta T_{min} = 30°C$

Solution. The program HXAREA.GMS may be used to solve the problem for the minimum countercurrent area.

(a) The listing given above for HXAREA.GMS will directly solve this part of the problem. Noting that the energy targets are 605 kW (hot utility) and 525 kW (cold utility) for $\Delta T_{min} = 20°C$, the area target is obtained as 1312.49 m^2. The result agrees with the area of 1312.57 m^2 obtained by PT in Example 1.4 since the heat transfer coefficients are uniform. The small difference in the areas obtained is due to the Paterson (1984) approximation used in MP (instead of the true LMTD used in PT).

(b) To find the minimum area requirement at ΔT_{min} of 13°C, it is required to change the MC_ps of the utilities in the listing. The MC_p for the hot and cold utility become 360 and 28 respectively (corresponding to $Q_{hu,min} = 360$ kW and $Q_{cu,min} = 280$). With only this change, HXAREA.GMS gives the area target as 1639.76 m^2. This agrees with the value of 1640.06 m^2 from PT (see Table 1.15), with the slight difference being due to the Paterson approximation used in MP.

(c) For Case Study 4S1t at $\Delta T_{min} = 30°C$, the following changes must be made in HXAREA.GMS.
   (i) Change the stream and utility data to that of Case Study 4S1t. The only differences between Case Studies 4S1 and 4S1t are in the heat transfer coefficients and in the MC_p of one hot stream (H2).
   (ii) Energy targets at ΔT_{min} of 30°C are $Q_{hu,min} = 1630$ kW and $Q_{cu,min} = 50$ kW. This gives the MC_ps of hot and cold utilities as 1630 and 5.
HXAREA.GMS gives the area target as 416.51 m^2. The power and accuracy of MP techniques is highlighted in this case where the heat transfer coefficients are unequal. The area estimated from the composite-curve-based uniform BATH formula is 466.43 m^2 (an overestimation of 12%) and that from the diverse BATH formula is 463.8 m^2 (an overestimation of 11.4%).

10.3 OPTIMAL HEN THROUGH SUPERSTRUCTURE

In Section 10.1, the minimum utility requirements are determined for a fixed HRAT by a transshipment LP model. The minimum number of units is then determined for a fixed utility consumption by an MILP model. These formulations are especially useful in the case of forbidden matches. In Section 10.2, the minimum network area is obtained for specified utilities. This is specially useful for unequal heat transfer coefficients and constrained matches. However, this decomposition into three separate targets for utilities, units, and area does not ensure that the total cost is minimized. Ideally, the factors affecting operating cost (usage of different utilities) and capital cost (number of units, heat transfer area, exchanger types, and materials of construction) must be considered simultaneously.

In this section, an MINLP formulation for the synthesis of an optimal HEN in terms of its total cost is presented based on the work of Yee and Grossmann (1990). The three-way tradeoff between energy, units, and area is considered by simultaneous optimization of these quantities. Owing to the high nonlinearity of the objective function and lack of robustness in MINLP solvers, it is difficult to obtain the global optimum with arbitrary initial guesses. Yee and Grossmann (1990) have suggested a guess generation procedure which solves a relaxed linearization of the MINLP. However, it is also possible to get the optimal solution by taking several runs with different bounds on the utility requirements as demonstrated below.

This formulation, which does not recognize the pinch division nor fixed approach temperatures (HRAT or EMAT), offers the following advantages over supertargeting using PT:

- A minimum-cost HEN structure is actually synthesized where utilities, units, and area are simultaneously optimized.
- Capital costs are calculated using the actual areas of the exchangers.
- It is possible to handle problems with multiple utilities, with different cost laws, with constraints on forbidden/preferred matches and with variable inlet/outlet temperatures.

10.3.1 The Simultaneous Optimization MINLP Model

The formulation is based on a network superstructure (Figure 10.6) which consists of a number of stages and allows for every potential match within each stage by the splitting of streams. The superstructure is simple enough to keep the large nonlinear combinatorial HENS problem tractable with available solution techniques. At the same time, it is powerful enough to be able to capture the general features of most networks.

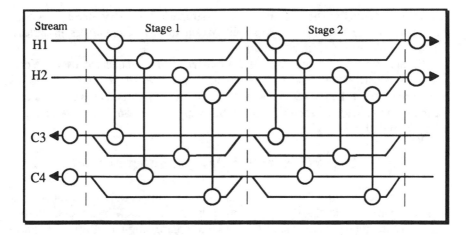

Figure 10.6 The superstructure comprises stages, within each of which heat exchanges occur between every hot stream and cold stream. For simplicity, heaters and coolers are placed at the ends (outlets) of the streams and isothermal mixing junctions are assumed.

The superstructure is conceptually similar to a spaghetti design. However, it differs in two important ways:

- The number of stages is typically much smaller than the number of enthalpy intervals. (In spaghetti designs, the number of stages and enthalpy intervals are necessarily equal.) The number of stages recommended by Yee and Grossmann (1990) is the maximum of the number of hot or cold streams. Although choosing a larger number of stages will allow for more combinations of stream matches, this number is usually sufficient for an optimal design.

- There is opportunity for crisscross heat transfer when streams have differing heat transfer coefficients. (Spaghetti designs are based on true countercurrency and vertical heat transfer.) The outlet temperatures of every stage are considered as variables for optimization.

A two-stage superstructure for a problem involving two hot and two cold streams (e.g., Case Study 4S1) is shown in Figure 10.6. It allows for a large number of series-parallel matching possibilities and sequences. However, there are certain restrictions on the superstructure. An isothermal mixing junction is assumed at the outlets of the stages. This simplification allows elimination of the mixing junction equations and the nonlinear heat balances for the individual stream splits. This results in reduction of the problem dimensionality, linearity in the constraints, robustness in the model, and relative ease in the solution (Yee et al., 1990a). For further simplicity, utilities are located at the ends of the superstructure for every stream. Also, the superstructure does not allow a split stream to have exchangers in series nor does it allow for stream bypassing.

Binary variables accounting for the existence of matches are used. Yee and Grossmann (1990) have recommended the use of the Chen (1987) approximation for the *LMTD*. However, experience suggests that accurate area estimates are provided by the Paterson (1984) approximation; hence, it is used in the formulation below.

10.3.2 GAMS Listing for Optimal Superstructure

The formulation for HENS using the simultaneous optimization model of Yee and Grossmann (1990) is presented below. The MINLP essentially minimizes the total annual cost of the network, comprising the utility costs and the capital costs of the exchangers. This formulation (and the previous "Area NLP" formulation) are robust and efficient, primarily due to the nonlinearities, associated with the area computation, being solely in the objective function and not in the constraints.

The GAMS listing that follows is specific to Case Study 4S1 but may be readily modified to solve other problems.

```
$TITLE HENS BY SUPERSTRUCTURE MINLP
* This program synthesizes the optimal HEN in terms of minimum total annual
* cost through simultaneous optimization of utilities, units, and area.
* Program Name: HXSUPER.GMS
*Program developed by Shwetal Patel, Yogesh Makwana, and Uday V. Shenoy
$onnestcom inlinecom // // eolcom ##
```

// A superstructure, similar to the spaghetti design, is assumed with a possibility of all potential exchanges between the various hot and cold streams in each stage. The number of stages, however, is not equal to the number of enthalpy intervals on the composite curves. Instead, it is chosen as the max(number of hot streams, number of cold streams). Forbidden matches and preferred matches can be handled by imposing constraints on the heat exchanged between streams. //

```
SETS
 JH hot streams /1*2/
 JC cold streams /1*2/
 JHU hot utilities /1*1/
 JCU cold utilities /1*1/
 DATA /TIN,TOUT,MCP,H/;
```

```
        TABLE HOTS(JH,DATA)  Hot streams data
        TIN    TOUT    MCP    H
1       175.00  45.00  10.00  0.2000
2       125.00  65.00  40.00  0.2000
;

        TABLE HUTIL(JHU,DATA)  Hot utilities data
        TIN    TOUT    MCP    H
1       180.00 179.00  0.00   0.2000
;

        TABLE COLDS(JC,DATA)  Cold streams data
        TIN    TOUT    MCP    H
1       20.00 155.00  20.00   0.2000
2       40.00 112.00  15.00   0.2000
;

        TABLE CUTIL(JCU,DATA)  Cold utilities data
        TIN    TOUT    MCP    H
1       15.00  25.00   0.0    0.2000
;

   TABLE  MHC(JH,JC)  Forbidden match data for hot and cold streams
        1   2
1       1   1
2       1   1
;

   TABLE  MSC(JHU,JC) Forbidden matches for hot utilities and cold streams
        1   2
1       1   1
;

   TABLE MHW(JH,JCU) Forbidden matches for hot streams and cold utilities
        1
1       1
2       1
;
SCALAR
NOI  number of stages in superstructure / 2 /;

SET
I    temperature locations NOI + 1 /1*3/;

PARAMETERS
CSTS(JHU)       cost of heating utility
CSTW(JCU)       cost of cooling utility
COEFFA          fixed cost of a unit
```

COEFFB	area cost coefficient for exchangers
COEFFC	cost exponent for exchangers
AF	annualization factor
HULIMIT	upper limit on hot utility to be given by user
AVHOT(JH)	heat content of hot stream
AVCOLD(JC)	heat content of cold stream
MAXDT(JH,JC)	upper bound on temperature difference
STAGE(I)	stages in the superstructure
FTB(I)	first temperature boundary in the superstructure
LTB(I)	last temperature boundary in the superstructure
AREA(JH,JC,I)	area of exchanger between streams JH and JC in interval I
AREAH(JHU,JC)	area of heater for hot utility JHU on cold stream JC
AREAC(JH,JCU)	area of cooler for cold utility JCU on hot stream JH;

CSTS('1') = 120;   CSTW('1') = 10;   AF   = (1.1**5)/5.;
COEFFA = 30000;   COEFFB = 750;   COEFFC = 0.81;
HULIMIT = 400;

AVHOT(JH) = HOTS(JH,'MCP')*(HOTS(JH,'TIN') - HOTS(JH,'TOUT')) ;
AVCOLD(JC) = COLDS(JC,'MCP')*(COLDS(JC,'TOUT')-COLDS(JC,'TIN'));
MAXDT(JH,JC) = MAX(0,COLDS(JC,'TIN') - HOTS(JH,'TIN'),
              COLDS(JC,'TIN') - HOTS(JH,'TOUT'),
              COLDS(JC,'TOUT') - HOTS(JH,'TIN'),
              COLDS(JC,'TOUT') - HOTS(JH,'TOUT')) ;
STAGE(I)$(ORD(I) LT CARD(I)) = 1;
FTB(I)$(ORD(I) EQ 1)   = 1;
LTB(I)$(ORD(I) EQ CARD(I))   = 1;

POSITIVE VARIABLES

TH(JH,I)	temperature of hot stream JH as it enters stage I
TC(JC,I)	temperature of cold stream JC as it leaves stage I
Q(JH,JC,I)	energy exchanged between JH and JC in stage I
QC(JH,JCU)	energy exchanged between JH and the cold utility
QH(JHU,JC)	energy exchanged between the hot utility and JC
DTHC(JH,JC,I)	approach between JH and JC at location I
DTCU(JH,JCU)	approach between JH and the cold utility
DTHU(JHU,JC)	approach between the hot utility and JC;

BINARY VARIABLES

NHC(JH,JC,I)	is one if the match between JH and JC exists in stage I
NIW(JH,JCU)	is one if the match between JH and JCU exists
NSJ(JHU,JC)	is one if the match between JHU and JC exists;

VARIABLE
TAC   total annual cost;

EQUATIONS

ENBALH(JH)
ENBALC(JC)
EBHST(JH,I)
EBCST(JC,I)
UTILH(JC,I)
UTILC(JH,I)
TINH(JH,I)
TINC(JC,I)
CONSTH(JH,I)
CONSTC(JC,I)
CONSTHL(JH,I)
CONSTCF(JC,I)
BOUNDQ(JH,JC,I)
BOUNDQH(JHU,JC)
BOUNDQC(JH,JCU)
DTHEQN(JH,JC,I)
DTCEQN(JH,JC,I)
DTHUEQN(JHU,JC,I)
DTCUEQN(JH,JCU,I)
FORBHC(JH,JC,I)
FORBHW(JH,JCU)
FORBSC(JHU,JC)
ENCON1
OBJ   Objective function for total annual cost;

// Overall energy balances for hot and cold streams are given below. //

ENBALH(JH)..
        (HOTS(JH,'TIN')-HOTS(JH,'TOUT'))*HOTS(JH,'MCP') =E=
        SUM((JC,I)$STAGE(I), Q(JH,JC,I)) +SUM(JCU,QC(JH,JCU));

ENBALC(JC)..
        (COLDS(JC,'TOUT')-COLDS(JC,'TIN'))*COLDS(JC,'MCP') =E=
        SUM((JH,I)$STAGE(I), Q(JH,JC,I)) +SUM(JHU,QH(JHU,JC));

// Energy balances for hot and cold streams in each stage are given next. //

EBHST(JH,I)$STAGE(I)..
        HOTS(JH,'MCP')*(TH(JH,I) - TH(JH,I+1)) =E= SUM(JC, Q(JH,JC,I));

EBCST(JC,I)$STAGE(I)..
        COLDS(JC,'MCP')*(TC(JC,I) - TC(JC,I+1)) =E= SUM(JH,Q(JH,JC,I));

// Utilities are assumed to be placed outside the superstructure. Hence, the utility load for each stream is determined based on its temperature at the outlet of the superstructure and its target temperature, as given by these equations. //

UTILH(JC,I)$FTB(I)..
        COLDS(JC,'MCP')*(COLDS(JC,'TOUT') - TC(JC,I)) =E=
        SUM(JHU,QH(JHU,JC));

UTILC(JH,I)$LTB(I)..
         HOTS(JH,'MCP')*(TH(JH,I) - HOTS(JH,'TOUT')) =E=
        SUM(JCU, QC(JH,JCU));

// The temperature of a hot stream at the first temperature boundary is equal to its inlet temperature. For a cold stream, the temperature at the last temperature boundary is equal to its inlet temperature. These are assigned below. //

TINH(JH,I)$FTB(I)..
        HOTS(JH,'TIN') =E= TH(JH,I);

TINC(JC,I)$LTB(I)..
        COLDS(JC,'TIN') =E= TC(JC,I);

// The following constraints ensure that temperatures continuously decrease from the first to the last temperature boundary. //

CONSTH(JH,I)$STAGE(I)..
        TH(JH,I) =G= TH(JH,I+1);
CONSTC(JC,I)$STAGE(I)..
        TC(JC,I) =G= TC(JC,I+1);
CONSTHL(JH,I)$LTB(I)..
        TH(JH,I) =G= HOTS(JH,'TOUT');
CONSTCF(JC,I)$FTB(I)..
        COLDS(JC,'TOUT') =G= TC(JC,I);

// The match existence indices are used to put bounds on the maximum possible heat exchange between two streams (or a stream and a utility). The upper bound is simply the smaller of the heat contents of the two streams involved in the match. //

BOUNDQ(JH,JC,I)$STAGE(I)..
        Q(JH,JC,I)-MIN(AVHOT(JH), AVCOLD(JC))*NHC(JH,JC,I) =L= 0;
BOUNDQC(JH,JCU)..
        QC(JH,JCU) - AVHOT(JH)*NIW(JH,JCU) =L= 0 ;
BOUNDQH(JHU,JC)..
        QH(JHU,JC) - AVCOLD(JC)*NSJ(JHU,JC) =L= 0 ;

// Since the superstructure assumes isothermal mixing at the stage outlets, the area calculations can be done using the temperature differences at the stage boundaries. When the heat load on a match is zero (i.e., it does not exist), the area requirement of the match is zero irrespective of the temperature differences. However, it is possible to get negative temperature differences for a match of zero load when the hot stream temperature is less than the cold stream temperature. To avoid numerical errors, the calculation of temperature differences is done with the binary variables for match existence. The following equations appropriately set the temperature differences when the match exists. //

DTHEQN(JH,JC,I)$STAGE(I)..
        DTHC(JH,JC,I) =L=
        TH(JH,I) - TC(JC,I) + MAXDT(JH,JC)*(1 - NHC(JH,JC,I));
DTCEQN(JH,JC,I)$STAGE(I)..
        DTHC(JH,JC,I+1) =L=
        TH(JH,I+1) - TC(JC,I+1) + MAXDT(JH,JC)*(1 - NHC(JH,JC,I));
DTHUEQN(JHU,JC,I)$FTB(I)..
        DTHU(JHU,JC) =L= (HUTIL(JHU,'TOUT') - TC(JC,I));
DTCUEQN(JH,JCU,I)$LTB(I)..
        DTCU(JH,JCU) =L= TH(JH,I) - CUTIL(JCU,'TOUT');

* If a match is forbidden, then its load must be set to zero.

FORBHC(JH,JC,I)$(MHC(JH,JC) EQ 0)..
        Q(JH,JC,I) =E= 0;
FORBHW(JH,JCU)$(MHW(JH,JCU) EQ 0)..
        QC(JH,JCU) =E= 0;
FORBSC(JHU,JC)$(MSC(JHU,JC) EQ 0)..
        QH(JHU,JC) =E= 0;

// To obtain the global optimum of the MINLP is not easy. One way to circumvent this difficulty is to solve the MINLP many times with different bounds on the utility requirements. //

ENCON1..
  SUM((JHU,JC),QH(JHU,JC)) =L= HULIMIT;

// The objective function to be minimized is the total annual cost target for the network. The Paterson (1984) approximation is used for LMTD. //

OBJ..
  TAC =E=   AF*(
    COEFFA*(SUM((JH,JC,I)$STAGE(I),NHC(JH,JC,I)) +
    SUM((JH,JCU),NIW(JH,JCU)) + SUM((JHU,JC),NSJ(JHU,JC))) +

    COEFFB*SUM((JH,JC,I)$STAGE(I),(Q(JH,JC,I)
    *(1/HOTS(JH,'H')+1/COLDS(JC,'H'))/
    ((2./3.)*(DTHC(JH,JC,I)*DTHC(JH,JC,I+1))**(0.5) +
    (DTHC(JH,JC,I) + DTHC(JH,JC,I+1))/6 +
    1E-6) + 1E-6)**COEFFC) +

    COEFFB*(SUM((JHU,JC),(QH(JHU,JC)
    *(1/COLDS(JC,'H')+1/HUTIL(JHU,'H')))/
((2./3.)*((HUTIL(JHU,'TIN')-COLDS(JC,'TOUT'))*DTHU(JHU,JC))**(0.5) +
    (HUTIL(JHU,'TIN')-COLDS(JC,'TOUT')+DTHU(JHU,JC))/6 +
    1E-6) + 1E-6)**COEFFC) +

    COEFFB*SUM((JH,JCU),(QC(JH,JCU)
    *(1/HOTS(JH,'H')+1/CUTIL(JCU,'H'))/
((2./3.)*((HOTS(JH,'TOUT')-CUTIL(JCU,'TIN'))*DTCU(JH,JCU))**(0.5) +
    (HOTS(JH,'TOUT')-CUTIL(JCU,'TIN')+DTCU(JH,JCU))/6 +
    1E-6) + 1E-6)**COEFFC)) +

    SUM((JHU,JC),QH(JHU,JC)*CSTS(JHU))+
    SUM((JH,JCU),QC(JH,JCU)*CSTW(JCU)) ;

MODEL AUTIL /ALL/ ;

* Upper/lower bounds and initialization

TH.UP(JH,I) = HOTS(JH,'TIN');
TH.LO(JH,I) = HOTS(JH,'TOUT');

```
TC.UP(JC,I) = COLDS(JC,'TOUT');
TC.LO(JC,I) = COLDS(JC,'TIN');
TH.L(JH,I) = HOTS(JH,'TIN');
TC.L(JC,I) = COLDS(JC,'TIN');
```

* Resetting some GAMS options

```
OPTION ITERLIM = 5000;
OPTION DOMLIM = 10;
```

```
SOLVE AUTIL USING MINLP MINIMIZING TAC;
```

* Calculating areas for units in superstructure

```
AREA(JH,JC,I)$STAGE(I) =
Q.L(JH,JC,I)*(1/HOTS(JH,'H')+1/COLDS(JC,'H'))/( (2./3.)*
(DTHC.L(JH,JC,I)*DTHC.L(JH,JC,I+1))**(0.5) +
(DTHC.L(JH,JC,I) + DTHC.L(JH,JC,I+1))/6 + 1E-6 ) + 1E-6;
```

```
AREAH(JHU,JC) =
QH.L(JHU,JC)*(1/COLDS(JC,'H')+1/HUTIL(JHU,'H'))/( (2./3.)*
((HUTIL(JHU,'TIN')-COLDS(JC,'TOUT'))*DTHU.L(JHU,JC))**(0.5) +
(HUTIL(JHU,'TIN')-COLDS(JC,'TOUT')+DTHU.L(JHU,JC))/6+1E-6 )+1E-6;
```

```
AREAC(JH,JCU) =
QC.L(JH,JCU)*(1/HOTS(JH,'H')+1/CUTIL(JCU,'H'))/( (2./3.)*
((HOTS(JH,'TOUT')-CUTIL(JCU,'TIN'))*DTCU.L(JH,JCU))**(0.5)+
(HOTS(JH,'TOUT')-CUTIL(JCU,'TIN')+DTCU.L(JH,JCU))/6 + 1E-6 ) + 1E-6;
```

```
DISPLAY AREA, AREAH, AREAC;
```

// It is possible to use the same formulation to obtain a minimum area target for the utility requirement obtained above. This is done by effectively replacing the total cost in the objective function by the area expression as shown below.//

```
CSTS('1') = 0;          CSTW('1') = 0;          AF      = 1.0;
COEFFA    = 0;          COEFFB    = 1.0;        COEFFC    = 1.0;
HULIMIT   = SUM((JHU,JC),QH.L(JHU,JC));
```

```
OPTION ITERLIM = 25000;
OPTION DOMLIM = 100;
```

SOLVE AUTIL USING MINLP MINIMIZING TAC;

* END OF GAMS LISTING FOR OPTIMAL SUPERSTRUCTURE

Example 10.3 Optimum HEN by Superstructure MINLP

a) Determine the cost-optimum network for Case Study 4S1 using a two-stage superstructure as in Figure 10.6. The nonconvex nature of the nonlinearities in the objective function may lead to the model having multiple local optima. In order to locate the global optimum, it is suggested that the upper bound on the hot utility consumption be varied.

   For comparison, also determine the minimum-area network for Case Study 4S1 using the same two-stage superstructure.

b) Determine the cost-optimum network for Case Study 4S1 which does not feature any stream splits using the two-stage superstructure (Figure 10.6).

c) Determine the cost-optimum network for Case Study 4S1 using the two-stage superstructure (Figure 10.6) when the match between streams H2 and C3 is forbidden.

d) Determine the minimum-area network for Case Study 4S1t for a hot utility consumption of 1980 kW.

Solution.   The program HXSUPER.GMS may be used to solve the problem for the optimum network based on minimum total annual cost as well as minimum area.

a) The listing given above for HXSUPER.GMS may be directly used. The upper limit on hot utility (HULIMIT) may be varied over a suitable range (say, 100 kW to 1000 kW in steps of 150 first, and 300 kW to 500 kW in steps of 50 next). The results from HXSUPER.GMS are given below.

HULIMIT (kW)	100	250	400	550	700	850	1000
Hot Utility (kW)	100	250	333.88	508.85	700	726.05	726.05
TAC (10^5 \$/yr)	3.760	2.389	2.354	2.441	2.535	2.531	2.531

HULIMIT (kW)	300	350	400	450	500
Hot Utility (kW)	300	333.88	333.88	333.88	500
TAC (10^5 \$/yr)	2.359	2.354	2.354	2.354	2.442

The optimum network (Figure 10.7) uses 333.88 kW (hot utility) and has a total annual cost (*TAC*) of 2.354 x 10^5 \$/yr. It involves two stream splits (note that the additional cost of stream splitting is not accounted for in the model).

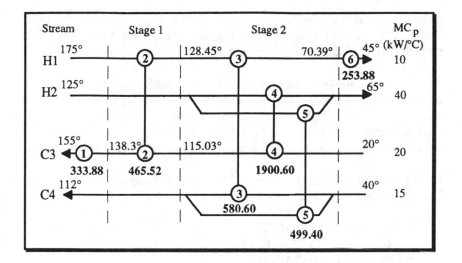

Figure 10.7 The cost-optimum network for Case Study 4S1 generated by the two-stage superstructure MINLP has two stream splits. (Temperature in °C and load in kW)

The total heat transfer area of the network in Figure 10.7 is 1717.19 m^2. The area distribution among the six units is shown in Table 10.1.

Table 10.1 Comparison of Superstructures for Minimum Cost and Minimum Area

Unit	Minimum Cost			Minimum Area		
	Load	Area	Capital Cost	Load	Area	Capital Cost
1	333.879	103.65	62.19	333.879	103.65	62.19
2	465.524	201.14	85.07	491.655	221.58	89.56
3	580.597	255.66	96.87	554.466	263.54	98.54
4	1900.597	816.31	201.26	1874.466	768.32	193.06
5	499.403	272.13	100.34	525.534	286.37	103.31
6	253.879	68.31	52.96	253.879	68.31	52.96
Total		1717.19	598.70		1711.76	599.62

Units: load in kW, area in m^2, and cost in 10^3 $.

The second part of HXSUPER.GMS solves the model for the minimum-area superstructure. The network has the same topology as in Figure 10.7

although the loads on the individual units are different. These are reported in Table 10.1 along with the individual areas, which total 1711.76 m^2. The capital cost is 5.996 x 10^5 \$, which leads to a negligibly higher *TAC* of 2.357 x 10^5 \$/yr.

The true minimum area target for Case Study 4S1 for a hot utility consumption of 333.879 kW is 1695.45 m^2 (from HXAREA.GMS) and 1695.57 m^2 (from PT using the procedure in Section 1.2.1). The superstructure yields a slightly higher area because it uses fewer stages (two, which is less than the number of enthalpy intervals on the composite curves).

b) To obtain the cost-optimum network with no stream splits, two more constraints need to be added. These are given below.

    NOSPLITSH(JC,I)$STAGE(I)..
        SUM(JH, NHC(JH,JC,I)) =L= 1;
    NOSPLITSC(JH,I)$STAGE(I)..
        SUM(JC, NHC(JH,JC,I)) =L= 1;

The constraints ensure that there is not more than one match per stream on any stage of the superstructure. Figure 10.8 shows the cost-optimum network for Case Study 4S1 without any splits, on setting HULIMIT = 550. It uses 508.85 kW (hot utility), has an area of 1767.34 m^2, and has a total annual cost (*TAC*) of 2.441 x 10^5 \$/yr.

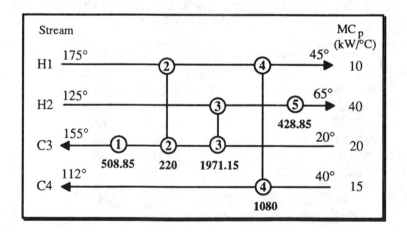

Figure 10.8 A cost-optimum network for Case Study 4S1 may be generated by the two-stage superstructure MINLP with additional constraints for no splits. (Temperature in °C and load in kW)

c) As the match H2-C3 is forbidden, MHC(2,1) is set to zero. Also, the utility consumption for this constrained case could be high, so HULIMIT is set to 2000 (say).   HXSUPER.GMS generates the network in Figure 1.2. Amazingly, the network with which it all started in this book is the cost-optimum one if the match H2-C3 is forbidden! It is also the minimum area network for the hot utility consumption of 1400 kW.

d) As pointed out earlier, the second part of HXSUPER.GMS solves the model for the minimum-area superstructure. The stream and utility data must be changed to correspond to Case Study 4S1t and the utility consumption set to 1980 kW. Figure 10.9 shows the minimum-area network based on a two-stage superstructure.   It has an area of 354.63 m^2, which is virtually the true minimum area target (354.39 m^2) from HXAREA.GMS.

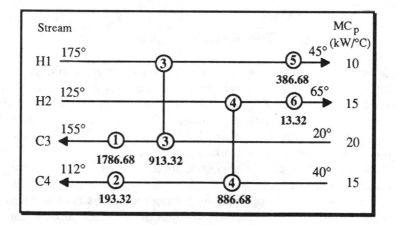

Figure 10.9 The above minimum-area network may be generated using a single-stage superstructure and evolved to a five-unit design with negligible area penalty. (Temperature in °C and load in kW)

Note that it is possible to generate the network in Figure 10.9 with less computational effort using a single-stage superstructure. Further, it may be observed that Figure 10.9 has one unit more than the Euler target, so an additional constraint may be imposed on the number of units as follows:

UNITCON..
    SUM((JH,JC,I)$STAGE(I),NHC(JH,JC,I)) +
    SUM((JH,JCU),NIW(JH,JCU)) + SUM((JHU,JC),NSJ(JHU,JC)) =E= 5;

Now, HXSUPER.GMS yields a network with an area of 354.71 m^2 (the penalty is indeed negligible!). The network is the same as that in Figure 4.26, so the diverse pinch design did virtually generate a minimum-area network for Case Study 4S1t.

10.4 NETWORK LOAD OPTIMIZATION

This section presents an NLP to minimize the area of a network for a given topology and fixed utility requirements. The network loads are the variables. The model is useful because it allows for nonisothermal mixing junctions, a limitation of the superstructure MINLP presented in the previous section. As a design strategy, the topology of the network may be obtained using the aforementioned MINLP, followed by area minimization using the NLP. The model can also be extended to minimize the capital cost (an MINLP formulation).

10.4.1 The Network Representation

In order to formulate the NLP for the network load optimization in a general manner, the following representation for a HEN may be conveniently exploited. For each unit, T_{in}, T_{out}, and MC_p are specified for the hot as well as cold stream passing through the exchanger. The network topology is further defined by supplying the order (from left to right) of the exchangers on each stream as they appear on the network (adjacency data).

These data are used to write constraints that ensure enthalpy balance on each exchanger and the continuity of temperatures on each stream. Networks with stream splits can be optimized by putting appropriate constraints on the temperature and MC_p at the mixing junction. The mixing can be nonisothermal. The objective function in this case is the total heat transfer area required by the network. The heat loads of the various exchangers are given to provide a starting point for the optimization.

10.4.2 GAMS Listing for Optimal Network Loads

The formulation for network load optimization is presented below. The NLP minimizes the overall countercurrent area required by a network where the variables are the heat loads of the various exchangers. A similar optimization was attempted by Mizsey and Rev (1991) using GAMS as well as a commercial flowsheet simulator. The GAMS formulation below is different from that of Mizsey and Rev (1991) in that the constraints and objective function are written in a general form and are not specific to a particular network. In other words, the model formulation is largely uncoupled from the network-specific data, which are given separately using the representation discussed in Section 10.4.1. Another major advantage is that essentially the same formulation may be used to perform reconciliation of data extracted in industrial problems (see next section).

The GAMS listing below is specific to Case Study 4S1 for the topology and utility requirements in Figure 10.7. However, it may be readily adapted to optimize the loads for the networks listed in Table 4.22 for Case Study 4S1 or, for that matter, any network whatsoever.

$TITLE NETWORK LOAD OPTIMIZATION NLP
* This program determines the minimum area required for a network of given
* topology when energy consumption is fixed. It may be used for networks
* with stream splits where the mixing junctions may or may not be isothermal.
* Program Name : HXNLO.GMS
* Program developed by Yogesh Makwana and Uday V. Shenoy
$onnestcom inlinecom // // eolcom ##

// As in previous formulations, sets for streams and exchangers are defined first. A set is defined for data of an exchanger, which includes inlet and outlet temperatures and MCp of the hot and cold streams on that exchanger. Note that the split streams are given as separate streams to simplify the formulation. Hence, for Figure 10.7, there are three hot and three cold streams. //

SETS
JH Hot streams /1*3/
JC Cold streams/1*3/
N   Exchangers /1*6/
HXDATA Exchanger data /HTIN,HTOUT,HMCP,CTIN,CTOUT,CMCP/;

TABLE HXINIT(N,HXDATA) Initial exchanger data

	HTIN	HTOUT	HMCP	CTIN	CTOUT	CMCP
1	180.00	179.00	333.88	138.306	155.0	20.00
2	175.00	128.45	10.00	115.03	138.306	20.00
3	128.45	70.388	10.00	40.00	112.0	8.064
4	125.00	65.00	31.68	20.00	115.03	20.00
5	125.00	65.00	8.32	40.00	112.0	6.936
6	70.388	45.00	10.00	15.00	25.0	25.388

;

// The user can fix some of the exchanger data by setting the FAITH index below to 1. It is essential to give faith as 1 for inlet and outlet temperatures of exchangers at the ends of the streams and for MCp of the nonsplit streams. The faith is set to 0 for split streams to allow nonisothermal mixing. //

TABLE FAITH(N,HXDATA)

	HTIN	HTOUT	HMCP	CTIN	CTOUT	CMCP
1	1	1	1	1	1	1
2	1	0	1	0	1	1
3	0	1	1	1	0	0
4	1	0	0	1	0	1
5	1	0	0	1	0	0
6	1	1	1	1	1	1

;

PARAMETERS

HORD(JH,N)	Order of exchanger from left to right on stream JH
CORD(JC,N)	Order of exchanger from left to right on stream JC
HH(N)	Heat transfer coefficient of the hot stream on exchanger N
HC(N)	Heat transfer coefficient of the cold stream on exchanger N
TOL	Tolerance on energy balance
LOADS(N)	Load on exchanger N
AREAS(N)	Area of exchanger N;

// The network topology is provided in terms of the adjacency of the exchangers on a stream. For example, the exchangers on stream H1 are 2, 3, and 6 in that order from left to right. These data are given in HORD and CORD for all the streams. Utilities are considered to be fixed; hence, no data are required for them. //

HORD('1','1') = 2;   HORD('1','2') = 3;   HORD('1','3') = 6;
HORD('2','1') = 4;
HORD('3','1') = 5;
CORD('1','1') = 1;   CORD('1','2') = 2;   CORD('1','3') = 4;
CORD('2','1') = 3;
CORD('3','1') = 5;

HH(N) = 0.2;   HC(N) = 0.2;
TOL = 0.1;

ALIAS (N,N1)
      (N,N2);

POSITIVE VARIABLES
HXCOMP(N,HXDATA) Computed data for exchangers during optimization;

VARIABLES
AREA Total area to be minimized;

EQUATIONS
ENBAL(N)
HSTRCON(JH,N)
CSTRCON(JC,N)
TEMPCON1(N)
TEMPCON2(N)
DTMINCON1(N)
DTMINCON2(N)
FAITHCON(N,HXDATA)
SPLITCON11
SPLITCON12
SPLITCON21
SPLITCON22
OBJ   Objective function for area;

// The heat lost by a hot stream must equal the heat gained by the cold stream
through an exchanger. Thus, the energy balance must be satisfied within a
certain tolerance. //

ENBAL(N)..
   ABS(HXCOMP(N,'HMCP')*(HXCOMP(N,'HTIN') - HXCOMP(N,'HTOUT'))
    - HXCOMP(N,'CMCP')*(HXCOMP(N,'CTOUT') - HXCOMP(N,'CTIN')))
    =L= TOL;

// The following equations ensure that the outlet temperature from an
exchanger equals the inlet temperature to an adjacent exchanger on a stream. //

HSTRCON(JH,N)$((HORD(JH,N) NE 0) AND
       (SUM(N1$(ORD(N1) EQ (ORD(N) + 1)),HORD(JH,N1)) NE 0))..
   SUM(N1$(ORD(N1) EQ HORD(JH,N)),HXCOMP(N1,'HTOUT')) =E=
   SUM(N1$(ORD(N1) EQ SUM(N2$(ORD(N2) EQ
       (ORD(N) + 1)),HORD(JH,N2))),HXCOMP(N1,'HTIN'));

CSTRCON(JC,N)$((CORD(JC,N) NE 0) AND
       (SUM(N1$(ORD(N1) EQ (ORD(N) + 1)),CORD(JC,N1)) NE 0))..
   SUM(N1$(ORD(N1) EQ CORD(JC,N)),HXCOMP(N1,'CTIN')) =E=
   SUM(N1$(ORD(N1) EQ SUM(N2$(ORD(N2) EQ
       (ORD(N) + 1)),CORD(JC,N2))),HXCOMP(N1,'CTOUT'));

// The inlet temperature of the hot stream on an exchanger must be greater than the outlet, and the reverse holds for the cold stream, as given below. //

TEMPCON1(N)..
      HXCOMP(N,'HTIN') =G= HXCOMP(N,'HTOUT');
TEMPCON2(N)..
      HXCOMP(N,'CTOUT') =G= HXCOMP(N,'CTIN');

// To ensure feasible exchangers, a lower bound of 0.1 is put on the EMAT. //

DTMINCON1(N)..
      HXCOMP(N,'HTIN') - HXCOMP(N,'CTOUT') =G= 0.1;
DTMINCON2(N)..
      HXCOMP(N,'HTOUT') - HXCOMP(N,'CTIN') =G= 0.1;

// The following equation utilizes the faith data to fix some of the variables at the initial values supplied. //

FAITHCON(N,HXDATA)$FAITH(N,HXDATA) ..
      HXCOMP(N,HXDATA) =E= HXINIT(N,HXDATA);

// The following equations model the nonisothermal mixing junction to obtain the correct sum of the MCp of the split streams and the appropriate temperature after mixing. SPLITCON11 and SPLITCON12 are for the split on stream H2, and SPLITCON21 and SPLITCON22 are for the split on stream C4 (Figure 10.7). //

SPLITCON11..
      HXCOMP('4','HMCP') + HXCOMP('5','HMCP') =E= 40;
SPLITCON12..
      (HXCOMP('4','HMCP')*HXCOMP('4','HTOUT') +
      HXCOMP('5','HMCP')*HXCOMP('5','HTOUT'))/
      (HXCOMP('4','HMCP') + HXCOMP('5','HMCP')) =E= 65;
SPLITCON21..
      HXCOMP('3','CMCP') + HXCOMP('3','CMCP') =E= 15;
SPLITCON22..
      (HXCOMP('3','CMCP')*HXCOMP('3','CTOUT')
      + HXCOMP('5','CMCP')*HXCOMP('5','CTOUT'))/
      (HXCOMP('3','CMCP') + HXCOMP('5','CMCP')) =E= 112;

// The objective function is simply minimization of the sum of the areas of the exchangers. The Paterson (1984) approximation is used for the LMTD. //

```
OBJ..
     AREA =E=  SUM(N, HXCOMP(N,'CMCP')*
     (HXCOMP(N,'CTOUT') -HXCOMP(N,'CTIN'))*(1/HH(N) + 1/HC(N))/(
     (2./3.)*((HXCOMP(N,'HTIN')-HXCOMP(N,'CTOUT'))
     *(HXCOMP(N,'HTOUT') - HXCOMP(N,'CTIN')))**(0.5) +
     (HXCOMP(N,'HTIN') - HXCOMP(N,'CTOUT') +
     HXCOMP(N,'HTOUT') - HXCOMP(N,'CTIN'))/6+ 1E-02));

MODEL NLO /ALL/;

* Initialization
HXCOMP.L(N,HXDATA) = HXINIT(N,HXDATA);

SOLVE NLO USING DNLP MINIMIZING AREA;
* Calculation of loads
LOADS(N) =
HXCOMP.L(N,'HMCP')*(HXCOMP.L(N,'HTIN') -HXCOMP.L(N,'HTOUT'));

* Calculation of areas
AREAS(N) = HXCOMP.L(N,'CMCP')*
  (HXCOMP.L(N,'CTOUT') -HXCOMP.L(N,'CTIN'))* (1/HH(N) + 1/HC(N))/(
  (2./3.)*((HXCOMP.L(N,'HTIN')-HXCOMP.L(N,'CTOUT'))
  *(HXCOMP.L(N,'HTOUT') - HXCOMP.L(N,'CTIN')))**(0.5) +
  (HXCOMP.L(N,'HTIN') - HXCOMP.L(N,'CTOUT') +
  HXCOMP.L(N,'HTOUT') - HXCOMP.L(N,'CTIN'))/6 + 1E-02);

* Calculation of the capital cost

PARAMETER
CCOST          Capital cost of the network
COEFFA         Fixed cost of an exchanger
COEFFB         Area cost coefficient for an exchanger
COEFFC         Cost exponent for an exchanger;

COEFFA = 30000;      COEFFB = 750;        COEFFC = 0.81;

CCOST =  SUM(N,COEFFA*(LOADS(N)/(LOADS(N) + 1E-6)) +
         COEFFB*(AREAS(N))**COEFFC);

DISPLAY LOADS, AREAS, CCOST;

* END OF GAMS LISTING FOR NETWORK LOAD OPTIMIZATION
```

Example 10.4 Network Load Optimization for Minimum Area by NLP

Determine the minimum heat transfer area required by a network for Case Study 4S1 if the topology and utility level are fixed as in Figure 10.7. The MC_p ratio of the stream splits as well as the heat loads of the process-to-process exchangers may be varied.

Solution. The problem may be solved by using the program HXNLO.GMS directly. The minimum area obtained (corresponding to the network in Figure 10.10) is 1699.22 m^2, which is less than the area of the network in Figure 10.7 (1717.19 m^2) and is close to the true minimum area target from HXAREA.GMS (1695.45 m^2). The stream splits in Figure 10.7 are isothermal, and their MC_p ratios are 31.68:8.32 (for stream H2) and 8.06:6.94 (for stream C4). Allowing for nonisothermal mixing junctions in Figure 10.10 (by splitting stream H2 in the MC_p ratio of 29.77:10.23 and stream C4 in the ratio of 7.5:7.5) reduces the network area from 1711.76 m^2 (Table 10.1) to 1699.22 m^2 and the total annual cost (*TAC*) from 2.354 x 10^5 \$/yr (Figure 10.7) to 2.349 x 10^5 \$/yr. However, it is important to note that the optimizations so far have ignored the additional cost of piping and maintenance for the stream splits, which is certain to cause significant errors.

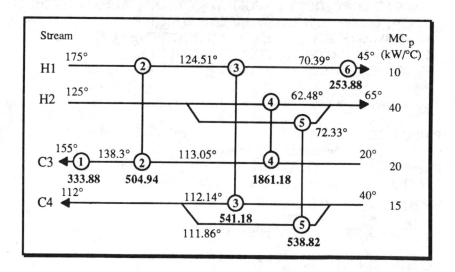

Figure 10.10 The above minimum-area network is evolved from Figure 10.7 through continuous optimization of the heat loads by maintaining the topology and utility levels but allowing for nonisothermal mixing junctions. (Temperature in °C and load in kW)

10.5 DATA RECONCILIATION OF HENs

When data are extracted for any industrial process, they often do not satisfy the basic mass and energy balances. This is essentially due to errors in the measurements, and reconciliation (Pai and Fisher, 1988) of the measured data is required to close the balances. In this section, an MP formulation is presented for the data reconciliation (DR) of heat exchanger networks.

10.5.1 Mathematical Model for Data Reconciliation

The data reconciliation of a HEN is a nonlinear problem. To understand the basics, the DR model for a single exchanger is discussed first. It is then extended to a network using the same representation as in Section 10.4.1. Consider a simple countercurrent exchanger, where typical measurements include the inlet temperature (T_{in}), the outlet temperature (T_{out}), and the heat capacity flow rate (MC_p) of the hot and cold streams. The mean (x_{mi}) and variance (v_i) for each measured quantity (denoted by subscript i) may be obtained by making several measurements under the same operating conditions. If the reconciled values are denoted by x_i, then the DR model considering the energy balance may be written as

$$\text{minimize } \Sigma_i \, (x_i - x_{mi})^2/v_i \qquad (10.1)$$
$$\text{subject to } f = MC_{p,h}\,(T_{in,h} - T_{out,h}) - MC_{p,c}\,(T_{out,c} - T_{in,c}) = 0$$

To extend the model to a network, the objective function must simply be the sum of terms as in Equation 10.1 but for every exchanger. In addition to an energy balance on each exchanger, additional constraints for ensuring the continuity of temperatures and heat capacity flow rates must be included.

10.5.2 GAMS Listing for HEN Data Reconciliation

The NLP for HEN data reconciliation is presented below. The objective function is along the lines of Equation 10.1, where the variables are the reconciled measurements x_i. As in the case of network load optimization, the constraints and objective function are written in a general form and are largely uncoupled from the network-specific data, which is given separately using the representation discussed in Section 10.4.1.

The GAMS listing below is specific to Case Study 4S1 for the network given in Figure 10.11. However, it may be easily modified to do data reconciliation for any other network.

$TITLE HEN DATA RECONCILIATION NLP
* This program does nonlinear data reconciliation for a network of heat
* exchangers. The measured quantities (average and variance for each) are the
* inputs to the program.
* Program Name: HXDR.GMS
* Program developed by Yogesh Makwana and Uday V. Shenoy
$onnestcom inlinecom // // eolcom ##

// Various sets, similar to those in HXNLO.GMS, are defined first. //

SETS
JH Hot streams /1*2/
JC Cold streams /1*2/
N Exchangers /1*4/
HXDATA Data of exchangers /HTIN,HTOUT,HMCP,CTIN,CTOUT,CMCP/;

// The average data measurements for the four exchangers are shown in Figure
10.11 and appear in the following table. //

TABLE MEAS(N,HXDATA)

	HTIN	HTOUT	HMCP	CTIN	CTOUT	CMCP
1	180	179	1400	83.5	157	20.2
2	126	96.5	40.2	40.5	111.5	14.6
3	175.5	44	10.3	21.5	83.5	20.2
4	96.5	67.5	40.2	15	25	132

;
// The variances for the above data are given in the following table. //

TABLE VARS(N,HXDATA)

	HTIN	HTOUT	HMCP	CTIN	CTOUT	CMCP
1	2.0	2.0	0.1	2.0	2.0	0.1
2	2.0	2.0	0.1	2.0	2.0	0.1
3	2.0	2.0	0.1	2.0	2.0	0.1
4	2.0	2.0	0.1	2.0	2.0	0.1

;
// As in the network load optimization program of the previous section, the
user is allowed to fix some variables at their initial values by giving the FAITH
as 1. This is equivalent to the variance being zero. If the variance itself is
specified as zero in the above table, it will cause a divide fault in the objective
function. For the problem under consideration, the data for the utilities are
assumed to be error-free and, hence, have FAITH equal to 1. //

```
TABLE FAITH(N,HXDATA)
     HTIN   HTOUT   HMCP   CTIN   CTOUT   CMCP
1     1      1       1       0       0       0
2     0      0       0       0       0       0
3     0      0       0       0       0       0
4     0      0       0       1       1       1
;
```

// The network topology is supplied in the same manner as in the previous section, by giving the order of exchangers as they appear (from left to right) on a stream. //

```
PARAMETER
HORD(JH,N)     Nth exchanger on hot stream JH
CORD(JC,N)     Nth exchanger on cold stream JC ;

HORD('1','1') = 3;
HORD('2','1') = 2;          HORD('2','2') = 4;
CORD('1','1') = 1;          CORD('1','2') = 3;
CORD('2','1') = 2;


ALIAS (N,N1)
        (N,N2);


POSITIVE VARIABLES
RECON(N,HXDATA)  Reconciled data;

VARIABLES
SUMERR  Sum of errors to be minimized;

EQUATIONS
ENBAL(N)
HSTRCON1(JH,N)
HSTRCON2(JH,N)
CSTRCON1(JC,N)
CSTRCON2(JC,N)
CONFID(N,HXDATA)
OBJ;
```

// The energy balance on each exchanger must be satisfied. //
```
ENBAL(N)..
    RECON(N,'HMCP')*(RECON(N,'HTIN') - RECON(N,'HTOUT'))
    - RECON(N,'CMCP')*(RECON(N,'CTOUT') - RECON(N,'CTIN')) =E= 0;
```

// The outlet temperature of a stream must equal the inlet of the next exchanger on that stream. Also the MCp must remain the same for all exchangers on a stream. This is ensured by the following four constraints. //

```
HSTRCON1(JH,N)$((HORD(JH,N) NE 0) AND
        (SUM(N1$(ORD(N1) EQ (ORD(N) + 1)),HORD(JH,N1)) NE 0))..
    SUM(N1$(ORD(N1) EQ HORD(JH,N)),RECON(N1,'HTOUT')) =E=
    SUM(N1$(ORD(N1) EQ SUM(N2$(ORD(N2) EQ (ORD(N) + 1)),
        HORD(JH,N2))),RECON(N1,'HTIN'));

HSTRCON2(JH,N)$((HORD(JH,N) NE 0) AND
        (SUM(N1$(ORD(N1) EQ (ORD(N) + 1)),HORD(JH,N1)) NE 0))..
    SUM(N1$(ORD(N1) EQ HORD(JH,N)),RECON(N1,'HMCP')) =E=
    SUM(N1$(ORD(N1) EQ SUM(N2$(ORD(N2) EQ (ORD(N) + 1)),
        HORD(JH,N2))),RECON(N1,'HMCP'));

CSTRCON1(JC,N)$((CORD(JC,N) NE 0) AND
        (SUM(N1$(ORD(N1) EQ (ORD(N) + 1)),CORD(JC,N1)) NE 0))..
    SUM(N1$(ORD(N1) EQ CORD(JC,N)),RECON(N1,'CTIN')) =E=
    SUM(N1$(ORD(N1) EQ SUM(N2$(ORD(N2) EQ (ORD(N) + 1)),
        CORD(JC,N2))),RECON(N1,'CTOUT'));

CSTRCON2(JC,N)$((CORD(JC,N) NE 0) AND
        (SUM(N1$(ORD(N1) EQ (ORD(N) + 1)),CORD(JC,N1)) NE 0))..
    SUM(N1$(ORD(N1) EQ CORD(JC,N)),RECON(N1,'CMCP')) =E=
    SUM(N1$(ORD(N1) EQ SUM(N2$(ORD(N2) EQ (ORD(N) + 1)),
        CORD(JC,N2))),RECON(N1,'CMCP'));
```

// The variable with FAITH equal to 1 must not change. //

```
CONFID(N,HXDATA)$FAITH(N,HXDATA)..
    RECON(N,HXDATA) =E= MEAS(N,HXDATA);
```

* The objective function is along the lines of Equation 10.1.
```
OBJ..
    SUMERR =E= SUM(N,
        (SQR(RECON(N,'HTIN') - MEAS(N,'HTIN'))/VARS(N,'HTIN')
        + SQR(RECON(N,'HTOUT') - MEAS(N,'HTOUT'))/VARS(N,'HTOUT')
        + SQR(RECON(N,'HMCP') - MEAS(N,'HMCP'))/VARS(N,'HMCP')
        + SQR(RECON(N,'CTIN') - MEAS(N,'CTIN'))/VARS(N,'CTIN')
        + SQR(RECON(N,'CTOUT') - MEAS(N,'CTOUT'))/VARS(N,'CTOUT')
        + SQR(RECON(N,'CMCP') - MEAS(N,'CMCP'))/VARS(N,'CMCP'))) ;
```

MODEL DR /ALL/;

* Initializing the variables to be reconciled.
RECON.L(N,HXDATA) = MEAS(N,HXDATA)

SOLVE DR USING DNLP MINIMIZING SUMERR;

* Displaying the reconciled variables.
MEAS(N,HXDATA) = RECON.L(N,HXDATA);
DISPLAY MEAS;

* END OF GAMS LISTING FOR HEN DATA RECONCILIATION

Example 10.5 Data Reconciliation of a HEN by NLP

Reconcile the measurements on the network shown in Figure 10.11 for Case Study 4S1. The mean for each measured quantity appears in Figure 10.11. Assume the variance to be 2.0 (for temperature measurements) and 0.1 (for heat capacity flow rate measurements).

Solution. The data reconciliation is done by directly using HXDR.GMS. The reconciled values are shown in bold in parenthesis next to each measurement in Figure 10.11 itself. The values after reconciliation are closer to those in Figure 1.2.

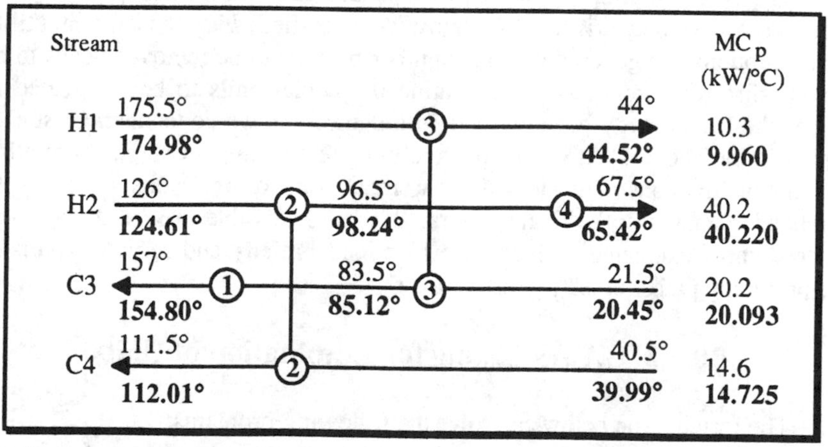

Figure 10.11 The measured plant data must be reconciled (values in bold) to close the energy balances prior to any analysis. (Temperature in °C and load in kW)

To understand the effect of the reconciliation, consider the energy balances on the four exchangers in Figure 10.11 before and after reconciliation.

1) $1400 (180 - 179) = 1400$ & $20.2 (157 - 83.5) = 1484.7$ (before)
   $1400 (180 - 179) = 1400$ & $20.093 (154.799 - 85.124) = 1399.98$

2) $40.2 (126 - 96.5) = 1185.9$ & $14.6 (111.5 - 40.5) = 1036.6$ (before)
   $40.22 (124.606 - 98.238) = 1060.52$ & $14.725(112.01 - 39.99) = 1060.49$

3) $10.3 (175.5 - 44) = 1354.45$ & $20.2 (83.5 - 21.5) = 1252.4$ (before)
   $9.96 (174.981 - 44.519) = 1299.40$ & $20.093 (85.124 - 20.454) = 1299.41$

4) $40.2 (96.5 - 67.5) = 1165.8$ & $132 (25 - 15) = 1320$ (before)
   $40.22 (98.238 - 65.418) = 1320.02$ & $132 (25 - 15) = 1320$.

10.6 EVOLUTION OF HENs

This section presents a formulation for the simplification of HENs by elimination of units, which is equivalent to loop breaking. The formulation is an MINLP when the network has stream splits and an MILP when there are no splits. The objective is to find the minimum energy penalty incurred when a specified number of units are removed. The user may specify the units to be removed.

10.6.1 Model for Loop Breaking

The representation of the network is the same as in the previous two sections. The network topology must be specified with the current utility consumptions. In general, only the number of units to be removed needs to be given since the program will determine the actual units to be eliminated to incur the least energy penalty. The formulation allows nonisothermal stream splits in the network. It is possible to eliminate stream splits by incorporating an inequality constraint such that, at most, one of the units on the split branches remains in the final network. It is also possible to seed the network with dummy exchangers which have zero load initially and serve the purpose of providing paths for ΔT_{min} restoration.

10.6.2 GAMS Listing for Elimination of Units

The formulation below can solve the following problems:
- minimizing the energy penalty when a specified number of units are removed from a network.
- minimizing the energy penalty when specific units are removed from a network.

- exploring the possibility of existence of an MER network of a desired topology. This is done by seeding the network with dummy exchangers at desired locations, and then setting the NREM parameter (that specifies the number of units to be removed) to zero.

Given below is the GAMS implementation of the loop-breaking model for the removal of units from the MER network for Case Study 4S1, shown in Figure 4.9 (ΔT_{min} = 20°C). It is easily adapted to other networks and cases.

```
$TITLE HEN EVOLUTION MIP;
* This program finds the minimum energy penalty that will be incurred on
* removal of a specified number of exchangers from a HEN. It is an MINLP
* formulation for networks with steam splits and an MILP when there are no
* splits.
* PROGRAM NAME : HXEVOL.GMS
* Program developed by Yogesh Makwana and Uday V. Shenoy
$onnestcom inlinecom // // eolcom ##
```

// The network representation is the same as in the load optimization and data reconciliation programs. The task is to find the least energy penalty when two of the seven units are removed from the network (Figure 4.9), with the EMAT remaining unchanged at 20°C. A dummy cooler with no load initially is placed on stream H2 for additional flexibility during the energy relaxation. Therefore, the number of exchangers to be removed becomes three. //

```
SETS
JH Hot streams /1*2/
JC Cold streams /1*2/
N Exchangers /1*8/
HXDATA Data of exchangers /HTIN,HTOUT,HMCP,CTIN,CTOUT,CMCP/;
```

* Data for the network to be evolved.
      TABLE MEAS(N,HXDATA) Data of current network

	HTIN	HTOUT	HMCP	CTIN	CTOUT	CMCP
1	180	179	500	130	155	20
2	180	179	105	105	112	15
3	175	125	10	105	130	20
4	125	40	20	20	105	20
5	125	90	20	58.333	105	15
6	125	97.5	10	40	58.333	15
7	97.5	45	10	15	25	52.5
8	65	65	40	15	25	0.0

;

TABLE FAITH(N,HXDATA) Variables not be changed

	HTIN	HTOUT	HMCP	CTIN	CTOUT	CMCP
1	1	1	0	0	1	1
2	1	1	0	0	1	1
3	1	0	1	0	0	1
4	1	0	0	1	0	1
5	1	0	0	0	0	1
6	0	0	1	1	0	1
7	0	1	1	1	1	0
8	0	1	1	1	1	0

;

SCALAR NHX Number of exchangers /8/;

PARAMETERS

HORD(JH,N)	Nth exchanger on hot stream JH
CORD(JC,N)	Nth exchanger on cold stream JC
EMAT	Exchanger minimum approach temperature
DTMAX	Maximum possible approach temperature
NREM	Number of exchangers to be removed
QH	Hot utility in the existing network
QC	Cold utility in the existing network
QMAX	Maximum heat load
HEATER(N)	Index for identifying heater
COOLER(N)	Index for identifying cooler
LOAD(N)	Load on exchanger N;

* Order of exchangers (from left to right) on each stream
HORD('1','1') = 3;   HORD('1','2') = 6;   HORD('1','3') = 7;
CORD('1','1') = 1;   CORD('1','2') = 3;   CORD('1','3') = 4;
CORD('2','1') = 2;   CORD('2','2') = 5;   CORD('2','3') = 6;

* Data for the problem
EMAT = 20;   DTMAX = 250;   NREM = 3;
QH = 605;   QC = 525;   QMAX = 3780;

* Indices for identification of heaters and coolers
HEATER('1') = 1;   HEATER('2') = 1;
COOLER('7') = 1;   COOLER('8') = 1;

ALIAS (N,N1);

POSITIVE VARIABLES
Q(N)   Load in exchanger N
CALC(N,HXDATA)   Variables for the network data;

BINARY VARIABLES
Y(N)   Binary variable for exchanger existence;

VARIABLES
QPENAL   Energy penalty;

EQUATIONS
XLOAD(N)
ENBAL(N)
HSTRCON1(JH,N)
CSTRCON1(JC,N)
TEMPCON1(N)
TEMPCON2(N)
DTMINCON1(N)
DTMINCON2(N)
FAITHCON(N,HXDATA)
SPLITCON1
SPLITCON2
SPLITCON3
SPLITCON4
EXIST(N)
UWANT
NOSPLITS
OBJ;

* The following two equations perform the energy balances on all exchangers.

XLOAD(N)..
   Q(N) =E= CALC(N,'HMCP')*(CALC(N,'HTIN') - CALC(N,'HTOUT'));

ENBAL(N)..
   Q(N) =E= CALC(N,'CMCP')*(CALC(N,'CTOUT') - CALC(N,'CTIN'));

// The following two equations maintain the continuity of temperatures in the
network during evolution. //

HSTRCON1(JH,N)$((HORD(JH,N) NE 0) AND (HORD(JH,N+1) NE 0))..
   SUM(N1$(ORD(N1) EQ HORD(JH,N)),CALC(N1,'HTOUT')) =E=

SUM(N1$(ORD(N1) EQ HORD(JH,N+1)),CALC(N1,'HTIN'));

CSTRCON1(JC,N)$((CORD(JC,N) NE 0) AND (CORD(JC,N+1) NE 0))..
  SUM(N1$(ORD(N1) EQ CORD(JC,N)),CALC(N1,'CTIN')) =E=
  SUM(N1$(ORD(N1) EQ CORD(JC,N+1)),CALC(N1,'CTOUT'));

// The temperature of a hot stream at the outlet must not be greater than the temperature at the inlet. Similar constraint exists for the cold streams. //

TEMPCON1(N)..
  CALC(N,'HTIN')   =G= CALC(N,'HTOUT');

TEMPCON2(N)..
  CALC(N,'CTOUT') =G= CALC(N,'CTIN');

// The approach temperatures at the exchangers must be greater than the specified EMAT. If the exchanger does not exist, the temperature difference is allowed to take negative values. //

DTMINCON1(N)$(HEATER(N) NE 1 AND COOLER(N) NE 1)..
  CALC(N,'HTIN') - CALC(N,'CTOUT') =G= EMAT - (1 - Y(N))*DTMAX;

DTMINCON2(N)$(HEATER(N) NE 1 AND COOLER(N) NE 1)..
  CALC(N,'HTOUT') - CALC(N,'CTIN') =G= EMAT - (1 - Y(N))*DTMAX;

// The variable with FAITH index equal to 1 must not change. //

FAITHCON(N,HXDATA)$(FAITH(N,HXDATA) EQ 1)..
  CALC(N,HXDATA) =E= MEAS(N,HXDATA);

// These four constraints model the split on stream H2. //

SPLITCON1..
  CALC('4','HMCP') + CALC('5','HMCP') =E= 40;

SPLITCON2..
  (CALC('4','HMCP')*CALC('4','HTOUT') +
  CALC('5','HMCP')*CALC('5','HTOUT'))/
  (CALC('4','HMCP') + CALC('5','HMCP')) =E= CALC('8','HTIN');

SPLITCON3..
  CALC('4','HMCP') =L= Y('4')*40;

```
SPLITCON4..
  CALC('5','HMCP') =L= Y('5')*40;
```

// The binary variables for existence of exchangers are assigned as follows. //

```
EXIST(N)..
  Q(N) =L= QMAX*Y(N);
```

// This constraint specifies the number of exchangers to be removed. //
```
UWANT..
  SUM(N,Y(N)) =E= NHX - NREM;
```

// The following constraint ensures that at least one of the two exchangers on
the split will be eliminated. //

```
NOSPLITS..
  Y('4') + Y('5') =L= 1;
```

// The objective is to minimize the penalty QPENAL on the hot and cold
utilities incurred when NREM units are removed from the network. //

```
OBJ..
  QPENAL =E=
  SUM(N$HEATER(N), CALC(N,'HMCP')*(CALC(N,'HTIN') -
      CALC(N,'HTOUT')))
  + SUM(N$COOLER(N), CALC(N,'CMCP')*(CALC(N,'CTOUT') -
      CALC(N,'CTIN'))) - QH - QC;

MODEL EVOLVE /ALL/;

* Initializing the variables and fixing bounds
CALC.L(N,HXDATA) = MEAS(N,HXDATA);
Q.L(N) =
  MIN(CALC.L(N,'HMCP')*(CALC.L(N,'HTIN') - CALC.L(N,'HTOUT')),
        CALC.L(N,'CMCP')*(CALC.L(N,'CTOUT') - CALC.L(N,'CTIN')) );
QPENAL.UP =
        2*SUM(N$(HEATER(N) NE 1 AND COOLER(N) NE 1), Q.L(N)) ;

OPTION ITERLIM = 10000;
OPTION DOMLIM = 500;

SOLVE EVOLVE USING MINLP MINIMIZING QPENAL;
```

DISPLAY CALC.L, Q.L;

* END OF GAMS LISTING FOR HEN EVOLUTION

Example 10.6 Evolution of a HEN by MINLP
Consider the MER network shown in Figure 4.9 for Case Study 4S1. Find the minimum energy penalty incurred for a specified EMAT of 20°C on removal of
a) one unit; and
b) two units and the split.

Solution. The minimum energy penalty for eliminating units from a network may be obtained by using HXEVOL.GMS. As Figure 4.9 involves a stream split, the formulation is an MINLP.

a) To remove one unit from Figure 4.9, the above listing for HXEVOL.GMS needs the following modifications:
(i)   The FAITH index for all variables on exchanger 8 are set to 1.
(ii) The NOSPLITS constraint is excluded.
(iii) NREM is set to 2 in order to account for the dummy cooler.
HXEVOL.LST yields the minimum energy penalty for the removal of one unit to be 137.5 kW. Unit 6 is eliminated, and the evolved network corresponds to the one in Figure 4.11.

b) To remove two units and the split, the listing given above is used as is. The minimum penalty is found to be 395 kW, and the units removed are 2, 3, and 5. The dummy cooler provided at the start now becomes active and has a load of 700 kW. The network is the same as the one obtained by pinch technology and shown in Figure 4.16. The power of mathematical programming is evident in this case because it leaves little doubt about the evolution being optimal in terms of the energy penalty.

The purpose of the GAMS programs presented in this chapter is to provide the user with a feel for MP techniques rather than an exhaustive coverage of all the HENS problems that may be tackled by mathematical programming. Although not discussed here, useful MP formulations can be developed for retrofitting (Ciric and Floudas, 1989, 1990b, 1990c; Yee and Grossmann, 1991), controllable HENs (Papalexandri and Pistikopoulos, 1994), and MENS (El-Halwagi and Manousiouthakis, 1990; El-Halwagi and Srinivas, 1992; Papalexandri et al., 1994; Srinivas and El-Halwagi, 1994).
It is clear from the discussion in this chapter that tools from operations research and mathematical programming have an important role to play in

OPERA. It is also possible that tools from artificial intelligence including knowledge-based expert systems, neural networks, and fuzzy logic will provide a framework for OPERA to support the manipulation of heuristic, ill-structured, and imprecise information. The development of such an intelligent system for synthesizing cost-effective and controllable exchanger networks has been recently described by Huang and Fan (1994).

It may be concluded that OPERA in the future should involve hybrid computer-aided design systems based on fundamental thermodynamic principles as well as mathematical programming strategies with necessary support from artificial intelligence tools.

QUESTIONS FOR DISCUSSION

1. The transshipment model was employed in Sections 10.1 and 10.2. Could the transportation model be used? Compare the transshipment and transportation models from operations research.

2. An NLP is easier to solve if the nonlinearities are placed solely in the objective function rather than the constraints. Comment.

3. Discuss the possibility of devising a formulation that accounts for the cost of stream splitting in the optimal superstructure approach in Section 10.3.

4. Compare and contrast the approaches based on pinch analysis and mathematical programming. List the strengths and weaknesses of both approaches.

5. Discuss possible areas for future research related to pinch analysis, mathematical programming, and hybrid systems. Consider the following specific examples: design of controllable HENs using pinch analysis and rigorous area targets for 1-2 shell-and-tube exchangers using mathematical programming.

PROBLEMS

10.A Optimal Heat Recovery in Unpinched Problems

For Case Study 4L1, determine the energy targets and the MER networks for the following cases using HXUTIL.GMS:
   a) for a ΔT_{min} of 10°C with no constraints on matches
   b) for a ΔT_{min} of 10°C with the match between streams H1 and C3 forbidden

Answers:
   a) $Q_{hu,min}$ = 127.68 kW and $Q_{cu,min}$ = 250.14 kW. A possible design
   appears in Figure 6 (Papoulias & Grossmann, 1983, p. 716):
   HU - C4 (127.68 kW)   H2 - C4 (747.84 kW)
   H1 - C3 (338.74 kW)   H2 - C3 (423.21 kW)
                                H1 - CU (250.14 kW)
   b) $Q_{hu,min}$ = 259.75 kW and $Q_{cu,min}$ = 382.21 kW. A possible design
   appears in Figure 7 (Papoulias & Grossmann, 1983, p. 716):
   HU - C4 (259.75 kW)   H2 - C4 (409.05 kW)
   H1 - C4 (206.72 kW)   H2 - C3 (762.00 kW)
                                H1 - CU (382.21 kW)

10.B Optimal Heat Recovery in Pinched Problems

For Case Study 7P1, determine the energy targets and the MER network
for a ΔT_{min} of 20°F with the match between streams H5 and C1 forbidden. It
is a pinched problem and must be split at the pinch while using
HXUTIL.GMS.

Answers:

A possible design appears in Figure 8 (Papoulias & Grossmann, 1983,
p. 717):

T_p = 420°F   *Above Pinch*   *Below Pinch*
                     H1 - C1a (3675 Btu/hr) H4 - C1   (5100 Btu/hr)
                     H2 - C1b (1540 Btu/hr) H1 - C1a (1182.5 Btu/hr)
                     H3 - C1c (495 Btu/hr) H6 - C1a (8750 Btu/hr)
                     HU - C1 (8390 Btu/hr) H3 - C1b (1417.5 Btu/hr)
                                             H1 - CU   (3017.5 Btu/hr)
                                             H5 - CU   (3600 Btu/hr)

10.C Rigorous Area Targets by NLP

Obtain the minimum heat transfer area requirements using
HXAREA.GMS for the following case studies discussed by Colberg and
Morari (1990):
   a) Case Study 4C1 for $Q_{hu,min}$ = 620 kW and $Q_{cu,min}$ = 230 kW
   b) Case Study 5N1 with no utility requirements
   c) Case Study 7C1 for $Q_{hu,min}$ = 244.125 kW and $Q_{cu,min}$ = 172.596 kW

Answers:
   a) 258.9 m^2;   b) 29.94 m^2;   c) 174.51 m^2.

10.D Optimum Pressure Drops for Shell-and-Tube Exchanger

Rework Example 6.7 using an NLP formulation.

Answers:

Same as in Example 6.7.

APPENDIX A

The Software Package HX

As mentioned in Chapter 1, the advances in pinch analysis during the last decade make software programs not merely convenient but essential. The accompanying student version of the PC-based software package HX provides a simple effective computing aid to reduce tedious hand-calculations during HENS targeting.

The four major capabilities of the HX student version are given below:

- Conventional point targeting for energy, area, units, shells, and cost at a particular value of ΔT_{min} for a grassroots design (see Sections 1.1 through 1.5);
- Conventional continuous targeting for energy, pinch, units, area, and cost over a specified range of ΔT_{min} for a grassroots design (see Section 1.6);
- Diverse pinch point targeting for energy and area at a particular value of Q''_{min} for a grassroots design (see Section 2.2); and
- Retrofit targeting for energy savings, extra area, cost savings, investment, and payback over a specified range of ΔT_{min} for an existing design based on fixed heat transfer coefficients (see Section 7.2.1).

During conventional targeting, options are available to specify nonuniform stream specifications (see Example 1.9) and individual stream ΔT contributions. Figure A.1 shows the four key features provided by HX along with the variety of plots through which the results may be viewed.

A.1 THE MENU STRUCTURE

A simplified overview of the HX software is provided through Figure A.2, which shows its menu structure. The various menu options are briefly explained below:

- File Open - allows an existing input file (in the format required by HX and specified in the next section) to be opened. The input file may be created with any standard text editor. The filename along with the path must be given as screen input. Note that the data file corresponding to the last case study run is automatically retrieved when the HX package is started.

- File Quit - allows for final exiting of the HX software package.

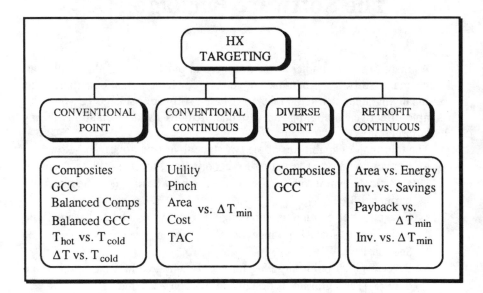

Figure A.1 Various plots are available to view the targeting results from the four major capabilities of the HX student version.

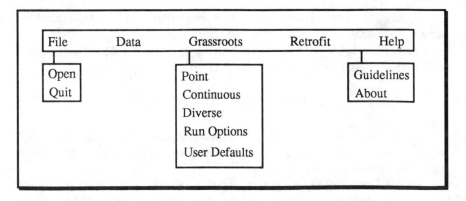

Figure A.2 The simple menu structure shown above is used by the HX student version. Highlighted characters in the menu denote the hot keys. To navigate through the main menu, the Alt-key must be pressed along with the highlighted character.

- Data - allows viewing of the stream, utility, and cost data for the active case study. If data on stream specifications and ΔT contributions exist, they are also displayed.
- Grassroots Point - allows targeting for energy (utility and pinch), area (countercurrent and 1-2 shell and tube), units (Euler minimum and MER), shells (approximate and accurate), and cost (operating, capital, and total annual cost) for a grassroots design.   The value of ΔT_{min} must be provided as screen input.
- Grassroots Continuous - allows range targeting for utility, pinch, area, units, operating cost, capital cost, and total annual cost for a grassroots design. The range of ΔT_{min} (lower limit, upper limit, and step size) must be provided as screen input.
- Grassroots Diverse - allows targeting for utility, pinch, and area for minimum flux specification based on diverse pinch concept. The value of Q''_{min} must be provided as screen input.
- Grassroots Run_Options - allows options to be set for nonuniform stream specifications and individual stream ΔT contributions. The space bar may be pressed to toggle the relevant option on or off. Data must be provided in the input data file under the keywords STREAM SPECS DATA and DELTA T DATA (see next section).
- Grassroots User_Defaults - allows user to set values for X_p (see Equation 1.19), the maximum area per shell (see Example 1.8), and the diverse coefficient (see Equation 2.11 and Problem 4.F).
- Retrofit - allows targeting for energy savings, extra area, cost savings, investment, and payback for an existing design based on fixed heat transfer coefficients. The current utility usage, the existing area, the area per match, the area per shell, and the desired payback must be provided as screen inputs. The range of ΔT_{min} (lower limit, upper limit, and step size) must also be specified as screen input.
- Help Guidelines - allows viewing of general information and instructions for use of the HX program.
- Help About - allows HX program information, version, and copyright to be displayed.

A.2 THE INPUT FILE STRUCTURE

The general format of the input data file to HX is outlined in this section. Note that the case studies for HENS in Appendix B are presented in a format consistent with the input data (.dat) files required by the HX software package. Comment lines at the beginning of a data file start with //. Consistency must be maintained in the units employed for the various quantities since the HX software does not perform any conversion of units. All pertinent data are classified using keywords as given below.

STREAM DATA

T_{in}, T_{out}, MC_p, h, Stream Label

(Note: The number of rows correspond to the number of process streams, both hot and cold.)

UTILITY DATA

T_{in}, T_{out}, MC_p, h, Utility Label

(Note: MC_p for a utility is typically given as zero and is calculated by HX. The number of rows correspond to the number of utilities, both hot and cold).

COST DATA

C_{hu}, C_{cu}

(Note: The entries for costs of unit loads of hot and cold utilities must correspond to the utilities under UTILITY DATA.)

a, b, c

(Note: These are cost coefficients for exchanger area in the case of uniform specifications.)

t, r

(Note: These are the plant life and rate of interest to be used in the annualization factor.)

STREAM SPECS DATA

(Note: Numbers are listed in a single row and correspond to identifiers used for the capital cost coefficients for various exchanger specifications given in a separate data file called 'HXSPEC.CST'. The number of entries must agree with the total number of streams, both process and utility.)

DELTA T DATA

(Note: Stream-dependent ΔT_{min} values are listed in a single row. The number of entries must agree with the total number of streams, both process and utility.)

Being a student version, the HX software accompanying this book contains the following limitations:

- A maximum of three hot streams and three cold streams can be specified under STREAM DATA.
- The multiple utility capability is not available, i.e., a maximum of one hot utility and one cold utility can be specified under UTILITY DATA.
- Isothermal phase changes for streams should be approximated by a small temperature change (say, a 1° temperature change so that MC_p may be set equal to the heat load).

- Straight line segments are employed to connect points on all the plots, so it is preferable to input a small step size when providing the range of ΔT_{min} during continuous targeting.
- Unit targeting is not available for multi-pinched problems.
- The student version does not support a mouse.

A.3 SOME DEMONSTRATION EXAMPLES

On the HX disk accompanying this book, the CS subdirectory contains data for the demonstration examples given below (as well as for some of the case studies given in Appendix B).

Example A.1 Determination of Conventional Point Targets

For the stream data in Case Study 4S1, determine the energy, units, area, shells, and cost targets for a grassroots design for $\Delta T_{min} = 20°C$.

Solution. The input data file 4S1.dat in the CS subdirectory contains the relevant data. It contains the STREAM DATA, the UTILITY DATA, and the COST DATA.

Step 1. Type "HX" at the DOS prompt.
Press the Enter key when the initial screen appears.

Step 2. Select the menu option "File Open."
Type "cs\4S1.dat" to load the input file.

Step 3. Select the menu option "Data." This allows viewing of the loaded data. The cursor keys may be used to scroll up and down, as well as left and right, on the screen. When done, press the Esc key.

Step 4. Select the menu option "Grassroots Point." Type 20 for Delta T_{min}.

Step 5. Within a few seconds, the following targeting results appear on the screen (The cursor keys may be used to scroll up and down.):
Hot utility = 605; cold utility = 525;
pinch temperature(s) = 115 (see Table 1.2).
Problem Table (as in Table 1.1).
NUmin = 5, NUmer = 7, Nloops = 2 (see Table 1.11).

Area and Shell Targets (see Tables 1.10 and 1.13b):

	Area Countercurrent	Area Shell & Tube	Shells Approx.	Shells Accurate
Below Pinch	1005.78	1176.17	7	7
Above Pinch	306.78	312.70	2	3
Total	1312.57	1488.86	9	10

Operating Cost = 77850

	Countercurrent	1-2 Shell & Tube
Capital Cost	574173.12	641592.62
Annualized Capital Cost	184942.31	206658.27
Total Annualized Cost	262792.31	284508.25

Step 6. Press "C" to view the composites (HCC and CCC as in Figure 1.4).
Press "G" to view the GCC (as in Figure 1.5).
Press "T" to view the driving force plot in terms of T_{hot} vs. T_{cold}.
Press "D" to view the driving force plot in terms of ΔT vs. T_{cold} (see Figure 4.22).
Press "B" to view the balanced composites (BHCC and BCCC as in Figure 1.6).
Press "U" to view the utility GCC with the process GCC (Figure 1.5).
Press the Esc key after viewing each of the graphs.
The above results hold for $X_p = 0.9$ (Equation 1.19). Next, let X_p be changed to 0.8.

Step 7. Select the menu option "Grassroots User_Defaults." Type 0.8 for X_p.

Step 8. Select the menu option "Grassroots Point." Type 20 for Delta T_{min}.
Within a few seconds, the targeting results appear on the screen.
There is no change in the energy and unit targeting results.

Area and Shell Targets:

	Area Countercurrent	Area Shell & Tube	Shells Approx.	Shells Accurate
Below Pinch	1005.78	1067.62	9	9
Above Pinch	306.78	312.70	3	4
Total	1312.57	1380.32	12	13

	Countercurrent	1-2 Shell & Tube
Capital Cost	574173.12	636672.25
Annualized Capital Cost	184942.31	205073.41
Total Annualized Cost	262792.31	282923.41

At this stage, the various plots may be viewed as in step 6.

Step 9. Select the menu option "File Quit" to end this HX run.

Example A.2 Determination of Conventional Continuous Targets

For the stream data in Case Study 4S1, determine the utility, pinch, area, units, and cost targets for a grassroots design over a range of ΔT_{min} from 5° to 50°C in steps of 5°C. Consider the following cases:

countercurrent exchangers of carbon-steel; and
nonuniform exchanger specifications as given in Example 1.9.

Solution. The input data file 4S1.dat in the CS subdirectory contains the
STREAM DATA, the UTILITY DATA, and the COST DATA as well as the
necessary STREAM SPECS DATA. Note that the identifiers (see Table 1.14)
correspond to the capital cost coefficients (a, b, and c) for various exchanger
specifications given in the separate data file called "HXSPEC.CST."

Step 1. Type "HX" at the DOS prompt.

Step 2. Select the menu option "File Open."
Type "cs\4S1.dat" to load the input file.

Step 3. Select the menu option "Data." The identifiers corresponding to the
nonuniform exchanger specifications appear under STREAM
SPECS DATA. There is no DELTA T DATA.

Step 4. Select the menu option "Grassroots Continuous."
Type 5 for lower limit, 50 for upper limit, and 5 for step size. This
sets the range for ΔT_{min}.

Step 5. The calculations are performed for the various ΔT_{min} values, and a
table of supertargeting results appears on the screen after a while
in the following format. The results correspond to countercurrent
exchangers of carbon-steel and agree with those in Table 1.15.

DTmin	Qhu	Tpinch	Area	Numer	Ocost	Ccost	TACost
5	200	122.5	2157.441	6	25200	708932	253548
-	---	-----	--------	--	-----	------	------
50	1655	100	951.415	7	214350	490614	372378

Step 6. Press "U" to view the utility vs. ΔT_{min} plot (see Figure 2.2). The two
curves correspond to the hot and cold utility consumptions.
Press "P" to view the pinch temperature vs. ΔT_{min} plot (see Figure
2.2).
Press "A" to view the area (countercurrent) vs. ΔT_{min} plot (see Figure
2.4).
Press "C" to view the costs (annual) vs. ΔT_{min} plot (see Figure 2.4).
The operating cost curve increases with ΔT_{min} whereas annualized
capital cost curve decreases with ΔT_{min}.
Press "T" to view the total annual cost vs. ΔT_{min} plot (see Figures 1.9
and 2.4).
Note the calculations are performed only at the ΔT_{min} values desired
by the user and not at the significant shifts (see Chapter 2). Also,

topology traps are not accurately detected by the HX student version.

Step 7. Select the menu option "Grassroots Run_Options." Use the spacebar to toggle the "Stream Specs Data" option on. Note that the Stream Specs Data option is active, but the Delta T Data option is not.

Step 8. Select the menu option "Grassroots Continuous" again.

Type 5 for lower limit, 50 for upper limit, and 5 for step size.

Step 9. The calculations are performed for the various ΔT_{min} values, and a table of supertargeting results for nonuniform exchanger specifications appears on the screen in the following format.

DTmin	Qhu	Tpinch	Area*	Numer	Ocost	Ccost	TACost
5	200	122.5	11412.94	6	25200	2218929	739922
-	---	-----	--------	--	-----	------	------
50	1655	100	4579.576	7	214350	1212071	604760

The above results may be compared with those in Table 1.15. Note that the areas reported above are fictitious ($A*$ as in Example 1.9).

At this stage, the various plots may be viewed as in step 6.

Step 10. Select the menu option "File Quit" to end this HX run.

Example A.3 Determination of Retrofit Continuous Targets

For the stream data in Case Study 4S1, determine the retrofit targets for an existing design (Figure 1.2) over a range of ΔT_{min} from 5° to 35°C in steps of 5°C. Given: Payback period = 2 years (approximate).

Solution. The input data file 4S1.dat in the CS subdirectory contains the necessary STREAM DATA and the COST DATA.

Step 1. Type "HX" at the DOS prompt.

Step 2. Select the menu option "File Open."

Type "cs\4S1.dat" to load the input file.

Step 3. Select the menu option "Retrofit."

Type 1400 (or 1320) for current hot (or cold) utility usage.

Type 615.11 for existing area.

Type 310 for area/match.

Type 310 for area/shell.

Type 2.01 for payback (yrs).

Type 5 for lower limit, 35 for upper limit, and 5 for step size. This sets the range for ΔT_{min}.

Step 4. The calculations are performed for the various ΔT_{min} values and the retrofit targeting results appear on the screen after a while. The results agree with those obtained in Problem 7.2 as indicated below:

α (existing) = 0.73778 (corresponding to ΔT_{min} = 42.714).

Retrofit curve on area-energy plot (as in Table 7.2).

Analysis based on energy saved and area required (as in Table 7.3, with two more columns for payback at $\Delta\alpha$ = 1 and $\Delta\alpha$ = 0.74).

Target ΔT_{min} for retrofitting = 20°C; target savings = 103350 \$/yr; and target investment = 207122.42 \$/yr (for payback = 2.01 yrs based on $\Delta\alpha$ = 1 criterion).

Step 5. Press "A" to view the area vs. energy (hot utility) plot (see Figure 7.2). The lowermost curve is the ideal target whereas the other two curves correspond to likely retrofit paths. The upper retrofit curve is based on $\Delta\alpha$ = $\alpha_{existing}$ = 0.74, and the lower retrofit curve is based on $\Delta\alpha$ = 1.

Press "I" to view the investment vs. annual savings plot (Figure 7.3).

Press "P" to view the payback vs. ΔT_{min} plot.

Press "N" to view the investment vs. ΔT_{min} plot.

The three plots (investment vs. savings, payback vs. ΔT_{min}, and investment vs. ΔT_{min}) allow establishment of targets based on economic criteria like payback and investment ceiling. Each of the three plots has two curves: the upper curve corresponds to $\Delta\alpha$ = $\alpha_{existing}$ = 0.74 whereas the lower curve corresponds to $\Delta\alpha$ = 1.

Step 6. Select the menu option "File Quit" to end this HX run.

Example A.4 Determination of Diverse Pinch Point Targets

For the stream data in Case Study 4S1t, determine the energy and area targets for a grassroots design for Q''_{min} = 8.1818 kW/m^2.

Solution. The input data file 4S1t.dat in the CS subdirectory contains the relevant STREAM DATA and UTILITY DATA. It also contains DELTA T DATA, which will be used later in the example.

Step 1. Type "HX" at the DOS prompt.

Step 2. Select the menu option "File Open."
Type "cs\4S1t.dat" to load the input file.

Step 3. Select the menu option "Grassroots Diverse."
Type 8.1818 for Q''_{min}.

Step 4. Within a few seconds, the following targeting results appear:
Hot utility = 1980; cold utility = 400;
pinch temperature(s) - shifted = 44.09 (see Table 2.14).
Problem table (diverse-type heat cascade as in Table 2.13).

Countercurrent area (diverse BATH formula) = 393.707.

Step 5. Press "C" to view the composites (HCC and CCC with shifted temperatures used on the ordinate).

Press "G" to view the GCC.

The above results hold for diverse coefficient $z = 1.0$ (i.e., the temperature shift is $Q"_{min}/h_j$). Next, let diverse coefficient $z = 0.5$ (i.e., the temperature shift is $Q"_{min}/\sqrt{h_j}$).

Step 6. Select the menu option "Grassroots User_Defaults." Type 0.5 for z.

Step 7. Select the menu option "Grassroots Diverse."

Type 8.1818 for $Q"_{min}$.

Step 8. Within a few seconds, the following targeting results appear:

Hot utility = 1695.9; cold utility = 115.9;

pinch temperature(s) - shifted = 38.3.

Problem table (diverse-type heat cascade).

Countercurrent area (diverse BATH formula) = 295.388.

At this stage, the various plots may be viewed as in step 5.

Can the above results be obtained by using stream-dependent ΔT_{min} values? The required values are given by 2 $Q"_{min}/h_j$ for $z = 1.0$. Thus, for the six streams, the values may be calculated as 81.82 (H1), 8.18 (H2), 81.82 (C3), 8.18 (C4), 4.09 (HU), and 8.18 (CU).

Step 9. Select the menu option "Data." The above stream-dependent ΔT_{min} values appear under DELTA T DATA. There is no STREAM SPECS DATA.

Step 10. Select the menu option "Grassroots Run_Options." Use the cursor keys to move to the "Delta T Data" option, and use the spacebar to toggle the option on.

Step 11. Select the menu option "Grassroots Point."

The targeting results appear on the screen. The energy targets are in agreement with those obtained earlier in step 4:

Hot utility = 1980; cold utility = 400;

pinch temperature(s) - shifted = 44.09.

Countercurrent area = 393.263 (see Table 2.25).

Step 12. Select the menu option "File Quit." This will end the HX run.

With proper understanding and some creativity on the part of the user, the HX software may be used for a variety of problems (e.g., single-component MENS, area for a single 1-2 heat exchanger, RPA, TAM, TSM, RPP, and process modifications).

Note that installation instructions and other useful information are included in the README.TXT file on the accompanying disk. Users are strongly urged to go through the README.TXT file and Examples A.1 - A.4 carefully before using the HX software on other problems.

APPENDIX B

HENS Case Studies from Literature

This appendix contains published case studies for HENS. All the case studies are presented in a format consistent with the input data (.dat) files required by the HX software package described in Appendix A. Each case study is identified by a code (e.g., in 4S1h, 4 stands for the number of streams in the problem, S is the starting letter in the last name of the first author of the publication in which the problem appears, and 1h is a tag to identify the particular data set). Also provided below are the references in which the case study appears and the units used. These items occupy the first few lines (commented by //) in the data file for each case study. Finally, all pertinent data are classified using keywords as discussed in Section A.2.

//Case Study 3T1
//Ref: Townsend & Linnhoff (1983)
//see Problems 8.A, 8.B
//Units: temp in °C, MCp in MW/°C
STREAM DATA

650,	550,	0.225,	0.0,	H1a
550,	450,	0.214,	0.0,	H1b
450,	350,	0.202,	0.0,	H1c
350,	260,	0.19111,	0.0,	H1d
260,	170,	0.18111,	0.0,	H1e
100,	264,	0.09512,	0.0,	C2a
264,	264.1,	348,	0.0,	C2b
264.1,	300,	0.07778,	0.0,	C2c
300,	350,	0.06,	0.0,	C2d
350,	400,	0.054,	0.0,	C2e
400,	650,	0.05,	0.0,	C2f
25,	155,	0.11462,	0.0,	C3a
155,	200,	0.3,	0.0,	C3b
200,	300,	0.142,	0.0,	C3c
300,	400,	0.147,	0.0,	C3d
400,	500,	0.154,	0.0,	C3e
500,	650,	0.15533,	0.0,	C3f

//Case Study 4A1
//Ref: Ahmad & Smith (1989)
//see Problems 1.B, 3.B
//Units: temp in °C, MCp in kW/°C,
//area in m^2
STREAM DATA

120,	60,	8,	0.2,	H1
160,	40,	10,	0.2,	H2
10,	100,	2,	0.2,	C3
80,	115,	60,	0.2,	C4

UTILITY DATA

180,	179,	0,	0.2,	HU
10,	20,	0,	0.2,	CU

//Case Study 4A2
//Ref: Ahmad & Smith (1989)
//Also: Trivedi et al. (1989b), (1990a)
//Also: Suaysompol & Wood (1991)
//Also: Wood et al. (1991)
//Also: Jezowski (1991a)
//see Problem 1.B
//see Problems 3.A, 4.A (data modified)
//see Problems 4.E, 5.C, 5.D

//see Problem 5.B (data modified)
//Units: temp in °C, MCp in kW/°C,
//area in m²
STREAM DATA
```
300,  80,  30,  0.4,  H1
200,  40,  45,  0.4,  H2
 40, 180,  40,  0.4,  C3
140, 280,  60,  0.4,  C4
```
UTILITY DATA
```
400, 399,   0,  0.4,  HU
 10,  11,   0,  0.4,  CU
```

//**Case Study 4C1**
//Ref: Colberg & Morari (1990)
//see Problems 1.A, 10.C
//Units: temp in K, MCp in kW/K,
//area in m²
STREAM DATA
```
395, 343,   4,  2.0,  H1
405, 288,   6,  0.2,  H2
293, 493,   5,  2.0,  C3
353, 383,  10,  0.2,  C4
```
UTILITY DATA
```
520, 519,   0,  2.0,  HU
278, 288,   0,  2.0,  CU
```

//**Case Study 4C2 (also called 4TC2)**
//Ref: Colbert (1982)
//Also: Linnhoff & Flower (1978)
//Also: Linnhoff & Hindmarsh (1983)
//Also: Trivedi et al. (1989b)
//Also: Jezowski (1991a)
//see Problems 5.B, 5.D
//Units: temp in °C, MCp in kW/°C
STREAM DATA
```
180,  40,   2,  0.0,  H1
150,  40,   4,  0.0,  H2
 60, 180,   3,  0.0,  C3
 30, 130, 2.6,  0.0,  C4
```

//**Case Study 4D1**
//Ref: Douglas (1988)

//see Problems 3.A, 4.A
//Units: temp in °F,
//MCp in 10**3 Btu/hr °F
STREAM DATA
```
250, 120,   1,  0.0,  H1
200, 100,   4,  0.0,  H2
 90, 150,   3,  0.0,  C3
130, 190,   6,  0.0,  C4
```

//**Case Study 4G1**
//Ref: Gundersen & Grossmann (1990)
//Also: Colberg & Morari (1990)
//Also: Jezowski (1991a)
//see Problems 1.A, 5.D
//Units: temp in K, MCp in kW/K,
//area in m², cost in $
STREAM DATA
```
423, 333,  20,  0.1,  H1
363, 333,  80,  0.1,  H2
293, 398,  25,  0.1,  C3
298, 373,  30,  0.1,  C4
```
UTILITY DATA
```
453, 452,   0,  0.1,  HU
283, 288,   0,  0.1,  CU
```
COST DATA
```
   0,  0
8600, 670, 0.83
   0,  0.0
```

//**Case Study 4L1 (also called 4SP1)**
//Ref: Lee et al. (1970)
//Also: Papoulias & Grossmann (1983)
//see Problems 2.B, 10.A
//Units: temp in °C, MCp in kW/°C
STREAM DATA
```
160,  93,  8.79,  0.0,  H1
249, 138, 10.55,  0.0,  H2
 60, 160,  7.62,  0.0,  C3
116, 260,  6.08,  0.0,  C4
```
UTILITY DATA
```
270, 269,   0,  0.0,  HU
 38,  82,   0,  0.0,  CU
```

//**Case Study 4L2**
//Ref: Linnhoff et al. (1982)
//see Problems 2.B, 3.A, 4.A, 4.D
//see Problem 9.A
//Units: temp in °C, MCp in kW/°C
STREAM DATA
  170,   60, 3.0, 0.0, H1
  150,   30, 1.5, 0.0, H2
   20, 135, 2.0, 0.0, C3
   80, 140, 4.0, 0.0, C4
UTILITY DATA
  200, 199,   0, 0.0, HU
   15, 16,   0, 0.0, CU

//**Case Study 4L3**
//Ref: Linnhoff et al. (1982)
//see Problems 3.C, 4.C
//Units: temp in °C, MCp in kW/°C
STREAM DATA
  180,   40, 20, 0.0, H1
  150,   40, 40, 0.0, H2
   60, 180, 30, 0.0, C3
   30, 130, 22, 0.0, C4
UTILITY DATA
  200, 199, 0, 0.0, HU(assumed)
  110, 111, 0, 0.0, VLP
   15, 16, 0, 0.0, CU(assumed)

//**Case Study 4L4**
//Ref: Linnhoff & Ahmad (1990)
//Also: Jezowski (1991a)
//see Problem 5.D
//Units: temp in °C, MCp in MW/°C
STREAM DATA
  170,   45, 3.0, 0.0, H1
  150,   30, 1.2, 0.0, H2
   20, 135, 2.0, 0.0, C3
   80, 150, 4.0, 0.0, C4

//**Case Study 4P1 (also called 4SP2)**
//Ref: Ponton & Donaldson (1974)
//see Problems 2.B, 3.F

//Units: temp in °F,
//MCp in 10**4 Btu/hr °F
STREAM DATA
  500, 110, 2, 0.0, H1
  430, 230, 5, 0.0, H2
  400, 110, 3, 0.0, H3
   25, 420, 7, 0.0, C4

//**Case Study 4S1**
//Ref: Shenoy (1995) - this book
//see Problems 3.H, 5.B, 5.D
//Units: temp in °C, MCp in kW/°C,
//area in m^2, cost in $, plant life in yr
STREAM DATA
  175, 45, 10, 0.2, H1
  125, 65, 40, 0.2, H2
   20, 155, 20, 0.2, C3
   40, 112, 15, 0.2, C4
UTILITY DATA
  180, 179, 0, 0.2, HU
   15, 25, 0, 0.2, CU
COST DATA
  120, 10
30000, 750, 0.81
   5, 0.1
STREAM SPECS DATA
  4, 1, 2, 4, 3, 3

//**Case Study 4S1h**
//Ref: Shenoy (1995) - this book
//see Problems 4.I, 5.E
//Units: temp in °C, MCp in kW/°C,
//area in m^2
STREAM DATA
  175, 45, 25, 0.2, H1
  125, 65, 40, 2.0, H2
   20, 155, 20, 0.2, C3
   40, 112, 15, 2.0, C4
UTILITY DATA
  180, 179, 0, 4.0, HU
   15, 25, 0, 2.0, CU

//**Case Study 4S1t**
//Ref: Shenoy (1995) - this book
//see Problem 4.H
//Units: temp in °C, MCp in kW/°C,
//area in m^2
STREAM DATA
 175, 45, 10, 0.2, H1
 125, 65, 15, 2.0, H2
  20, 155, 20, 0.2, C3
  40, 112, 15, 2.0, C4
UTILITY DATA
 180, 179,   0, 4.0, HU
  15, 25,   0, 2.0, CU

//**Case Study 4T1**
//Ref: Trivedi et al. (1990a)
//see Problems 3.B, 4.B
//Units: temp in °C, MCp in kW/°C
STREAM DATA
 250, 70, 3.5, 0.0, H1
 170, 70, 6.0, 0.0, H2
  60, 160, 5.0, 0.0, C3
 110, 260, 4.0, 0.0, C4

//**Case Study 5A1**
//Ref: Ahmad et al. (1990)
//see Problem 4.F
//Units: temp in °C, MCp in kW/°C,
//area in m^2
STREAM DATA
 159, 77, 22.85, 1.0, H1
 267, 80,   2.04, 0.4, H2
 343, 90,   5.38, 5.0, H3
  26, 127,   9.33, 0.1, C4
 118, 265, 19.61, 5.0, C5
UTILITY DATA
 300, 299, 0.0, 0.5, HU
  20, 60, 0.0, 2.0, CU

//**Case Study 5J1**
//Ref: Jezowski & Friedler (1992)
//see Problem 1.F

//Units: temp in °C, MCp in kW/°C
STREAM DATA
 249, 138, 10.55, 0.0, H1
 126, 80, 16.50, 0.0, H2
 160, 93, 8.79, 0.0, H3
  60, 160, 7.62, 0.0, C4
 116, 260, 6.08, 0.0, C5
UTILITY DATA
 290, 289, 0.0, 0.0, HU
 150, 151, 0.0, 0.0, CU1
 130, 131, 0.0, 0.0, CU2
  38, 39, 0.0, 0.0, CU3

//**Case Study 5L1**
//Ref: Linnhoff (1986)
//see Problem 3.D
//Units: temp in °C, MCp in kW/°C
STREAM DATA
1000, 50, 0.7, 0.0, H1
 540, 50, 1.0, 0.0, H2
 300, 500, 2.25, 0.0, C3
 100, 500, 1.4, 0.0, C4
 100, 101, 275, 0.0, C5
UTILITY DATA
1500, 160, 0.0, 0.0, HU
   0,   5, 0.0, 0.0, CU(assumed)

//**Case Study 5M1 (also called 5SP1)**
//Ref: Masso & Rudd (1969)
//see Problem 2.B
//Units: temp in °F, MCp in Btu/hr °F
STREAM DATA
 480, 250, 3.15, 0.0, H1
 400, 150, 2.52, 0.0, H2
 200, 400, 2.42, 0.0, C3
 100, 400, 2.16, 0.0, C4
 150, 360, 2.45, 0.0, C5

//**Case Study 5N1**
//Ref: Nishimura (1980)
//Also: Colberg & Morari (1990)
//see Problems 1.A, 10.C

//Units: temp in K, MCp in kW/K,
//area in m^2
STREAM DATA
443, 293, 0.5, 2.0000, H1
416, 393, 2.0, 0.2857, H2
438, 408, 0.5, 0.0645, H3
448, 423, 1.0, 0.0408, H4
273, 434, 1.0, 2.0000, C5

//**Case Study 5T1**
//Ref: Tjoe & Linnhoff (1986)
//see Problem 7.A
//Units: temp in °C, MCp in kW/°C,
//area in m^2, cost in £
STREAM DATA
159, 77, 228.5, 0.4, H1
267, 80, 20.4, 0.3, H2
343, 90, 53.8, 0.25, H3
26, 127, 93.3, 0.15, C4
118, 265, 196.1, 0.5, C5
UTILITY DATA
400, 399, 0.0, 0.5, HU (assumed)
20, 25, 0.0, 2.0, CU (assumed)
COST DATA
63.36, 0
8600, 670, 0.83
5, 0.1 (assumed)

//**Case Study 6C1**
//Ref: Ciric & Floudas (1989)
//see Problem 2.B
//Units: temp in °C, MCp in kW/°C
STREAM DATA
500, 350, 10, 0.0, H1
450, 350, 12, 0.0, H2
400, 320, 8, 0.0, H3
300, 480, 9, 0.0, C4
340, 420, 10, 0.0, C5
340, 400, 8, 0.0, C6
UTILITY DATA
540, 539, 0, 0.0, HU
300, 320, 0, 0.0, CU

//**Case Study 6G1**
//Ref: Gundersen & Grossmann (1990)
//Also: Rev & Fonyo (1991)
//see Problem 2.D
//Units: temp in °C, MCp in kW/°C,
//area in m^2
STREAM DATA
300, 200, 10, 0.1, H1
200, 190, 100, 1.0, H2
190, 170, 50, 1.0, H3
160, 180, 50, 0.1, C4
180, 190, 100, 1.0, C5
190, 230, 25, 1.0, C6
UTILITY DATA
350, 349, 0, 4.0, HU
20, 50, 0, 2.0, CU

//**Case Study 6L1 (also called 6SP1)**
//Ref: Lee et al. (1970)
//see Problem 2.B
//Units: temp in °F, MCp in Btu/hr °F
STREAM DATA
520, 300, 2.38, 0.0, H1
440, 150, 2.80, 0.0, H2
390, 150, 3.36, 0.0, H3
100, 430, 1.60, 0.0, C4
200, 400, 2.63, 0.0, C5
180, 350, 3.27, 0.0, C6

//**Case Study 6M1**
//Ref: Makwana & Shenoy (1993)
//see Problem 2.A
//Units: temp in °C, MCp in kW/°C,
STREAM DATA
113, 30, 3, 0.0, H1
113, 100, 2, 0.0, H2
60, 30, 3, 0.0, H3
25, 60, 4, 0.0, C4
50, 110, 2.6, 0.0, C5
100, 110, 4.4, 0.0, C6

//**Case Study 7C1**
//Ref: Colberg & Morari (1990)
//see Problems 1.A, 10.C
//Units: temp in K, MCp in kW/K,
//area in m^2
STREAM DATA

626,	586,	9.802,	1.25,	H1
620,	519,	2.931,	0.05,	H2
528,	353,	6.161,	3.20,	H3
497,	613,	7.179,	0.65,	C4
389,	576,	0.641,	0.25,	C5
326,	386,	7.627,	0.33,	C6
313,	566,	1.690,	3.20,	C7

UTILITY DATA

650,	649,	0.000,	3.5,	HU
293,	308,	0.000,	3.5,	CU

//**Case Study 7M1 (also called 7SP1)**
//Ref: Masso & Rudd (1969)
//see Problem 2.B
//Units: temp in °F, MCp in Btu/hr °F
STREAM DATA

520,	300,	2.38,	0.0,	H1
440,	150,	2.80,	0.0,	H2
390,	150,	3.36,	0.0,	H3
100,	430,	1.60,	0.0,	C4
350,	410,	1.98,	0.0,	C5
200,	400,	2.63,	0.0,	C6
180,	350,	3.27,	0.0,	C7

//**Case Study 7M2 (also called 7SP2)**
//Ref: Masso & Rudd (1969)
//see Problem 2.B
//Units: temp in °F, MCp in Btu/hr °F
STREAM DATA

590,	400,	2.376,	0.0,	H1
471,	200,	1.577,	0.0,	H2
533,	150,	1.320,	0.0,	H3
200,	400,	1.600,	0.0,	C4
100,	430,	1.600,	0.0,	C5
300,	400,	4.128,	0.0,	C6
150,	280,	2.624,	0.0,	C7

//**Case Study 7P1 (also called 7SP4)**
//Ref: Papoulias & Grossmann (1983)
//see Problem 10.B
//Units: temp in °F, MCp in Btu/hr °F
STREAM DATA

675,	150,	15,	0.0,	H1
590,	450,	11,	0.0,	H2
540,	115,	4.5,	0.0,	H3
430,	345,	60,	0.0,	H4
400,	100,	12,	0.0,	H5
300,	230,	125,	0.0,	H6
60,	710,	47,	0.0,	C1

UTILITY DATA

800,	799,	0.0,	0.0,	HU
80,	140,	0.0,	0.0,	CU

//**Case Study 7S1**
//Ref: Shenoy (1995) - this book
//see Problem 7.E
//Units: temp in °C, MCp in kW/°C,
//area in m^2, cost in $, plant life in yr
STREAM DATA

101,	40,	95.6,	0.464,	LN
193,	30,	64.8,	0.402,	K
158,	30,	62.6,	0.427,	HN
346,	30,	56.1,	0.397,	HD
263,	50,	115.4,	0.431,	LD
419,	80,	174.8,	0.518,	B
30,	390,	573.0,	0.434,	C

UTILITY DATA

520,	519,	0,	5.0,	HU
5,	10,	0,	2.5,	CU

COST DATA
120, 10
10000, 750, 0.81
  5, 0.1

//**Case Study 7T1**
//Ref: Trivedi et al. (1990a)
//see Problem 2.B
//Units: temp in °C, MCp in kW/°C

STREAM DATA
```
160, 110,   7.03,  0.0,  H1
249, 138,   8.44,  0.0,  H2
227, 106,  11.81,  0.0,  H3
271, 146,   5.60,  0.0,  H4
 96, 160,   9.14,  0.0,  C5
116, 217,   7.27,  0.0,  C6
140, 250,  18.00,  0.0,  C7
```

//Case Study 8A1
//Ref: Ahmad & Polley (1990)
//see Problem 7.D
//Units: temp in °C, MCp in kW/°C
//area in m^2, cost in $
STREAM DATA
```
140,  40, 517.000, 0.2, H1
160, 120, 825.000, 1.5, H2
210,  45,  46.662, 0.8, H3
260,  60, 110.000, 0.7, H4
280, 210, 392.854, 1.0, H5
350, 170,  61.116, 0.5, H6a
170,  45,  52.800, 0.4, H6b
380, 160, 160.006, 0.4, H7a
160,  80, 137.500, 0.3, H7b
 10, 130, 430.837, 0.5, C8a
130, 270, 550.000, 0.7, C8b
270, 385, 908.699, 0.9, C8c
```
UTILITY DATA
```
500, 499, 0.0, 0.50, HU (assumed)
-20, -15, 0.0, 0.50, CU (assumed)
```
COST DATA
```
57.6,  9.6
8600,  670, 0.83
   5,  0.1 (assumed)
```

//Case Study 8F1
//Ref: Farhanieh & Sunden (1990)
//see Problem 7.C
//Units: temp in °C, MCp in kW/°C
//area in m^2, cost in SEK
//heat transfer coeffts vary linearly
//between supply and target temps.

STREAM DATA
```
109,  40,  42.03, (0.50, 0.50), H1
295,  45,   3.24, (0.12, 0.08), H2
271,  60,  10.89, (0.12, 0.08), H3
287,  60,   1.98, (0.11, 0.07), H4
361, 190,  31.58, (0.10, 0.06), H5
 66, 340,  40.51, (0.06, 0.09), C6
 10,  95,   4.70, (0.30, 0.30), C7
137, 138,   1000, (0.50, 0.50), C8
```
UTILITY DATA
```
500, 499, 0.0,  0.50, HU (assumed)
-20, -15, 0.0,  0.50, CU (assumed)
```
COST DATA
```
1600, 400
   0, 12500, 0.65
   5, 0.1 (assumed)
```

//Case Study 9A1
//Ref: Ahmad & Smith (1989)
//see Problems 1.B, 2.B
//Units: temp in °C, MCp in kW/°C,
//area in m^2
STREAM DATA
```
327,  30, 100, 1.0, H1
220, 160, 160, 1.0, H2
220,  60,  60, 1.0, H3
160,  45, 200, 1.0, H4
100, 300, 100, 1.0, C5
 35, 164,  70, 1.0, C6
 80, 125, 175, 1.0, C7
 60, 170,  60, 1.0, C8
140, 300, 200, 1.0, C9
```
UTILITY DATA
```
330, 329, 0, 1.0, HU
 15,  40, 0, 1.0, CU
```
DELTA T DATA
20, 20, 20, 20, 20, 20, 20, 20, 20, 20, 15

//Case Study 9A2 (Aromatics Plant)
//Ref: Ahmad & Linnhoff (1989)
//Also: Linnhoff & Ahmad (1990)
//Also: Suaysompol & Wood (1991)

//Also: Wood et al. (1991)
//Also: Smith & Delaby (1991)
//see Problems 2.B, 3.A, 5.C
//see Problem 8.C (data modified)
//Units: temp in °C, MCp in MW/°C,
//area in m^2, cost in $, plant life in yr
STREAM DATA

327,	40,	0.1,	0.0005,	H1
220,	160,	0.16,	0.0005,	H2
220,	60,	0.06,	0.0005,	H3
160,	60,	0.4,	0.0005,	H4
100,	300,	0.1,	0.0005,	C5
35,	164,	0.07,	0.0005,	C6
85,	138,	0.35,	0.0005,	C7
60,	170,	0.06,	0.0005,	C8
140,	300,	0.2,	0.0005,	C9

UTILITY DATA

330,	329,	0,	0.0005,	HU
15,	40,	0,	0.0005,	CU

COST DATA
75000, 10000
0, 700, 0.83
5, 0.05

//**Case Study 9H1**
//Ref: Hall et al. (1990)
//see Problems 1.C, 1.D, 1.E, 2.B
//Units: temp in °C, MCp in kW/°C,
//area in m^2
STREAM DATA

120,	65,	50,	0.50,	H1
80,	50,	300,	0.25,	H2
135,	110,	290,	0.30,	H3
220,	95,	20,	0.18,	H4
135,	105,	260,	0.25,	H5
65,	90,	150,	0.27,	C6
75,	200,	140,	0.25,	C7
30,	210,	100,	0.15,	C8
60,	140,	50,	0.45,	C9

UTILITY DATA

250,	249,	0,	0.35,	HU
15,	16,	0,	0.20,	CU

COST DATA
120, 10
30800, 750, 0.81
6, 0.1
STREAM SPECS DATA
3, 3, 4, 4, 3, 4, 3, 3, 4, 3, 3

//**Case Study 9P1 (Aromatics Plant)**
//Ref: Polley & Panjeh Shahi (1991)
//see Problem 6.F
//Units: temp in °C, MCp in kW/°C,
//area in m^2
STREAM DATA

327,	40,	100,	0.5,	H1
220,	160,	160,	0.5,	H2
220,	60,	60,	0.5,	H3
160,	45,	400,	0.5,	H4
100,	300,	100,	0.5,	C5
35,	164,	70,	0.5,	C6
85,	138,	350,	0.5,	C7
60,	170,	60,	0.5,	C8
140,	300,	200,	0.5,	C9

UTILITY DATA

330,	230,	0,	1.0,	HU
10,	30,	0,	2.5,	CU

//**Case Study 9T1 (Aromatics Plant)**
//Ref: Tjoe & Linnhoff (1986, 1987)
//see Problem 7.B
//Units: temp in °C, MCp in kW/°C,
//area in m^2, cost in £
STREAM DATA

327,	30,	100,	0.8,	H1
220,	160,	160,	0.5,	H2
220,	60,	60,	2.0,	H3
160,	45,	200,	0.4,	H4
100,	300,	100,	5.0,	C5
35,	164,	70,	1.0,	C6
80,	125,	175,	0.5,	C7
60,	170,	60,	0.2,	C8
140,	300,	200,	0.8,	C9

UTILITY DATA
500, 495,   0,   0.2, HU (assumed)
−20, −10,   0,   0.2, CU (assumed)
COST DATA
57.6, 0
  0, 252, 1.0
  5, 0.1 (assumed)

//**Case Study 10L1**
//Ref: Linnhoff & Ahmad (1986)
//Also: Trivedi et al. (1989b)
//see Problem 5.A
//Units: temp in °C, MCp in kW/°C
STREAM DATA
  85,   45,   156.3,   0.0, H1
120,   40,   50.0,   0.0, H2
125,   35,   23.9,   0.0, H3
  56,   46, 1250.0,   0.0, H4
  90,   85, 1500.0,   0.0, H5
227,   75,   50.0,   0.0, H6
  40,   55,   466.7,   0.0, C7
  55,   65,   600.0,   0.0, C8
  65, 165,   195.0,   0.0, C9
  10, 170,   81.3,   0.0, C10

//**Case Study 10O1**
//Ref: O'Young & Linnhoff (1989)
//see Problems 3.G, 4.G
//Units: temp in °C, MCp in kW/°C,
//area in m^2
STREAM DATA
327,   30,   0.10,   0.05, H1
220, 160,   0.25,   0.05, H2
220,   60,   0.02,   0.05, H3
160,   45,   0.34,   0.05, H4
100, 300,   0.20,   0.05, C5
  35, 164,   0.07,   0.05, C6
  80, 125,   0.175,   0.05, C7
  60, 170,   0.06,   0.05, C8
140, 300,   0.20,   0.05, C9
  10,   50,   0.30,   0.05, C10

//**Case Study 10P1 (also called 10SP1)**
//Ref: Pho & Lapidus (1973)
//see Problem 2.B
//Units: temp in °F, MCp in Btu/hr °F
STREAM DATA
520, 300,   2.38,   0.0, H1
480, 280,   2.00,   0.0, H2
440, 150,   2.80,   0.0, H3
390, 150,   3.36,   0.0, H4
320, 200, 1.667,   0.0, H5
240, 431, 1.153,   0.0, C6
100, 430,   1.60,   0.0, C7
200, 400, 2.635,   0.0, C8
180, 350, 3.276,   0.0, C9
140, 320, 1.445,   0.0, C10

//**Case Study 28A1**
//Ref: Ahmad & Smith (1989)
//see Problem 1.B
//Units: temp in °C, MCp in kW/°C,
//area in m^2
STREAM DATA
230,   80, 30, 0.4, H1
200,   40, 45, 0.4, H2
  40, 180, 40, 0.4, C3
140, 280, 60, 0.4, C4
110,   45, 0.1, 0.4, H5
115,   40, 0.1, 0.4, H6
105,   40, 0.1, 0.4, H7
110,   42, 0.1, 0.4, H8
117,   48, 0.1, 0.4, H9
103,   50, 0.1, 0.4, H10
170, 270, 0.1, 0.4, C11
175, 265, 0.1, 0.4, C12
180, 275, 0.1, 0.4, C13
168, 277, 0.1, 0.4, C14
181, 267, 0.1, 0.4, C15
110,   45, 0.1, 0.4, H16
115,   40, 0.1, 0.4, H17
105,   40, 0.1, 0.4, H18
110,   42, 0.1, 0.4, H19
117,   48, 0.1, 0.4, H20

```
103,  50, 0.1,  0.4,  H21
170, 270, 0.1,  0.4,  C22
175, 265, 0.1,  0.4,  C23
180, 275, 0.1,  0.4,  C24
168, 277, 0.1,  0.4,  C25
181, 267, 0.1,  0.4,  C26
115,  42, 0.1,  0.4,  H27
117,  43, 0.1,  0.4,  H28
```
UTILITY DATA
```
300, 230,  0,  0.4,  HU
 10,  50,  0,  0.4,  CU
```

References

Note: References with an asterisk (after the year of publication) are not cited in the text of this book, and are provided for further reading.

Aguirre, P.A., Pavani, E.O. and Irazoqui, H.A. (1989) "Comparative analysis of pinch and operating line methods for heat and power integration", *Chem. Eng. Sci.*, **44**(4), 803-816.

Aguirre, P.A., Pavani, E.O. and Irazoqui, H.A. (1990) "Optimal synthesis of heat-and-power systems with multiple steam levels", *Chem. Eng. Sci.*, **45**, 117-129.

Ahmad, S. (1985) "Heat exchanger networks: cost tradeoffs in energy and capital", Ph.D. Thesis, University of Manchester Institute of Science and Technology, U.K.

Ahmad, S. and Linnhoff, B. (1984)* "Overall cost targets for heat exchanger networks", *IChemE 11th Annual Res. Meeting*, April, Bath, U.K.

Ahmad, S. and Petela, E. (1987)* "Supertarget: applications software for oil refinery retrofit", *AIChE Annual Meeting*, March 29-April 2, Houston, Texas.

Ahmad, S. and Shah, J.V. (1987)* "Supertarget: a software interface for pinch technologists", *AIChE Annual Meeting*, March 29-April 2, Houston, Texas.

Ahmad, S. and Linnhoff, B. (1989) "Supertargeting: Different process structures for different economics", *ASME J. Energy Resources Tech.*, September, **111** (3), 131-136; also, Ahmad, S. and Linnhoff, B. (1986) "Supertarget: optimisation of a chemical solvents plant - different process structures for different economics", *ASME Winter Annual Meeting*, December, Anaheim, Calif. *AES*, **2-1**, 15-21.

Ahmad, S. and Smith, R. (1989) "Targets and design for minimum number of shells in heat exchanger networks", *Trans. IChemE. Chem. Eng. Res. Des.*, **67**(5), September, 481-494; also, correspondence (1990), *Trans. IChemE. Chem. Eng. Res. Des.*, **68**, Part A, May, 299-301.

Ahmad, S. and Polley, G.T. (1990) "Debottlenecking of heat exchanger networks", *Heat Recovery Systems & CHP*, **10**(4), 369-385.

Ahmad, S. and Hui, D.C.W. (1991)* "Heat recovery between areas of integrity", *Comp. and Chem. Eng.*, **15**(12), 809-832.

Ahmad, S., Linnhoff, B. and Smith, R. (1988) "Design of multipass heat exchangers: an alternative approach", *ASME J. Heat Transfer*, May, **110**, 304-309.

Ahmad, S., Polley, G.T. and Petela, E.A. (1989)* "Retrofit of heat exchanger networks subject to pressure drop constraints", *AIChE Spring Meeting*, April, Houston, Paper No. 34a.

Ahmad, S., Linnhoff, B. and Smith, R. (1990) "Cost optimum heat exchanger networks - 2. Targets and design for detailed capital cost models", *Comp. and Chem. Eng.*, **14**(7), 751-767.

Akselvoll, K. and Loken, P.A. (1987)* "Automatic heat exchanger network synthesis", *CEF87 - The use of computers in chem. eng.*, Taormina, Italy.

Ali, Z. (1993) "Multiple utilities and multiple pinches in heat exchanger networks", B.Tech. Seminar Report, Indian Institute of Technology, Bombay.

Andrecovich, M.J. and Westerberg, A.W. (1985a)* "A simple synthesis method based on utility bounding for heat integrated distillation sequences", *AIChE J.*, **31**(3), 363.

Andrecovich, M.J. and Westerberg, A.W. (1985b)* "An MILP formulation for heat integrated distillation sequence synthesis", *AIChE J.*, **31**(9), 1461-1474.

Ashton, G.J., Linnhoff, B. and Obeng, E.D.A. (1988)* "Understanding process integration II", *IChemE Symp. Ser. No. 109*, 221.

Barton, J. (1989) "Pinch technology improves olefin heat recovery", *Hydrocarbon Processing*, February, 47-48.

Beautyman, A.C. and Cornish, A.R.H. (1984)* "The design of flexible heat exchanger networks", *1st U.K. National Heat Transfer Conf.*, July, Leeds, 547-565.

Bell, K.J. (1963) "Final report of the co-operative research program on shell-and-tube heat exchangers", *University of Delaware Eng. Exptl. Statn. Bulletin 5.*

Bell, K.J. (1978)* "Estimate S and T exchanger design fast", *Oil Gas J.*, December, **59**.

Bell, K.J. (1981) "Preliminary design of shell and tube heat exchangers", *in Heat Exchangers: Thermal-Hydraulic Fundamentals and Design*, Kakac, S., Bergles, A.E. and Mayinger, F. (eds.), Hemisphere Publishing Corp., Washington D.C., 559.

Bell, K.J. (1983) *Heat Exchanger Data Handbook*, Volume 3, Hemisphere Publishing Corp., Washington D.C.

Bell, K.J. (1984)* "Thermal design of heat-transfer equipment", *in Chemical Engineers' Handbook*, 6th ed., Perry, R.H. and Green, D. (eds.), McGraw-Hill, New York, 10-24.

Bell, K.J. (1987)* "Process heat-exchanger design: Qualitative factors in selection and applications", *in Recent Developments in Chemical Process* *and Plant Design*, Liu, Y.A., McGee, H.A. and Epperly, W.R. (eds.), John Wiley, New York, 41-69.

Benstead, R. and Sharman, F.W. (1990)* "Heat pumps and pinch technology", *Heat Recovery Systems & CHP*, **10**(4), 387-398.

Bhatwadekar, S. (1993) "Retrofitting by pinch technology", B.Tech. Project Report, Indian Institute of Technology, Bombay.

Boland, D. (1983) *The Chemical Engineer*, March, 24.

Boland, D. and Linnhoff, B. (1979)* "The preliminary design of networks for heat exchange by systematic methods", *The Chemical Engineer*, April 9-15, 222-228.

Bowman, R.A. (1936) "Mean temperature difference correction in multipass exchangers", *Ind. Eng. Chem.*, **28**, 541.

Bowman, R.A., Mueller, A.C. and Nagle, W.M. (1940) "Mean temperature difference in design", *Trans. ASME*, **62**, 283-294.

Brooke, A., Kendrick, D. and Meeraus, A. (1992), *GAMS: A User's Guide, Release 2.25*, The Scientific Press, San Francisco.

Calandranis, J. and Stephanopoulos, G. (1986)* "Structural operability analysis of heat exchanger networks", *Trans. IChemE. Chem. Eng. Res. Des.*, **64**(5), 347-364.

Calandranis, J. and Stephanopoulos, G. (1988)* "A structural approach to the design of control systems in heat exchanger networks", *Comp. and. Chem. Eng.*, **12**(7), 651-669.

Carlsson, A., Franck, P.A. and Berntsson, T. (1992)* "Retrofit of heat exchanger networks", *The 1992 IChemE Research Event*.

Carlsson, A., Franck, P.A. and Berntsson, T. (1993)* "Design better heat exchanger retrofits", *Chem. Eng. Prog.*, March, 87-96.

Cena, V., Mustacchi, C. and Natali, F. (1977)* "Synthesis of heat exchanger networks: a non-iterative approach", *Chem. Eng. Sci.*, **32**, 1227-1231.

Cerda, J. (1980)* "Transportation models for the optimal synthesis of heat exchanger networks", Ph. D. Thesis, Carnegie Mellon University, Pittsburgh.

Cerda, J., and Westerberg A.W. (1983)* "Synthesizing heat exchanger networks having restricted stream/stream matches using transportation problem formulations", *Chem. Eng. Sci.*, **38**(10), 1723-1740.

Cerda, J. and Galli, M.R. (1990)* "Synthesis of flexible heat exchanger networks - II. Nonconvex networks with large temperature variations", *Comp. and Chem. Eng.*, **14**(2), 213-225.

Cerda, J., Westerberg, A.W., Mason, D. and Linnhoff, B. (1983)* "Minimum utility usage in heat exchanger network synthesis - a transportation problem", *Chem. Eng. Sci.*, **38**(3), 373-387.

Cerda, J., Galli, M.R., Camussi, N. and Isla, M.A. (1990)* "Synthesis of flexible heat exchanger networks - I. Convex networks", *Comp. and Chem. Eng.*, **14**(2), 197-211.

Challand, T.B. and O'Reilly (1981)* "New engineering software for energy conservation projects", *Proc. 2nd Intl. Conf. Eng. Software, London*, 306-316.

Challand, T.B., Colbert, R.W. and Venkatesh, C.K. (1981) "Computerised heat exchanger networks", *Chem. Eng. Prog.*, July, **77**(7), 65-71.

Chang, C.-T., Chu, K.-K. and Hwang, J.-R. (1994)* "Application of the generalized stream structure in HEN synthesis", *Comp. and. Chem. Eng.*, **18**(4), 345-368.

Chato, J.C. and Damianides, ˙C (1986), *Int. J. Heat Mass Transfer*, **29**, 1079-1086.

Chen, J.J.J. (1987) "Letter to the editor: Comments on improvement on a replacement for the logarithmic mean", *Chem. Eng. Sci.*, **42**, 2488-2489.

Chen, B. Shen, J., Sun, Q. and Hu, S. (1989)* "Development of an expert system for synthesis of heat exchanger networks", *Comp. and. Chem. Eng.*, **13**(11/12), 1221-1227.

Cheng, W.B. and Mah, R.S.H. (1980)* "Interactive synthesis of cascaded refrigeration systems", *Ind. Eng. Chem. Proc. Des. Dev.*, **19**, 410.

Churchill, S.W. and Usagi, R. (1972) "A general expression for the correlation of rates of transfer and other phenomena", *AIChE J.*, **18**, 1121.

Ciric, A.R. and Floudas, C.A. (1989) "A retrofit approach for heat exchanger networks", *Comp. and Chem. Eng.*, **13**(6), 703-715.

Ciric, A.R. and Floudas, C.A. (1990a)* "Application of the simultaneous match-network optimization approach to the pseudo-pinch problem", *Comp. and Chem. Eng.*, **14**(3), 241-250.

Ciric, A.R. and Floudas, C.A. (1990b) "A comprehensive optimization model of the heat exchanger network retrofit problem", *Heat Recovery Systems & CHP*, **10**(4), 407-422.

Ciric, A.R. and Floudas, C.A. (1990c) "A mixed integer nonlinear programming model for retrofitting heat exchanger networks", *Ind. Eng. Chem. Res.*, **29**, 239-251.

Ciric, A.R. and Floudas, C.A. (1991)* "Heat exchanger network synthesis without decomposition", *Comp. and Chem. Eng.*, **15**, 385-396.

Clayton, R.W. (1986) *Energy Efficiency Office R&D reports RD/13/15 and RD/14/14*.

Colberg, R.D. (1989)* "Area, cost and resilience targets for heat exchanger networks", Ph.D. Thesis, California Institute of Technology, Pasadena.

Colberg, R.D. and Morari, M. (1990) "Area and capital cost targets for heat exchanger network synthesis with constrained matches and unequal heat transfer coefficients", *Comp. and Chem. Eng.*, **14**(1), 1-22.

Colberg, R.D., Morari, M. and Townsend, D.W. (1989) "A resilience target for heat exchanger network synthesis", *Comp. and Chem. Eng.*, **13**(7), 821-837.

Colbert, R.W. (1982) "Industrial heat exchange networks", *Chem. Eng. Prog.*, July, **78**(7), 47-54.

Colburn, A.P. (1933), "Mean temperature difference and heat transfer coefficient in liquid heat exchangers", *Ind. Eng. Chem.*, **25**(8), 873-877.

Colmenares, T.R. and Seider, W.D. (1987)* "Heat and power integration of chemical processes", *AIChE J.*, **33**(6), 898-915.

Colmenares, T.R. and Seider, W.D. (1989)* "Synthesis of utility systems integrated with chemical processes", *Ind. Eng. Chem. Res.*, **28**, 84-93.

Davison, R. (1992)* "Pinch helps energy efficiency drive", *European Chem. News*, 27 July.

Delaby, O. (1989)* "Minimum area in heat exchanger networks", M.Sc. Thesis, University of Manchester Institute of Science and Technology, U.K.

Dhole, V.R. and Linnhoff, B. (1992) "Setting targets for distillation" (3 parts), *Process Eng.*, June, 33-34, 37-38, 39-40.

Dhole, V.R. and Linnhoff, B. (1993a) "Total site targets for fuel, cogeneration, emissions, and cooling", *Comp. and Chem. Eng.*, **17** Supplement, S101-109; also, Dhole, V.R. and Linnhoff, B. (1992) "Total site targets for fuel, cogeneration, emissions, and cooling", *European Symp. on Comp. Applications in Process Eng. (ESCAPE - 2)*, October, Toulouse, France.

Dhole, V.R. and Linnhoff, B. (1993b) "Distillation column targets", *Comp. and Chem. Eng.*, **17**(5/6), 549-560; also, Dhole, V.R. and Linnhoff, B. (1992) "Distillation column targets", *European Symp. on Comp. Applications in Process Eng. (ESCAPE - 1)*, May 24-28, Elsinore, Denmark.

Dhole, V.R. and Zheng, J.P. (1993)* "Applying combined pinch and exergy analysis to closed cycle gas turbine system design", *ASME Cogen Turbo Power Conf.*, September 21-22, Bournemouth, U.K.

Dhole, V.R. and Linnhoff, B. (1994)* "Overall design of low temperature processes", *Comp. and Chem. Eng.*, **18** Supplement, S105-111; also, Dhole, V.R. and Linnhoff, B. (1993) "Overall design of subambient processes", *European Symp. on Comp. Applications in Process Eng. (ESCAPE - 3)*, July, Graz, Austria.

Dhole, V.R. and Mavromatis, S. (1994)* "Design and monitoring of site-wide utility systems for operational variations", *The 1994 IChemE Research Event*, 5-6 January, University College, London.

Diachendt, M.M. and Grossmann, I.E. (1994)* "A preliminary screening procedure for MINLP heat exchanger network synthesis using aggregated models", *Trans. IChemE. Chem. Eng. Res. Des.*, **72**, Part A, 357-363.

Dixon, A.G. (1987)* "Teaching heat exchanger network synthesis using interactive microcomputer graphics", *Chem. Eng. Educ.*, Summer, 118-121, 156.

Dolan, O.B., Bagajewicz, M.J. and Cerda, J. (1985)* "Designing heat exchanger networks for existing chemical plants", *Comp. and Chem. Eng.*, **9**, 483-498.

Dolan, W.B., Cummings, P.T. and Levan, M.D (1989)* "Process optimization via simulated annealing: Application to network design", *AIChE J.*, **35**(5), 725-736.

Dolan, W.B., Cummings, P.T. and Levan, M.D (1990)*, *Comp. and Chem. Eng.*, **14**, 1039-1050.

Domingos, J.D. (1969) "Analysis of complex assemblies of heat exchangers", *Int. J. Heat Mass Transfer*, **12**, 537.

Donaldson, R.A.B. (1976)* "Studies in computer aided design of complex heat exchange networks", Ph. D. Thesis, University of Edinburgh, Scotland.

Douglas, J.M. (1988), *Conceptual design of chemical processes*, McGraw Hill, New York.

Dunford, H.A. and Linnhoff, B. (1981)* "Energy savings by appropriate integration of distillation columns into overall processes", *Cost Savings in Distillation Symposium*, IChemE, Leeds, U.K.

Duran, M.A. and Grossmann, I.E. (1986)* "Simultaneous optimization and heat integration of chemical processes", *AIChE J.*, **32**(1), 123-138; also, Duran, M.A. and Grossmann, I.E. (1985) "Simultaneous optimization and heat integration of chemical processes", *AIChE National Meeting, Houston, Texas.

Duvedi, A. (1993) "Interfacing heat exchanger design and network synthesis", B.Tech. Project Report, Indian Institute of Technology, Bombay.

Eastwood, A.R. and Linnhoff, B. (1985)* "CHP and process integration", *51st Autumn Meeting*, November 12-13, Inst. Gas Engineers, London, U.K, Communication 1272, 1-26.

Edgar, T.F and Himmelblau, D.M. (1989), *Optimization of chemical processes*, McGraw Hill, New York.

El-Halwagi, M.M. and Manousiouthakis, V. (1989) "Synthesis of mass exchange networks", *AIChE J.*, **35**(8), 1233-1244.

El-Halwagi, M.M. and Manousiouthakis, V. (1990a) "Automatic synthesis of mass-exchange networks with single-component targets", *Chem. Eng. Sci.*, **45**(9), 2813-2831.

El-Halwagi, M.M. and Manousiouthakis, V. (1990b) "Simultaneous synthesis of mass-exchange and regeneration networks", *AIChE J.*, **36**(8), 1209-1219.

El-Halwagi, M.M. and Srinivas, B.K. (1992) "Synthesis of reactive mass-exchange networks", *Chem. Eng. Sci.*, **47**(8), 2113-2119.

Elshout, R.V. and Hohmann, E.C. (1979)* "The heat exchanger network simulator", *Chem. Eng. Prog.*, **75**(3), 72-77.

Engel, P. and Morari, M. (1988)* "Limitations of the primary loop breaking method for synthesis of heat exchanger networks", *Comp. and Chem. Eng.*, **12**(4), 307-310.

Fair, J.R. (1987)* "Energy-efficient separation process design", *in Recent Developments in Chemical Process and Plant Design*, Liu, Y.A., McGee, H.A. and Epperly, W.R. (eds.), John Wiley, New York, 71-99.

Farhanieh, B. and Sunden, B. (1990) "Analysis of an existing heat exchanger network and effects of heat pump installations - Case Study", *Heat Recovery Systems & CHP*, **10**(3), 285-296.

Farhanieh, B. and Sunden, B. (1992a)* "Analysis of heat pump integration into an optimized heat exchanger network with high temperature levels of the heat sink and heat source", *Recent Adv. in Heat Transfer (Proc. of First Baltic Heat Transfer Conf., Goteberg, Sweden, Aug 26-28, 1991)*, eds. Sunden B. and Zukauskus, A., 1000-1016.

Farhanieh, B. and Sunden, B. (1992b)* "Pinch design method assists retrofit design and predicts its economical profitability", *Recent Adv. in Heat Transfer (Proc. of First Baltic Heat Transfer Conf., Goteberg, Sweden, Aug 26-28, 1991)*, eds. Sunden B. and Zukauskus, A., 1025-1039.

Floudas, C.A. and Grossmann, I.E. (1986)* "Synthesis of flexible heat exchanger networks for multiperiod operation", *Comp. and Chem. Eng.*, **10**, 153.

Floudas, C.A. and Grossmann, I.E. (1987a)* "Synthesis of flexible heat exchanger networks with uncertain flowrates and temperatures", *Comp. and Chem. Eng.*, **11**(4), 319-336.

Floudas, C.A. and Grossmann, I.E. (1987b)* "Automatic generation of multiperiod heat exchanger network configuration", *Comp. and Chem. Eng.*, **11**(2), 123-142.

Floudas, C.A. and Ciric, A.R. (1988)* "Global optimum issues on heat exchanger network synthesis", *Third Intl. Symp. on Process Systems Eng.*, Sydney, Australia, 104-110; also, Ciric, A.R. and Floudas, C.A. (1988) "Global optimum search in heat exchanger networks - II. Simultaneous optimization of network configuration and process stream matches", *AIChE Annual Meeting*, Washington D.C.

Floudas, C.A. and Ciric, A.R. (1989)* "Strategies for overcoming uncertainties in heat exchanger network synthesis", *Comp. and Chem. Eng.*, **13**, 1133-1152.

Floudas, C.A., Ciric, A.R. and Grossmann, I.E. (1986) "Automatic synthesis of optimum heat exchanger network configurations", *AIChE J.*, **32**(2), 276-290.

Flower, J.R. and Linnhoff, B. (1979)* "Thermodynamic analysis in the design of process networks", *Comp. and Chem. Eng.*, **3**, 283-291.

Flower, J.R. and Linnhoff, B. (1980)* "A thermodynamic combinatorial approach to the design of optimum heat exchanger networks", *AIChE J.*, **26**(1), 1-9.

Fogler, H.S. (1992), *Elements of chemical reaction engineering*, Prentice-Hall, New Jersey.

Fonyo, Z. (1974) "Thermodynamic analysis of rectification (2 parts)", *Int. Chem. Eng.*, **14**, 18-27 and 203-210.

Fonyo, Z., Rev, E. and Mizsey, P. (1991)* "Heat exchanger network synthesis at unequal heat transfer conditions", *4th Intl. Symp. Process Systems Eng. (PSE-91)*, Montebello, Quebec, Canada.

Forder, G.J. and Hutchison, H.P. (1969) "The analysis of chemical process flowsheets", *Chem. Eng. Sci.*, **24**, 771-778.

Fraser, D.M. (1989a) "The use of minimum flux instead of minimum approach temperature as a design specification for heat exchanger networks", *Chem. Eng. Sci.*, **44**(5), 1121-1127.

Fraser, D.M. (1989b) "The application of a minimum flux specification to the design of heat exchanger networks", *Dechema-Monographs-VCH Verlagsgesellschaft*, **116**, 253-260; also, Fraser, D.M. (1989) "The use of a minimum flux specification to the design of heat exchanger networks", *Eur Symp. on the Use of Computers in the Chem. Industry*, Erlangan, 23-26 April.

Fraser, D.M. (1991)* "Minimum flux values to use in heat exchanger network design", *AIChE Annual Meeting*, Los Angeles, November.

Fraser, D.M. and Gillespie, N.E. (1992) "The application of pinch technology to retrofit energy integration of an entire oil refinery", *Trans. IChemE. Chem. Eng. Res. Des.*, **70**, Part A, July, 395-406.

Galli, M.R. (1989)* "Optimal synthesis of flexible heat exchanger networks", Ph.D. Thesis, University of Litoral, Santa Fe, Argentina.

Galli, M.R. and Cerda, J. (1991)* "Synthesis of flexible heat exchanger networks - III. Temperature and flowrate variations", *Comp. and Chem. Eng.*, **15**(1), 7-24.

Gautam, R. and Smith, J.A. (1985)* "A computer aided system for process synthesis and optimization", *AIChE Meeting*, Houston, March, Paper No 146.

Gibbs, N.E. (1969) "A cycle generation algorithm for finite undirected linear graphs", *JACM*, **16**(4), 564-568.

Glavic, P. and Novak, Z. (1991)* "Improved pinch techniques for process design", *Vestn. Slov. Kem. Drus.*, **38**(3), 271-286.

Glavic, P. and Novak, Z. (1993)* "Completely analyze energy-integrated processes", *Chem. Eng. Prog.*, February, 49-60.

Glavic, P., Kravanja, Z. and Homsak, M. (1988a) "Modeling of reactors for process heat integration", *Comp. and Chem. Eng.*, **12**(2/3), 189-194.

Glavic, P., Kravanja, Z. and Homsak, M. (1988b) "Heat integration of reactors - I. Criteria for the placement of reactors into process flowsheets", *Chem. Eng. Sci.*, **43**(3), 593-608.

Glavic, P., Kravanja, Z. and Homsak, M. (1990) "Putting the pinch on reactors", *Chem. Eng.*, June, 106-121.

Glinos, K. and Malone, M.F. (1989)* "Net work consumption in distillation - Short-cut evaluation and applications to synthesis", *Comp. and Chem. Eng.*, **13**(3), 295-305.

Gorsek, A., Glavic, P. and Sencar, P. (1992)* "Optimal process design for specialty products", ESCAPE-1, *Comp. and Chem. Eng.*, S321-S328.

Govind, R., Mocsny, D., Cosson, P. and Klei, J. (1986)* "Exchanger network synthesis on a microcomputer", *Hydrocarbon Processing*, July, **65**, 53-57.

Gremouti, I.D. (1991)* "Integration of batch processes for energy savings and debottlenecking", M.Sc. Thesis, University of Manchester Institute of Science and Technology, U.K.

Grimes, L.E. (1980)* "The synthesis and evolution of networks of heat exchanger that feature the minimum number of units", M.S. Thesis, Carnegie Mellon University, Pittsburgh.

Grimes, L.E., Rychener, M.D. and Westerberg, A.W. (1982) "The synthesis and evolution of networks of heat exchange that feature the minimum number of units", *Chem. Eng. Commun.*, **14**, 339 -360.

Grossmann, I.E. (1985)* "Mixed integer programming approach for the synthesis of integrated process flowsheets", *Comp. and Chem. Eng.*, **9**, 463-482.

Grossmann, I.E. and Sargent, R.W.H. (1976)* "Optimum design of heat exchanger networks", *Comp. and Chem. Eng.*, **2**, 1-7.

Grossmann, I.E. and Morari, M. (1983)* "Operability, resiliency and flexibility - process design objectives for a changing world", *2nd Intl. Conf. Foundations Comput. Aided Process Des.*, Snowmass.

Grossmann, I.E. and Floudas, C.A. (1987)* "Active constraint strategy for flexibility analysis in chemical processes", *Comp. and Chem. Eng.*, **11**, 675.

Golwelker, S. (1994) "Energy integration of batch processes: Pinch technology approach", M.Tech. Thesis, Indian Institute of Technology, Bombay.

Gundersen, T. (1991)* "Achievements and future challenges in industrial design applications of process systems engineering", *4th Intl. Symp. Process Systems Eng. (PSE-91)*, Montebello, Quebec, Canada.

Gundersen, T. and Naess, L. (1988) "The synthesis of cost optimal heat exchanger networks: an industrial review of the state of the art", *Comp. and Chem. Eng.*, **12**(6), 503-530.

Gundersen, T. and Grossmann, I.E. (1990) "Improved optimization strategies for automated heat exchanger network synthesis through physical insights", *Comp. and Chem. Eng.*, **14**(9), 925-944; also, Gundersen, T. and Grossmann, I.E. (1988) "Improved optimization strategies for automated heat exchanger network synthesis through physical insights", *AIChE Annual Meeting*, Washington D.C., Paper No 81g.

Gundersen, T., Sagli, B. and Kiste, K. (1991)* "Problems in sequential and simultaneous strategies for heat exchanger network synthesis", in *Computer-Oriented Process Eng.* eds. Puigjaner L. and Espuna, A., Elsevier Science Publishers, Amsterdam.

Gupta, A. and Manousiouthakis, V. (1993) "Minimum utility cost of mass exchange networks with variable single component supplies and targets", *Ind. Eng. Chem. Res.*, **32**, 1937-1950.

Gupta, A. and Manousiouthakis, V. (1994)* "Waste reduction through multicomponent mass exchange network synthesis", *Comp. and Chem. Eng.*, **18** Supplement, S585-590;

Hall, S.G. (1986) "Capital cost targets for heat exchanger networks: differing materials of construction and different heat exchanger types", M.Sc. Thesis, University of Manchester Institute of Science and Technology, U.K.

Hall, S.G. and Morgan, S.W. (1994)* "Heat exchanger databases: Accelerate process design and costing", *Chem. Eng.*, July, 139-144.

Hall, S.G., Ahmad, S. and Smith, R. (1990) "Capital cost targets for heat exchanger networks comprising mixed materials of construction, pressure ratings, and exchanger types", *Comp. and Chem. Eng.*, **14**(3), 319-335; also, Hall, S.G., Ahmad, S. and Smith, R. (1988) "Capital cost targets for heat exchanger networks comprising mixed materials of construction", *AIChE Spring Meeting*, March 6-10, New Orleans, Paper No 38a.

Hall, S.G., Parker, S.J. and Linnhoff, B. (1992) "Process integration of utility systems", *IEA Workshop on Process Integration*, January 28-29, Gothenburg, Sweden.

Hama, A. (1984)* "Computer-aided synthesis of heat exchanger networks - An iterative approach to global optimum networks", *1st U.K. National Heat Transfer Conf.*, July, Leeds, 567-598.

Heggs, P.J. (1989), *Heat Recovery Systems & CHP*, **9**, 367-375.

Hillenbrand, J.B. Jr. and Westerberg, A.W. (1984)* "Synthesis of evaporation systems using minimum utility insights", *AIChE Annual Meeting*, San Francisco, Calif.

Himsworth, J.R. and Cooper, A.C.G. (1993)* "Supercharged heat exchanger networks", *Trans. IChemE. Chem. Eng. Res. Des.*, March, **71**, 203-211.

Hindmarsh, E. and Townsend, D.W. (1984)* "Use of complex columns to enhance heat integration of distillation columns into total flowsheets", *AIChE Annual Meeting*, San Francisco, Calif., Paper No 88a.

Hindmarsh, E., Boland, D. and Townsend, D.W. (1985)* "Maximizing energy savings for heat engines in process plants", *Chem. Eng.*, Feb. 4, **92**, 38-47.

Hlavacek, V. (1976)* *Comp. and Chem. Eng.*, **2**, 67.

Ho, F-G and Keller, G.E. II (1987) "Process integration", *in Recent Developments in Chemical Process and Plant Design*, Liu, Y.A., McGee, H.A. and Epperly, W.R. (eds.), John Wiley, New York, 101-126.

Hohmann, E.C. (1971) "Optimum networks for heat exchanger", Ph. D. Thesis, University of Southern California, U.S.A.

Hohmann, E. (1984)* "Heat exchange technology, network synthesis", *Kirk-Othmer Encyclopedia of Chemical Tech.*, pp 521-545.

Hohmann, E.C. and Lockhart, F.J. (1976) "Optimum heat exchanger network synthesis", *AIChE 82nd National Meeting*, Atlantic City, NJ, Paper No 22a.

Hohmann, E.C. and Sander, M.T. (1982)* "A new approach to the synthesis of multicomponent separation schemes", *Chem. Eng. Commun.*, **17**, 273.

Holger-Martin (1990)* "Simple new formulae for efficiency and mean temperature difference in heat exchangers", *Chem. Eng. Tech.*, **13**, 237-241.

Homsak, M. and Glavic, P. (1991)* "Thermodynamic analysis of inappropriately placed energetic units", in *Computer-Oriented Process Eng.* eds. Puigjaner L. and Espuna, A., Elsevier Science Publishers, Amsterdam, 363-368.

Huang, F. and Elshout, R.V. (1976)* "Optimizing the heat recovery of crude units", *Chem. Eng. Prog.*, July, **72**(7), 68-74.

Huang, Y.L. and Fan, L.T. (1994) "HIDEN: A hybrid intelligent system for synthesizing highly controllable exchanger networks. Implementation of a distributed strategy for integrating process design and control", *Ind. Eng. Chem. Res.*, **33**, 1174-1187.

Hui, C.W. and Ahmad, S. (1994a)* "Minimum cost heat recovery between separate plant regions", *Comp. and Chem. Eng.*, **18**(8), 711-728.

Hui, C.W. and Ahmad, S. (1994b) "Total site heat integration using the utility system", *Comp. and Chem. Eng.*, **18**(8), 729-742.

Hwa, C.S. (1965)* "Mathematical formulation and optimization for heat exchanger networks using separable programming", *AIChE - IChemE Symp. Ser.*, **4**, 101-106.

Irazoqui, H.A. (1986) "Optimal thermodynamic synthesis of thermal energy recovery systems", *Chem. Eng. Sci.*, **41**(5), 1243-1255.

Itoh, J., Shiroko, K. and Umeda, T. (1982)* "Extensive applications of T-Q diagram to heat integrated system synthesis", *Proc. Intl. Symp. Process Systems Eng. (PSE-82)*, August, Kyoto, Japan, 92.

Jegede, F.O. (1990) Ph.D. Thesis, University of Manchester Institute of Science and Technology, U.K.

Jegede, F.O. and Polley, G.T. (1992a) "Optimum heat exchanger design", *Trans. IChemE. Chem. Eng. Res. Des.*, **70**, Part A, 133-141.

Jegede, F.O. and Polley, G.T. (1992b) "Capital cost targets for networks with non-uniform heat exchanger specifications", *Comp. and Chem. Eng.*, **16**(5), 477-495.

Jezowski, J. (1989)*, *Hungarian J. Ind. Chem.*, **17**, 295-310.

Jezowski, J. (1990) "A simple synthesis method for heat exchanger networks with minimum number of matches", *Chem. Eng. Sci.*, **45**(7), 1928-1932.

Jezowski, J. (1991a) "A note on the use of dual temperature approach in heat exchanger network synthesis", *Comp. and Chem. Eng.*, **15**(5), 305-312.

Jezowski, J. (1991b)* "On match calculation in heat exchanger network synthesis", *Inzynieria Chemiczna I Procesowa*, **2**, 203-215.

Jezowski, J. (1992a) "The pinch design method for tasks with multiple pinches", *Comp. and Chem. Eng.*, **16**(2), 129-133.

Jezowski, J. (1992b)* "SYNHEN: Microcomputer directed package of programs for heat exchanger network synthesis", *Comp. and Chem. Eng.*, **16**(7), 691-706.

Jezowski, J. (1994-1995)*, "Heat exchanger network grassroot and retrofit design. The review of the state-of-the-art.", I. Heat exchanger network targeting and insight based methods of synthesis, *to appear in Hungarian J. Ind. Chem.*; II. Heat exchanger network synthesis by mathematical methods and approaches for retrofit design, *to appear in Hungarian J. Ind. Chem.*

Jezowski, J. and Friedler, F. (1991)* "A note on targeting in the design of cost optimal heat exchanger networks", *Chem. Biochem. Eng. Q.*, **5**(1-2), 1-9.

Jezowski, J. and Friedler, F. (1992) "A simple approach for maximum heat recovery calculations", *Chem. Eng. Sci.*, **47**(6), 1481-1494.

Jones, S.A. (1987)* "Methods for the generation and evaluation of alternative heat exchanger networks", Ph.D. Thesis, ETH, Zurich.

Jones, P.S. (1991)* "Targeting and design for heat exchanger networks under multiple base case operation", Ph.D. Thesis, University of Manchester Institute of Science and Technology, U.K.

Jones, S.A. and Rippin, D.W.T. (1985)* "The generation of heat load distributions in heat exchanger network synthesis", *Proc. Intl. Conf. Process Systems Eng. (PSE-85)*, 155-177.

Jones, P.S. and Kotjabasakis, E. (1990)* "Multiple base cases: Targeting for heat exchanger networks", *17th IChemE Annual Res. Meeting*, April 5, Swansea.

Jones, D.A., Yilmaz, A.N. and Tilton, B.E. (1986) "Synthesis techniques for retrofitting heat recovery systems", *Chem. Eng. Prog.*, **82**(7), 28-33; also Jones, D.A., Yilmaz, A.N. and Tilton, B.E. (1985) "Practical synthesis techniques for retrofitting heat recovery systems", *AIChE Annual Meeting*, Chicago, Paper No 35c.

Kalitventzeff, B. (1993)* "Mixed integer non-linear programming and its application to the management of utility networks", *Private communication*.

Kalitventzeff, B. and Marechal, F. (1988)* "The management of a utility network", *Proc. Systems Eng.*, Aug 28-Sept 2, Sydney.

Karp, A., Smith, R. and Ahmad, S. (1990)* "Pinch technology: a primer", *EPRI Report CU-6775*, Electric Power Research Institute, Palo Alto, Calif.

Kelahan, R.C. and Gaddy, J.L. (1977)* "Synthesis of heat exchanger networks by mixed integer optimization", *AIChE J.*, **23**, 816-822; also, Kelahan, R.C. and Gaddy, J.L. (1976) "Synthesis of heat exchange networks by mixed integer optimization", *AIChE National Meeting*, Atlantic City, NJ, Paper No 22c.

Kemp, I.C. (1986)* "Analysis of separation systems by process integration", *J. Separation Proc. Technol.*, **7**, 9-23.

Kemp, I.C. (1988) "Letter to the editor", *Trans. IChemE. Chem. Eng. Res. Des.*, **66**(6), 569.

Kemp, I.C. (1990a) "Applications of the time-dependent cascade analysis in process integration", *Heat Recovery Systems & CHP*, **10**(4), 423-435.

Kemp, I.C. (1990b) "Process integration: process change and batch processes", *ESDU Data Item 90033*, ESDU International plc, London.

Kemp, I.C. (1991) "Some aspects of the practical application of pinch technology methods", *Trans. IChemE. Chem. Eng. Res. Des.*, **69**, Part A, 471-479.

Kemp, I.C. and Macdonald, E.K. (1987)* "Process integration", *ESDU Data Item 87030*, ESDU International plc, London.

Kemp, I.C. and Macdonald, E.K. (1988) "Application of pinch technology to separation, reaction and batch processes", *IChemE Symp. Ser. No. 109 Understanding Process Integration II*, 239-258.

Kemp, I.C. and Deakin, A.W. (1989a)* "Application of process integration to utilities, combined heat and power and heat pumps", *ESDU Data Item 89001*, ESDU International plc, London.

Kemp, I.C. and Deakin, A.W. (1989b) "The cascade analysis for energy and process integration of batch processes", 1. Calculation of energy targets, *Trans. IChemE. Chem. Eng. Res. Des.*, **67**(5), 495-509; 2. Network design and process scheduling, *Trans. IChemE. Chem. Eng. Res. Des.*, **67**(5), 510-516; 3. A case study, *Trans. IChemE. Chem. Eng. Res. Des.*, **67**(5), 517-525.

Kern, D.Q. (1950), *Process Heat Transfer*, McGraw Hill, New York.

Kesler, M.G. and Parker, R.O. (1969)* "Optimal networks of heat exchange", *Chem. Eng. Prog. Symp. Ser. No. 92*, **65**, 111-120.

King, C.J. (1980), *Separation processes*, McGraw Hill, New York.

Kobayashi, S., Umeda, T. and Ichikawa, A. (1971)* "Synthesis of optimal heat exchange systems - an approach by the optimal assignment problem in linear programming", *Chem. Eng. Sci.*, **26**, 1367-1380.

Korner H. (1988)* "Optimal use of energy in the chemical industry", *Chem. Ing. Tech.*, **60**(7), 511-518.

Kotjabasakis, E. and Linnhoff, B. (1986) "Sensitivity tables for the design of flexible processes (1) - how much contingency in heat exchanger networks is cost effective?", *Trans. IChemE. Chem. Eng. Res. Des.*, **64**(3), 197-211.

Kotjabasakis, E. and Linnhoff, B. (1987a). "Better system design reduces heat-exchanger fouling costs", *Oil and Gas J.*, September, 49-56; also Kotjabasakis, E. and Linnhoff, B. (1987) "An optimal overdesign strategy for fouling", *IChemE. Symposium on Process Optimisation*, April 7-10, Nottingham.

Kotjabasakis, E. and Linnhoff, B. (1987b) "Flexible heat exchanger network design: comments on the problem definition and on suitable solution techniques", *IChemE Symposium on Innovation in Process Energy Utilisation*, September 16-18, Bath.

Kotjabasakis, E. and Linnhoff, B. (1988a)* "Sensitivity tables for the design of flexible processes (2) - a case study", *IChemE Symp. Ser. No. 109 Understanding Process Integration II*, 181-204.

Kotjabasakis, E. and Linnhoff, B. (1988b)* "Sensitivity tables for the design of flexible heat exchanger networks: systems with variable physical properties", *AIChE Spring Meeting*, March 6-10, New Orleans, Paper No 39b.

Kotjabasakis, E. and Gremouti, I.D. (1992) "Practical aspects of process integration and their implications for design", *IEA Workshop on Process Integration*, January 28-29, Gothenburg, Sweden.

Krajnc, M. and Glavic, P. (1992)* "Energy integration of mechanical heat pumps with process fluid as working fluid", *Trans. IChemE. Chem. Eng. Res. Des.*, **70**, Part A, 407-420.

Kravanja, Z. and Glavic, P. (1989)* "Heat integration of reactors - II. Total flowsheet integration", *Chem. Eng. Sci.*, **44**(11), 2667-2682.

Lambert, A.J.D. (1994)* "Minimization of number of units in heat exchanger networks using a lumped approach", *Comp. and Chem. Eng.*, **18**(1), 71-74.

Lang, Y.-D., Biegler, L.T. and Grossmann, I.E. (1988)* "Simultaneous optimization and heat integration with process simulators", *Comp. and Chem. Eng.*, **12**(4), 311-327.

Lee, In-Beum and Reklaitis, G.V. (1989)* "Towards the synthesis of global optimum heat exchange networks", *Chem. Eng. Commun.*, **75**, 57-88.

Lee, K.F., Masso, A.H and Rudd, D.F. (1970) "Branch and bound synthesis of integrated process design", *Ind. Eng. Chem. Fundam.*, **9**, 48-58.

Lee, K.L., Morabito, M. and Wood, R.M. (1989) "Refinery heat integration using pinch technology", *Hydrocarbon Processing*, April, 49-53.

Li, Y. and Motard, R.L. (1986)* "Optimal pinch approach temperature in heat exchanger networks", *Ind. Eng. Chem. Fundam.*, **25**, 577-581.

Linnhoff, B. (1979) "Thermodynamic analysis in the design of process networks", Ph.D. Thesis, University of Leeds, U.K.

Linnhoff, B. (1982)* "Interpreting exergy analysis: a case study", *IChemE Jubilee Symposium*, London.

Linnhoff, B. (1983) "New concepts in thermodynamics for better chemical process design", *Proc. Royal Soc.*, March, **386**(1790), 1-33.

Linnhoff, B. (1986) "The process/utility interface", *Second International Meeting for National Use of Energy*, March 10-12, Liege, Belgium.

Linnhoff, B. (1989) "Pinch technology for the synthesis of optimal heat and power systems", *ASME J. Energy Resources Tech.*, September, 111(3), 137-147; also, Linnhoff, B. (1986) "Pinch technology for the synthesis of optimal heat and power systems", *ASME Winter Annual Meeting*, December, Anaheim, Calif. *AES*, **2-1**, 23-35.

Linnhoff, B. (1993a) "Pinch analysis - A state-of-the-art overview", *Trans. IChemE. Chem. Eng. Res. Des.*, **71**, Part A5, 503-522.

Linnhoff, B. (1993b)* "Total site integration and emissions targeting by pinch analysis", *J. Israel Inst. of Chem. Engrs.*, Professor William Resnick Memorial Issue, April, 81-87.

Linnhoff, B. (1993c)* "Pinch analysis and exergy - A comparison", *Energy Systems and Ecology Conf.*, July 5-9, Cracow, Poland.

Linnhoff, B. and Flower, J.R. (1978) "Synthesis of heat exchanger networks", I. Systematic generation of energy optimal networks, *AIChE J.*, 24(4), 633-642; II. Evolutionary generation of networks with various criteria of optimality, *AIChE J.*, **24**(4), 642-654.

Linnhoff, B. and Flower, J.R. (1979)* "A thermodynamic approach to practical process network design", *AIChE 72nd Annual Meeting*, November 25-29, San Francisco, Paper No 28b.

Linnhoff, B. and Smith, R. (1979)* "The thermodynamic efficiency of distillation", *IChemE Symp. Ser.*, **56**.

Linnhoff, B. and Turner, J.A. (1980) "Simple concepts in process synthesis give energy savings and elegant designs", *The Chemical Engineer*, December, 742-746.

Linnhoff, B. and Carpenter, K.J. (1981)* "Energy conservation by exergy analysis - The quick and simple way", *Second World Congress of Chem. Eng.*, October, Montreal, Canada.

Linnhoff, B. and Turner, J.A. (1981) "Heat recovery networks: new insights yield big savings", *Chem. Eng.*, November 2, **88**, 56-70.

Linnhoff, B. and Townsend, D.W. (1982)* "Designing total energy systems", *Chem. Eng. Prog.*, **78**, 72-80.

Linnhoff, B. and Hindmarsh, E. (1983) "The pinch design method for heat exchanger networks", *Chem. Eng. Sci.*, **38**(5), 745-763; also, Linnhoff, B.

and Hindmarsh, E. (1982) "The pinch design method for heat exchanger networks", *Understanding Process Integration Conference of IChemE*, March 22-24, Lancaster.

Linnhoff, B. and Senior, P.R. (1983)* "Energy targets clarify scope for better heat integration", *Process Eng.*, March, 29-33.

Linnhoff, B. and Parker, S. (1984) "Heat exchanger networks with process modifications", *IChemE 11th Annual Res. Meeting*, April, Bath, U.K.

Linnhoff, B. and Vredeveld, D.R. (1984) "Pinch technology has come of age", *Chem. Eng. Prog.*, July, **80**(7), 33-40; also, Linnhoff, B. and Vredeveld, D.R. (1983) "Retrofit projects through process synthesis", *AIChE Diamond Jubilee Meeting*, October 30 - November 4, Washington, Paper No 5f.

Linnhoff, B. and Ahmad, S. (1986) "Supertargeting, or the optimisation of heat exchanger networks prior to design", *IIIrd World Congress on Chem. Eng.*, September, Tokyo.

Linnhoff, B. and Kotjabasakis, E. (1986) "Process optimization: downstream paths for operable process design", *Chem. Eng. Prog.*, May, **82**(5), 23-28; also, Linnhoff, B. and Kotjabasakis, E. (1984) "Design of operable heat exchanger networks", *1st U.K. National Heat Transfer Conf.*, July, Leeds, Vol. 1, 599-618.

Linnhoff, B. and Witherell, W.D. (1986) "Pinch technology guides retrofit", *Oil and Gas J.*, April 7, 54-65.

Linnhoff, B. and Eastwood, A.R. (1987a)* "Overall site optimisation by pinch technology", *Trans. IChemE. Chem. Eng. Res. Des.*, September, **65**, 408-414; Linnhoff, B. and Eastwood, A. (1987), "Experience from the application of pinch technology to ethylene plants", *AIChE National Meeting*, March 29- April 2, Houston, Texas, Paper No 70a.

Linnhoff, B. and Eastwood, A. (1987b)* "Process integration using pinch technology", *108th ASME Winter Meeting*, December, Boston, USA.

Linnhoff, B. and Lenz, W. (1987)* "Thermal integration and process optimisation", *Chem. Ing. Tech.*, November, **59**(11), 851-857.

Linnhoff, B. and O'Young, D.L. (1987) "The three components of cross pinch heat flow in constrained heat exchanger networks", *AIChE Annual Meeting*, November 15-20, New York, Paper No 91.

Linnhoff, B. and Sahdev, V. (1987)* "Pinch technology", *Ullmann's Encyclopedia of Industrial Chemistry*, Vol. B3, 5th ed., 13-1 - 13-6.

Linnhoff, B. and Smith, R. (1987)* "Design of flexible plant - how much capital for how much flexibility", *Chem. Ing. Tech.*, **59**, 166-167; also, Linnhoff, B. and Smith, R. (1985) "Design of flexible plant - how much capital for how much flexibility?", *AIChE National Meeting*, March 24-28, Houston, Texas, Paper No 24f.

Linnhoff, B. and de Leur, J. (1988) "Appropriate placement of furnaces in the integrated process", *IChemE Symp. Ser. No. 109 Understanding Process Integration II*, 259-282.

Linnhoff, B. and Polley, G. (1988a)* "Stepping beyond the pinch", *The Chemical Engineer*, February, 25-32.

Linnhoff, B. and Polley, G.T. (1988b)* "Total process design through pinch technology", *AIChE Spring Meeting*, March 6-10, New Orleans, Paper No 38c.

Linnhoff, B. and Smith, R. (1988)* "The pinch principle", *Mech. Eng.*, February, 70-73.

Linnhoff, B. and Ahmad, S. (1989) "Supertargeting: Optimum synthesis of energy management systems", *ASME J. Energy Resources Tech.*, September, 111(3), 121-130; also, Linnhoff, B. and Ahmad, S. (1986) "Supertargeting: Optimal synthesis of energy management systems", *ASME Winter Annual Meeting*, December, Anaheim, Calif. *AES*, **2-1**, 1-13.

Linnhoff, B. and Alanis, F.J. (1989)* "A system's approach based on pinch technology to commercial power station design", *ASME Winter Annual Meeting*, December 10-15, San Francisco, Calif.

Linnhoff, B., and Ahmad, S. (1990) "Cost optimum heat exchanger networks - 1. Minimum energy and capital using simple models for capital cost", *Comp. and Chem. Eng.*, **14**(7), 729-750.

Linnhoff, B. and Alanis, F.J. (1991)* "Integration of a new process into an existing site: A case study in the application of pinch technology", *ASME J. of Eng. for Gas Turbines and Power*, **113**, 159-169; also, Linnhoff, B. and Alanis, F.J. (1987)* "Integration of a new process into an existing site - a case study", *108th ASME Winter Meeting*, December, Boston, USA.

Linnhoff, B., and Dhole, V.R. (1992) "Shaftwork targets for low-temperature process design", *Chem. Eng. Sci.*, **47**(8), 2081-2091; also, Linnhoff, B., and Dhole, V.R. (1989) "Shaft work targeting for subambient plants", *AIChE Spring Meeting*, April, Houston, Paper No 34d.

Linnhoff, B., and Dhole, V.R. (1993) "Targeting for CO_2 emissions for total sites", *Chem. Eng. Technol.*, **16**, 252-259; also, Linnhoff, B., and Dhole, V.R. (1992) "Targeting for CO_2 emissions for total sites", *VDI Jahrestreffen*, October, Vienna.

Linnhoff, B., Mason, D.R. and Wardle, I. (1979) "Understanding heat exchanger networks", *Comp. and Chem. Eng.*, **3**, 295-302.

Linnhoff, B., Townsend, D.W., Boland, D., Hewitt, G.F., Thomas, B.E.A., Guy, A.R. and Marsland, R.H. (1982), *User guide on process integration for the efficient use of energy*, The Institution of Chemical Engineers, Rugby, U.K.; available in the U.S. through Pergamon Press, Inc., Elmsford, N.Y.

Linnhoff, B., Dunford, H. and Smith, R. (1983) "Heat integration of distillation columns into overall processes", *Chem. Eng. Sci.*, **38**(8), 1175-1188: also, Dunford, H.A. and Linnhoff, B. (1981) "Energy savings by appropriate integration of distillation columns into overall processes", *Cost Savings in Distillation Symposium*, July 9-10, Leeds, Paper No 10.

Linnhoff, B., Ashton, G.J. and Obeng, E.D.A. (1988a) "Process integration of batch processes", *IChemE Symp. Ser. No. 109 Understanding Process Integration II*, 221-238; also, Linnhoff, B., Ashton, G.J. and Obeng, E.D.A. (1987) "Process integration of batch processes", *AIChE Annual Meeting*, November 15-20, New York, Paper No 92.

Linnhoff, B., Kotjabasakis, E. and Smith, R. (1988b)* "Flexible heat exchanger network design: problem definition and one method of approach", *AIChE Annual Meeting*, November 27-December 2, Washington, Paper No 79d.

Linnhoff, B., Polley, G.T. and Sahdev, V. (1988c)* "General process improvements through pinch technology", *Chem. Eng. Prog.*, June, 51-58.

Linnhoff, B., Smith, R. and Williams, J.D. (1990) "The optimisation of process changes and utility selection in heat integrated processes", *Trans. IChemE. Chem. Eng. Res. Des.*, **68**, Part A, 221-236.

Lipowicz, M. (1986)* "Heat-exchanger software", *Chem. Eng.*, August 4, 73.

Liu, Y.A. (1987) "Process synthesis: Some simple and practical developments", *in Recent Developments in Chemical Process and Plant Design*, Liu, Y.A., McGee, H.A. and Epperly, W.R. (eds.), John Wiley, New York, 147-260.

Liu, Y.A., Pehler, F.A. and Cahela, D.R. (1985) "Studies in chemical process design and synthesis. Part VII: systematic synthesis of multipass heat exchanger networks", *AIChE J.*, **31**, 487-491.

Makwana, Y. and Shenoy, U.V. (1993) "A new algorithm for continuous energy targeting and topology trap determination of heat exchanger networks", *Technical report*, Indian Institute of Technology, Bombay.

Malek, N. and Glavic, P. (1994)* "Theoretical bases of separation sequence heuristics", *Comp. and Chem. Eng.*, **18** Supplement, S143-147.

Marechal, F. and Kalitventzeff, B. (1989) "SYNEP1: a methodology for energy integration and optimal heat exchanger network synthesis", *Comp. and Chem. Eng.*, **10**, 603-611.

Marselle, D.F., Morari, M. and Rudd, D.F. (1982)* "Design of resilient processing plants II. Design and control of energy management systems", *Chem. Eng. Sci.*, **37**, 259-270.

Masso, A. H. and Rudd, D.F. (1969) "The synthesis of system designs. II. Heuristic structuring", *AIChE J.*, **15**, 10-17.

Mathisen, K.W., Morari, M. and Skogestad, S. (1994)* "Dynamic models for heat exchangers and heat exchanger networks", *Comp. and Chem. Eng.*, **18** Supplement, S459-463.

McAvoy, T.J. (1987)* "Integration of process design and process control", *in Recent Developments in Chemical Process and Plant Design*, Liu, Y.A., McGee, H.A. and Epperly, W.R. (eds.), John Wiley, New York, 289-324.

Menzies, M.A. and Johnson, A.I. (1972)* "Synthesis of optimal energy recovery network using discrete methods", *Can. J. Chem. Eng.*, **50**, 290.

Mizsey, P. and Fonyo, Z. (1990) "Toward a more realistic overall process synthesis - The combined approach", *Comp. and Chem. Eng.*, **14**(11), 1213-1236.

Mizsey, P. and Rev, E. (1991) "The use of flowsheet simulators in heat exchanger network synthesis", *Hungarian J. Ind. Chem.*, **19**, 293-299.

Mizsey, P. and Rev, E. (1992)*, *Hungarian J. Ind. Chem.*, **20**, 91-97.

Morgan, S.W. (1992)* "Use process integration to improve process designs and the design process", *Chem. Eng. Prog.*, September, 62-68.

Morton, R. J. and Linnhoff, B. (1984)* "Individual process improvements in the context of site-wide interactions", *IChemE 11th Annual Res. Meeting*, April, Bath, U.K.

Muraki, M. and Hayakawa, T. (1982)* "Practical synthesis method for heat exchanger network", *J. Chem. Eng. Japan*, **15**(2), 136-141.

Natarajan, U. and Shenoy, U.V. (1992) "Modeling varying-heat-transfer-area batch processes", *Heat Transf. Eng.*, **13**(1), 34-41.

Nilsson, K. and Sunden, B. (1992)* "A combined method for the thermal and structural design of industrial energy systems", *Recent Adv. in Heat Transfer (Proc. of First Baltic Heat Transfer Conf., Goteberg, Sweden, Aug 26-28, 1991)*, eds. Sunden B. and Zukauskus, A., 1017-1024.

Nishida, N., Kobayashi, S. and Ichikawa, A. (1971) "Optimal synthesis of heat exchange systems: necessary conditions for minimum heat transfer area and their application to systems synthesis", *Chem. Eng. Sci.*, **26**, 1841.

Nishida, N., Liu, Y.A. and Lapidus, L. (1977) "Studies in chemical process design and synthesis: III. A simple and practical approach to the optimal synthesis of heat exchanger networks", *AIChE J.*, **23**(1), 77-93; also, Nishida, N., Liu, Y.A. and Lapidus, L. (1977) "Studies in chemical process design and synthesis : III. A simple and practical approach to the optimal synthesis of heat exchanger networks", *AIChE National Meeting*, Atlantic City, NJ, Paper No 22b.

Nishida, N., Stephanopoulos, G. and Westerberg, A.W. (1981)* "A review of process synthesis", *AIChE J.*, **27**, 321-351.

Nishimura, H. (1980) "A theory for the optimal synthesis of heat exchange systems", *J. Optimization Theory Appl.*, **30**, 423-450.

Nishio, M., Itoh, J., Shiroko, K. and Umeda, T. (1981)* "Thermodynamic approach to steam and power system design", *Ind. Eng. Chem. Res. Des. Dev.*, **19**, 306.

Obeng, E.D.A. and Ashton, G.J. (1988) "On pinch technology based procedures for the design of batch processes", *Trans. IChemE. Chem. Eng. Res. Des.*, May, **66**(3), 255-259.

O'Young, D.L. (1989)* "Constrained heat exchanger networks: targeting and design", Ph. D. Thesis, University of Manchester Institute of Science and Technology, U.K.

O'Young, D.L. and Linnhoff, B. (1989) "Degrees of freedom analysis and a systematic procedure for the design and evolution of constrained heat exchanger networks", *AIChE Spring Meeting*, April, Houston, Paper No 32e.

O'Young, D.L., Jenkins, D.M. and Linnhoff, B. (1988) "The constrained problem table for heat exchanger networks", *IChemE Symp. Ser.*, **109**, 75-116; also, O'Young, D.L., Jenkins, D.M. and Linnhoff, B. (1988) "The constrained problem table for heat exchanger networks", *Understanding Process Integration II Conference of IChemE*, March 22-23, UMIST, Manchester.

Pai, D.C.C. and Fisher, G.D. (1988) "Application of Broyden's method to reconciliation of nonlinearly constrained data", *AIChE J.*, **34**(5), 873-876.

Palen, J.W. and Taborek, J. (1969) "Solution of shell-side flow pressure drop and heat transfer by stream analysis method", *CEP Symp. Ser.* 92, **65**, Heat Transfer - Philadelphia, 53-63.

Panjeh Shahi, M.H. (1991) Ph.D. Thesis, University of Manchester Institute of Science and Technology, U.K.

Papalexandri, K.P. and Pistikopoulos, E.N. (1993)* "An MINLP retrofit approach for improving the flexibility of heat exchanger networks", *Ann. Oper. Res.*, 42.

Papalexandri, K.P. and Pistikopoulos, E.N. (1994a) "Synthesis of cost optimal and controllable heat exchanger networks", *Trans. IChemE. Chem. Eng. Res. Des.*, **72**, Part A, 350-356.

Papalexandri, K.P. and Pistikopoulos, E.N. (1994b) "Synthesis and retrofit design of operable heat exchanger networks", 1. Flexibility and structural controllability aspects, *Ind. Eng. Chem. Res.*, **33**(7), 1718-1737; 2. Dynamics and control structure considerations, *Ind. Eng. Chem. Res.*, **33**(7), 1738-1755.

Papalexandri, K.P., Pistikopoulos, E.N. and Floudas, C.A. (1994) "Mass exchange networks for waste minimization: A simultaneous approach", *Trans. IChemE. Chem. Eng. Res. Des.*, **72**, Part A, 279-294.

Papoulias, S.A. (1982)* "Studies in the optimal synthesis of chemical processing and energy systems", Ph.D. Thesis, Carnegie Mellon University, Pittsburgh.

Papoulias, S.A. and Grossmann, I.E. (1983) "A structural optimization approach in process synthesis - II. Heat recovery networks", *Comp. and Chem. Eng.*, 7(6), 707-721.

Parker, S.J. (1989) "Supertargeting for multiple utilities", Ph.D. Thesis, University of Manchester Institute of Science and Technology, U.K.

Patel, K.S. (1993) "An algorithm for identification of loops using graph theory", private communication.

Paterson, W.R. (1984) "A replacement for the logarithmic mean", *Chem. Eng. Sci.*, **39**, 1635-1636.

Paton, K. (1969) "An algorithm for finding a fundamental set of cycles of a graph", *Comm. ACM*, **12**(9), 514-518.

Pavani, E.O., Aguirre, P.A. and Irazoqui, H.A. (1990) "Optimal synthesis of heat-and-power systems: the operating line method", *Intl. J. Heat Mass Transfer*, **33**(12), 2683-2699.

Pehler, F.A. and Liu, Y.A. (1981) "Thermodynamic availability analysis in the synthesis of energy-optimum and minimum-cost heat exchanger networks", *AIChE National Meeting*, August, Detroit; also, Pehler, F.A. and Liu, Y.A. (1983) *ACS Symp. Ser. No. 235 - Efficiency and Costing: Second Law Analysis of Processes*, Giaggioli, R.A. (ed.), ACS, Washington D.C., 161-178.

Pehler, F.A. and Liu, Y.A. (1984) "Studies in chemical process design and synthesis: VI. A thermoeconomic approach to the evolutionary synthesis of heat exchanger networks", *Chem. Eng. Commun.*, **25**, 295-310.

Peters, M.S. and Timmerhaus, K.D. (1981), *Plant design and economics for chemical engineers*, McGraw Hill, New York.

Pethe, S., Singh, R., Bhargava, S. and Knopf, F.C. (1988)* "HEN - interactive software for the synthesis of heat exchanger networks", Louisiana State University.

Pethe, S., Singh, R. and Knopf, F.C. (1989) "A simple technique for locating loops in heat exchanger networks", *Comp. and Chem. Eng.*, **13**(7), 859-860.

Pho, T.K. and Lapidus, L. (1973) "Topics in computer-aided design. II. Synthesis of optimal heat exchanger networks by tree searching algorithms", *AIChE J.*, **19**(6), 1182-1189.

Polley, G.T. and Linnhoff, B. (1988)* "Interface between conceptual design of a process and detailed design of equipment", *One day seminar : Process Heat Exchangers*, November 29, BHRA, Cranfield, U.K.

Polley, G.T. and Panjeh Shahi, M.H. (1990) "Process integration retrofit subject to pressure drop constraint", *Computer Applications in Chem. Eng.*, ed. Bussemaker, H.Th. and Iedema, P.D., Elsevier, Amsterdam, 31-37; also, *Conchem '90: European Symposium on Computer Applications in Chemical Engineering*, May 7-9, The Hague, Netherlands.

Polley, G.T. and Panjeh Shahi, M.H. (1991) "Interfacing heat exchanger network synthesis and detailed heat exchanger design", *Trans. IChemE. Chem. Eng. Res. Des.*, **69**, Part A, 445-457.

Polley, G.T., Panjeh Shahi, M.H. and Jegede, F.O. (1990) "Pressure drop considerations in the retrofit of heat exchanger networks", *Trans. IChemE. Chem. Eng. Res. Des.*, **68**, Part A, 211-220.

Polley, G.T., Panjeh Shahi, M.H. and Picon Nunez, M. (1991) "Rapid design algorithms for shell-and-tube and compact heat exchangers", *Trans. IChemE. Chem. Eng. Res. Des.*, **69**, Part A, 435-444.

Polley, G.T., Reyes Athie, C.M. and Gough, M. (1992)*, *Heat Recovery Systems & CHP*, **12**, 191-202.

Ponton, J.W. and Donaldson, R.A.B. (1974) "A fast method for the synthesis of optimal heat exchanger networks", *Chem. Eng. Sci.*, **29**, 2375-2377.

Ptacnik, R. and Klemes, J. (1988)*, *Comp. and Chem. Eng.*, **12**, 231-235.

Qassim, R.Y. and Silveira, C.S. (1988) "Heat exchanger network synthesis: the goal programming approach", *Comp. and Chem. Eng.*, **12**(11), 1163-1165.

Raghavan, S. (1977)* "Heat exchanger network synthesis: a thermodynamic approach", Ph.D. Thesis, Purdue University, Lafayette.

Raman, Raghu (1985), *Chemical process computations*, Chapter 5 - Heat Transfer, Elsevier Applied Science Publishers, London.

Ranade, S.M. (1988) "New insights on optimal integration of heat pumps in industrial sites", *Heat Recovery Systems & CHP*, **8**(3), 255-263.

Ranade, S.M. and Sullivan, M.O. (1988) "Optimal integration of industrial heat pumps", *IChemE Symp. Ser.*, **109**, 303-325.

Ranade, S.M., Jones, D.H. and Zapata-Suarez, A. (1989)* "Impact of utility costs on pinch designs", *Hydrocarbon Processing*, July, 39-43.

Rastogi, S.K., Shankara, J. and Srinivasan, J. (1988)* "Synthesis of optimal heat exchanger networks", in *Heat Transfer Equipment Design* eds. Shah, R. K., Subbarao, E. C. and Mashelkar, R. A., Hemisphere Publishing Corp., New York.

Rathore, R.N.S. and Powers, G.J. (1975)* "A forward branching scheme for the synthesis of energy recovery systems", *Ind. Eng. Chem. Proc. Des. Dev.*, **14**, 175-181.

Ratnam, R. and Patwardhan, V.S. (1991)* "Sensitivity analysis for heat exchanger networks", *Chem. Eng. Sci.*, **46**(2), 451-458.

Ratnam, R. and Patwardhan, V.S. (1993)* "Analyze single- and multiple-pass heat exchangers", *Chem. Eng. Prog.*, June, 85-91.

Reppich, M. and Kohoutek, J. (1994)* "Optimal design of shell-and-tube heat exchangers", *Comp. and Chem. Eng.*, **18** Supplement, S295-299.

Rev, E. and Fonyo, Z. (1986a) "Hidden and pseudo pinch phenomena and relaxation in the synthesis of heat-exchange networks", *Comp. and Chem. Eng.*, **10**(6), 601-607.

Rev, E. and Fonyo, Z. (1986b) "Additional pinch phenomena providing improved synthesis of heat exchanger networks", *Hungarian J. Ind. Chem.*, **14**,181-201.

Rev, E. and Fonyo, Z. (1991) "Diverse pinch concept for heat exchange network synthesis: the case of different heat transfer conditions", *Chem. Eng. Sci.*, **46**(7), 1623-1634.

Rev, E. and Fonyo, Z. (1993) "Comments on diverse pinch concept for heat exchanger network synthesis", *Chem. Eng. Sci.*, **48**(3), 627-628.

Rossiter, A.P., Seetharam, R.V. and Ranade, S.M. (1988)* "Scope for industrial heat pump applications in the United States", *Heat Recovery Systems & CHP*, **8**(3), 279-287.

Rossiter, A.P., Spriggs, H.D. and Klee, H. Jr. (1993)* "Apply process integration to waste minimization", *Chem. Eng. Prog.*, January, 30-36.

Saboo, A.K. and Morari, M. (1984)* "Design of resilient processing plants. 4. Some new results on heat exchanger network synthesis", *Chem. Eng. Sci.*, **39**, 579-592.

Saboo, A.K., Morari, M. and Woodcock, D.C. (1985)* "Design of resilient processing plants: VIII. A resilience index for heat exchanger networks", *Chem. Eng. Sci.*, **40**(8), 1553-1565.

Saboo, A.K., Morari, M. and Colberg, R.D. (1986)* "RESHEX: an interactive software package for the synthesis and analysis of resilient heat exchanger networks", I. Program description and application, *Comp. and Chem. Eng.*, **10**(6), 577-589; II. Discussion of area targeting and network synthesis algorithms, *Comp. and Chem. Eng.*, **10**(6), 591-599.

Saboo, A.K., Morari, M. and Colberg, R.D. (1987a)* "Resilience analysis for heat exchanger networks - I. Temperature dependent heat capacities", *Comp. and Chem. Eng.*, **11**, 399.

Saboo, A.K., Morari, M. and Colberg, R.D. (1987b)* "Resilience analysis for heat exchanger networks - II. Stream splits and flowrate variations", *Comp. and Chem. Eng.*, **11**, 457.

Sachdeva, N. (1993) "Retrofitting of heat exchanger networks using pinch technology", M.Tech. Thesis, Indian Institute of Technology, Bombay.

Sagli, B., Gundersen, T. and Yee, T.F. (1990) "Topology traps in evolutionary strategies for heat exchanger network synthesis", *Computer Applications in Chem. Eng.*, ed. Bussemaker, H.Th. and Iedema, P.D., Elsevier, Amsterdam, 51-57.

Scenna, N.J. and Aguirre, P.A. (1993)* "A thermodynamic methodology for process synthesis and its application to dual-purpose desalination plants.

Extraction vs. Back Pressure turbines", *Trans. IChemE. Chem. Eng. Res. Des.*, January, **71**, 77-84.

Schembecker, G., Schuttenhelm, W. and Simmrock, K.H. (1994)* "Cooperating knowledge integrating systems for the synthesis of energy-integrated distillation processes", *Comp. and Chem. Eng.*, **18** Supplement, S131-135.

Shah, J.V. and Westerberg, A.W. (1975) "Evolutionary synthesis of heat exchanger networks", *AIChE Annual Meeting*, Los Angeles, Calif, Paper No 60c.

Shenoy, U.V., Kapoor, B. and Makwana, Y. (1993) "A criterion based on constant slope for design of multipass heat exchangers", *Technical report*, Indian Institute of Technology, Bombay.

Shiroko, K. and Umeda, T. (1983)* "A practical approach for the optimum design of heat exchange system", *Proc. Econ. Int.*, **3**, 44-49.

Shokoya, C.G. (1992) "Retrofit of heat exchanger networks for debottlenecking and energy savings", Ph.D. Thesis, University of Manchester Institute of Science and Technology, U.K.

Shokoya, C.G. and Kotjabasakis, E. (1991)* "A new targeting procedure for the retrofit of heat exchanger networks", *Intl. Conference*, Athens, Greece, June, 265-279; also, Shokoya, C.G. and Kotjabasakis, E. (1990) "Retrofit of heat exchanger networks - targeting", *17th IChemE Annual Res. Meeting*, April 5, Swansea.

Siirola, J.J (1974)* "Status of heat exchanger network synthesis", *AIChE National Meeting*, Tulsa, Okla., Paper No 42a.

Silangwa, M. (1986) "Evaluation of various surface area efficiency criteria in heat exchanger network retrofits", M.Sc. Thesis, University of Manchester Institute of Science and Technology, U.K.

Sinnott, R.K. (1989), *An introduction to chemical engineering design - Chemical engineering series by J. M. Coulson and J. F. Richardson, Volume 6*, Pergamon Press, Oxford.

Smith, R. (1992) "Environmental consequences of process integration", *IEA Workshop on Process Integration*, January 28-29, Gothenburg, Sweden.

Smith, R. and Linnhoff, B. (1987)* "Process integration using pinch technology", *ATEE Symposium on Energy Management in Industry*, March 31 - April 3, Paris.

Smith, R. and Linnhoff, B. (1988) "The design of separators in the context of overall processes", *Trans. IChemE. Chem. Eng. Res. Des.*, May, **66**(3), 195-228.

Smith, R. and Jones, P.S. (1990) "The optimal design of integrated evaporation systems", *Heat Recovery Systems & CHP*, **10**(4), 341-368.

Smith, R. and Delaby, O. (1991) "Targeting Flue Gas Emissions", *Trans. IChemE. Chem. Eng. Res. Des.*, **69**, No. A6, 492-505.

Smith, R. and Petela, E.A. (1991-1992)* "Waste minimization in the process industries", 1. The problem, *The Chemical Engineer*, **506**, 31 October 1991, 24-25; 2. Reactors, *The Chemical Engineer*, **509-510**, 12 December 1991, 17-23; 3. Separation and recycle systems, *The Chemical Engineer*, **513**, 13 February 1992, 24-28; 4. Process operations, *The Chemical Engineer*, **517**, 9 April 1992, 21-23; 5. Utility waste, *The Chemical Engineer*, **523**, 16 July 1992, 32-35.

Smith, R. and Petela, E.A. (1993)* "The interface between the chemist and the chemical engineer as a source of waste", *The Chemist's Contribution to Waste Minimization Conf.*, The Royal Society of Chemistry, University of Lancaster, June 30.

Smith, R. and Omidkhah Nasrin, M. (1993)* "Trade-offs and interactions in reaction and separation systems", I. Reactions with no selectivity losses, *Trans. IChemE. Chem. Eng. Res. Des.*, **71**, No. A5, 467-473; II. Reactors with selectivity losses, *Trans. IChemE. Chem. Eng. Res. Des.*, **71**, No. A5, 474-478.

Smith, R., Petela, E.A. and Spriggs, H.D. (1990) "Minimization of environmental emissions through improved process integration", *Heat Recovery Systems & CHP*, **10**(4), 329-339.

Srinivas, B.K. and El-Halwagi, M.M. (1994a) "Synthesis of reactive mass-exchange networks with general nonlinear equilibrium functions", *AIChE J.*, **40**(3), 463-472.

Srinivas, B.K. and El-Halwagi, M.M. (1994b) "Synthesis of combined heat and reactive mass-exchange networks", *Chem. Eng. Sci.*, **49**(13), 2059-2074.

Steinmetz, F.J. and Chaney, M.O. (1985) "Total plant process energy integration", *Chem. Eng. Prog.*, **81**, July, 27-32.

Steinmeyer, D. (1992) "Save energy, without entropy", *Hydrocarbon Processing*, **71**, October, 55-95.

Stephanopoulos, G., Linnhoff, B. and Sophos, A. (1982)* "Synthesis of heat integrated distillation sequences", *Understanding Process Integration Conference of IChemE*, March 22-24, Lancaster, U.K.

Su, J.L. (1979)* "A loop breaking evolutionary method for the synthesis of heat exchanger networks", M.S. Thesis, Washington University, St. Louis, Missouri.

Su, J.L. and Motard, R.L. (1984) "Evolutionary synthesis of heat exchanger networks", *Comp. and Chem. Eng.*, **8**, 67-80.

Suaysompol, K. (1991)* Ph.D. Thesis, University of New South Wales, Australia.

Suaysompol, K. and Wood, R.M. (1991) "The flexible pinch design method for heat exchanger networks", I. Heuristic guidelines for free hand designs, *Trans. IChemE. Chem. Eng. Res. Des.*, **69**, Part A, 458-464; II. FLEXNET -

heuristic searching guided by the A* algorithm, *Trans. IChemE. Chem. Eng. Res. Des.*, **69**, Part A, 465-470.

Suaysompol, K. and Wood, R.M. (1993)* "Estimation of the installed cost of heat exchanger networks", *Intl. J. Production Economics*, **29**, 303-312.

Sunden, B. (1988)* "Analysis of the heat recovery in two crude distillation units", *Heat Recovery Systems & CHP*, **8**(5), 483-488.

Sunden, B., Thersthol, H. and Wernersson, C. (1985)* "Improvements of the energy system at an oil refinery by optimization of the heat exchanger network", *4th Intl. Chem. Eng. Conf. (Chempor '85)*, Coimbra-Portugal, 42/1-42/9.

Taborek, J. (1979) "Evolution of heat exchanger design techniques", *Heat Transf. Eng.*, **1**, 15.

Taborek, J. (1983) "Shell-and-tube heat exchangers", *Heat Exchanger Data Handbook*, Section 3.3, Hemisphere Publishing Corp., Washington D.C.

Terrill, D.L. and Douglas, J.M. (1987)*, *Ind. Eng. Chem. Res.*, **26**, 685-691; also, Terrill, D.L. and Douglas, J.M. (1987)* "Optimization and operability of heat exchanger networks", *AIChE Meeting*, Chicago, Illinois, Paper No 35b.

Tinker, T. (1951) "Shellside characteristics of shell and tube heat exchangers", *General Discussion on Heat Exchangers* (IMechE), 97-116.

Tjoe, T.N. (1986)* Ph.D. Thesis, University of Manchester Institute of Science and Technology, U.K.

Tjoe, T.N. and Linnhoff, B. (1984)* "Heat exchanger network retrofits", *IChemE 11th Annual Res. Meeting*, April, Bath, U.K.

Tjoe, T.N. and Linnhoff, B. (1986) "Using pinch technology for process retrofit", *Chem. Eng.*, April 28, 47-60; also, Linnhoff, B. and Tjoe, T.N. (1985) "Pinch technology retrofit: setting targets for existing plant", *AIChE National Meeting*, March 24-28, Houston, Texas, Paper No 88.

Tjoe, T.N. and Linnhoff, B. (1987) "Achieving the best energy saving retrofit", *AIChE Annual Meeting*, March 29-April 4, Houston, Texas, Paper No 17d.

Townsend, D.W. (1980)* "Second law analysis in practice", *The Chemical Engineer*, October, 628-633.

Townsend, D.W. (1989) "Surface area and capital cost targets for heat exchanger networks", Ph.D. Thesis, University of Manchester Institute of Science and Technology, U.K.

Townsend, D.W. and Linnhoff, B. (1982) "Designing total energy systems by systematic methods", *The Chemical Engineer*, March, 91-97.

Townsend, D.W. and Linnhoff, B. (1983) "Heat and power networks in process design", I. Criteria for placement of heat engines and heat pumps in process networks, *AIChE J.*, **29**(5), 742-748; II. Design procedure for equipment selection and process matching, *AIChE J.*, **29**(5), 748-771.

Townsend, D.W. and Linnhoff, B. (1984) "Surface area targets for heat exchanger networks", *IChemE 11th Annual Res. Meeting*, April, Bath, U.K.

Triantafyllou, C. and Smith, R. (1992)* "The design and optimisation of fully thermally coupled distillation columns", *Trans. IChemE. Chem. Eng. Res. Des.*, **70**, Part A, 118-132.

Trivedi, K.K. (1988) "The pinch design method for the synthesis of heat exchanger networks: the constrained case", *AIChE Annual Meeting*, Washington D.C., Paper No 81b.

Trivedi, K.K. and Roach, J.R. (1986)* "Overall capital - energy tradeoffs in heat exchanger networks", *Chemeca 86, 14th Australian Chemical Eng. Conf.*, Adelaide.

Trivedi, K.K., Roach, J.R. and O'Neill, B.K. (1987a) "Shell targeting in heat exchanger networks", *AIChE J.*, **33**(12), 2087-2090.

Trivedi, K.K., O'Neill, B.K., Roach, J.R. and Wood, R.M. (1987b)* "Loop breaking and energy relaxation in heat exchanger networks", *Chemeca 87, 15th Australian Chemical Conf.*, Melbourne.

Trivedi, K.K., O'Neill, B.K., Roach, J.R. and Wood, R.M. (1988)* "The synthesis of heat exchanger networks using an improved dual-temperature difference design procedure", *Proc. 3rd Intl. Conf. Process Systems Eng. (PSE-88)*, Sydney, 95-103.

Trivedi, K.K., O'Neill, B.K. and Roach, J.R. (1989a) "Synthesis of heat exchanger networks featuring multiple pinch points", *Comp. and Chem. Eng.*, **13**(3), 291-294.

Trivedi, K.K., O'Neill, B.K., Roach, J.R. and Wood, R.M. (1989b) "A new dual-temperature design method for the synthesis of heat exchanger networks", *Comp. and Chem. Eng.*, **13**, 667-685.

Trivedi, K.K., O'Neill, B.K., Roach, J.R. and Wood, R.M. (1990a) "Systematic energy relaxation in MER heat exchanger networks, *Comp. and Chem. Eng.*, **14**(6), 601-611.

Trivedi, K.K., O'Neill, B.K., Roach, J.R. and Wood, R.M. (1990b) "A best-first search strategy for energy relaxation in MER heat exchanger networks", *Eng. Optim.*, **16**(3), 165-189.

Umeda, T., Itoh, J. and Shiroko, K. (1978) "Heat exchange system synthesis", *Chem. Eng. Prog.*, **74**, 70-76.

Umeda, T., Harada, T. and Shiroko, K. (1979a) "A thermodynamic approach to the synthesis of heat integration systems in chemical processes", *Comp. and Chem. Eng.*, **3**, 273-282.

Umeda T., Niida, K. and Shiroko, K. (1979b)* "A thermodynamic approach to heat integration in distillation columns", *AIChE J.*, **25**, 423.

Vishwanathan, M. and Evans, L.B. (1987)* "Studies in the heat integration of chemical process plants", *AIChE J.*, **33**(11), 1781-1790.

Viswanathan, J. and Grossmann, I.E. (1990)*, *Comp. and Chem. Eng.*, **14**, 769-782.

Wallin, E., Franck, P.A. and Berntsson, T. (1990)* "Heat pumps in industrial processes - an optimization methodology", *Heat Recovery Systems & CHP*, **10**(4), 437-446.

Wang, Y.P. and Smith, R. (1994) "Wastewater minimization", *Chem. Eng. Sci.*, **49**(7), 981-1006.

Wang, Y.P., Chen, Z.H. and Groll, M. (1990)* "A new approach to heat exchanger network synthesis", *Heat Recovery Systems & CHP*, **10**(4), 399-405.

Wells, G.L. and Hodgkinson, M.G. (1977)* *Process Eng.*, 59-63.

Wells, G. and Rose, L.M. (1986)* *The art of chemical process design*, Elsevier, Amsterdam.

Westerberg, A.W. (1983)* "The heat path diagram for energy management", *AIChE Diamond Jubilee Meeting*, October-November, Washington D.C., Paper No 22f.

Westerberg, A.W. (1987)* "Process synthesis: A morphological view", *in Recent Developments in Chemical Process and Plant Design*, Liu, Y.A., McGee, H.A. and Epperly, W.R. (eds.), John Wiley, New York, 127-145.

Westerberg, A.W. and Andrecovich, M.J. (1985)* "Utility bounds for nonconstant QΔT for heat-integrated distillation sequence synthesis", *AIChE J.*, **31**(9), 1475-1479.

Westerberg, A.W. and Grossmann, I.E. (1985)* "Process synthesis techniques in the process industries and their impact on energy use", Report sponsored by EPRI, December 1, Dept. of Chemical Eng., Carnegie Mellon University, Pittsburgh.

Westerberg, A.W., Hutchison, H.P., Motard, R.L. and Winter P. (1979), *Process flowsheeting*, Cambridge University Press, Cambridge.

Whistler, A.M. (1948) "Heat exchangers as money makers", *Petroleum Refiner*, **27**(1), 83-86.

Winter, P. (1992)* "Computer-aided process engineering: The evolution continues", *Chem. Eng. Prog.*, February, 76-83.

Wood, R.M., Wilcox, R.J. and Grossmann, I.E. (1985) "A note on the minimum number of units for heat exchanger network synthesis", *Chem. Eng. Commun.*, **39**, 371-380.

Wood, R.M., Suaysompol, K., O'Neill, B.K., Roach, J.R. and Trivedi, K.K. (1991) "A new option for heat exchanger network design", *Chem. Eng. Prog.*, September, 38-43.

Yee, T.F. and Grossmann, I.E. (1989)* "A simultaneous optimization approach for heat exchanger network synthesis", *AIChE Annual Meeting*, San Francisco, Paper No 136f.

Yee, T.F. and Grossmann, I.E. (1990) "Simultaneous optimization models for heat integration - II. Heat exchanger network synthesis", *Comp. and Chem. Eng.*, **14**(10), 1165-1184.

Yee, T.F. and Grossmann, I.E. (1991) "A screening and optimization approach for the retrofit of heat exchanger networks", *Ind. Eng. Chem. Res.*, **30**, 146-162; also, Yee, T.F. and Grossmann, I.E. (1988) "A screening and optimization approach for the retrofit of heat exchanger networks", *AIChE Annual Meeting*, Washington D.C, Paper 81d.

Yee, T.F., Grossmann, I.E. and Kravanja, Z. (1990a) "Simultaneous optimization models for heat integration - I. Area and energy targeting and modeling of multi-stream exchangers", *Comp. and Chem. Eng.*, **14**(10), 1151-1164.

Yee, T.F., Grossmann, I.E. and Kravanja, Z. (1990b) "Simultaneous optimization models for heat integration - III. Process and heat exchanger network optimization", *Comp. and Chem. Eng.*, **14**(11), 1185-1200.

Zhelev, T.K. and Boyadzhiev, Khv. B. (1988)* "A method for the optimal synthesis of heat exchanger systems" (3 parts), *Intl. Chem. Eng.*, **28**(3), 543-558.

Zhelev, T.K. and Kotjabasakis, E. (1994)* "System control of energy efficient batch processes", *Comp. and Chem. Eng.*, **18** Supplement, S471-S475; also, Zhelev, T.K. and Kotjabasakis, E. (1993)* "System control of energy efficient batch processes", *European Symp. on Comp. Applications in Process Eng. (ESCAPE - 3)*, July, Graz, Austria.

Zhu, X.X., O'Neill, B.K., Roach, J.R. and Wood, R.M. (1994)* "Kirchhoff's law and loop-breaking for the design of heat exchanger networks", *accepted for publication in Chem. Eng. Commun.*

Index

A

B

C

M

Mass Exchanger Network Synthesis (MENS), 493-500, 521-524, 582

Mass Separating Agents (MSAs), 493-494, 497, 500

Match area distribution matrix, 52-53, 55, 166, 289-291, 354

Mathematical Programming (MP), 16, 308, 500, 516, 525-526, 582-583

Maximum Energy Recovery. *See* MER design.

Maximum Possible Remaining (MPR) concept, 148-152, 154-155

MC_p criterion, 118, 125, 230, 498

MC_p-ratio heuristic, 214-215

MER design, 7, 12, 33, 117-125, 127-130, 154-158, 161-164, 179, 195, 401-402, 452-457, 526-541

MILP, 247, 308, 499, 525-526, 536, 576

Minimum allowable stack temperature, 132

Minimum flux specification, 64, 71, 95-108, 205

Minimum Number of Units (MNU), 32-33, 189, 191, 194-197, 239-240, 246, 248-249, 504-505, 527

Minimum temperature difference. *See* ΔT_{min}.

Minimum utility, 8, 10, 16, 80, 88, 92, 97, 111, 148, 498

MINLP, 499, 525-526, 550, 552, 576-577, 581-582

Mirror image matches, 240-242

Mixed Integer Linear Programming. *See* MILP.

Mixed Integer Non-Linear Programming. *See* MINLP.

Mixing junction, 123, 195-197, 221, 551, 564, 568, 570

Modification of distillation column, 482-484, 518, 520

Modification of process, 409, 415, 517-519

Modification of reactor, 513-515

Modification of utility, 415-417

Multiple pinches, 71, 73, 89, 115-116, 125-130, 161, 228, 539-540

Multiple utilities, 57-62, 69, 125-130, 161, 165-166, 176-178, 225, 253, 497, 539-541

N

Network representation, 564-566, 572-573, 577-578

NLP, 111-112, 209, 300, 525-526, 541-542, 549, 564-565, 569-572, 575, 583-584

Non-Linear Programming. *See* NLP.

Non-uniform exchanger specifications, 42-46, 48-50, 54-57, 68-69, 86, 486

Number criterion, 118, 125

Number of shells, 30, 36-41, 221, 258-259, 263-264

Number of transfer units N, 30-31, 258-259, 325, 463, 466

O

Onion model, 516, 520

OPERA, 1-2, 417, 432, 516, 520, 583

Operating cost, 41, 85, 219, 415